MW01623006

HANDBOOK OF REACTIVE CHEMICAL HAZARDS

Handbook of Reactive Chemical Hazards

L. Bretherick, B.Sc., F.R.I.C.
Project Leader
New Technology Division
Sunbury Research Centre
British Petroleum Company, Ltd.

CRC PRESS, Inc.
18901 Cranwood Parkway · Cleveland, Ohio 44128

First published 1975

International Standard Book Number 0-408-70643-0

Printed in the United States

FOREWORD

During the two years in which I chaired the Enquiry Committee on Safety and Health of people at their place of work, what was certainly borne upon us was the tremendous speed of advance of modern technology and the importance of keeping up to date with the safety factors surrounding new developments.

This handbook, the product of Mr. L. Bretherick, certainly adds a new dimension to this whole problem.

It is very clear from its exposition that every single contingency that can at present be contemplated in the whole field of laboratory safety has been covered, and there is no doubt that from time to time there will require to be reviews in the light of chemical development. Sufficient to say, however, that within these pages lies an enormous amount of research work in this whole area of laboratory work in the chemical industry, that there has been collected a tremendous amount of information on hazards in laboratories and that Mr. Bretherick has become recognised as an expert in this field.

Safety Officers who are responsible for safety in industrial laboratories will find this book absolutely essential reading if they are to do their own work to the satisfaction of themselves and of the company which they serve.

I know of no previous book containing so much information as this one, and there is no doubt at all that if the book is diligently used there will be found within its pages answers to many of the problems in relation to safety in the use of chemicals.

I can only say, having spent so much time in this whole field, that the development of the modern technologies is the area in which safety research needs to be constantly updated. In the older industries already there is sufficient information to enable us to conduct our affairs in such a manner as to provide adequate safety and a proper healthy environment for the workers in those fields. It is in the modern fields that the problems arise, because with great frequency there are new discoveries that require fresh thought in relation to safety hazards.

Mr. Bretherick has undoubtedly produced for all those in this field a handbook that will enable the greatest possible safety precautions to accompany the developing sciences. That Mr. Bretherick should have been able to produce what is a monumental contribution to safety is a matter of great credit to him and the enormous amount of research that was required in order to produce this book.

I therefore recommend most strongly a careful survey of the entries in this book relevant to the particular activities of companies in this field, and feel very strongly that Safety Officers require this constantly at their elbow in order to provide the right kind of environment in which work in laboratories can be conducted with greater safety than is possible without the research benefits that have come from this book.

It is absolutely essential in the modern world that safety factors of those employed in research must be adequately safeguarded, and this means that the maximum amount of information must be available at all times, based upon present knowledge. It is because I recognise the value of the work that this book embraces in the whole difficult field of modern research and technology that I unhesitatingly in this foreword make the strongest recommendation for its close attention and use in laboratories.

In our travels as a Committee of Enquiry, we found that many countries in the world look to the United Kingdom for a lead in these matters. This book produces further evidence of the care and attention required in this highly critical area of laboratory research and, therefore, will be welcomed both in the United Kingdom and overseas.

I am certain that as a result of this work a great many people engaged in actual research, and the companies concerned, will find this of considerable help and assistance in safeguarding all those who are engaged in breaking through the frontiers of knowledge.

Robens

LORD ROBENS OF WOLDINGHAM

November, 1974

PREFACE

Although I had been aware during most of my career as a preparative chemist of a general lack of information relevant to the reactive hazards associated with the use of chemicals, the realisation that this book needed to be compiled came soon after my reading *Chemistry & Industry* for June 6th, 1964. This issue contained an account of an unexpected violent laboratory explosion involving chromium trioxide and acetic anhydride, a combination which I knew to be extremely hazardous from close personal experience 16 years previously.

This hazard had received wide publicity in the same journal in 1948, but during the intervening years had apparently lapsed into relative obscurity. It was then clear that currently existing arrangements for communicating 'well-known' reactive chemical hazards to practising chemists and students were largely inadequate. I resolved to try to meet this obvious need for a single source of information with a logically arranged compilation of available material. After a preliminary assessment of the over-all problems involved, work began in late 1964.

By late 1971, so much information had been uncovered but remained to be processed that it was apparent that the compilation would never be finished on the spare-time basis then being used. Fortunately I then gained the support of my employers, the British Petroleum Company, Ltd., and have now been able to complete this compilation as a supporting research objective since January, 1972.

The detailed form of presentation adopted has evolved steadily since 1964 to meet the dual needs for information on reactive chemical hazards in both specific and general terms, and the conflicting practical requirements of completeness and brevity. A comprehensive explanation of how this has been attempted, with suggestions on using this Handbook to best advantage, is given in the Introduction.

In an attempt to widen the scope of this work, unpublished information has been sought from many sources, both by published appeals and by correspondence. In this latter area, the contribution made by a friend, the late Mr. A. Kruk-Schuster, of Laboratory Chemicals Disposal Company, Ltd., Billericay, has been outstanding. During 1965–1968 his literature work and global letter campaign to 2000 University chemistry departments and industrial institutions yielded some 300 contributions.

The coverage attempted in this Handbook is wide, but is certainly incomplete because of the difficulties in retrieving relevant information from original literature when it does not appear in the indices of either primary or abstract journals. Details of such new material known to users of this Handbook and within the scope given in the Introduction will be welcomed for inclusion in supplementary or revised editions of this work.

L.B.

ACKNOWLEDGMENTS

It is a pleasure to acknowledge the contributions of unpublished information on hazardous chemical reactions which have been made by both close colleagues and distant correspondents during the course of several years. Many of the earlier contributors must have despaired of ever seeing their help acknowledged in print (as private communications) in the text, but I hope that they will feel the long delay has been worthwhile. Thanks are also due to those whose contributions could not be incorporated.

I should like to record my special indebtedness to the following, who all provided substantial quantities of information from their own files or experience:

Mr R. N. Beer, BP Chemicals International Limited, Grangemouth
Dr A. F. Crowther, ICI Pharmaceuticals Limited, Macclesfield
Mr K. Everett, Safety Officer, University of Leeds
Dr C. S. Foote, University of California, Los Angeles
Dr H. Feuer, Ohio State University, Columbus
Mr. C. H. R. Gentry, Mr H. Sutcliffe } Mullard Radio Valve Company, Mitcham
Dr P. R. Huddleston, Trent Polytechnic, Nottingham
Dr E. W. Jenkins, Centre for Studies in Science Education, University of Leeds
Mr A. Kruk-Schuster, Laboratory Chemicals Disposal Co. Ltd., Billericay
Dr E. Levens, Douglas Aircraft Co. Inc., Santa Monica
Dr D. G. Muir, Mr H. G. Moss } BDH Chemicals Limited, Poole
Dr T. Poles, Engelhard Sales Limited, Cinderford
Dr C. H. Tilford, Warner-Lambert Research Institute, Morris Plains

Thanks are also due to the British Chemical Industries Safety Council, London, and the Manufacturing Chemists Association Inc.,

Washington, whose secretaries, Mr D. E. Pickbourne and Mr M. F. Crass, Jr., respectively, gave free use of their regularly published safety material.

I am indebted to colleagues, library and clerical staff at the BP Research Centre for much general encouragement and assistance, and to my employers, the British Petroleum Company Limited, without whose close support since 1971 this work would not have been completed. My former colleagues at May and Baker Limited, Dagenham, were largely responsible for the original development of my personal interest in laboratory safety matters.

The assistance rendered during preparation of this book by the Publisher's staff and their IUPAC nomenclature consultant, Dr R. S. Cahn, is also gratefully acknowledged.

Finally, I would record my heartfelt gratitude to my wife, who has cheerfully during several years converted miles of illegibility into a coherent typescript. To her I dedicate this work.

CONTENTS

INTRODUCTION

THIS SHOULD BE READ THROUGH CAREFULLY TO GAIN FULL BENEFIT FROM THE FOLLOWING SECTIONS.

Aims of the Handbook

This compilation has been prepared to give the research student or practising chemist access to a wide selection of documented information to allow him or her readily to assess the likely reaction hazard-potential associated with an existing or proposed chemical compound or reaction system.

A secondary, longer-term purpose is to present the information in a way which will, as far as possible, bring out the causes of, and interrelationships between, apparently disconnected facts and incidents. This should encourage an increased awareness of potential chemical reactivity hazards in school, college and university teaching laboratories and help to dispel the considerable ignorance of such matters which appears to be widespread in this important area of safety training during formative years.

Scope and Source Coverage

This Handbook includes all information which had become available to the author by June 1973 on the reactivity hazards of individual elements or compounds, either alone or in combination. Appropriate source references are included to give access to more expansive information than that compressed into the text entries.

A wide variety of possible sources of published information has been scanned to ensure maximum coverage. Primary sources have largely been restricted to journals known to favour or specialise in publication of safety matters, and the textbook series specialising in synthetic procedures.

Secondary sources have been a fairly wide variety of both specialised and general textbooks (notably those of Mellor and Sidgwick in the inorganic area), part of *Chemical Abstracts*, and various safety summaries, data sheets and case histories which cover the industrial area. Details of all the secondary sources and the primary textbooks will be found in Appendix 1. Textbook references are characterised by the absence of an author's initials.

Information on toxic hazards has been specifically excluded because this is collectively available elsewhere and its attempted inclusion would have delayed publication by several more years. For similar reasons, no attempt has been made to include details of all flammable materials capable of burning explosively when mixed with air and ignited, or any incidents related to this frequent cause of accidents.

However, to focus attention on the potential hazards always associated with the use of flammable substances, some 550 gases and liquids with flash points below 25°C and/or autoignition temperatures below 225°C have been included, with figures for flash point, autoignition temperature and explosive limits in air where known. Those elements or compounds which ignite on exposure to air are also included.

General Arrangement

The information herein on reactive hazards is of two main types, specific or general, and these have been arranged differently in their respective separate sections. Specific information on instability of individual chemical compounds and on hazardous interactions of elements and/or compounds is contained in the main formula-based second section of the Handbook.

See ETHYL PERCHLORATE, $C_2H_5ClO_4$
NITRIC ACID, HNO_3 : Acetone

General information relating to types or groups of elements or compounds possessing similar structural or hazardous characteristics is contained in the smaller alphabetically based first section:

See ACYL NITRATES
PYROPHORIC METALS

Individual materials of variable composition and materials which cannot conveniently be formulated are also included in this section:

See BLEACHING POWDER
CELLULOSE NITRATE

Both theoretical and practical hazard topics, some indirectly related to the main theme of this book, are also included:

See DISPOSAL
EXPLOSIBILITY
GAS CYLINDERS
OXYGEN ENRICHMENT

Specific Formula Entries

1. A single unstable compound of known composition is placed in the main (formula) section on the basis of its molecular formula expressed in the Hill system used by *Chemical Abstracts* (C and H if present, then all other elements alphabetically). The use of this indexing basis permits a compound to be located if its structure can be drawn, irrespective of whether it can be named or not. The name corresponding to the indexing formulation appears in capitals to the left of the formula. References to the information source are given, followed by a statement of the observed hazard, with any relevant explanation. Cross-reference to similar compounds, preferably as a group entry, completes the item:

See TRIFLUOROACETYL NITRITE, $C_2F_3NO_3$

2a. Where two or more elements or compounds are involved in a reactive hazard, and an intermediate or product of reaction is identifiable as being responsible for the hazard, both reacting substances are normally cross-referred to the identified product. The well-known reaction of ammonia and iodine to give explosive nitrogen triiodide-ammonia is of this type. The entries

AMMONIA, H_3N: Halogens
IODINE, I_2 : Ammonia

are referred back to the main entry under the identified material

NITROGEN TRIIODIDE-AMMONIA, $I_3N{\cdot}H_3N$

No attempt has been made, however, to list all combinations of reactants which can lead to the formation of a particular unstable main entry compound.

2b. In a multi-reactant system where no identification of an unstable product was possible, one of the reactants had to be selected as primary reactant to prepare and index the main entry, with the other material(s) as secondary reactant(s). No strictly logical basis of choice for this is obvious.

However, it emerged during the compilation phase that most two-component reaction hazard systems of this type involve a fairly obvious oxidant material as one of the reactants. Where this situation was recognised, the oxidant has normally been selected as primary (indexing) reactant with its name in capitals, secondary reactant(s) following in lower case type:

See POTASSIUM PERMANGANATE, $KMnO_4$: Acetic acid, etc.

In the markedly fewer cases where an obvious reducant has been involved as one reactant, this has normally been selected as primary reactant:

See LITHIUM TETRAHYDROALUMINATE, AlH_4Li: 3,5-Dibromocyclopentene

In the relatively few cases where neither (or none) of the reactants can be recognised as oxidant or reducant, the choice was made which appeared to give the more informative main entry text:

See CHLOROFORM, $CHCl_3$: Acetone, etc.

Where some hazard has been noted during the preparation of a specific compound, but without it being possible to identify a specific cause, an entry for that compound states 'preparative hazard', and back-refers to the reactants involved in the preparation:

See SULPHUR DIOXIDE, O_2S

Occasionally, departures from these considerations have been made where such action appeared advantageous in bringing out a relationship between formally unrelated compounds or hazards. In all multi-component cases, however, the secondary reactants (except air and water) appear as formula entries back-referred to the main entry text, so the latter is accessible from either primary or secondary reactants:

See DIMETHYL SULPHOXIDE, C_2H_6OS: Acyl halides (main entry)
ACETYL CHLORIDE, C_2H_3ClO: Dimethyl sulphoxide (back reference)

It may be noted here that interaction of obvious oxidants and reducants without very careful control of reaction conditions may be catastrophic:

See ROCKET PROPELLANTS
DIBENZOYL PEROXIDE, $C_{14}H_{10}O_4$: Lithium tetrahydroaluminate

The presence of reducant and oxidant functions in the same molecule naturally raises the possibility of a violent intramolecular redox reaction being initiated:

See HYDRAZINIUM NITRATE, $H_5N_2O_3$

Grouping of Reactants

There are advantages to be gained in grouping together elements or compounds showing similar reactivities, because this tends to bring out the relationships between structure and reactivity more clearly than separate treatment. This course has been adopted widely for primary reactants (see next section), and for secondary reactants where one primary reactant has been separately involved in hazardous reactions with a large number of secondary materials. Where possible, the latter have been collected together under a suitable general group title indicative of the composition or characteristics of the collected materials:

See CHLORINE, Cl_2 : Hydrocarbons
HYDROGEN PEROXIDE, H_2O_2 : Metals, Metal oxides, Metal salts
HYDROGEN SULPHIDE, H_2S: Oxidants

This arrangement means, however, that some practice will be necessary on the user's part in deciding into what group an individual secondary reactant falls before the longer-term advantages of the system become apparent. The group titles listed in Appendix 2 will be of use in this connection.

General Group Entries

In some cases literature references relevant to well-defined hazard topics or groups of hazardous compounds have been found, and these are given, with a condensed version of appropriate information, at the beginning of the topic or group entry.

Cross-references to related group or sub-group entries are also included, with a group list of individually indexed main-section entries which lie within the structural or functional scope of the group entry, these lists being arranged in order of formulae. Compounds which are closely similar to, but not in strict conformity with, the group definition are indicated by an initial asterisk.

The group entries thus serve as sub-indexes for each structurally based group of hazardous compounds. Conversely, each individual compound entry is back-referred to the group entry, and thence to all its strict structural analogues and related congeners included in this Handbook.

These features should be useful in attempts to estimate the stability or reactivity of a compound or reaction system which does not appear in this Handbook. The effects on stability or reactivity of changes in the molecular structure to which the destabilising group(s) is attached are in some cases discussed in the group entry. Otherwise such information may be gained from comparison of the information available from the individual compound entries collectively listed in the group entry.

Care is, however, necessary in extrapolating from described hazardous compounds or systems to others in which the user of this Handbook may be interested. Due allowance must be made for changes in elemental reactivity up or down the columns of the Periodic Table, and for the effects of variation in chain length, branching and point of group-attachment in organic systems. Purity of materials, possible catalytic effects (positive or negative) of impurities, and scale of operations may all have a direct bearing upon a particular reaction rate. Examples of all these effects will be found throughout this compilation.

Nomenclature

With the direct encouragement and assistance of the Publishers, an attempt has been made in this Handbook to use chemical names which conform to the most recent recommendations of IUPAC. While this has not been an essential part of the compilation, because each name has a corresponding molecular formula adjacent, it seems nonetheless desirable to minimise possible confusion by adopting the unambiguous system of nomenclature presented in the IUPAC publications.

Where the IUPAC name for a compound is very different from the previously used trivial name, the latter is included in parentheses (and in quotes where no longer an acceptable name). Generally, retained trivial names have not been used as main entry titles, but they have been used in the entry texts. Rarely, on the grounds of brevity, names not strictly conforming to IUPAC principles but recommended for chemicals used in industry in BS 2474: 1965 (amended 1970) have been used. The prefix *mixo-* , to represent the mixtures of isomers sometimes used as industrial materials, is a case in point.

No attempt has been made to use only one of the nomenclature systems acceptable to IUPAC, but no non-preferred or deprecated names are believed still to lurk in the text.

In connection with the group titles adopted for the alphabetical section, it has been necessary in some cases to devise group names (particularly in the inorganic field) to indicate in a very general way the chemical structures involved in various classes, groups or sub-groups of compounds.

For this purpose, all elements have been considered either as METALS or NON-METALS, and, of the latter, HALOGENS, HYDROGEN, NITROGEN, OXYGEN and SULPHUR were selected as specially important. Group names have then been coined from suitable combinations of these, such as the simple

METAL OXIDES, NON-METAL SULPHIDES
N-HALOGEN COMPOUNDS, NON-METAL HALIDES
METAL NON-METALLIDES, COMPLEX HYDRIDES

or the more complex

METAL OXOHALOGENATES
AMMINECHROMIUM PEROXOCOMPLEXES
OXOSALTS OF NITROGENOUS BASES
METAL OXONON-METALLATES

Organic group entries are fairly conventional, such as

HALOALKENES
NITROSO COMPOUNDS

Where necessary, such group names are explained in the appropriate group entry, of which a full listing is given in Appendix 2.

Cross-reference System

The cross-reference system adopted in this Handbook plays a large part in providing maximum access to, and use of, the rather heterogeneous collection of information herein. The significance of the four types of cross-reference which have been used is as follows.

See . . . . cross-refers to a directly related material or entry.
See also. . . . . cross-refers to an indirectly related material or entry.
See other . . . cross-refers to strict analogues of the compound or type.

See related . . cross-refers to related compounds (congeners) which are fairly closely but not strictly analogous structurally.

Information Content of Individual Entries

A conscious effort has been made throughout this compilation to exclude all fringe information not directly relevant to the involvement of chemical reactivity in the various incidents, with just enough detail present to allow the reader to judge the relevance or otherwise of the quoted references to his or her particular reactivity problems.

It must be stressed that this book can do no more than serve as a guide to much more detailed information available via the quoted references. In all but a few cases it cannot relieve the student or chemist of their moral obligation to themselves and others to equip themselves with the fullest possible information from the technical literature resources which are available, before attempting any experimental work with materials known, or suspected, to be hazardous or potentially so.

THE ABSENCE OF A MATERIAL OR A COMBINATION OF MATERIALS FROM THIS HANDBOOK CANNOT BE TAKEN TO IMPLY THAT NO HAZARD EXISTS. LOOK THEN FOR ANALOGOUS MATERIALS USING THE GROUP ENTRY SYSTEM.

One aspect which, although it is excluded from all entry texts is nevertheless of vital importance, is that of the potential for damage, injury or death associated with the various materials and reaction systems dealt with in this Handbook.

Though some of the incidents have involved little or no damage (see CAN OF BEANS), others have involved personal injuries, often of unexpected severity (see SODIUM PRESS), and material damage is often immense (PERCHLORIC ACID, $ClHO_4$: Cellulose derivatives, ref. 1, involved damage to 116 buildings and a loss approaching $3 M). The death-roll associated with reactive chemical hazards has ranged from 1 or 2 (see TETRAFLUOROETHYLENE, C_2F_4 : Iodine pentafluoride) to some 600 with 2000 injured in the incident at Oppau in 1923 (see AMMONIUM NITRATE, $H_4N_2O_3$, ref. 4).

This potential for destruction emphasises again the need to gain detailed knowledge in this area.

CLASS, GROUP AND TOPIC SECTION

(Entries arranged in alphabetical order)

EXPLANATORY NOTES

All temperatures are expressed in degrees Celsius, pressures in bars or mbars and volumes in m^3, litres or ml.

The following abbreviations have been widely used.

Fl.P., meaning Flash Point, a suffix (o) indicating an open-cup rather than the usual closed-cup value. See the entry FLASH POINTS.

E.L., meaning Explosive Limits in air at ambient temperature and usually determined by a standard method. In some cases, where the upper limit has not been recorded, a blank space has been left before %. Where temperatures above ambient were used, or where a calculated result is given, this is indicated. See the entry FLAMMABILITY.

A.I.T., meaning Auto-Ignition Temperature in air. See the entry AUTOIGNITION TEMPERATURE.

MCA-SD... , referring to the appropriate MCA data sheet.

ACETYLENIC COMPOUNDS

Davidsohn, W. B. *et al., Chem. Rev.,* 1967, **67**, 74
The presence of the endothermic acetylene group confers explosive instability on a wide range of acetylenic compounds (notably when halogen is also present) and derivatives of metals (especially of heavy metals). The class includes the separately treated groups:

ACETYLENIC PEROXIDES
ETHOXYETHYNYL ALCOHOLS
HALOACETYLENE DERIVATIVES
METAL ACETYLIDES

and the individually indexed compounds:

ACETYLENE, C_2H_2
PROPIOLALDEHYDE, C_3H_2O
PROPYNE, C_3H_4
METHOXYACETYLENE, C_3H_4O
2-PROPYN-1-OL (PROPARGYL ALCOHOL), C_3H_4O
2-PROPYNE-1-THIOL, C_3H_4S
1,3-BUTADIYNE, C_4H_2
1-BUTEN-3-YNE, C_4H_4
ETHYNYL VINYL SELENIDE, C_4H_4Se
1- or 2-BUTYNE, C_4H_6
ETHOXYACETYLENE, C_4H_6O
3-METHOXYPROPYNE, C_4H_6O
2-BUTYNE-1, 4-DIOL, $C_4H_6O_2$
2-BUTYNE-1-THIOL, C_4H_6S
DIMETHYLAMINOACETYLENE, C_4H_7N
1,3-PENTADIYNE, C_5H_4
2-METHYL-1-BUTEN-3-YNE, C_5H_6
2-PENTEN-4-YN-3-OL, C_5H_6O
2-PROPYNYL VINYL SULPHIDE, C_5H_6S
1-PENTYNE, C_5H_8
1-ETHOXY-2-PROPYNE, C_5H_8O
2-METHYL-3-BUTYN-2-OL, C_5H_8O
TRIETHYNYLARSINE, C_6H_3As
TRIETHYNYLPHOSPHINE, C_6H_3P
1,3-HEXADIEN-5-YNE, C_6H_6
4,5-HEXADIEN-2-YN-1-OL, C_6H_6O
3-METHYL-2-PENTEN-4-YN-1-OL, C_6H_8O
tert-BUTYLNITROACETYLENE, $C_6H_9NO_2$
1-, 2- or 3-HEXYNE, C_6H_{10}

BUTOXYACETYLENE, $C_6H_{10}O$
2-HEPTYN-1-OL, $C_7H_{12}O$
1-ETHOXY-3-METHYL-1-BUTYN-3-OL, $C_7H_{12}O_2$
PHENOXYACETYLENE, C_8H_6O
DIETHYL ACETYLENEDICARBOXYLATE, $C_8H_{10}O_4$
1-DIETHYLAMINO-1-BUTEN-3-YNE, $C_8H_{13}N$
1-, 2-, 3- or 4-OCTYNE, C_8H_{14}
2-NONEN-4,6,8-TRIYN-1-AL, C_9H_4O
BENZYLOXYACETYLENE, C_9H_8O
3-BUTYN-1-YL *p*-TOLUENESULPHONATE, $C_{11}H_{12}O_3S$
1,3,5-TRIETHYNYLBENZENE, $C_{12}H_6$
DIBUTYL-3-METHYL-3-BUTEN-1-YNYLBORANE, $C_{13}H_{23}B$
BIS(DIBUTYLBORINO)ACETYLENE, $C_{18}H_{36}B_2$

See also PEROXIDISABLE COMPOUNDS

ACETYLENIC PEROXIDES

Milas, N. A. *et al., Chem. Eng. News,* 1959, **37**(37), 66; *J. Amer. Chem. Soc.,* 1952, **75**, 1472; *J. Amer. Chem. Soc.,* 1953, **76**, 5970
The importance of strict temperature control to prevent explosion during preparation (second and third references) of acetylenic peroxides is stressed. Use of inert solvent to prevent undue thickening and consequent poor temperature control is recommended.

ACYL AZIDES

1. Smith, P. A. S., *Org. React.,* 1946, **3**, 373–375
2. Houben-Weyl, 1952, Vol. 8, 680
3. Lieber, E. *et al., Chem. Rev.,* 1965, **65**, 377

Azides of low molecular weight (more than 25% nitrogen content) should not be isolated from solution since the concentrated material is likely to be dangerously explosive [1]. The concentration of such solutions (prepared below 10°C) should be below 10% [2].

Carbonyl azides are explosive compounds, some exceptionally so; suitable handling precautions are necessary [3]. Individually indexed compounds are:

TRIFLUOROMETHANESULPHONYL AZIDE, $CF_3N_3O_2S$

AZIDODITHIOFORMIC ACID, CHN_3S_2
CARBONYL DIAZIDE, CN_6O
1,2-DIAZIDOCARBONYLHYDRAZINE, $C_2H_2N_8O_2$
ACETYL AZIDE, $C_2H_3N_3O$
BIS(AZIDOTHIOCARBONYL) DISULPHIDE, $C_2N_6S_4$
4-AZIDOCARBONYL-1,2,3-THIADIAZOLE, C_3HN_5OS
ETHYL AZIDOFORMATE, $C_3H_5N_3O_2$
N-BUTYLAMIDOSULPHURYL AZIDE, $C_4H_{10}N_4O_2S$
GLUTARYL DIAZIDE, $C_5H_6N_6O_2$
PIVALOYL AZIDE, $C_5H_9N_3O$
tert-BUTYL AZIDOFORMATE, $C_5H_9N_3O_2$
PHENYLPHOSPHONIC DIAZIDE, $C_6H_5N_6OP$
PHENYLTHIOPHOSPHONIC DIAZIDE, $C_6H_5N_6PS$
p-BROMOBENZOYL AZIDE, $C_7H_4BrN_3O$
p-CHLOROBENZOYL AZIDE, $C_7H_4ClN_3O$
BENZOYL AZIDE, $C_7H_5N_3O$
N-AZIDOCARBONYLAZEPINE, $C_7H_6N_4O$
p-TOLUENESULPHONYL AZIDE, $C_7H_7N_3O_2S$
PHTHALOYL DIAZIDE, $C_8H_4N_6O_2$
3-PHENYLPROPIONYL AZIDE, $C_9H_9N_3O$
6-QUINOLINECARBONYL AZIDE, $C_{10}H_6N_4O$
SULPHURYL AZIDE CHLORIDE, ClN_3O_2S
FLUOROTHIOPHOSPHORYL DIAZIDE, FN_6PS
AMIDOSULPHURYL AZIDE, $H_2N_4O_2S$
POTASSIUM AZIDOSULPHATE, KN_3O_3S
POTASSIUM AZIDODISULPHATE, $KN_3O_6S_2$
SULPHURYL DIAZIDE, N_6O_2S
DISULPHURYL DIAZIDE, $N_6O_5S_2$

ACYL CHLORIDES

Aromatic Hydrocarbons
Trifluoromethanesulphonic acid
See TRIFLUOROMETHANESULPHONIC ACID, CHF_3OS: Acyl chlorides, etc.

ACYL NITRATES

A thermally unstable class, tending to violent decomposition or explosion on heating.

See ACETYL NITRATE, $C_2H_3NO_4$
BUTYRYL NITRATE, $C_4H_7NO_4$

m-NITROBENZOYL NITRATE, $C_7H_4N_2O_6$
BENZOYL NITRATE, $C_7H_5NO_4$
See also NITRIC ACID, HNO_3 : Phthalic anhydride
SODIUM NITRITE, $NNaO_2$: Phthalic acid

ACYL NITRITES

Ferrario, E., *Gazz. Chim. Ital.*, 1901, **40**(2), 98–99
Francesconi, L. *et al., Gazz. Chim. Ital.*, 1895, **34**(1), 442
The stabilities of propionyl and butyryl nitrites are greater than that of acetyl nitrite, butyryl nitrite being the least explosive of these homologues. Benzyl nitrite is also unstable.
See ACETYL NITRITE, $C_2H_3NO_3$
TRIFLUOROACETYL NITRITE, $C_2F_3NO_3$

ALKALI-METAL DERIVATIVES OF HYDROCARBONS

Sidgwick, 1950, 68, 75
Alkali-metal derivatives of aliphatic or aromatic hydrocarbons, such as methyllithium, ethylsodium or phenylpotassium, are the most reactive towards moisture and air, immediately igniting in the latter. Derivatives of benzyl-type compounds, such as benzylsodium, are of slightly lower activity, usually, but not always, igniting in air. Derivatives of hydrocarbons with definitely acidic hydrogen atoms (acetylene, phenylacetylene, cyclopentadiene, fluorene), though readily oxidised, are usually relatively stable in ambient air. Sodium phenylacetylide if moist with ether ignites; derivatives of triphenylmethane also when dry.

ALKALI METALS

Handling and Uses of the Alkali Metals (Advances in Chemistry Series No. 19), Washington, American Chemical Society, 1957
The collected papers of a symposium at Dallas, April 1956, covering

the handling, use and hazards of lithium, sodium, potassium, their alloys, oxides and hydrides, in 19 chapters.

ALKANETHIOLS

Ethylene oxide
See ETHYLENE OXIDE, C_2H_4O: Alkanethiols

ALKENES

Oxides of nitrogen
See NITROGEN OXIDE, NO: Dienes, Oxygen

ALKENYL NITRATES

Tetrafluorohydrazine
See TETRAFLUOROHYDRAZINE, F_4N_2: Alkenyl nitrates

ALKYLALUMINIUM ALKOXIDES AND HYDRIDES

Though substitution of a hydrogen atom or an alkoxy group for one alkyl group in a trialkylaluminium tends to increase stability and reduce reactivity and the tendency to ignition, these compounds are still of high potential hazard, the hydrides being used industrially as powerful reducants.
See ALKYLALUMINIUM DERIVATIVES (references 1, 3, 6)

Individually indexed compounds of this relatively small sub-group of commercially available compounds are:

DIETHYLALUMINIUM HYDRIDE, $C_4H_{11}Al$
TETRAMETHYLDIALUMINIUM DIHYDRIDE, $C_4H_{14}Al_2$
DIPROPYLALUMINIUM HYDRIDE, $C_6H_{15}Al$

ETHOXYDIETHYLALUMINIUM, $C_6H_{15}AlO$
DIISOBUTYLALUMINIUM HYDRIDE, $C_8H_{19}Al$
ETHOXYDIISOBUTYLALUMINIUM, $C_{10}H_{23}AlO$

ALKYLALUMINIUM DERIVATIVES

1. Mirviss, S. B. *et al., Ind. Eng. Chem.*, 1961, **53**(1), 53A–56A
2. Heck, W. B. *et al., Ind. Eng. Chem.*, 1962, **54**(12), 35–38
3. Kirk-Othmer, 1963, Vol. 2, 38, 40
4. Houben-Weyl, 1970, Vol. 13(4), 19
5. 'Aluminum Alkyls', Brochure PB 3500/1 New 568, New York, Ethyl Corp., 1968
6. 'Aluminum Alkyls', New York, Texas Alkyls, 1971
7. Knap, J. E. *et al., Ind. Eng. Chem.*, 1957, **49**, 875

This main class of ALKYLALUMINIUM DERIVATIVES is divided into 3 groups for convenience:

TRIALKYLALUMINIUMS
ALKYLALUMINIUM ALKOXIDES and HYDRIDES
ALKYLALUMINIUM HALIDES

Individual compounds are indexed under their appropriate group or sub-group headings, but since most of the available compounds are liquids with similar hazardous properties, these will be described collectively here. This aspect of the class has been extensively reviewed and documented [1–7].

Compounds with alkyl groups of C_4 and below all ignite immediately on exposure to air, unless diluted with a hydrocarbon solvent to 10–20% concentration [6]. Even these may ignite on prolonged exposure to air, owing to exothermic autoxidation, which becomes rapid if solutions are spilled [2, 6]. Compounds with C_5–C_{14} alkyl groups (safe at 20–30% concentration) smoke in air but do not burn unless ignited externally or if the air is very moist. Contact with air containing more oxygen than normal (21%) will cause explosive oxidation.

Fires involving alkylaluminium compounds are difficult to control and must be treated appropriately to particular circumstances [1, 5, 6], usually with dry-powder extinguishers. Halocarbon fire extinguishants (carbon tetrachloride, chlorobromomethane, etc.), water or water-based foam must not be applied to alkylaluminium fires. Carbon dioxide is ineffective unless dilute solutions are involved [5, 6].

Alcohols
Alkylaluminium derivatives up to C_4 react explosively with methanol or ethanol, and triethylaluminium also with 2-propanol [1].

Halocarbons
With the exception of chlorobenzene and 1,2-dichloroethane, halocarbon solvents are unsuitable diluents, since carbon tetrachloride and chloroform may react violently with alkylaluminium derivatives [1].

Oxidants
In view of the generally powerful reducing properties of alkylaluminium derivatives, deliberate contact with known oxidants must be under careful control with appropriate precautions [3].

Water
Interaction of alkylaluminium derivatives up to C_9 chain length with liquid water is explosive [1] and violent shock effects have been noted [4].

Suitable handling and disposal techniques have been detailed for both laboratory [1, 2, 5–7] and manufacturing [5,6] scales of operation.

See other ALKYLMETAL HALIDES
ALKYLMETALS

ALKYLALUMINIUM HALIDES

Three main structural sub-groups can be recognised: (1) Dialkylaluminium halides; (2) Alkylaluminium dihalides; (3) Trialkyl - dialuminium trihalides (equimolar complexes of an aluminium trihalide and a trialkylaluminium).

While this is generally a very reactive group of compounds, similar in reactivity characteristics to trialkylaluminium compounds, increase in size of the alkyl groups and the number of halogens present tends to reduce pyrophoricity.

See ALKYLALUMINIUM DERIVATIVES (references 1, 2)

Individually indexed compounds of the ALKYLALUMINIUM HALIDE group, many of which are commercially available in bulk, are:

METHYLALUMINIUM DIIODIDE, CH_3AlI_2
ETHYLALUMINIUM BROMIDE IODIDE, C_2H_5AlBrI
ETHYLALUMINIUM DIBROMIDE, $C_2H_5AlBr_2$
ETHYLALUMINIUM DICHLORIDE, $C_2H_5AlCl_2$
DIMETHYLALUMINIUM BROMIDE, C_2H_6AlBr
TRIMETHYLDIALUMINIUM TRICHLORIDE, $C_3H_9Al_2Cl_3$
DIETHYLALUMINIUM BROMIDE, $C_4H_{10}AlBr$
DIETHYLALUMINIUM CHLORIDE, $C_4H_{10}AlCl$
DIETHYLALUMINIUM THIOCYANATE: see the trimer, $C_{15}H_{30}Al_3N_3S_3$
DIISOBUTYLALUMINIUM CHLORIDE, $C_8H_{18}AlCl$
* HEXAETHYLTRIALUMINIUM TRITHIOCYANATE, $C_{15}H_{30}Al_3N_3S_3$

ALKYLBORANES

1. Sidgwick, 1951, 371
2. Mirviss, S. B. *et al., Ind. Eng. Chem.*, 1961, **53**(1), 53A

Trimethyl- and triethyl-borane ignite in air, and tributylborane ignites in a thinly diffuse layer, as when poured on cloth [1]. Generally, the pyrophoric tendency of trialkylboranes decreases with increasing branching on the 2- and 3-carbon atoms [2].

ALKYL CHLORITES

See SILVER CHLORITE, $AgClO_2$: Alkyl iodides

ALKYL HYDROPEROXIDES

Swern, 1970, Vol. 1, 19; 1971, Vol. 2, 1, 29

Most alkyl monohydroperoxides are liquid, the explosivity of the lower members (possibly due to traces of the dialkyl peroxides) decreasing with increasing chain length.

See METHYL HYDROPEROXIDE, CH_4O_2
ETHYL HYDROPEROXIDE, $C_2H_6O_2$

ALLYL HYDROPEROXIDE, $C_3H_6O_2$
ISOPROPYL HYDROPEROXIDE, $C_3H_8O_2$
tert-BUTYL HYDROPEROXIDE, $C_4H_{10}O_2$
2-PHENYL-2-PROPYL HYDROPEROXIDE, $C_9H_{12}O_2$
1,2,3,4-TETRAHYDRO-1-NAPHTHYL HYDROPEROXIDE, $C_{10}H_{12}O_2$
2-METHYL-1-PHENYL-2-PROPYL HYDROPEROXIDE, $C_{10}H_{14}O_2$
α-BENZENEDIAZOBENZYL HYDROPEROXIDE, $C_{13}H_{12}N_2O_2$

See also HYDRAZONES

ALKYLMETAL HALIDES

This highly reactive class of organometallic compounds includes the group of

ALKYLALUMINIUM HALIDES

as well as the individually indexed compounds:

METHYLMAGNESIUM IODIDE, CH_3IMg
METHYLZINC IODIDE, CH_3IZn
ETHYLMAGNESIUM IODIDE, C_2H_5IMg
DIMETHYLBISMUTH CHLORIDE, C_2H_6BiCl
DIMETHYLANTIMONY CHLORIDE, C_2H_6ClSb
DIETHYLBISMUTH CHLORIDE, $C_4H_{10}BiCl$

ALKYLMETALS

This reactive and usually pyrophoric class includes the groups

DIALKYLZINCS
DIPLUMBANES
TRIALKYLALUMINIUMS
TRIALKYLBISMUTHS

as well as the individually indexed compounds:

POLY(METHYLENEMAGNESIUM), $(CH_2Mg)_n$
METHYLSILVER, CH_3Ag
* METHYLBISMUTH OXIDE, CH_3BiO
METHYLCOPPER, CH_3Cu
METHYLPOTASSIUM, CH_3K
METHYLLITHIUM, CH_3Li
METHYLSODIUM, CH_3Na
VINYLLITHIUM, C_2H_3Li

DIMETHYLBERYLLIUM, C_2H_6Be
DIMETHYLCADMIUM, C_2H_6Cd
DIMETHYLMERCURY, C_2H_6Hg
DIMETHYLMAGNESIUM, C_2H_6Mg
DIMETHYLZINC, C_2H_6Zn
PROPYLCOPPER(I), C_3H_7Cu
TRIMETHYLGALLIUM, C_3H_9Ga
TRIMETHYLANTIMONY, C_3H_9Sb
TRIMETHYLTHALLIUM, C_3H_9Tl
BUTYLLITHIUM, C_4H_9Li
DIETHYLBERYLLIUM, $C_4H_{10}Be$
DIETHYLCADMIUM, $C_4H_{10}Cd$
DIETHYLMAGNESIUM, $C_4H_{10}Mg$
DIETHYLZINC, $C_4H_{10}Zn$
* BIS-DIMETHYLSTIBINYL OXIDE, $C_4H_{12}OSb_2$
TETRAMETHYLLEAD, $C_4H_{12}Pb$
TETRAMETHYLPLATINUM, $C_4H_{12}Pt$
TETRAMETHYLDISTIBANE, $C_4H_{12}Sb_2$
TETRAMETHYLTIN, $C_4H_{12}Sn$
DIMETHYL-1-PROPYNYLTHALLIUM, C_5H_9Tl
TRIVINYLBISMUTH, C_6H_9Bi
BIS(DIMETHYLTHALLIUM) ACETYLIDE, $C_6H_{12}Tl_2$
DIISOPROPYLBERYLLIUM, $C_6H_{14}Be$
TRIETHYLBISMUTH, $C_6H_{15}Bi$
TRIETHYLGALLIUM, $C_6H_{15}Ga$
TRIETHYLANTIMONY, $C_6H_{15}Sb$
TETRAVINYLLEAD, $C_8H_{12}Pb$
DIMETHYL-PHENYLETHYNYLTHALLIUM, $C_{10}H_{11}Tl$

See also ARYLMETALS

ALKYL NITRATES

Alone,
or Lewis acids

1 Boschan, R. *et al., Chem. Rev.,* 1955, **55**, 505
2. Slavinskaya, R. A., *J. Gen. Chem. (USSR).* 1957, 27, 844
3. Boschan, R. *et al., J.Org. Chem.*, 1960, **25**, 2012
4. Smith, 1966, Vol 2, 458

The class has been reviewed [1]. Ethyl [1], isopropyl, butyl, benzyl and triphenylmethyl nitrates [3] in contact with sulphuric acid [2], tin(IV) chloride [2, 3] or boron trifluoride [3] interact violently (after an induction period of up to several hours) with gas evolution. An autocatalytic mechanism is proposed.

Although pure alkyl nitrates are stable in storage, traces of oxides of nitrogen sensitise them to decomposition, and may cause explosion on heating or storage at ambient temperature [4].

The most important applications of alkyl nitrates are based on their explosive properties. Methyl and ethyl nitrates are too volatile for widespread use; propyl nitrate has been used as a liquid monopropellant. Ethylene dinitrate, glyceryl trinitrate and pentaerythrityl tetranitrate are widely used as explosives, the two former adsorbed on or blended with other compounds [1].

See METHYL NITRATE, CH_3NO_3
ETHYL NITRATE, $C_2H_5NO_3$
ISOPROPYL NITRATE, $C_3H_7NO_3$
PROPYL NITRATE, $C_3H_7NO_3$
2,2'-OXYDI(ETHYL NITRATE), $C_4H_8N_2O_7$
* 2-DIETHYLAMMONIOETHYL NITRATE NITRATE, $C_6H_{15}N_3O_6$
BENZYL NITRATE, $C_7H_7NO_3$
2(*trans*-1-AZIDO-1,2-DIHYDROACENAPHTHYL) NITRATE, $C_{12}H_8N_4O_3$

See also LITHIUM AZIDE, LiN_3: Alkyl nitrates, etc.

ALKYLNON-METAL HALIDES

This small class of flammable, air-sensitive and usually pyrophoric compounds includes the individual entries for:

TRICHLORO(METHYL)SILANE, CH_3Cl_3Si
DICHLORO(METHYL)SILANE, CH_4Cl_2Si
TRICHLORO(VINYL)SILANE, $C_2H_3Cl_3Si$
TRICHLORO(ETHYL)SILANE, $C_2H_5Cl_3Si$
IODO DIMETHYLARSINE, C_2H_6AsI
DICHLORODIMETHYLSILANE, $C_2H_6Cl_2Si$
DICHLORO(ETHYL)SILANE, $C_2H_6Cl_2Si$
BUTYLDICHLOROBORANE, $C_4H_9BCl_2$
DICHLORODIETHYLSILANE, $C_4H_{10}Cl_2Si$

See related HALOSILANES

ALKYLNON-METAL HYDRIDES

Several of the partially lower-alkylated derivatives of non-metal hydrides are pyrophoric in air. Individually indexed compounds are:

METHYLPHOSPHINE, CH_5P
METHYLSILANE, CH_6Si
DIMETHYLARSINE, C_2H_7As
DIMETHYLPHOSPHINE, C_2H_7P
ETHYLPHOSPHINE, C_2H_7P
1,1-DIMETHYLDIBORANE, $C_2H_{10}B_2$
1,2-DIMETHYLDIBORANE, $C_2H_{10}B_2$
DIETHYLARSINE, $C_4H_{11}As$
DIETHYLPHOSPHINE, $C_4H_{11}P$

See also the fully alkylated class, ALKYLNON-METALS

ALKYLNON-METALS

Several of the fully lower-alkylated non-metals are pyrophoric in air. Individually indexed compounds are:

TRIMETHYLARSINE, C_3H_9As
TRIMETHYLBORANE, C_3H_9B
TRIMETHYLPHOSPHINE, C_3H_9P
ETHYLDIMETHYLPHOSPHINE, $C_4H_{11}P$
TETRAMETHYLDIARSANE, $C_4H_{12}As_2$
* BIS-DIMETHYLARSINYL OXIDE, $C_4H_{12}As_2O$
* BIS-DIMETHYLARSINYL SULPHIDE, $C_4H_{12}As_2S$
TETRAMETHYLSILANE, $C_4H_{12}Si$
DIETHYLMETHYLPHOSPHINE, $C_5H_{13}P$
TRIETHYLARSINE, $C_6H_{15}As$
TRIETHYLBORANE, $C_6H_{15}B$
TRIETHYLPHOSPHINE, $C_6H_{15}P$
* DIETHOXYDIMETHYLSILANE, $C_6H_{16}O_2Si$
* BIS-TRIMETHYLSILYL OXIDE, $C_6H_{18}OSi_2$
TETRAETHYLDIARSANE, $C_8H_{20}As_2$
TRIISOPROPYLPHOSPHINE, $C_9H_{21}P$
mixo-TRIBUTYLBORANE, $C_{12}H_{27}B$
TRIBUTYLPHOSPHINE, $C_{12}H_{27}P$
DIBUTYL-3-METHYL-3-BUTEN-1-YNYLBORANE, $C_{13}H_{23}B$
BIS(DIBUTYLBORINO)ACETYLENE, $C_{18}H_{36}B_2$
TRIBENZYLARSINE, $C_{21}H_{21}As$

See also the partially alkylated class, ALKYLNON-METAL HYDRIDES

ALKYL PERCHLORATES

1. Hofmann, K. A. *et al., Ber.*, 1909, **42**, 4390; Meyer, J. *et al., Z. Anorg. Chem.*, 1936, **228**, 341; Schumacher, 1960, 214
2. Burton, H. *et al., Analyst*, 1955, **80**, 4
3. Hoffman, D. M., *J. Org. Chem.*, 1971, **36**, 1716

Methyl, ethyl and propyl perchlorates, readily formed from the alcohol and anhydrous perchloric acid, are highly explosive oils, sensitive to shock, heat and friction. Many of the explosions which have occurred on contact of hydroxylic compounds with concentrated perchloric acid or anhydrous metal perchlorates are attributable to the formation and decomposition of perchlorate esters.

Safe procedures for preparation of solutions of 14 *sec*-alkyl perchlorates are described. Hot evaporation of solvent caused explosions in all cases [3].

See TRICHLOROMETHYL PERCHLORATE, CCl_4O_4
METHYL PERCHLORATE, CH_3ClO_4
ETHYL PERCHLORATE, $C_2H_5ClO_4$
3-CHLORO-2-HYDROXYPROPYL PERCHLORATE, $C_3H_6Cl_2O_5$
PROPYL PERCHLORATE, $C_3H_7ClO_4$
2(2-HYDROXYETHOXY)ETHYL PERCHLORATE, $C_4H_9ClO_6$

ALKYNES

Oxides of nitrogen

See NITROGEN OXIDE, NO: Dienes, etc.

ALLOYS (INTERMETALLIC COMPOUNDS)

The individually indexed alloys or intermetallic compounds are:

ALUMINIUM–COPPER–ZINC ALLOY, Al–Cu–Zn
ALUMINIUM AMALGAM, Al–Hg
ALUMINIUM–MAGNESIUM ALLOY, Al–Mg
ALUMINIUM–NICKEL ALLOYS, Al–Ni
ALUMINIUM–TITANIUM ALLOYS, Al–Ti
BISMUTH PLUTONIDE, BiPu
COPPER–ZINC COUPLE, Cu–Zn
SODIUM GERMANIDE, GeNa
ZINC AMALGAM, Hg–Zn
POTASSIUM–SODIUM ALLOY, K–Na
LITHIUM–TIN ALLOYS, Li–Sn
SODIUM–ANTIMONY ALLOY, Na–Sb
LEAD–ZIRCONIUM ALLOY, Pb–Zr

ALLYL COMPOUNDS

Several allyl compounds are notable for their high flammability and reactivity, including:

1-BROMO-2-PROPENE, C_3H_5Br
1-CHLORO-2-PROPENE, C_3H_5Cl
1-IODO-2-PROPENE, C_3H_5I
2-PROPEN-1-OL, C_3H_6O
2-PROPEN-1-THIOL, C_3H_6S
ALLYLAMINE, C_3H_7N
3-CYANOPROPENE, C_4H_5N
ALLYL ISOTHIOCYANATE, C_4H_5NS
ALLYL FORMATE, $C_4H_6O_2$
ALLYL ETHYL ETHER, $C_5H_{10}O$
DIALLYL ETHER, $C_6H_{10}O$
DIALLYL SULPHATE, $C_6H_{10}O_4S$
DIALLYL SULPHIDE, $C_6H_{10}S$
DIALLYLAMINE, $C_6H_{11}N$
DIALLYL PHOSPHITE, $C_6H_{11}O_3P$
DIALLYL PEROXODICARBONATE, $C_8H_{10}O_6$
ALLYL BENZENESULPHONATE, $C_9H_{10}O_3S$
TRIALLYL PHOSPHATE, $C_9H_{15}O_4P$

See other PEROXIDISABLE COMPOUNDS

AMMINECHROMIUM PEROXOCOMPLEXES

This class of compounds, previously described as 'amine perchromates', is characterised by the presence of basic nitrogen and peroxo ligands within the same coordination sphere. This creates a high tendency towards explosive decomposition, which sometimes apparently occurs spontaneously. Individually indexed compounds are:

* DIMETHYLETHEROXODIPEROXOCHROMIUM(VI), $C_2H_6CrO_6$
1,2-DIAMINOETHANEAQUADIPEROXOCHROMIUM(IV), $C_2H_{10}CrN_2O_5$
1,2-DIAMINOETHANEAMMINEDIPEROXOCHROMIUM(IV), $C_2H_{11}CrN_3O_4$
1,2-DIAMINOPROPANEAQUADIPEROXOCHROMIUM(IV), $C_3H_{12}CrN_2O_5$
1,2-DIAMINO-2-METHYLPROPANEOXODIPEROXOCHROMIUM(VI), $C_4H_{12}CrN_2O_5$
DIETHYLENETRIAMINEDIPEROXOCHROMIUM(IV), $C_4H_{13}CrN_3O_4$

1,2-DIAMINO-2-METHYLPROPANEAQUADIPEROXOCHROMIUM (IV), $C_4H_{14}CrN_2O_5$
OXODIPEROXODIPYRIDINECHROMIUM(VI), $C_{10}H_{10}CrN_2O_5$
OXODIPEROXODIPIPERIDINECHROMIUM(VI), $C_{10}H_{22}CrN_2O_5$
DIANILINEOXODIPEROXOCHROMIUM(VI), $C_{12}H_{14}CrN_2O_5$
TRIAMMINEDIPEROXOCHROMIUM(IV), $CrH_9N_3O_4$

See related AMMINEMETAL OXOSALTS

AMMINECOBALT(III) AZIDES

Joyner, T. B., West States Sect. Combust. Inst., 1967, Paper WSS/CI-67-15 (*Chem. Abs.*, 1970, **72**, 113377h)

The explosive properties of a series of five amminecobalt(III) azides were examined in detail. Compounds are hexaammine-triazide; pentaammine azido-diazide; *cis*- and *trans*-tetraammine-diazido-azide; triammine-triazide.

See related METAL AZIDES

AMMINEMETAL OXOSALTS

1. Mellor, 1941, Vol. 2, 341–364, 404
2. Tomlinson, W. R. *et al., J. Amer. Chem. Soc.*, 1949, **71**, 375
3. Wendlandt, W. W. *et al., Thermal Properties of Transition Metal Complexes,* Barking, Elsevier, 1967
4. Bretherick, L., *J. Chem. Educ.*, 1970, **47**, A204
5. Ray, P., *Chem. Rev.*, 1961, **61**, 313

Metal compounds containing both coordinated ammonia, hydrazine or similar nitrogenous donors, and coordinated or ionic perchlorate, chlorate, nitrate, nitrite, nitro, permanganate or other oxidising groups, will decompose violently under various conditions of impact, friction or heat [1, 2].

From tabulated data for 17 such compounds of Co and Cr, it is considered that oxygenated *N*-coordination compounds cover a wide range of explosive types; many may explode powerfully with little provocation, and should be considered extremely dangerous, as some are sensitive enough to propagate explosion under water. The same considerations may be expected to apply to ammines of silver, gold, cadmium, lead and zinc which contain oxidising radicals [2]. The topic has been reviewed [3] and possible hazards in published student preparations were emphasised [4]. Some of the derivatives of metal biguanide and guanylurea complexes [5] are of this class.

Individually indexed compounds in this large class are:

PENTAAMMINETHIOCYANATOCOBALT(III) PERCHLORATE, $CH_{15}Cl_2CoN_6O_8S$
PENTAAMMINETHIOCYANATORUTHENIUM(III) PERCHLORATE, $CH_{15}Cl_2N_6O_8RuS$
TETRAAMMINEDITHIOCYANATOCOBALT(III) PERCHLORATE, $C_2H_{12}ClCoN_6O_4S_2$
BIS-1,2-DIAMINOETHANEDICHLOROCOBALT(III) CHLORATE, $C_4H_{16}Cl_3CoN_4O_3$
BIS-1,2-DIAMINOETHANEDICHLOROCOBALT(III) PERCHLORATE, $C_4H_{16}Cl_3CoN_4O_4$
cis-BIS-1,2-DIAMINOETHANEDINITROCOBALT(III) IODATE, $C_4H_{16}CoIN_6O_7$
BIS(1,2-DIAMINOETHANE)HYDROXOOXORHENIUM(V) DIPERCHLORATE, $C_4H_{17}Cl_2N_4O_{10}Re$
PENTAAMMINEPYRAZINERUTHENIUM(III) DIPERCHLORATE, $C_4H_{19}Cl_2N_7O_8Ru$
PENTAAMMINEPYRIDINERUTHENIUM(III) DIPERCHLORATE, $C_5H_{20}Cl_2N_6O_8Ru$
BIS-1,2-DIAMINOPROPANE-*cis*-DICHLOROCHROMIUM(III) PERCHLORATE, $C_6H_{20}Cl_3CrN_4O_4$
HEXAUREAGALLIUM(III) PERCHLORATE, $C_6H_{24}Cl_3GaN_{12}O_{18}$
TRIS-1,2-DIAMINOETHANECOBALT(III) NITRATE, $C_6H_{24}CoN_9O_9$
HEXAUREACHROMIUM(III) NITRATE, $C_6H_{24}CrN_{15}O_{15}$
BIS-DIETHYLENETRIAMINECOBALT(III) PERCHLORATE, $C_8H_{26}Cl_3CoN_6O_{12}$
DIPYRIDINESILVER(I) PERCHLORATE, $C_{10}H_{10}AgClN_2O_4$
1,4,8,11-TETRAAZACYCLOTETRADECANENICKEL(II) PERCHLORATE, $C_{10}H_{24}Cl_2NiO_8$
TETRAACRYLONITRILECOPPER(I) PERCHLORATE, $C_{12}H_{12}ClCuN_4O_4$
TETRAACRYLONITRILECOPPER(II) PERCHLORATE, $C_{12}H_{12}Cl_2CuN_4O_8$
DI[TRIS-1,2-DIAMINOETHANECOBALT(III)] TRIPEROXODISULPHATE, $C_{12}H_{24}Co_2N_{12}O_{24}S_6$
DI[TRIS-1,2-DIAMINOETHANECHROMIUM(III)] TRIPEROXODISULPHATE, $C_{12}H_{24}Cr_2N_{12}O_{24}S_6$
BROMO-5,7,7,12,14,14-HEXAMETHYL-1,4,8,11-TETRAAZA-4, 11-CYCLOTETRADECADIENEIRON(II) PERCHLORATE, $C_{16}H_{32}BrClFeN_4O_4$
IODO-5,7,7,12,14,14-HEXAMETHYL-1,4,8,11-TETRAAZA-4, 11-CYCLOTETRADECADIENEIRON(II) PERCHLORATE, $C_{16}H_{32}ClFeIN_4O_4$

CHLORO-5,7,7,12,14,14-HEXAMETHYL-1,4,8,11-TETRAAZA-4,11-CYCLOTETRADECADIENEIRON(II) PERCHLORATE, $C_{16}H_{32}Cl_2FeN_4O_4$

DICHLORO-5,7,7,12,14,14-HEXAMETHYL-1,4,8,11-TETRAAZA-4,11-CYCLOTETRADECADIENEIRON(III) PERCHLORATE, $C_{16}H_{32}Cl_3FeN_4O_4$

HEXAMETHYLENETETRAMMONIUM TETRAPEROXOCHROMATE (V) (?), $C_{18}H_{48}Cr_4N_{12}O_{32}$

DIACETONITRILE-5,7,7,12,14,14-HEXAMETHYL-1,4,8,11-TETRAAZA-4,11-CYCLOTETRADECADIENEIRON(II) PERCHLORATE, $C_{20}H_{36}Cl_2FeN_6O_8$

ACETONITRILEIMIDAZOLE-5,7,7,12,14,14-HEXAMETHYL-1,4,8,11-TETRAAZA-4,11-CYCLOTETRADECADIENEIRON(II) PERCHLORATE, $C_{21}H_{39}Cl_2FeN_7O_8$

4-[2-(4-HYDRAZINO-1-PHTHALAZINYL)HYDRAZINO]-4-METHYL-2-PENTANONE (4-HYDRAZINO-1-PHTHALAZINYL) HYDRAZONEDINICKEL(II) TETRAPERCHLORATE, $C_{22}H_{28}N_{12}Ni_2Cl_4O_{16}$

TRIS-2,2′-BIPYRIDINESILVER(II) PERCHLORATE, $C_{30}H_{24}AgCl_2N_6O_8$

TRIS-2,2′-BIPYRIDINECHROMIUM(II) PERCHLORATE, $C_{30}H_{24}Cl_2CrN_6O_8$

* TRIS-2,2′-BIPYRIDINECHROMIUM (0), $C_{30}H_{24}CrN_6$

HEXAPYRIDINEIRON(II) TRIDECACARBONYLTETRAFERRATE (2−), $C_{43}H_{30}Fe_5N_6O_{13}$

OXYBIS (*N,N*-DIMETHYLACETAMIDETRIPHENYLSTIBONIUM) DIPERCHLORATE, $C_{44}H_{50}Cl_2N_2O_{11}Sb$

TETRAAMMINECADMIUM(II) PERMANGANATE, $CdH_{12}Mn_2N_4O_8$

DIHYDRAZINECOBALT(II) CHLORATE, $Cl_2CoH_8N_4O_6$

* HEXAAQUACOBALT(II) PERCHLORATE, $Cl_2CoH_{12}O_{14}$

PENTAAMMINEPHOSPHINECOBALT(III) PERCHLORATE, $Cl_2CoH_{17}N_5O_{10}P$

* BISHYDROXYLAMINEZINC(II) CHLORIDE, $Cl_2H_6N_2O_2Zn$

BISHYDRAZINENICKEL(II) PERCHLORATE, $Cl_2H_8N_4NiO_8$

* BISHYDRAZINETIN(II) CHLORIDE, $Cl_2H_8N_4Sn$

TETRAAMMINEBIS-DINITROGENOSMIUM DIPERCHLORATE, $Cl_2H_{12}N_8O_8Os$

PENTAAMMINECHLOROCOBALT(III) PERCHLORATE, $Cl_3CoH_{15}N_5O_8$

PENTAAMMINEAQUACOBALT(III) CHLORATE, $Cl_3CoH_{17}N_5O_{10}$

HEXAAMMINECOBALT(III) CHLORITE, $Cl_3CoH_{18}N_6O_6$

HEXAAMMINECOBALT(III) CHLORATE, $Cl_3CoH_{18}N_6O_9$

HEXAAMMINECOBALT(III) PERCHLORATE, $Cl_3CoH_{18}N_6O_{12}$

DIAMMINENITRATOCOBALT(II) NITRATE, $CoH_6N_4O_6$

* TRIAMMINETRINITROCOBALT(III), $CoH_9N_6O_6$

TRIHYDRAZINECOBALT(II) NITRATE, $CoH_{12}N_8O_6$

* AMMONIUM HEXANITROCOBALTATE (3−), $CoH_{12}N_9O_{12}$

PENTAAMMINENITRATOCOBALT(III) NITRITE, $CoH_{15}N_8O_7$
HEXAAMMINECOBALT(III) IODATE, $CoH_{18}I_3N_6O_9$
HEXAAMMINECOBALT(III) PERMANGANATE, $CoH_{18}Mn_3N_6O_{12}$
HEXAHYDROXYLAMINECOBALT(III) NITRATE, $CoH_{18}N_9O_{15}$
HEXAAMMINECOBALT (3+) HEXANITROCOBALTATE (3−), $Co_2H_{18}N_{12}O_{12}$
HEXAAMMINECHROMIUM(III) NITRATE, $CrH_{18}N_9O_9$
TETRAAMMINECOPPER(II) NITRITE, $CuH_{12}N_6O_4$
TETRAAMMINECOPPER(II) NITRATE, $CuH_{12}N_6O_6$
DIAMMINEPALLADIUM(II) NITRATE, $H_6N_4O_6Pd$
TRIAMMINENITRATOPLATINUM(II) NITRATE, $H_9N_5O_6Pt$
TETRAAMMINEZINC PEROXODISULPHATE, $H_{12}N_4O_8S_2Zn$
TETRAAMMINEPALLADIUM(II) NITRATE, $H_{12}N_6O_6Pd$
TRIHYDRAZINENICKEL(II) NITRATE, $H_{12}N_8NiO_6$
TETRAAMMINEHYDROXONITRATOPLATINUM(IV) NITRATE, $H_{13}N_7O_{10}Pt$

See related AMMINECHROMIUM PEROXOCOMPLEXES
[14] DIENE-N_4 IRON COMPLEXES

AMMONIUM PERCHLORATES

Datta, R. L. *et al.*, *J. Chem. Soc.*, 1919, **115**, 1006–1010
Many perchlorate salts of amines explode in the range 215–310°C, and some on impact at ambient temperature.
See other OXOSALTS OF NITROGENOUS BASES

APROTIC SOLVENTS

1. Buckley, A., *J. Chem. Educ.*, 1965, **42**, 674
2. Banthorpe, D. V., *Nature*, 1967, **215**, 1296

Many aprotic (non-hydroxylic) solvents are not inert towards other reagents and care is necessary when using untried combinations of solvents and reagents for the first time.

A further potential hazard which should be considered is that some aprotic solvents, notably dimethyl sulphoxide [1] and *N,N*-dimethylformamide [2], may greatly promote the toxic properties of solutes because of their unique ability readily to penetrate synthetic rubber protective gloves and the skin.

See METALS: Halocarbons
FORMAMIDE, CH_3NO
DIMETHYL SULPHOXIDE, C_2H_6OS
N,N-DIMETHYLFORMAMIDE, C_3H_7NO

AQUA REGIA

Fawcett, H. H. *et al., Chem. Eng. News*, **33**, 897, 1406, 1622, 1844
Aqua regia (nitric and hydrochloric acids, 1:4), which had been used for cleaning purposes, was stored in screw-capped winchesters. Internal pressure developed overnight, one bottle being shattered. Aqua regia decomposes with evolution of gas and should not be stored in tightly closed bottles. Undersized stoppers are suggested. It is also a powerful oxidant.

ARENEDIAZO ARYL SULPHIDES

See DIAZONIUM SULPHIDES AND DERIVATIVES

ARENEDIAZOATES

Houben-Weyl, 1965, Vol. 10 (3), 563–564
Alkyl and aryl arenediazoates ('diazoethers'), Ar—N=N—O—R, are generally unstable and even explosive compounds. They are produced by interaction of alcohols with (explosive) bis(arenendiazo) oxides, or of *p*-blocked phenols with diazonium salts. The thio analogues are similar.

See METHYL 4-BROMOBENZENEDIAZOATE, $C_7H_7BrN_2O$
METHYL 2-NITROBENZENEDIAZOATE, $C_7H_7N_3O_3$
METHYL BENZENEDIAZOATE, $C_7H_8N_2O$
4-NITROPHENYL 2-CARBOXYBENZENEDIAZOATE, $C_{13}H_9N_3O_5$

See related DIAZONIUM SULPHIDES

ARENEDIAZONIUMOLATES

This group of internal diazonium salts (previously called diazooxides) are, like many other internal diazonium salts, explosively unstable and shock-sensitive materials.

Individually indexed compounds are:

4-CHLORO-2,5-DINITROBENZENEDIAZONIUM-6-OLATE, $C_6HClN_4O_5$
3,4-DIFLUORO-2-NITROBENZENEDIAZONIUM-6-OLATE, $C_6HF_2N_3O_3$
3,6-DIFLUORO-2-NITROBENZENEDIAZONIUM-4-OLATE, $C_6HF_2N_3O_3$
4,6-DINITROBENZENEDIAZONIUM-2-OLATE, $C_6H_2N_4O_5$
3-BROMO-2,7-DINITRO-5-BENZO[b]THIOPHENEDIAZONIUM-4-OLATE, $C_8HBrN_4O_5S$

See also DIAZONIUM CARBOXYLATES
5-DIAZONIOTETRAZOLIDE, CN_6
BENZENEDIAZONIUM-4-SULPHONATE, $C_6H_4N_2O_3S$

ARYL CHLOROFORMATES

Water

Muir, G.D., private comm., 1968

During preparation of aryl chloroformates, it is essential to keep the reaction mixture really cold during water washing to prevent vigorous decomposition. Phenyl and naphthyl chloroformates may be distilled, but benzyl chloroformate is considered too unstable.

ARYLMETALS

This reactive group includes the individually indexed compounds:

CYCLOPENTADIENYLGOLD, C_5H_5Au
CYCLOPENTADIENYLSODIUM, C_5H_5Na
m-DILITHIOBENZENE, $C_6H_4Li_2$
p-DILITHIOBENZENE, $C_6H_4Li_2$
PHENYLSILVER, C_6H_5Ag
PHENYLLITHIUM, C_6H_5Li
TOLYLCOPPER (*o*-, *m*-, *p*-), C_7H_7Cu

NAPHTHYLSODIUM, $C_{10}H_7Na$
DIPHENYLMERCURY, $C_{12}H_{10}Hg$
DIPHENYLMAGNESIUM, $C_{12}H_{10}Mg$
DIPHENYLTIN, $C_{12}H_{10}Sn$
TRIPHENYLALUMINIUM, $C_{18}H_{15}Al$

ARYLTHALLIC ACETATE PERCHLORATES

See PERCHLORIC ACID, $ClHO_4$: Ethylbenzene, etc.

AUTOIGNITION TEMPERATURE

1. Hilado, C. J. *et al., Chem. Eng.*, 1972, **79**(19), 75–80
2. *Fire Hazard Properties of Flammable Liquids, Gases, Solids,* No. 325M, Boston, Nat. Fire Prot. Ass., 1969

Autoignition temperature (AIT) is the temperature at which a material in contact with air undergoes oxidation at a sufficiently high rate to initiate combustion without an external ignition source.

Though only those compounds with unusually low AITs (below 225°C) have been included in this Handbook, the reference above is to a compilation of data for over 300 organic compounds, which also includes the theoretical background and discussion of the effect of variations in test methods upon AIT values obtained [1]. Further AIT data are given in the tabulated publication [2].

AUTOXIDATION

1. Davies, 1961, 11
2. Ingold, K. U., *Chem. Rev.*, 1961, **61**, 563

Autoxidation (interaction of a substance with molecular oxygen at below 120°C without flame [1]) has often been involved in the generation of hazardous materials from reactive compounds exposed to air. Methods of inhibiting autoxidation of organic compounds in the liquid phase have been reviewed [2].

See PEROXIDISABLE COMPOUNDS

AZIDES

Many compounds of both organic and inorganic derivation, which contain the azide function, are unstable or explosive under appropriate conditions of initiation. The large number of compounds has been subdivided for convenience on the basis of structure. The inorganic groups are:

ACYL AZIDES
METAL AZIDES
NON-METAL AZIDES

and the organic groups:

ACYL AZIDES
2-AZIDOCARBONYL COMPOUNDS
ORGANIC AZIDES

2-AZIDOCARBONYL COMPOUNDS

Boyer, J. H. *et al., Chem. Rev.*, 1954, **54**, 33
Certain 2-azidocarbonyl compounds and congeners have long been known as unstable substances.

* AZIDOACETONITRILE, $C_2H_2N_4$
AZIDOACETALDEHYDE, $C_2H_3N_3O$
AZIDOACETONE, $C_3H_5N_3O$
AZIDOACETONE OXIME, $C_3H_6N_4O$
* DIAZIDOMALONONITRILE, C_3N_8
TETRAAZIDO-*p*-BENZOQUINONE, $C_6N_{12}O_2$

AZO COMPOUNDS

The common structural characteristic of this group of unstable compounds is the $-N{=}N-$ group linked to two different carbon atoms. Individually indexed compounds are:

* METHYLDIAZENE, CH_4N_2
AZO-*N*-CHLOROFORMAMIDINE, $C_2H_4Cl_2N_6$
AZO-*N*-NITROFORMAMIDINE, $C_2H_4N_8O_4$
* ISOPROPYLDIAZENE, $C_3H_8N_2$
DIMETHYL AZODIFORMATE, $C_4H_6N_2O_4$

DIETHYL AZODIFORMATE, $C_6H_{10}N_2O_4$
AZOISOBUTYRONITRILE, $C_8H_{12}N_4$
3,4-DIMETHYL-4-(3,4-DIMETHYL-5-ISOXAZOLYLAZO)-ISOXAZOLIN-5-ONE, $C_{10}H_{12}N_4O_3$
* POTASSIUM AZODISULPHONATE (2−), $K_2N_2O_6S_2$

BIS-ARENEDIAZO OXIDES

1. Bamberger, E., *Ber.*, 1896, **29**, 451
2. Kaufmann, T. *et al.*, *Ann.*, 1960, **634**, 77

Action of alkalies on diazonium solutions, or of acids on alkali diazoates to give a final pH of 5–6, causes these compounds ('diazoanhydrides') to separate as oils or solids. Many of these are violently explosive (some exceeding nitrogen trichloride in effect), sensitive to friction and heat or contact with aromatic hydrocarbons.

See BIS-2,4,5-TRICHLOROBENZENEDIAZO OXIDE, $C_{12}H_4Cl_6N_4O$
BIS-*p*-CHLOROBENZENEDIAZO OXIDE, $C_{12}H_8Cl_2N_4O$
BIS-BENZENEDIAZO OXIDE, $C_{12}H_{10}N_4O$
BIS-TOLUENEDIAZO OXIDE, $C_{14}H_{14}N_4O$

See related BIS-ARENEDIAZO SULPHIDES

BIS-ARENEDIAZO SULPHIDES

Some of the products of interaction of diazonium salts with sulphides may have this structure.

See DIAZONIUM SULPHIDES

BLEACHING POWDER

1. Mellor, 1956, Vol. 2, Suppl. 1,564–567
2. *Accid. Bull.* No. 30, Washington, Amer. Rail Assoc. Bur. Explosives, 1921
3. Gill, A. H., *Ind. Eng. Chem.*, 1924, **16**, 577
4. 'Leaflet No. 6', London, Inst. Chem., 1941

Bleaching powder is effectively a mixture of calcium hypochlorite, calcium hydroxide and a non-hygroscopic form of calcium chloride [1] and may therefore be regarded as a less active form of oxidant than undiluted calcium hypochlorite. There is a long history of explosions, many apparently spontaneous, involving bleaching powder. On storage or heating, several modes of decomposition are possible, one involving formation of chlorate which may increase the hazard potential.

Of the three possible routes for thermal decomposition, that involving liberation of oxygen predominates as the water content decreases, and at 150°C the decomposition becomes explosive [1].

Material which has been stored for a long time is liable to explode on exposure to sunlight, or on overheating of tightly packed material in closed containers [2]. The spontaneous explosion of material packed in drums was attributed to catalytic liberation of oxygen by iron and manganese oxides present in the lime used for manufacture [3]. Traces of metallic cobalt, iron, magnesium or nickel may also catalyse explosive decomposition [1].

When the lever-lid of a 6-month-old tin of bleaching powder was being removed, it flew off with explosive violence, possibly due to rust-catalysed slow liberation of oxygen [4].

Bis (2-chloroethyl) sulphide
Mellor, 1956, Vol. 2, Suppl. 1, 567
Interaction is very exothermic and ignition may occur, particularly in presence of water.

Wood
ABCM Quart. Safety Summ., 1933, **4**, 15
A mixture of sawdust and bleaching powder ignites when moistened.

See other OXIDANTS

CAN OF BEANS

Foote, C. S., private comm., 1965
An unopened can of beans, placed in an oven initially at 110°C, later reset to 150°C, exploded, causing extensive damage. Comments were judged to be superfluous.

CARBONACEOUS DUSTS

Explosion and Ignition Hazards, Report 6597, Washington, US Bureau of Mines, 1965

Hazards of 241 industrial dusts which may explode or burn because of their carbon content are defined, covering particle size and chemical composition in 10 categories.

CARBONYLMETALS

Many members of this reactive class are air-sensitive and pyrophoric (not always immediately), individually indexed compounds being:

CARBONYLPOTASSIUM, CKO
CARBONYLLITHIUM, CLiO
CARBONYLSODIUM, CNaO
TETRACARBONYLNICKEL, C_4NiO_4
PENTACARBONYLIRON, C_5FeO_5
HEXACARBONYLCHROMIUM, C_6CrO_6
HEXACARBONYLMOLYBDENUM, C_6MoO_6
HEXACARBONYLTUNGSTEN, C_6O_6W
OCTACARBONYLDICOBALT, $C_8Co_2O_8$
NONACARBONYLDIIRON, $C_9Fe_2O_9$
DODECACARBONYLTETRACOBALT, $C_{12}Co_4O_{12}$
DODECACARBONYLTRIIRON, $C_{12}Fe_3O_{12}$
DODECACARBONYLDIVANADIUM, $C_{12}O_{12}V_2$

CELLULOSE

Calcium oxide

See CALCIUM OXIDE, CaO: Water

Oxidants

See BLEACHING POWDER: Wood
PERCHLORATES; Organic matter
PERCHLORIC ACID, $ClHO_4$: Cellulose and derivatives
SODIUM CHLORATE, $ClNaO_3$: Paper, etc; Wood
MAGNESIUM DIPERCHLORATE, Cl_2MgO_8 : Cellulose, etc.
FLUORINE, F_2 : Miscellaneous materials
NITRIC ACID, HNO_3 : Cellulose

SODIUM NITRITE, $NNaO_2$: Wood
SODIUM NITRATE, $NNaO_3$: Fibrous material
SODIUM PEROXIDE, Na_2O_2 : Fibrous materials

CELLULOSE NITRATE

1. Kirk-Othmer, 1964, Vol. 4, 627–628
2. Anon., *Accidents,* 1961 (46), 12
3. *ABCM Quart. Safety Summ.,* 1963, **34**, 13
4. *MCA Case History No. 1614*

Cellulose nitrate is very easily ignited and burns very rapidly or explosively, depending on the degree of confinement and state of subdivision. Unless very pure and stabilised, it deteriorates in storage and may ignite spontaneously [1].

Removal of the emulsion coating from celluloid film base gives unstable material which may spontaneously ignite on prolonged storage in an enclosed space [2].

Unused but aged centrifuge tubes, made of 'Nitrocellulose' plasticised with dibutyl phthalate, ignited and exploded while being steam-sterilised in an autoclave at 125°C. The violent decomposition was attributed to the age of the tubes, the high temperature and the presence of steam [3].

During hacksaw cutting of a pipe containing cellulose nitrate residues, a violent explosion occurred [4].

Amines

1. *ABCM Quart. Safety Summ.,* 1956, **27**, 2
2. Thurlow, G.K. *et al.,* private comm., 1973

Cellulose nitrate of high surface area (dry or alcohol-wet guncotton or scrap) spontaneously ignited in contact with various amines used as curing agents for epoxide resins. These included: 1,2-diaminoethane, *N*-hydroxyethyl-1,2-diaminoethane, diethylenetriamine, triethylenetetramine, *N*-2-hydroxyethyltriethylenetetramine, tetraethylenepentamine, 2-hydroxyethylamine, 2-hydroxyethyl-dimethylamine, 2-hydroxypropylamine, 3-dimethylaminopropylamine, 3-diethylaminopropylamine, 3-dibutylaminopropylamine, morpholine, diethylamine. Ethylamine and dibutylamine caused charring but not ignition [1].

Similar results were obtained during an investigation of the compatibility of cellulose nitrate with a range of amine and amide components used in paint manufacture. Preliminary small-scale (12g) tests in which ethyl acetate solutions of the nitrate and other components were mixed in a lagged boiling tube showed large exotherms (which boiled the solvent off) with 1,4-diazabicyclo [2.2.2]octane, 2,4,6-tris(dimethylaminomethyl)phenol, morpholine and benzyldimethylamine. Smaller exotherms were shown by dodecanamine, dodecyldimethylamine (both fat-derived, containing homologues) and a polyamide resin, Versamid 140.

Subsequent tests in which small portions of these undiluted liquid amines and dried cellulose nitrate linters were contacted (with a little added butyl acetate for the solid phenol) under various conditions gave ignition with the first three amines, and exotherms to 110°C with foaming decomposition with the remaining four.

Other amine resin components showed slight or no exotherms in either test [2].

CHLORINATED RUBBER

Metal oxides or hydroxides

1. Anon., *Chem. Trade J.*, 1962, **151**, 672
2. *Euro. Chem. News*, 1963 (May 24th), 29
3. 'Report GCS 27130', London, ICI, 1963
4. *ABCM Quart. Safety Summ.*, 1963, **34**, 12

Intimate mixtures of chlorinated rubber and zinc oxide or powdered zinc, with or without hydrocarbon or chlorinated solvents, react violently or explosively when heated at about 216°C. If in milling such mixtures local overheating occurs, a risk of a violent reaction exists. Such risks can be minimised by controlling milling temperatures, by cooling, or by using a mixture of maximum possible fluidity [1]. Similar reactions have been observed with antimony or lead oxides, or aluminium, barium or zinc hydroxides [2]. The full report [3] has been abstracted [4].

CHLORITE SALTS

Mellor 1941, Vol. 2, 284; 1956, Vol. 2, Suppl. 1, 573–575

Many of the salts which have been prepared are explosive and sensitive to heat or impact. These include chlorites of copper (violent on impact), hydrazine (monochlorite, inflames when dry), lead, mercury (spontaneously explosive dry), nickel (explodes at 100°C but not on impact), silver (at 105°C or on impact), sodium, tetramethylammonium and thallium (which shows detonator properties). Several other chlorites not isolated and unstable in solution include ammonium chlorite and its mono-, di- and trimethyl- derivatives. The metal salts are powerful oxidants.

N-CHLORONITROAMINES

Grakauskas, V. *et al., J. Org. Chem.,* 1972, **37**, 334
N-Chloronitroamines and the derived *N*-chloro-*N*-nitrocarbamates are explosive compounds and decompose rapidly on storage.
See also *N*-FLUORO-*N*-NITROBUTYLAMINE, $C_4H_9FN_2O_2$
See other *N*-HALOGEN COMPOUNDS
N-NITRO COMPOUNDS

COMMERCIAL ORGANIC PEROXIDES

This group of compounds is widely used in industry as a radical source for initiation of polymerisation. They are available from several manufacturers in a very wide range of formulations in various diluents to minimise operational hazards. These have been classified into six hazard levels and of the many materials available, the few below (all dry and unformulated except for *, which is water-wetted) are included in the highest risk category. This specifies the material as being sensitive to friction or mechanical shock equivalent to the dissipation of 1kg-m or less of energy within the sample.

See BIS(3-CARBOXYPROPIONYL) PEROXIDE, $C_8H_{10}O_8$
O,O-tert-BUTYL HYDROGENMONOPEROXYMALEATE, $C_8H_{12}O_5$
* ACETYL CYCLOHEXANESULPHONYL PEROXIDE, $C_8H_{14}O_5S$
DIISOPROPYL PEROXYDICARBONATE, $C_8H_{14}O_6$
BIS(1-HYDROPEROXYCYCLOHEXYL) PEROXIDE, $C_{12}H_{22}O_6$
BIS(2,4-DICHLOROBENZOYL) PEROXIDE, $C_{14}H_6Cl_4O_4$

COMPLEX HYDRIDES

Gaylord, 1956

This class of highly reactive compounds includes several which have found extensive use as reducants in preparative chemistry.

ALUMINIUM TETRAHYDROBORATE, AlB_3H_{12}

* ALUMINIUM DICHLORIDE HYDRIDE DIETHYLETHERATE, $AlCl_2H \cdot C_4H_{10}O$

CAESIUM HEXAHYDROALUMINATE(3–), $AlCs_3H_6$

COPPER(I) TETRAHYDROALUMINATE, $AlCuH_4$

LITHIUM TETRAHYDROALUMINATE, AlH_4Li

SODIUM TETRAHYDROALUMINATE, AlH_4Na

POTASSIUM HEXAHYDROALUMINATE (3–), AlH_6K_3

MAGNESIUM TETRAHYDROALUMINATE, Al_2H_8Mn

MANGANESE(II) TETRAHYDROALUMINATE, Al_2H_8Mn

CERIUM(III) TETRAHYDROALUMINATE, Al_3CeH_{12}

LITHIUM TETRAHYDROBORATE, BH_4Li

SODIUM TETRAHYDROBORATE, BH_4Na

HYDRAZINE-MONOBORANE, BH_7N_2

BERYLLIUM TETRAHYDROBORATE, B_2BeH_8

HYDRAZINE-BISBORANE, $B_2H_{10}N_2$

AMMONIUM [AMINYLENIUMBIS(TRIHYDROBORATE)](1–), $B_2H_{12}N_2$

URANIUM(III) TETRAHYDROBORATE, $B_3H_{12}U$

HAFNIUM(IV) TETRAHYDROBORATE, $B_4H_{16}Hf$

ZIRCONIUM(IV) TETRAHYDROBORATE, $B_4H_{16}Zr$

BERYLLIUM TETRAHYDROBORATE-TRIMETHYLAMINE, $C_3H_{17}B_2BeN$

BIS(DIMETHYLAMINOBORANE)ALUMINIUM TETRAHYDROBORATE, $C_4H_{22}AlB_3N_2$

HEPTAKIS(DIMETHYLAMINO)TRIALUMINIUM TRIBORON PENTAHYDRIDE, $C_{14}H_{47}Al_3B_3N_7$

LITHIUM TETRAHYDROGALLATE, GaH_4Li

SODIUM TETRAHYDROGALLATE, GaH_4Na

See other REDUCANTS

CRYOGENIC LIQUIDS

1. Zabetakis, M. G., *Safety with Cryogenic Fluids,* London, Heywood, 1967
2. *Cryogenics Safety Manual–A Guide to Good Practice,* London, Safety Panel, British Cryogenics Council, 1970

Two reference works are available which together cover this important technological field.

The first is a monograph concisely presenting principles of safety applicable to cryogenics in 6 chapters: Introduction; Physiological Hazards; Physical Hazards; Chemical Hazards; Laboratory Safety; Plant and Test Site Safety. There are 3 Appendices: Physical Constants and Conversion Factors; Safety Data Sheets (for air, argon, carbon dioxide, carbon monoxide, ethylene, fluorine, helium, hydrogen, krypton, methane, neon, nitrogen, oxygen, ozone, xenon); Disaster Investigation (Explosions).

The second was produced for the guidance of those concerned with operation and maintenance of plant for producing, storing and handling commercial gases which liquefy at relatively low temperatures. This illustrated book covers both general safety requirements and specific safety requirements for: Air Separation Plants Producing Oxygen, Nitrogen and Argon; Liquefied Natural Gas; Hydrogen Separation Plants; Ethylene and Ethane.

CRYSTALLINE HYDROGEN PEROXIDATES

Castrantas, 1965, 4
Emeléus, 1960, 432
Kirk-Othmer, 1966, Vol. 11, 395

A few compounds will crystallise out with hydrogen peroxide in the crystal lattice, analogous to crystalline hydrates. These represent a form of concentrated hydrogen peroxide, which may react violently in close contact (grinding or heating) with oxidisable materials.

See SODIUM BORATE HYDROGEN PEROXIDATE, $BNaO_2 \cdot H_2O_2$
UREA HYDROGEN PEROXIDATE, $CH_4N_2O \cdot H_2O_2$
SODIUM CARBONATE HYDROGEN PEROXIDATE, $CNa_2O_3 \cdot 1.5H_2O_2$
TRIETHYLAMINE HYDROGEN PEROXIDATE, $C_6H_{15}N \cdot 4H_2O_2$
POTASSIUM FLUORIDE HYDROGEN PEROXIDATE, $FK \cdot H_2O_2$
SODIUM PYROPHOSPHATE HYDROGEN PEROXIDATE, $Na_4O_7P_2 \cdot 2H_2O_2$

See also HYDROGEN PEROXIDE, H_2O_2: Nitric acid, etc.

CYANO COMPOUNDS

Metal cyanides are readily oxidised and those of some heavy metals show instability. Many organic nitriles are unusually reactive under appropriate circumstances, and *N*-cyano derivatives are reactive or unstable. The class includes the groups:

3-CYANOTRIAZENES
METAL CYANIDES AND CYANOCOMPLEXES

as well as the individually indexed compounds:

DISILVER CYANAMIDE, CAg_2N_2
HYDROGEN CYANIDE, CHN
CYANAMIDE, CH_2N_2
CYANOGEN AZIDE, CN_4
ACETONITRILE, C_2H_3N
* METHYL ISOCYANIDE, C_2H_3N
GLYCOLONITRILE, C_2H_3NO
CHLOROCYANOACETYLENE, C_3ClN
2-CHLOROACRYLONITRILE, C_3H_2ClN
MALONONITRILE, $C_3H_2N_2$
ACRYLONITRILE, C_3H_3N
CYANOACETIC ACID, $C_3H_3NO_2$
2-CHLORO-1-CYANOETHANOL, C_3H_4ClNO
* ETHYL ISOCYANIDE, C_3H_5N
PROPIONITRILE, C_3H_5N
2-AMINOPROPIONITRILE, $C_3H_6N_2$
MESOXALONITRILE, C_3N_2O
PHOSPHORUS TRICYANIDE, C_3N_3P
DICYANODIAZOMETHANE, C_3N_4
DIAZIDOMALONONITRILE, C_3N_8
2-CYANO-1,2,3-TRIS(DIFLUOROAMINO)PROPANE, $C_4H_4F_6N_4$
1- and 3-CYANOPROPENE, C_4H_5N
1-CYANO-2-PROPEN-1-OL, C_4H_5NO
2-CYANO-2-PROPYL NITRATE, $C_4H_6N_2O_3$
BUTYRONITRILE, C_4H_7N
ISOBUTYRONITRILE, C_4H_7N
DIMETHYLAMINOACETONITRILE, $C_4H_8N_2$
DICYANOACETYLENE, C_4N_2
PIVALONITRILE, C_5H_9N
CYANODIETHYLGOLD: see the tetramer, $C_{20}H_{40}Au_4N_4$
3-DIMETHYLAMINOPROPIONITRILE, $C_5H_{10}N_2$
1,4-DICYANO-2-BUTENE, $C_6H_6N_2$
BIS(ACRYLONITRILE)NICKEL(0), $C_6H_6N_2Ni$
BIS(2-CYANOETHYL)AMINE, $C_6H_9N_3$
o-NITROBENZONITRILE, $C_7H_4N_2O_2$
N-CYANO-2-BROMOETHYLBUTYLAMINE, $C_7H_{13}BrN$
PHENYLACETONITRILE, C_8H_7N
AZOISOBUTYRONITRILE, $C_8H_{12}N_4$
N-CYANO-2-BROMOETHYLCYCLOHEXYLAMINE, $C_9H_{15}BrN$

TETRAACRYLONITRILECOPPER(I) PERCHLORATE, $C_{12}H_{12}ClCuN_4O_4$
TETRAACRYLONITRILECOPPER(II) PERCHLORATE, $C_{12}H_{12}Cl_2CuN_4O_8$
TETRACYANOOCTAETHYLTETRAGOLD, $C_{20}H_{40}Au_4N_4$

3-CYANOTRIAZENES

Bretschneider, H. *et al.*, *Monatsh.*, 1950, **81**, 981
Many aromatic 3-cyanotriazenes are shock-sensitive, explosive compounds.
See other HIGH-NITROGEN COMPOUNDS
TRIAZENES

CYCLIC PEROXIDES

Swern, 1970, Vol. 1, 37; 1972, Vol. 3, 67, 81
Generally produced *inter alia* during peroxidation of aldehydes or ketones, the lower members are often violently explosive. Dimeric and trimeric ketone peroxides are the most dangerous classes of organic peroxide, exploding on heating, touching or friction.

See 3,6-DIMETHYL-1,2,4,5-TETRAOXANE, $C_4H_8O_4$
3,3,6,6-TETRAMETHYL-1,2,4,5-TETRAOXANE, $C_6H_{12}O_4$
3,6-DIETHYL-3,6-DIMETHYL-1,2,4,5-TETRAOXANE, $C_8H_{16}O_4$
3,3,6,6-TETRAKIS(BROMOMETHYL)-9,9-DIMETHYL-1,2,4,5,7,8-HEXAOXAONANE, $C_9H_{14}Br_4O_6$
3,3,6,6,9,9-HEXAMETHYL-1,2,4,5,7,8-HEXAOXAONANE, $C_9H_{18}O_6$
3,6,9-TRIETHYL-1,2,4,5,7,8-HEXAOXAONANE, $C_9H_{18}O_6$
1,4-EPIDIOXY-2-*p*-MENTHENE (ASCARIDOLE), $C_{10}H_{20}O_4$
3,6-DI(SPIROCYCLOHEXANE)TETRAOXANE, $C_{12}H_{20}O_4$
3,6,9-TRIETHYL-3,6,9-TRIMETHYL-1,2,4,5,7,8-HEXAOXAONANE, $C_{12}H_{24}O_6$
9,10-EPIDIOXYANTHRACENE, $C_{14}H_8O_2$
TRI(SPIROCYCLOPENTANE)-1,1,4,4,7,7-HEXAOXAOXONANE, $C_{15}H_{24}O_6$

DEVARDA'S ALLOY

1. Cameron, W.G., *Chem. & Ind.*, 1948, 158
2. Chaudhuri, B.B., ibid., 462

The analytical use of the alloy to reduce nitrates is usually accompanied by the risk of a hydrogen explosion, particularly if heating is effected by flame. Use of a safety screen, and flameless heating, coupled with displacement of hydrogen by an inert gas, are recommended precautions [1]. The explosion was later attributed to gas pressure in a restricted system [2].
See other ALLOYS (INTERMETALLIC COMPOUNDS)

DIACYL PEROXIDES

Swern, 1970, Vol. 1, 70
Most of the isolated diacyl (including sulphonyl) peroxides are solids with relatively low decomposition temperatures, and are explosive, and sensitive to shock, heat or friction. Several, particularly the lower members, will detonate on the slightest disturbance. Autocatalytic, self-accelerating decomposition, which is promoted by tertiary amines, is involved. Individually indexed compounds are:

BISFLUOROFORMYL PEROXIDE, $C_2F_2O_4$
* *O*-TRIFLUOROACETYL-*S*-FLUOROFORMYL THIOPEROXIDE, $C_3F_4O_3S$
BISTRICHLOROACETYL PEROXIDE, $C_4Cl_6O_4$
BISTRIFLUOROACETYL PEROXIDE, $C_4F_6O_4$
DIACETYL PEROXIDE, $C_4H_6O_4$
POTASSIUM BENZENESULPHONYLPEROXOSULPHATE, $C_6H_5KO_7S_2$
DIPROPIONYL PEROXIDE, $C_6H_{10}O_4$
POTASSIUM BENZOYLPEROXOSULPHATE, $C_7H_5KO_6S$
PHTHALOYL PEROXIDE, $C_8H_4O_4$
DICROTONOYL PEROXIDE, $C_8H_{10}O_4$
BIS(3-CARBOXYPROPIONYL) PEROXIDE, $C_8H_{10}O_8$
DIISOBUTYRYL PEROXIDE, $C_8H_{14}O_4$
ACETYL CYCLOHEXANESULPHONYL PEROXIDE, $C_8H_{14}O_5S$
DI-2-FUROYL PEROXIDE, $C_{10}H_6O_6$
DI-2-METHYLBUTYRYL PEROXIDE, $C_{10}H_{18}O_4$
DIBENZENESULPHONYL PEROXIDE, $C_{12}H_{10}O_6S_2$
DIHEXANOYL PEROXIDE, $C_{12}H_{22}O_4$
BIS-2,4-DICHLOROBENZOYL PEROXIDE, $C_{14}H_6Cl_4O_4$
BIS-*o*-AZIDOBENZOYL PEROXIDE, $C_{14}H_8N_6O_4$
2,2 -BIPHENYLDICARBONYL PEROXIDE, $C_{14}H_8O_4$
DIBENZOYL PEROXIDE, $C_{14}H_{10}O_4$
DICYCLOHEXYLCARBONYL PEROXIDE, $C_{14}H_{22}O_4$
DI-1-NAPHTHOYL PEROXIDE, $C_{22}H_{14}O_4$

See PEROXYCARBONATE ESTERS

DIALKYL HYPONITRITES

Partington, J. R. *et al., J. Chem. Soc.*, 1932, **135**, 2593
The violence of the explosion when the ethyl ester was heated at 80°C was not so great as previously reported. The propyl and butyl esters explode if heated rapidly, but decompose smoothly if heated gradually.
See related AZO COMPOUNDS

DIALKYLMAGNESIUMS

Sidgwick, 1950, 233
This series, either as the free alkyls or as their etherates, is extremely reactive, igniting in air or carbon dioxide and reacting violently or explosively with alcohols, ammonia or water.
See DIMETHYLMAGNESIUM, C_2H_6Mg
DIETHYLMAGNESIUM, $C_4H_{10}Mg$
DIPHENYLMAGNESIUM, $C_{12}H_{10}Mg$
See other ALKYLMETALS

DIALKYL PEROXIDES

1. Castrantas, 1965, 12; Swern, 1970, Vol. 1, 38, 54
2. Davies, 1961, 75

The high and explosive instability of the lower dialkyl peroxides and 1,1-bis-peroxides decreases rapidly with increasing chain length and degree of branching, the di-*tert*-alkyl derivatives being amongst the most stable class of peroxides [1]. Though many 1,1-bisperoxides have been reported, few have been purified because of the higher explosion hazards compared with the monofunctional peroxides [2].
See DIMETHYL PEROXIDE, $C_2H_6O_2$
ETHYL METHYL PEROXIDE, $C_3H_8O_2$
DIETHYL PEROXIDE, $C_4H_{10}O_2$
DIPROPYL PEROXIDE, $C_6H_{14}O_2$
DI-*tert*-BUTYL PEROXIDE, $C_8H_{18}O_2$
2,2-DI(*tert*-BUTYLPEROXY)BUTANE, $C_{12}H_{26}O_4$

DIALKYLZINCS

Sidgwick, 1950, 266
The dialkylzincs up to the dibutyl derivative readily ignite and burn in air. The higher alkyls fume but do not always ignite.

DIAZIRINES

Schmitz, E. *et al., Org. Synth.*, 1965, **45**, 85
Diazirine and several of its 3-substituted homologues, formally cyclic azo compounds, are explosive on heating or impact.

See DIAZIRINE, CH_2N_2
3-METHYLDIAZIRINE, $C_2H_4N_2$
3-PROPYLDIAZIRINE, $C_4H_8N_2$
3,3-PENTAMETHYLENEDIAZIRINE, $C_6H_{10}N_2$
PHENYLCHLORODIAZIRINE, $C_7H_5ClN_2$

See related AZO COMPOUNDS

DIAZO COMPOUNDS

In this group of reactive and unstable compounds the common structural feature is two nitrogen atoms attached to the same carbon atom. Individually indexed compounds are:

DIAZOMETHYLLITHIUM, $CHLiN_2$
DIAZOMETHYLSODIUM, CHN_2Na
DIAZOMETHANE, CH_2N_2
ISODIAZOMETHANE, CH_2N_2
DINITRODIAZOMETHANE, CN_4O_4
DIAZOACETONITRILE, C_2HN_3
METHYL DIAZOACETATE, $C_3H_4N_2O_2$
DICYANODIAZOMETHANE, C_3N_4
DIAZOCYCLOPENTADIENE, $C_5H_4N_2$
2-DIAZOCYCLOHEXANONE, $C_6H_8N_2O$
2-BUTEN-1-YL DIAZOACETATE, $C_6H_8N_2O_2$
tert-BUTYL DIAZOACETATE, $C_6H_{10}N_2O_2$
tert-BUTYL 2-DIAZOACETOACETATE, $C_8H_{12}N_2O_3$
1-DIAZOINDENE, $C_9H_6N_2$
1,1-BENZOYLPHENYLDIAZOMETHANE, $C_{14}H_{10}N_2O$

DIAZONIUM CARBOXYLATES

Several of these internal salts, prepared by diazotisation of anthranilic acids, are explosive in the solid state, or react violently with various materials.

See 4-IODOBENZENEDIAZONIUM-2-CARBOXYLATE, $C_7H_3IN_2O_2$
BENZENEDIAZONIUM-2-CARBOXYLATE, $C_7H_4N_2O_2$
4-HYDROXYBENZENEDIAZONIUM-3-CARBOXYLATE, $C_7H_4N_2O_3$
3,6-DIMETHYLBENZENEDIAZONIUM-2-CARBOXYLATE, $C_9H_8N_2O_2$
4,6-DIMETHYLBENZENEDIAZONIUM-2-CARBOXYLATE, $C_9H_8N_2O_2$

DIAZONIUM PERCHLORATES

Hofmann, K. A. *et al., Ber.*, 1906, **39**, 3146; ibid., 1910, **43**, 2624
Burton, H. *et al., Analyst,* 1955, **80**, 4
Extremely explosive, shock-sensitive materials when dry, some even when damp. The salt derived from diazotised *p*-phenylenediamine was considered to be more explosive than any other substance known in 1910.

See *m*-NITROBENZENEDIAZONIUM PERCHLORATE, $C_6H_4ClN_3O_6$
p-AMINOBENZENEDIAZONIUM PERCHLORATE, $C_6H_6ClN_3O_4$
o-TOLUENEDIAZONIUM PERCHLORATE, $C_7H_7ClN_2O_4$

DIAZONIUM SALTS

1. Houben-Weyl, 1965, Vol. 10(3), 32–38
2. Doyle, W. H., *Loss Prevention*, 1969, 3, 14

A few diazonium salts are unstable in solution, and many are in the solid state. Of these, the azides, chromates, nitrates, perchlorates (outstandingly), picrates, sulphides, triiodides and xanthates are noted as being explosive, and sensitive to friction, shock, heat and radiation. In view of their technical importance, diazonium salts are often isolated as their zinc chloride (or other) double salts, and although these are considerably more stable, some incidents involving explosive decomposition have been recorded.

During bottom discharge of an undefined diazonium chloride preparation, operation of a valve initiated explosion of the friction-sensitive chloride which had separated from solution. The latter did not occur with the corresponding sulphate [2].

Separately treated groups are:

ARENEDIAZONIUMOLATES
DIAZONIUM CARBOXYLATES
DIAZONIUM PERCHLORATES
DIAZONIUM SULPHATES
DIAZONIUM SULPHIDES AND DERIVATIVES
DIAZONIUM TETRAHALOBORATES
DIAZONIUM TRIIODIDES

Individually indexed compounds are:

TETRAZOLE-5-DIAZONIUM CHLORIDE, $CHClN_6$
5-DIAZONIOTETRAZOLIDE, CN_6
3,4,5-TRIIODOBENZENEDIAZONIUM NITRATE, $C_6H_2I_3N_3O_3$
BENZENEDIAZONIUM-4-SULPHONATE, $C_6H_4N_2O_3S$
4-NITROBENZENEDIAZONIUM NITRATE, $C_6H_4N_4O_5$
4-NITROBENZENEDIAZONIUM AZIDE, $C_6H_4N_6O_2$
BENZENEDIAZONIUM CHLORIDE, $C_6H_5ClN_2$
BENZENEDIAZONIUM NITRATE, $C_6H_5N_3O_3$
MERCURY 2-NAPHTHALENEDIAZONIUM TRICHLORIDE, $C_{10}H_7Cl_3HgN_2$
DI(BENZENEDIAZONIUM) ZINC TETRACHLORIDE, $C_{12}H_{10}Cl_4N_4Zn$
BIS-5-CHLOROTOLUENEDIAZONIUM ZINC TETRACHLORIDE, $C_{14}H_{12}Cl_6N_4Zn$

DIAZONIUM SULPHATES

Bersier, P. *et al.*, *Chem. Ing. Tech.*, 1971, **43**(24), 1311–1315
During investigation after the violent explosion of a 6-chloro-2, 4-dinitrobenzenediazonium sulphate preparation made in nitrosylsulphuric acid, it was found that above certain minimum concentrations some diazonium sulphates prepared in sulphuric acid media could be brought to explosive decomposition by local application of thermal shock. Classed as dangerous were the diazonium derivatives of 6-chloro-2, 4-dinitroaniline (at 1.26 mmol/g, very dangerous at 1.98 mmol/g); 6-bromo-2,4-dinitroaniline (very dangerous at 1.76 mmol/g); 2,4-dinitroaniline (2.0 mmol/g). Classed as suspect were the diazonium derivatives above at lower concentrations, and those of 2-chloro-5-trifluoromethylaniline (at 1.84 mmol/g); 2,6-dichloro-4-nitroaniline

(0.80 mmol/g); 2-methanesulphonyl-4-nitroaniline (0.80 mmol/g); 2-cyano-4-nitroaniline (1.04 mmol/g). A further 11 derivatives were not found to be unstable. Details of several stability testing methods are given.

DIAZONIUM SULPHIDES AND DERIVATIVES

1. Graebe, C. *et al., Ber.*, 1882, **15**, 1683
2. Bamberger, E. *et al., Ber.*, 1896, **29**, 272
3. Nawiasky, P. *et al., Chem. Eng. News,* 1945, **23**, 1247
4. Hodgson, H. H., *Chem. & Ind.*, 1945, 362
5. Tomlinson, W. R., *Chem. Eng. News,* 1957, **29**, 5473
6. Hollingshead, R. G. W. *et al., Chem. & Ind.*, 1953, 1179
7. Anon., *Angew. Chem. (Nachr.)*, 1962, **10**, 278
8. Parham, W. E. *et al., Org. Synth.*, 1967, **47**, 107
9. *BCISC Quart. Safety Summ.*, 1969, **40**, 17
10. Zemlyanskii, N.I. *et al.*, *Zh. Obsch. Khim.*, 1970, **40**, 1976–1978 (*Chem. Abs.*, 1971, **74**, 53204d)

There is a long history of the preparation of explosive solids or oils from interaction of diazonium salts with solutions of various sulphides and related derivatives. Such products have arisen from benzene- and toluene-diazonium salts with hydrogen, ammonium or sodium sulphides [1,5]; 2- or 3-chlorobenzene-, 4-chloro-2-methylbenzene-, 2- or 4-nitrobenzene- or 1- or 2-naphthalene-diazonium solutions with hydrogen sulphide, sodium hydrogensulphide or sodium mono-, di- or polysulphides [1–4, 7].

4-Bromobenzenediazonium solutions gave with hydrogen sulphide at –5°C a product which exploded under water at 0°C [2], and every addition of a drop of 3-chlorobenzenediazonium solution to sodium disulphide solution at 0°C caused a violent explosion [4]. In general, these compounds appear to be bis(arenediazo) sulphides or hydrogensulphides, since some of the corresponding disulphides are considerably more stable [2].

Interaction of 2-, 3- or 4-chlorobenzenediazonium salts with *O*-alkyldithiocarbonate ('xanthate') solutions [8] or thiophenolate solutions [9] produces explosive products, possibly arenediazo aryl sulphides. The intermediate diazonium *O*-ethyldithiocarbonate produced during the preparation of *m*-thiocresol can be dangerously explosive under the wrong conditions [8]. The product of interaction of 2-chlorobenzenediazonium chloride and sodium 2-chlorothiophenolate exploded violently on heating to 100°C, and the oil precipitated from interaction of potassium thiophenolate with 3-chlorobenzenediazonium chloride exploded during mixing of the solutions [9].

Interaction of substituted arenediazonium salts with potassium *O, O*-diphenyldithiophosphate gave a series of solid diazonium salts which decomposed explosively when heated dry [10].

The unique failure of diazotised anthranilic acid solutions to produce any explosive sulphide derivatives under a variety of conditions has been investigated and discussed [6].

Individually detailed compounds of this type relevant to the above include:

m-THIOCRESOL, C_7H_8S
BIS(*p*-NITROBENZENEDIAZO) SULPHIDE, $C_{12}H_8N_6O_4S$
DI(BENZENEDIAZO) SULPHIDE, $C_{12}H_{10}N_4S$

DIAZONIUM TETRAHALOBORATES

1. Olah, G. A. *et al., J. Org. Chem.,* 1961, **26**, 2053
2. Doak, G. O. *et al., Chem. Eng. News,* 1967, **45**(53), 8
3. Fletcher, T. L., *Chem. & Ind.,* 1972, 370
4. Pavlath, A. E. *et al., Aromatic Fluorine Compounds,* 15, New York, Reinhold, 1962

Solid diazonium tetrachloroborates decompose very vigorously, sometimes explosively, on heating in absence of solvent. Dry *o*-nitrobenzenediazonium tetrachloroborate is liable to explode spontaneously during storage at ambient temperature [1].
Hazards involved in drying off diazonium tetrafluoroborates have been discussed. Decomposition temperature of any new salt should be checked first on a small sample, and only if it is above 100°C should the bulk be dried off and stored. Salts which show signs of decomposition at or below ambient temperature must be kept moist and used immediately [2]. The need to use an inert solvent in any deliberate thermal decomposition is stressed in the later publication [3], which draws attention to an erroneous reference to use of tetrahydrofuran which could be hazardous. The presence of nitro substituent groups

may greatly increase the decomposition temperature, so that preparative decomposition may become violent or even explosive [4].
See also 3-PYRIDINEDIAZONIUM TETRAFLUOROBORATE, $C_5H_4BF_4N_3$

DIAZONIUM TRIIODIDES

Carey, J. G. *et al., Chem. & Ind.*, 1960, 97
The products produced by interaction of diazonium salts and iodides are unstable and liable to be explosive in the solid state. They are usually the triiodides, but monoiodides have been isolated under specific conditions from diazotised aniline and *o*-toluidine. Products prepared from diazotised *o*-, *m*- or *p*-nitroanilines, *m*-chloro-, -methoxy- or -methylaniline are too unstable to isolate, decomposing below 0°C.

Isolated compounds were:

p-CHLOROBENZENEDIAZONIUM TRIIODIDE, $C_6H_4ClI_3N_2$
* BENZENEDIAZONIUM IODIDE, $C_6H_5IN_2$
BENZENEDIAZONIUM TRIIODIDE, $C_6H_5I_3N_2$
* *m*-TOLUENEDIAZONIUM IODIDE, $C_7H_7IN_2$
* *o*-TOLUENEDIAZONIUM IODIDE, $C_7H_7IN_2$
p-TOLUENEDIAZONIUM TRIIODIDE, $C_7H_7I_3N_2$
o-METHOXYBENZENEDIAZONIUM TRIIODIDE, $C_7H_7I_3N_2O$
p-METHOXYBENZENEDIAZONIUM TRIIODIDE, $C_7H_7I_3N_2O$
2,4-DIMETHYLBENZENEDIAZONIUM TRIIODIDE, $C_8H_9I_3N_2$

See other DIAZONIUM SALTS
IODINE COMPOUNDS

(DIBENZOYLDIOXYIODO)BENZENES

Plesnicar, B. *et al., Angew. Chem. (Intern. Ed.)*, 1970, **9**, 797
Compounds of the general formula $XArI(O_2COArY)_2$, where X = H, *p*-Cl or *o*-CH_3 and Y = *m*-Cl or *p*-NO_2, are extremely powerful oxidants, unstable when dry, and will explode during manipulation at ambient temperature (particularly with metal spatulae), or on heating to 80–120°C. The group exceeds the oxidising power of organic peroxy acids.

See *p*-CHLORO(BIS-*p*-NITROBENZOYLDIOXYIODO)BENZENE, $C_{20}H_{12}ClIN_2O_{10}$
p-CHLORO(BIS-*m*-CHLOROBENZOYLDIOXYIODO)BENZENE, $C_{20}H_{12}Cl_3IO_6$

(BIS-*m*-CHLOROBENZOYLDIOXYIODO)BENZENE, $C_{20}H_{13}Cl_2IO_6$
(BIS-*p*-NITROBENZOYLDIOXYIODO)BENZENE, $C_{20}H_{13}IN_2O_{10}$
o-METHOXY(BIS-*p*-NITROBENZOYLDIOXYIODO)BENZENE, $C_{21}H_{15}IN_2O_{11}$

See other DIACYL PEROXIDES
IODINE COMPOUNDS
OXIDANTS

DICHROMATE SALTS OF NITROGENOUS BASES

Gibson, G. M., *Chem. & Ind.*, 1966, 553
The dichromates of 1-phenylbiguanide, its *p*-chloro-*p*-methyl- and 1-naphthyl- analogues all decompose violently at *ca.* 130°C.

See also DIPYRIDINIUM DICHROMATE, $C_{10}H_{12}Cr_2N_2O_7$
DIANILINIUM DICHROMATE, $C_{12}H_{16}Cr_2N_2O_7$

See other OXOSALTS OF NITROGENOUS BASES

[14] DIENE-N_4 IRON COMPLEXES

Goedken, V. L. *et al.*, *J. Amer. Chem. Soc.*, 1972, **94**, 3397
The macroheterocyclic 5,7,7,12,14,14-hexamethyl-1,4,8,11-tetraaza-4,11-cyclotetradeca diene (abbreviated to [14] diene-N_4) forms cationic complexes with iron (II) or (III), also containing acetonitrile, imidazole, phenanthroline or halogen ligands. When the anion is perchlorate, the products are explosive, sensitive to heat and impact, and some appear to decompose on storage (1 week) and become sensitive to slight disturbance.

Specific compounds include:

BROMO-5,7,7,12,14,14-HEXAMETHYL-1,4,8,11-TETRAAZA-4,11-CYCLOTETRADECADIENEIRON(II) PERCHLORATE, $C_{16}H_{32}BrClFeN_4O_4$
IODO-5,7,7,12,14,14-HEXAMETHYL-1,4,8,11-TETRAAZA-4,11-CYCLOTETRADECADIENEIRON(II) PERCHLORATE, $C_{16}H_{32}ClFeIN_4O_4$
CHLORO-5,7,7,12,14,14-HEXAMETHYL-1,4,8,11-TETRAAZA-4,11-CYCLOTETRADECADIENEIRON(II) PERCHLORATE, $C_{16}H_{32}Cl_2FeN_4O_4$

DICHLORO-5,7,7,12,14,14-HEXAMETHYL-1,4,8,11-TETRAAZA-4,11-CYCLOTETRADECADIENEIRON(III) PERCHLORATE, $C_{16}H_{32}Cl_3FeN_4O_4$

DIACETONITRILE-5,7,7,12,14,14-HEXAMETHYL-1,4,8,11-TETRAAZA-4,11-CYCLOTETRADECADIENEIRON(II) PERCHLORATE, $C_{20}H_{36}Cl_2FeN_6O_8$

ACETONITRILEIMIDAZOLE-5,7,7,12,14,14-HEXAMETHYL-1,4,8,11-TETRAAZACYCLODODECADIENEIRON(II) PERCHLORATE, $C_{21}H_{39}Cl_2FeN_7O_8$

5,7,7,12,14,14-HEXAMETHYL-1,4,8,11-TETRAAZA-4,11-CYCLOTETRADECADIENE-1,10-PHENANTHROLINEIRON(II) PERCHLORATE, $C_{28}H_{40}Cl_2FeN_6O_8$

BIS(5,7,7,12,14,14-HEXAMETHYL-1,4,8,11-TETRAAZA-4,11-CYCLOTETRADECADIENE)HYDROXODIIRON(II) TRIPERCHLORATE, $C_{32}H_{65}Cl_3Fe_2N_8O_{13}$

BIS[AQUA-5,7,7,12,14,14-HEXAMETHYL-1,4,8,11-TETRAAZA-4,11-CYCLOTETRADECADIENEIRON(II)] OXIDE TETRAPERCHLORATE, $C_{32}H_{68}Cl_4Fe_2N_8O_{19}$

See other AMMINEMETAL OXOSALTS

DIENES

The 1,2- and 1,3-dienes (vicinal and conjugated, respectively) are more reactive than the separated dienes. Individually indexed compounds are:

PROPADIENE, C_3H_4
1,2-BUTADIENE, C_4H_6
1,3-BUTADIENE, C_4H_6
CYCLOPENTADIENE, C_5H_6
2-METHYL-1,3-BUTADIENE (ISOPRENE), C_5H_8
1,3-PENTADIENE, C_5H_8
1,4-PENTADIENE, C_5H_8
1,3-CYCLOHEXADIENE, C_6H_8
1,4-CYCLOHEXADIENE, C_6H_8
2,3-DIMETHYL-1,3-BUTADIENE, C_6H_{10}
1,4-HEXADIENE, C_6H_{10}
1,5-HEXADIENE, C_6H_{10}
2-METHYL-1,3-PENTADIENE, C_6H_{10}
4-METHYL-1,3-PENTADIENE, C_6H_{10}
* 1,3,5,7-CYCLOOCTATETRAENE, C_8H_8
* 1,3,5-CYCLOOCTATRIENE, C_8H_{10}
4-VINYLCYCLOHEXENE, C_8H_{12}
1,7-OCTADIENE, C_8H_{14}

See other PEROXIDISABLE COMPOUNDS

Oxides of nitrogen

See NITROGEN OXIDE ('NITRIC OXIDE'), NO: Dienes, Oxygen

DIFFERENTIAL THERMAL ANALYSIS

1. Corignan, Y. P. *et al., J. Org. Chem.*, 1967, **32**, 285
2. Prugh, R. W., *Chem. Eng. Prog.*, 1967, **63**(11), 53

Thermal stabilities of 40 explosive or potentially explosive *N*-nitroamines, aminium nitrates and guanidine derivatives were studied in relation to structure; 14 of the compounds decomposed violently when the exotherm occurred [1]. The value of the DTA technique in assessing reactive hazards of compounds or reaction mixtures is discussed [2].

DIFLUOROAMINO COMPOUNDS

Freeman, J. P., *Advan. Fluorine Chem.*, 1970, **6**, 321, 325

All organic compounds containing one or more difluoroamino groups should be treated as explosive oxidants and excluded from contact with strong reducing agents. If the ratio of CH_2 to NF_2 groups is below 5:1, the compound will be impact-sensitive. Direct combustion for elemental analysis is unsafe, but polarography is applicable.
See other *N*-HALOGEN COMPOUNDS

DIPLUMBANES

Sidgwick, 1950, 595

The higher homologues of the hexaalkyldiplumbane series may explosively disproportionate during distillation.
See other ALKYLMETALS

DISPOSAL

1. *Laboratory Waste Disposal Manual,* Washington, MCA, 2nd Ed., 1969 (revised, November 1972).

2. Gaston, P.J., *The Care, Handling and Disposal of Dangerous Chemicals,* Inst. Sci. Tech., Aberdeen, Northern Publishers, 1965
3. Voegelein, J. F., *J. Chem. Educ.,* 1966, **43**, A151–157
4. Teske, J. W., *J. Chem. Educ.,* 1970, **47**, A291–295

The problems of devising effective methods for the disposal of surplus reactive chemicals or hazardous chemical wastes or residues have been well covered in several publications. A comprehensive, classified and tabulated guide to disposal procedures [1] covers the same ground as, but in more detail than, an earlier publication [2], and includes details of facilities, reagents, protective clothing and equipment required. Further publications have described the practical solutions to disposal problems adopted by an explosives research laboratory [3] and an American university [4], the latter including details of the use of explosives to rupture corroded and unusable cylinders of compressed gases or liquids. Many of the techniques described, however, need a remote area for safe operations, which may not be accessible to laboratories in urban locations. In such cases, the services of a specialist chemical disposal contractor may be the most practical solution. It is possible to minimise the disposal problem by careful stock room procedures, adequately durable labelling arrangements for samples and materials in storage, and careful segregation of materials for disposal.

DUST EXPLOSIONS

1. Rep. RI 7132, Dorsett, H. G. *et al.,* Washington, US Bureau of Mines, 1968
2. Hartmann, I., *Ind. Eng. Chem.,* 1948, **40**, 752
3. Palmer, K. N., *Dust Explosions and Fires*, London, Chapman and Hall, 1973

Laboratory dust explosion data are presented for 73 chemical compounds, 29 drugs, 27 dyes and 46 pesticides, including ignition temperatures of clouds and layers, minimum igniting energy, explosion-limiting concentrations and pressures, and rates rise at various dust concentrations. Explosibility indices are computed where possible and variation of explosibility parameters with chemical composition is discussed. General means of minimising dust explosion hazard are reviewed [1].

Of the 17 dusts investigated earlier, the four metals examined—aluminium, magnesium, titanium and zirconium—were among the most hazardous. Aluminium and magnesium showed the maximum rates of pressure rise and final pressures, magnesium having a low minimum explosive concentration. Ignition of zirconium dust often occurred spontaneously, apparently due to static electric discharges, and undispersed layers of the dust could be ignited by less than 1μJ compared with 15μJ for a dispersed dust. All except aluminium ignite in carbon dioxide atmospheres [2].

Recently a comprehensive account of practical and theoretical aspects of laboratory and plant-scale dust explosions and fires has appeared.

See CARBONACEOUS DUSTS
METAL DUSTS

See also TITANIUM, Ti: Air
ZINC, Zn: Air

1,2-EPOXIDES

The three lower members of this class are bulk industrial chemicals and their high reactivities have been involved in several large-scale incidents. Individually indexed compounds are:

ETHYLENE OXIDE, C_2H_4O
1-CHLORO-2,3-EPOXYPROPANE, C_3H_5ClO
PROPYLENE OXIDE, C_3H_6O
3,4-EPOXYBUTENE, C_4H_6O
1,2-EPOXYBUTANE, C_4H_8O
EPOXYETHYLBENZENE, C_8H_8O
2-*tert*-BUTYL-3-PHENYLOXAZIRANE, $C_{11}H_{15}NO$

ETHERS

1. Jackson, H. L. *et al., J. Chem. Educ.*, 1970, **47**, A175
2. Davies, A. G., *J.R. Inst. Chem.*, 1956, **80**, 386–389
3. Dasler, W. *et al., Ind. Eng. Chem. (Anal. Ed.)* 1946, **18**, 52
4. *Laboratory Waste Disposal Manual,* 145, Washington, MCA, 1969

There is a long history of laboratory and plant fires and explosions involving the high flammability and/or tendency to peroxide formation in these widely used solvents, diisopropyl ether being the most notorious. Methods of controlling peroxide hazards in the use of ethers have been reviewed [1] and information on storage, handling, purification [2,3] and disposal [4] have been detailed.

See the individual entries:

TETRAHYDROFURAN, C_4H_8O
m- and *p*-DIOXAN, $C_4H_8O_2$
DIETHYL ETHER, $C_4H_{10}O$
1,1- and 1,2-DIMETHOXYETHANE, $C_4H_{10}O_2$
DIISOPROPYL ETHER, $C_6H_{14}O$
BIS(2-METHOXYETHYL) ETHER, $C_6H_{14}O_3$

See other PEROXIDISABLE COMPOUNDS

ETHOXYETHYNYL ALCOHOLS

1. Arens, J. F., *Adv. Org. Chem.*, 1960, **2**, 126
2. Brandsma, 1971, 12, 78

Vigorous decompositions or violent explosions have been observed on several occasions during careless handling (usually overheating) of ethoxyethynyl alcohols (structures not stated) [1]. The explosions noted [2] when magnesium sulphate was used to dry their ethereal solutions were attributed to the slight acidity of the salt causing exothermic rearrangement of the alcohols to acrylic esters and subsequent explosive reactions. Glassware used for distillation must be pretreated with ammonia gas to remove traces of acid [2].

See 1-ETHOXY-3-METHYL-1-BUTYN-3-OL, $C_7H_{12}O_2$
See other ACETYLENIC COMPOUNDS

EXPLOSIBILITY

1. Lothrop, W. C. *et al., Chem. Rev.*, 1949, **14**, 419–445
2. Tomlinson, W. R. *et al., J. Chem. Educ.*, 1950, **27**, 606–609
3. Coffee, R. D., *J. Chem. Educ.*, 1972, **49**, A343–349
4. Van Dolah, R. W., *Ind. Eng. Chem.*, 1961, **53**(7), 50A–53A
5. Stull, R. D., 'Prediction of Real Chemical Hazards,' 65th AIChE Meeting, New York, 1973

This may be defined as the tendency of a chemical system (involving one or more compounds) to undergo violent or explosive decomposition under appropriate conditions of reaction or initiation. It is obviously of great practical interest to be able to predict which compound or reaction systems are likely to exhibit explosibility, and much work has been devoted to this end.

Early work [1] on the relationship between structure and performance of 176 organic explosives (mainly nitro or nitrate ester compounds) was summarised and extended in general terms to multi-component systems [2]. The contribution of various structural factors (bond-groupings) was discussed in terms of heats of decomposition and oxygen balance of the compound or compounds involved in the system. Materials or systems approaching zero oxygen balance are the most powerfully explosive (give maximum heat release). Bond groupings (below) known to confer explosibility were classed as 'plosophors', and explosibility-enhancing groups as 'auxoploses' by analogy with dyestuffs terminology.

The latter groups (ether, nitrile or oximino) tend to increase the proportion of nitrogen and/or oxygen in the molecule towards (or past) zero oxygen balance.

BOND GROUPINGS *as in*	*CLASS ENTRY*
$-C\equiv C-$	ACETYLENIC COMPOUNDS
$-C\equiv C-$Metal	METAL ACETYLIDES
$-C\equiv C-X$	HALOACETYLENE DERIVATIVES
N = N \ / C / \	DIAZIRINES
$>CN_2$	DIAZO COMPOUNDS
$\geq C-N=O$	NITROSO COMPOUNDS
$\geq C-NO_2$	NITROALKANES, *C*-NITRO and POLYNITROARYL COMPOUNDS
$>C(NO_2)_2$	POLYNITROALKYL COMPOUNDS
$\geq C-O-N=O$	ACYL OR ALKYL NITRITES

BOND GROUPINGS	*as in*	*CLASS ENTRY*
$\gtrless C=O-NO_2$		ACYL OR ALKYL NITRATES
$>C-C<$ bridged by O		1,2-EPOXIDES
$>C=N-O-$Metal		METAL FULMINATES or *aci*-NITRO SALTS
NO_2 / C F / NO_2		FLUORODINITROMETHYL COMPOUNDS
$>N-$Metal		*N*-METAL DERIVATIVES
$>N-N=O$		NITROSO COMPOUNDS
$>N-NO_2$		*N*-NITRO COMPOUNDS
$\gtrless C-N=N-C\lessgtr$		AZO COMPOUNDS
$\gtrless C-N=N-O-C\lessgtr$		ARENEDIAZOATES
$\gtrless C-N=N-S-C\lessgtr$		ARENEDIAZO ARYL SULPHIDES
$\gtrless C-N=N-O-N=N-C\lessgtr$		BIS-ARENEDIAZO OXIDES
$\gtrless C-N=N-S-N=N-C\lessgtr$		BIS-ARENEDIAZO SULPHIDES
$\gtrless C-N=N-N(R)-C\lessgtr$ R(R = H,−CN,−OH,−NO)		TRIAZENES
$-N=N-N=N-$		HIGH-NITROGEN COMPOUNDS TETRAZOLES
$\gtrless C-O-O-H$		ALKYLHYDROPEROXIDES, PEROXYACIDS
$\gtrless C-O-O-C\lessgtr$		PEROXIDES (CYCLIC, DIACYL, DIALKYL), PEROXYESTERS
$-O-O-$Metal		METAL PEROXIDES, PEROXOACID SALTS

BOND GROUPINGS	*as in*	*CLASS ENTRY*
–O–O–Non-metal		PEROXOACIDS
$N \rightarrow Cr-O_2$		AMMINECHROMIUM PEROXOCOMPLEXES
$-N_3$		AZIDES (ACYL, HALOGEN, NON-METAL, ORGANIC)
$\gt C-N_2^+ O^-$ (ring)		ARENEDIAZONIUMOLATES
$\gt C-N_2^+ S^-$		DIAZONIUM SULPHIDES AND DERIVATIVES, 'XANTHATES'
$N^+-H\ Z^-$		HYDRAZINIUM SALTS, OXOSALTS OF NITROGENOUS BASES
$-N^+-OH\ Z^-$		HYDROXYLAMMONIUM SALTS
$\gt C-N_2^+ Z^-$		DIAZONIUM CARBOXYLATES or SALTS
$[N \rightarrow Metal]^+ Z^-$		AMMINEMETAL OXOSALTS
Ar–Metal–X X–Ar–Metal		HALO-ARYLMETALS
N–X		HALOGEN AZIDES *N*-HALOGEN COMPOUNDS *N*-HALOIMIDES
$-NF_2$		DIFLUOROAMINO COMPOUNDS *N,N,N*-TRIFLUOROALKYLAMIDINES
–O–X		ALKYL PERCHLORATES CHLORITE SALTS HALOGEN OXIDES HYPOHALITES PERCHLORIC ACID (no class entry: *see* $ClHO_4$) PERCHLORYL COMPOUNDS

Although the semi-empirical approach outlined above is of some value in assessing potential explosibility hazards, more fundamental work has recently been done to institute a quantitative basis for such assessment.

A combination of thermodynamical calculations with laboratory thermal stability and impact sensitivity determinations has enabled a system to be developed which indicates the relative potential of a given compound or reaction system for sudden energy release, and the relative magnitude of the latter [3].

A similar treatment, specifically for compounds expected to be explosive, has also been separately developed [4].

A further computational technique which takes account of both thermodynamic and kinetic considerations has permitted of the development of a system which provides a numerical Reaction Hazard Index (RHI) for each compound, which is a real, rather than a potential, indication of hazard. The RHIs calculated for 80 compounds are in fairly close agreement with the relative hazard values (assessed on the basis of experience) assigned on the NFPA Reactivity Rating scale for these same compounds [5].

See also OXYGEN BALANCE

EXPLOSIVES

1. Federoff, 1960
2. *Explosives, Propellants and Pyrotechnic Safety Covering Laboratory, Pilot Plant and Production Operations:* Manual AD272–424, Washington, US Naval Ordnance Laboratory, 1962
3. Kirk-Othmer, 1965, Vol. 8, 581
4. Urbanski, 1964–1967, Vols. 1, 2, 3

Explosive materials intended as such are outside the scope of this Handbook, but several specialist reference works on them contain much information relevant to safety practices for unstable materials [1–4].

FIRE

1. Williamson, J. J., *The Chartered Insurance Institute Handbook,* No. 8, London, Pitman, 1966
2. Crees, J. C. Vol. 23A, Boreham Wood, Fire Research Station, 1972
3. Bahme, 1972

The book covers the CII syllabus for general fire hazards, which, together with means for their prevention, are presented simply and concisely. Constructional features, management considerations, heating, lighting, etc., are considered in detail. An alphabetical list of hazardous materials and an extensive bibliography are included [1].

An annual compilation of references to the scientific literature on fire is classified under headings: Occurrence of Fire; Fire Hazards; Initiation and Development of Combustion; Fire Precautions; Fire Resistance; Fire Fighting; Nuclear Energy; General. Name and Subject Index. The latest available volume is reference 2.

A specialist account of problems associated with fires involving reactive and hazardous materials is also available [3].

FIRE EXTINGUISHERS

'Technical Booklet No. 6', London, FPA, 1965
An illustrated guide to the selection, installation, maintenance and use of various types of portable fire extinguishers.

FLAMMABILITY

1. Coward, H. F. *et al., Limits of Flammability of Gases and Vapors,* Bull. 503, Washington, US Bureau of Mines, 1952
2. Zabetakis, M. G., *Flammability Characteristics of Combustible Gases and Vapors,* Bull. 627, Washington, US Bureau of Mines, 1965

The hazards associated with flammability characteristics of combustible gases and vapours are excluded from detailed consideration in this Handbook, since the topic is adequately covered in standard reference works on combustion, including the two sources of much of the data on flammability limits [1,2].

However, to reinforce the constant need to consider flammability problems in laboratory and plant operations, the explosive limits have been included (where known) for those individual substances with flash point below 25°C. With the few noted exceptions, explosive limits quoted are those at ambient temperature and are expressed in % by volume in air.

See also OXYGEN ENRICHMENT

FLASH POINTS

1. *Flash Points,* Poole, BDH Ltd., 1962
2. 'Catalogue KL4', Colnbrook, Koch-Light Laboratories Ltd., 1973
3. *Flash Point Index of Trade Name Liquids,* 325A, Boston, Nat. Fire Prot. Assoc., 1972
4. *Properties of Flammable Liquids,* 325M, Boston, Nat. Fire Prot. Assoc., 1969
5. Prugh, R. W., *J. Chem. Educ.,* 1973, **50**, A85–89

Flash point is defined as the temperature at which the concentration of vapour above a flammable liquid (or volatile solid) in contact with air is sufficient to exceed the lower explosive limit and so produce an ignitable mixture which will propagate flame.

There is usually a fair correlation between flash point and probability of involvement in fire if an ignition source is present in the vicinity of a source of the vapour; materials with low flash points being more likely to be so involved than those with higher flash points. While no attempt has been made to include in this Handbook details of all known combustible materials, it has been thought worthwhile to include substances with flash points below 25°C, a likely maximum ambient temperature in many laboratories in warm temperature zones. These materials have been included to draw attention to the high probability of fire if such flammable materials are handled with insufficient care to prevent contact of their vapours with an ignition source (stirrer motor, hot-plate, energy controller, flame, etc.).

The figures for flash points quoted in the text entries are closed cup values except where indicated by (o) and most are reproduced by permission of the two companies concerned [1,2].

A comprehensive listing of flash points for commercial liquids and formulated mixtures is also available [3,4].

Recently a method for estimating approximate flash-point temperatures based upon the boiling point and molecular structure of a given compound has been published. After calculation of the stoicheiometric concentration in air, a nomograph is used to estimate the flash point to within 11°C [5].

FLUORINATED COMPOUNDS

See LITHIUM TETRAHYDROALUMINATE, AlH_4Li: Fluoroamides
SODIUM, Na: Fluorinated compounds
: Halocarbons (Ref. 6)

FLUORINATION

Grakauskas, V., *J. Org. Chem.*, 1970, **35**, 723; 1969, **34**, 2835
Safety precautions applicable to direct liquid-phase fluorination of aromatic compounds are discussed.
For further references see FLUORINE, F_2

FLUORODINITROMETHYL COMPOUNDS

Kamlet, M. J. *et al., J. Org. Chem.*, 1968, **33**, 3073
Witucki, E. F. *et al., J. Org. Chem.*, 1972, **37**, 152
Adolph, H. J., ibid., 749
Gilligan, W. H., ibid., 3947
Several of this group are explosives of moderate-to-considerable sensitivity to impact or friction and need careful handling. Fluorodinitromethane and fluorodinitroethanol are also vesicant.
See FLUORODINITROMETHYL AZIDE, CFN_5O_4
FLUORODINITROMETHANE, $CHFN_2O_4$
2-FLUORO-2,2-DINITROETHANOL, $C_2H_3FN_2O_5$
2-FLUORO-2,2-DINITROETHYLAMINE, $C_2H_4FN_3O_4$
BIS(2-FLUORO-2,2-DINITROETHYL)AMINE, $C_4H_5F_2N_5O_8$

FOAM RUBBER

Oxygen
See OXYGEN (Gas), O_2 : Polymers

FRICTIONAL IGNITION OF GASES

Powell, F., *Ind. Eng. Chem.*, 1969, **61**(12), 29
The ignition of flammable gases and vapours by friction or impact is reviewed, with 82 references.

FULLER'S EARTH

Turpentine
See TURPENTINE: Diatomaceous earth

GAS CYLINDERS

1. Whalley, E. W. F. (UKAEA), London, HMSO, 1965
2. *Properties of Gases,* Wall chart 57012, Poole, BDH Chemicals Ltd., 1973
3. Braker, W. *et al., Matheson Gas Data Book,* E. Rutherford, N.J., Matheson Gas Products, 5th Ed., 1971

In a report on safety in use of gas cylinders, the nature of hazards associated with gas cylinders and their contents and statutory requirements are discussed. Appendices outline model safeguards for storage, handling, use and transport of gas cylinders [1]. An inexpensive wall chart summarises the important properties of 116 gases and volatile liquids [2] and the *Gas Data Book* [3] gives comprehensive details of handling techniques and cylinder equipment necessary for safe use of 130 gases.

GOLD COMPOUNDS

Gold compounds exhibit a tendency to decompose violently with separation of the metal. Individually indexed gold compounds are:

GOLD(III) CHLORIDE, $AuCl_3$
TRIAMMINEGOLD TRIHYDROXIDE, $AuH_{12}N_3O_3$
GOLD(III) OXIDE, Au_2O_3
GOLD(III) SULPHIDE, Au_2S_3
GOLD NITRIDE-AMMONIA, $Au_3N{\cdot}H_3N$
TRIGOLD DISODIUM HEXAAZIDE, $Au_3N_{18}Na_2$
GOLD(I) CYANIDE, CAuN
DIGOLD ACETYLIDE, C_2Au_2
DIMETHYLGOLD SELENOCYANATE, C_3H_6AuNSe
DIMETHYLGOLD(III) AZIDE: see the dimer $C_4H_{12}Au_2N_6$
TETRAMETHYLDIGOLD DIAZIDE, $C_4H_{12}Au_2N_6$
CYCLOPENTADIENYLGOLD(I), C_5H_5Au
CYANODIETHYLGOLD: see the tetramer $C_{20}H_{40}Au_4N_4$
DIMETHYLTRIMETHYLSILOXOGOLD : see the dimer $C_{10}H_{30}Au_2O_2Si_2$
1,2-DIAMINOETHANEBIS-TRIMETHYLGOLD, $C_8H_{26}Au_2N_2$
TETRAMETHYLBIS-(TRIMETHYLSILOXO)DIGOLD, $C_{10}H_{30}Au_2O_2Si_2$
TETRACYANOCTAETHYLTETRAGOLD, $C_{20}H_{40}Au_4N_4$

See also PLATINUM COMPOUNDS
See other HEAVY METAL DERIVATIVES
METAL AZIDES

HALOACETYLENE DERIVATIVES

1. Whiting, M. C. *Chem. Eng. News*, 1972, **50**(23), 86
2. Brandsma, 1971, 99

The tendency towards explosive decomposition noted for dihalo-2,4-hexadiyne derivatives appears to be associated with the coexistence of halo- and acetylene functions in the same molecule, rather than with its being a polyacetylene. Haloacetylenes should be used with exceptional precautions [1]. Explosions may occur during distillation of bromoacetylenes when bath temperatures are too high, or if air is admitted to a hot vacuum-distillation residue [2].

See LITHIUM BROMOACETYLIDE, C_2BrLi
DIBROMOACETYLENE, C_2Br_2
LITHIUM CHLOROACETYLIDE, C_2ClLi
SODIUM CHLOROACETYLIDE, C_2ClNa
DICHLOROACETYLENE, C_2Cl_2
BROMOACETYLENE, C_2HBr
CHLOROACETYLENE, C_2HCl
FLUOROACETYLENE, C_2HF
DIIODOACETYLENE, C_2I_2
SILVER TRIFLUOROMETHYLACETYLIDE, C_3AgF_3
CHLOROCYANOACETYLENE, C_3ClN
LITHIUM TRIFLUOROMETHYLACETYLIDE, C_3F_3Li
3,3,3-TRIFLUOROPROPYNE, C_3HF_3
1-BROMO-2-PROPYNE, C_3H_3Br
1-CHLORO-2-PROPYNE, C_3H_3Cl
1-IODO-1,3-BUTADIYNE, C_4HI
1,4-DICHLORO-2-BUTYNE, $C_4H_4Cl_2$
1-IODO-3-PENTEN-1-YNE, C_5H_5I
1,6-DICHLORO-2,4-HEXADIYNE, $C_6H_4Cl_2$
2,4-HEXADIYNYLENE BISCHLOROSULPHITE, $C_6H_4Cl_2O_4S_2$
TETRA(CHLOROETHYNYL)SILANE, C_8Cl_4Si
2,4-HEXADIYNYLENE BISCHLOROFORMATE, $C_8H_4Cl_2O_4$
1-IODO-3-PHENYL-2-PROPYNE, C_9H_7I
1-BROMO-1,2-CYCLOTRIDECADIEN-4,8,10-TRIYNE, $C_{13}H_9Br$

HALOALKENES

Of the lower members of this reactive class, the more lightly substituted are of high flammability and many are classed as peroxidisable compounds. Individually indexed members are:

BROMOTRIFLUOROETHYLENE, C_2BrF_3
CHLOROTRIFLUOROETHYLENE, C_2ClF_3
TETRACHLOROETHYLENE, C_2Cl_4
TETRAFLUOROETHYLENE, C_2F_4
cis- or *trans*-1,2-DICHLOROETHYLENE, $C_2H_2Cl_2$
1,1-DIFLUOROETHYLENE, $C_2H_2F_2$
BROMOETHYLENE (VINYL BROMIDE), C_2H_3Br
CHLOROETHYLENE (VINYL CHLORIDE), C_2H_3Cl
FLUOROETHYLENE (VINYL FLUORIDE), C_2H_3F
HEXAFLUOROPROPENE, C_3F_6
1,3-DICHLOROPROPENE, $C_3H_4Cl_2$
2,3-DICHLOROPROPENE, $C_3H_4Cl_2$
1-BROMO-2-PROPENE (ALLYL BROMIDE), C_3H_5Br
1-CHLORO-1-PROPENE, C_3H_5Cl
2-CHLOROPROPENE, C_3H_5Cl
1-CHLORO-2-PROPENE (ALLYL CHLORIDE), C_3H_5Cl
1-IODO-2-PROPENE (ALLYL IODIDE), C_3H_5I
1,1,4,4-TETRACHLOROBUTATRIENE, C_4Cl_4
1,1,4,4-TETRAFLUOROBUTATRIENE, C_4F_4
2-CHLORO-1,3-BUTADIENE, C_4H_5Cl
1-BROMO-2-BUTENE, C_4H_7Br
4-BROMO-1-BUTENE, C_4H_7Br
2-CHLORO-2-BUTENE, C_4H_7Cl
3-CHLORO-1-BUTENE, C_4H_7Cl
3-CHLORO-2-METHYL-1-PROPENE, C_4H_7Cl

See other PEROXIDISABLE COMPOUNDS

HALOARYL COMPOUNDS

Though normally not very reactive, haloaryl compounds if sufficiently activated by other substituents may undergo violent reactions. Individually indexed compounds in this small class are:

1,2,4,5-TETRACHLOROBENZENE, $C_6H_2Cl_4$
1-CHLORO-2,4-DINITROBENZENE, $C_6H_3ClN_2O_4$
1-FLUORO-2,4-DINITROBENZENE, $C_6H_3FN_2O_4$
p-CHLORONITROBENZENE, $C_6H_4ClNO_2$
* 2,6-DINITROBENZYL BROMIDE, $C_7H_5BrN_2O_4$
2-CHLORO-4-NITROTOLUENE, $C_7H_6ClNO_2$
4-CHLORO-2-METHYLPHENOL, C_7H_7ClO
2-IODO-3,5-DINITROBIPHENYL, $C_{12}H_7IN_2O_4$

HALO-ARYLMETALS

The name adopted for this class of highly reactive (and in some circumstances self-reactive) compounds is intended to cover both arylmetal halides (halogen bonded to the metal) and haloaryl metals (halogen attached to the aryl nucleus). Individually indexed compounds are:

PENTAFLUOROPHENYLALUMINIUM DIBROMIDE, $C_6AlBr_2F_5$
PENTAFLUOROPHENYLLITHIUM, C_6F_5Li
m- or *p*-BROMOPHENYLLITHIUM, C_6H_4BrLi
m- or *p*-CHLOROPHENYLLITHIUM, C_6H_4ClLi
p-FLUOROPHENYLLITHIUM, C_6H_4FLi
o-, *m*- or *p*-TRIFLUOROMETHYLPHENYLMAGNESIUM BROMIDE, $C_7H_4BrF_3Mg$
o-, *m*- or *p*-TRIFLUOROMETHYLPHENYLLITHIUM, $C_7H_4F_3Li$
BIS(PENTAFLUOROPHENYL)ALUMINIUM BROMIDE, $C_{12}AlBrF_{10}$
BIS(CYCLOPENTADIENYL)BIS(PENTAFLUOROPHENYL)ZIRCONIUM, $C_{22}H_{10}F_{10}Zr$

See also ORGANOLITHIUM REAGENTS

HALOCARBONS

This generic name has been adopted to designate a range of halogenated aliphatic hydrocarbons widely used in research and industrial operations, often as solvents or diluents. Few are completely inert chemically but, in general, reactivity decreases with increasing substitution of halogen atoms (particularly of fluorine) for hydrogen atoms in the saturated or unsaturated parent hydrocarbons. For other reactants which have been involved, see:

METALS: Halocarbons
PENTABORANE(9), B_5H_9 : Reactive solvents
CALCIUM DISILICIDE, $CaSi_2$: Carbon tetrachloride
FLUORINE, F_2 : Halocarbons
DISILANE, H_6Si_2 : Non-metal halides
DINITROGEN TETRAOXIDE, N_2O_4: Halocarbons
OXYGEN (Liquid), O_2: Halocarbons

Unsaturated halocarbons are separately listed as:

HALOALKENES

and individually indexed saturated halocarbons are:

BROMOTRICHLOROMETHANE, $CBrCl_3$
BROMOTRIFLUOROMETHANE, $CBrF_3$
CARBON TETRABROMIDE, CBr_4
DICHLORODIFLUOROMETHANE, CCl_2F_2
TRICHLOROFLUOROMETHANE, CCl_3F
CARBON TETRACHLORIDE, CCl_4
CARBON TETRAFLUORIDE, CF_4
BROMOFORM, $CHBr_3$
CHLORODIFLUOROMETHANE, $CHClF_2$
CHLOROFORM, $CHCl_3$
IODOFORM, CHI_3
DIBROMOMETHANE, CH_2Br_2
DICHLOROMETHANE, CH_2Cl_2
DIIODOMETHANE, CH_2I_2
BROMOMETHANE, CH_3Br
CHLOROMETHANE, CH_3Cl
FLUOROMETHANE, CH_3F
IODOMETHANE, CH_3I
PENTACHLOROETHANE, C_2HCl_5
1,1,1,2-TETRACHLOROETHANE, $C_2H_2Cl_4$
1-CHLORO-1,1-DIFLUOROETHANE, $C_2H_3ClF_2$
1,1,1-TRICHLOROETHANE, $C_2H_3Cl_3$
1,1,2-TRICHLOROETHANE, $C_2H_3Cl_3$
1,1-DICHLOROETHANE, $C_2H_4Cl_2$
1,2-DICHLOROETHANE, $C_2H_4Cl_2$
1,1-DIFLUOROETHANE, $C_2H_4F_2$
BROMOETHANE, C_2H_5Br
CHLOROETHANE, C_2H_5Cl
FLUOROETHANE, C_2H_5F
IODOETHANE, C_2H_5I
3-BROMO-1,1,1-TRICHLOROPROPANE, $C_3H_4BrCl_3$
1,1-DICHLOROPROPANE, $C_3H_6Cl_2$
1,2-DICHLOROPROPANE, $C_3H_6Cl_2$
1-BROMOPROPANE, C_3H_7Br
2-BROMOPROPANE, C_3H_7Br
1-CHLOROPROPANE, C_3H_7Cl
2-CHLOROPROPANE, C_3H_7Cl
1-IODOPROPANE, C_3H_7I
2-IODOPROPANE, C_3H_7I
mixo-DICHLOROBUTANE, $C_4H_8Cl_2$
1-BROMOBUTANE, C_4H_9Br
2-BROMOBUTANE, C_4H_9Br
1-BROMO-2-METHYLPROPANE, C_4H_9Br
2-BROMO-2-METHYLPROPANE, C_4H_9Br
1-CHLOROBUTANE, C_4H_9Cl
2-CHLOROBUTANE, C_4H_9Cl

1-CHLORO-2-METHYLPROPANE, C_4H_9Cl
2-CHLORO-2-METHYLPROPANE, C_4H_9Cl
2-IODOBUTANE, C_4H_9I
1-IODO-2-METHYLPROPANE, C_4H_9I
2-IODO-2-METHYLPROPANE, C_4H_9I
1-BROMO-3-METHYLBUTANE, $C_5H_{11}Br$
2-BROMOPENTANE, $C_5H_{11}Br$
1-CHLORO-3-METHYLBUTANE, $C_5H_{11}Cl$
2-CHLORO-2-METHYLBUTANE, $C_5H_{11}Cl$
1-CHLOROPENTANE, $C_5H_{11}Cl$
2-IODOPENTANE, $C_5H_{11}I$
PERFLUOROHEXYL IODIDE, $C_6F_{13}I$

HALOGEN AZIDES

Metals,
or Phosphorus

Dehnicke, K., *Angew. Chem. (Intern. Ed.)*, 1967, **6**, 240

A comprehensive review covering stability relationships and reactions of these explosive compounds and their derivatives. Bromine, chlorine and iodine azides all explode violently in contact with magnesium, sodium, zinc or white phosphorus.

See BROMINE AZIDE, BrN_3
* CYANOGEN AZIDE, CN_4
CHLORINE AZIDE, ClN_3
FLUORINE AZIDE, FN_3
IODINE AZIDE, IN_3

See other AZIDES
N-HALOGEN COMPOUNDS

N-HALOGEN COMPOUNDS

1. Kovacic, P. *et al.*, *Chem. Rev.*, 1970, **70**, 640
2. Petry, R. C. *et al.*, *J. Org. Chem.*, 1967, **32**, 4034
3. Freeman, J. P. *et al.*, *J. Amer. Chem. Soc.*, 1969, **91**, 4778

Many compounds containing one or more N–X bonds show unstable or explosive properties (and are also oxidants), and this topic has been reviewed [1]. Difluoroamino compounds, ranging from difluoroamine and tetrafluorohydrazine to polydifluoroamino compounds, are notably explosive and suitable precautions have been detailed [2, 3]. Within this class fall the separate groups:

DIFLUOROAMINO COMPOUNDS
HALOGEN AZIDES
N-HALOIMIDES
N,N,N'-TRIFLUOROALKYLAMIDINES

as well as the individually indexed compounds:

TETRAFLUOROAMMONIUM TETRAFLUOROBORATE, BF_8N
BROMOAMINE, BrH_2N
NITROGEN TRIBROMIDE HEXAAMMONIATE, $Br_3N \cdot H_{18}N_6$
NITROSYL TRIBROMIDE, Br_3NO
TETRAFLUORODIAZIRIDINE, CF_4N_2
3-DIFLUOROAMINO-1,2,3-TRIFLUORODIAZIRIDINE, CF_5N_3
PENTAFLUOROGUANIDINE, CF_5N_3
BIS(DIFLUOROAMINO)DIFLUOROMETHANE, CF_6N_2
1-DICHLOROAMINOTETRAZOLE, $CHCl_2N_5$
N,N-DICHLOROMETHYLAMINE, CH_3Cl_2N
N-FLUOROIMINODIFLUOROMETHANE, CNF_3
PERFLUORO-*N*-CYANODIAMINOMETHANE, $C_2F_5N_3$
PERFLUORO-1-AMINOMETHYLGUANIDINE, $C_2F_8N_4$
PERFLUORO-*N*-AMINOMETHYLTRIAMINOMETHANE, $C_2F_{10}N_4$
1-BROMOAZIRIDINE, C_2H_4BrN
N-BROMOACETAMIDE, C_2H_4BrNO
1-CHLOROAZIRIDINE, C_2H_4ClN
N-CHLOROACETAMIDE, C_2H_4ClNO
AZO-*N*-CHLOROFORMAMIDINE, $C_2H_4Cl_2N_6$
1,2-BIS(DIFLUOROAMINO)ETHANOL, $C_2H_4F_4N_2O$
1,2-BIS(DIFLUOROAMINO)-*N*-NITROETHYLAMINE, $C_2H_4F_4N_4O_2$
DIMETHYL *N,N*-DICHLOROPHOSPHORAMIDATE, $C_2H_6Cl_2NO_3P$
2,4,6-TRIS(DICHLOROAMINO)-1,3,5-TRIAZINE (HEXACHLOROMELAMINE), $C_3Cl_6N_6$
2,4,6-TRIS(BROMOAMINO)-1,3,5-TRIAZINE, $C_3H_3Br_3N_6$
2,4,6-TRIS(CHLOROAMINO)-1,3,5-TRIAZINE, $C_3H_3Cl_3N_6$
2-CYANO-1,2,3-TRIS(DIFLUOROAMINO)PROPANE, $C_4H_4F_6N_4$
DI-1,2-BIS(DIFLUOROAMINO)ETHYL ETHER, $C_4H_6F_8N_4O$
4,4-BIS(DIFLUOROAMINO)-3-FLUOROIMINO-1-PENTENE, $C_5H_6F_5N_3$
TETRAKIS-(*N,N*-DICHLOROAMINOMETHYL)METHANE $C_5H_8Cl_8N_4$
N-BROMOTETRAMETHYLGUANIDINE, $C_5H_{12}BrN_3$
N-CHLOROTETRAMETHYLGUANIDINE, $C_5H_{12}ClN_3$
2,6-DIBROMOBENZOQUINONE-4-CHLOROIMINE, $C_6H_2Br_2ClNO$
2,6-DICHLOROBENZOQUINONE-4-CHLOROIMINE, $C_6H_2Cl_3NO$
1-CHLOROBENZOTRIAZOLE, $C_6H_4ClN_3$
N,N-DICHLOROANILINE, $C_6H_5Cl_2N$
NITROGEN CHLORIDE DIFLUORIDE, ClF_2N

CHLOROAMINE, ClH_2N
NITROSYL CHLORIDE, $ClNO$
N-CHLOROSULPHINYLIMIDE, $ClNOS$
NITRYL CHLORIDE, $ClNO_2$
NITROGEN TRICHLORIDE, Cl_3N
FLUOROAMINE, FH_2N
NITROSYL FLUORIDE, FNO
NITRYL FLUORIDE, FNO_2
DIFLUOROAMINE, F_2HN
DIFLUORODIAZENE, F_2N_2
NITROGEN TRIFLUORIDE, F_3N
TETRAFLUOROHYDRAZINE, F_4N_2
DIIODOAMINE, HI_2N
NITROGEN TRIIODIDE-SILVER AMIDE, $I_3N{\cdot}AgH_2N$
NITROGEN TRIIODIDE-AMMONIA, $I_3N{\cdot}NH_3$

HALOGEN OXIDES

The various compounds arising from union of oxygen with one or more halogens are a class of generally unstable but powerful oxidants, individually indexed compounds being:

BROMINE PERCHLORATE, $BrClO_4$
PERBROMYL FLUORIDE, $BrFO_3$
BROMINE DIOXIDE, BrO_2
BROMINE TRIOXIDE, BrO_3
CHLORYL HYPOFLUORITE, $ClFO_3$
PERCHLORYL FLUORIDE, $ClFO_3$
FLUORINE PERCHLORATE, $ClFO_4$
CHLORINE DIOXYGEN TRIFLUORIDE, ClF_3O_2
CHLORINE DIOXIDE, ClO_2
CHLORINE TRIOXIDE, see dimeric Cl_2O_6
DICHLORINE OXIDE, Cl_2O
DICHLORINE TRIOXIDE, Cl_2O_3
CHLORINE PERCHLORATE, Cl_2O_4
CHLORYL PERCHLORATE, Cl_2O_6
PERCHLORYL PERCHLORATE (DICHLORINE HEPTOXIDE), Cl_2O_7
IODINE DIOXYGEN TRIFLUORIDE, F_3IO_2
IODINE(V) OXIDE, I_2O_5

HALOGENS

The reactivity hazard of this group of oxidants towards other materials decreases progressively from fluorine, which reacts violently with most materials under appropriate conditions (except for the metals on which resistant fluoride films form), through chlorine and bromine to iodine. Astatine may be expected to continue this trend.

N-HALOIMIDES

Alcohols,
or Amines,
or Diallyl sulphide,
or Hydrazine,
or Xylene

Martin, R. H., *Nature,* 1951, **168**, 32

Many of the reactions of several *N*-chloro- and *N*-bromoimides are extremely violent or explosive. Those observed include *N*-chlorosuccinimide with aliphatic alcohols or benzylamine or hydrazine hydrate; *N*-bromosuccinimide with aniline, diethyl sulphide or hydrazine hydrate; or 3-nitro-*N*-bromophthalimide with tetrahydrofurfuryl alcohol; 1,3-dichloro-5,5-dimethyl-2,4-imidazolidindione with xylene (violent explosion).

Other *N*-haloimides which may react similarly are:

POTASSIUM 1,3-DIBROMO-2,4-DIKETO-1,3,5-TRIAZINE-6-OLATE, $C_3Br_2KN_3O_3$
SODIUM 1,3-DICHLORO-2,4-DIKETO-1,3,5-TRIAZINE-6-OLATE, $C_3Cl_2N_3NaO_3$
1,3,5-TRICHLORO-1,3,5-TRIAZINETRIONE, $C_3Cl_3N_3O_3$

N-Haloamides may be expected to react similarly.

See *N*-BROMOACETAMIDE, C_2H_4BrNO
N-CHLOROACETAMIDE, C_2H_4ClNO

See other *N*-HALOGEN COMPOUNDS

HALOSILANES

Schumb, W. C. *et al., Inorg. Synth.,* 1939, **1**, 46

When heated, the vapours of the higher chlorosilanes (hexachlorodisilane to dodecachloropentasilane) ignite in air. Other halo- and alkylhalo-silanes ignite without heating or have low flash points.

See POLYDIBROMOSILANE, $(Br_2Si)_n$
TRIBROMOSILANE, Br_3HSi
TRICHLOROSILANE, Cl_3HSi
HEXACHLORODISILANE, Cl_6Si_2
OCTACHLOROTRISILANE, Cl_8Si_3
DECACHLOROTETRASILANE, $Cl_{10}Si_4$
DODECACHLOROPENTASILANE, $Cl_{12}Si_5$
TRIFLUOROSILANE, F_3HSi

See other NON-METAL HALIDES

See related ALKYLNON-METAL HALIDES
NON-METAL HYDRIDES

HAZARDOUS MATERIALS

Cloyd, D. R. *et al.*, *Handling Hazardous Materials*, NASA Technology Survey SP-5032, Washington, NASA, 1965
A survey of hazards and safety procedures involved in handling rocket fuels and oxidisers, including liquid hydrogen, pentaborane, fluorine, chlorine trifluoride, ozone, dinitrogen tetraoxide, hydrazine, methylhydrazine and 1,1-dimethylhydrazine.

HEAVY METAL DERIVATIVES

This class of compounds showing explosive instability deals with heavy metals bonded to elements other than nitrogen and contains the separately treated groups:

GOLD COMPOUNDS
METAL ACETYLIDES
METAL FULMINATES
METAL OXALATES
PLATINUM COMPOUNDS

as well as the individually indexed compounds:

SILVER PEROXOCHROMATE, $AgCrO_5$
SILVER HYPONITRITE, $Ag_2N_2O_2$
SILVER CYANIDE, CAgN
DISILVER KETENDIIDE, C_2Ag_2O
'ETHANE HEXAMERCARBIDE', $C_2H_2Hg_6O_4$

MERCURY(II) CYANIDE, C_2HgN_2
DIMERCURY DICYANIDE OXIDE, $C_2Hg_2N_2O$
LEAD DITHIOCYANATE, $C_2N_2PbS_2$
SILVER TRIFLUOROMETHYLACETYLIDE, C_3AgF_3
CYCLOPENTADIENYLSILVER PERCHLORATE, $C_5H_5AgClO_4$
LEAD 2,4,6-TRINITRORESORCINOLATE, $C_6HN_3O_8Pb$
SILVER BENZO-1,2,3-TRIAZOL-1-OLATE, $C_6H_4AgN_3O$
SILVER PHENYLSELENONATE, $C_6H_5AgO_3Se$
COPPER DIPICRATE, $C_{12}H_4CuN_6O_{14}$
MERCURY DIPICRATE, $C_{12}H_4HgN_6O_{14}$
LEAD DIPICRATE, $C_{12}H_4N_6O_{14}Pb$
ZINC DIPICRATE, $C_{12}H_4N_6O_{14}Zn$

See also METAL AZIDES
METAL CYANIDES AND CYANOCOMPLEXES
N-METAL DERIVATIVES

HIGH-NITROGEN COMPOUNDS

This class heading is intended to include not only those compounds containing a high total proportion of nitrogen (up to 87%) but also those containing high local concentrations in substituent groups (notably azide and diazonium) within the molecule.

Many organic molecular structures containing several chain-linked atoms of nitrogen are unstable or explosive and the tendency is exaggerated by attachment of azide or diazonium groups, or a high-nitrogen heterocyclic nucleus. Closely related but separately treated classes or groups include:

AZIDES (in several sections)
DIAZO COMPOUNDS
DIAZONIUM SALTS
HYDRAZINIUM SALTS
N-NITRO COMPOUNDS
TETRAZOLES
TRIAZENES

Individually indexed compounds in this class are:

5-AMINO-1,2,3,4-THIATRIAZOLE, CH_2N_4S
METHYLDIAZENE, CH_4N_2
1-AMINO-3-NITROGUANIDINE, $CH_5N_5O_2$
DINITRODIAZOMETHANE, CN_4O_4
DIAZOACETONITRILE, C_2HN_3
AZIDOACETONITRILE, $C_2H_2N_4$

5-METHOXY-1,2,3,4-THIATRIAZOLE, $C_2H_3N_3OS$
1,6-BIS(5-TETRAZOLYL)HEXAZ-1,5-DIENE, $C_2H_4N_{14}$
1,2-DIMETHYLNITROSOHYDRAZINE, $C_2H_7N_3O$
DIAZIDOMETHYLENECYANAMIDE, C_2N_8
DIAZIDOMETHYLENEAZINE, C_2N_{14}
BIS(1,2,3,4-THIATRIAZOL-5-YLTHIO)METHANE, $C_3H_2N_6S_4$
DIAZIDOMALONONITRILE, C_3N_8
2,4,6-TRIAZIDO-1,3,5-TRIAZINE, C_3N_{12}
1,3-BIS(5-AMINO-1,3,4-TRIAZOL-2-YL)TRIAZENE, $C_4H_7N_{11}$
POTASSIUM 3,5-DINITRO-2(1-TETRAZENYL)PHENOLATE, $C_6H_5KN_6O_5$
TETRAAZIDO-*p*-BENZOQUINONE, $C_6N_{12}O_2$
1,3-DIPHENYLTRIAZENE, $C_{12}H_{11}N_3$
1,5-DIPHENYL-1,4-PENTAZDIENE, $C_{12}H_{11}N_5$
1,3,6,8-TETRAPHENYLOCTAAZTRIENE, $C_{24}H_{20}N_8$

HYDRAZINIUM SALTS

Mellor, 1940, Vol. 8, 327; 1967, Vol. 8, Suppl. 2.2, 84–86
Salvadori, J., *Gazz. Chim. Ital.*, 1907, **37**(2), 32
Levi, G. R., *Gazz. Chim. Ital.*, 1923, [2], **53**, 105

Several salts are explosively unstable, including hydrazinium azide (explodes on rapid heating or on initiation by a detonator even when damp), chlorate (explodes at m.p. 80°C), chlorite (also highly flammable when dry), hydrogenselenate, hydrogensulphate (decomposes explosively when melted), nitrate, nitrite and the highly explosive perchlorate and diperchlorates used as propellants.

See PROPELLANTS

HYDRAZONES

Air

1. Swern, 1971, Vol. 2, 19
2. Busch, M. *et al.*, *Ber.*, 1914, **47**, 3277

Alkyl- and aryl-hydrazones of aldehydes and ketones readily peroxidise in solution and rearrange to diazo hydroperoxides [1], some of which are explosively unstable [2].

See α-BENZENEDIAZOBENZYL HYDROPEROXIDE, $C_{13}H_{12}N_2O_2$

HYDROGENATION CATALYSTS

1. Augustine, R.L., *Catalytic Hydrogenation*, 23, 28, London, Arnold, 1965
2. Poles, T., private comm., 1973

Many hydrogenation catalysts are sufficiently active to effect rapid interaction of hydrogen and/or solvent vapour with air, causing ignition or explosion. This is particularly so where hydrogen is adsorbed on the catalyst either before a hydrogenation (Raney cobalt, nickel, etc.) or after a hydrogenation during separation of catalyst from the reaction mixture. Exposure to air of such a catalyst should be avoided until complete purging with an inert gas, such as nitrogen, has been effected.

With catalysts of high activity and readily reducible substrates, control of the exotherm may be required to prevent runaway reactions, particularly at high pressures [1].

Platinum-metal catalysts are preferably introduced to the reactor or hydrogenation system in the form of a water-wet paste or slurry. The latter is charged to the empty reactor: air is removed by purging with nitrogen or by several evacuations alternating with nitrogen filling: the reaction mixture is charged, after which hydrogen is admitted. The same procedure applies where it is mandatory to charge the catalyst in the dry state, but in this case the *complete* removal of air before introduction of the reaction mixture and/or hydrogen is of vital importance.

Platinum, palladium and rhodium catalysts are non-pyrophoric as normally manufactured. Iridium and, more particularly, ruthenium catalysts may exhibit pyrophoricity in their fully reduced form and for this reason are usually manufactured in the unreduced form and reduced *in situ.*

Spent catalysts should be purged from hydrogen and washed free from organics with water before storage in the water-wet condition. Under no circumstances should any attempt be made to dry a spent catalyst [2].

Specialist advice on safety and other problems in the use of catalysts and associated equipment is freely available from Engelhard Industries Ltd. at Cinderford, Gloucester, where a model high-pressure hydrogenation laboratory with full safety facilities is maintained.

For individually indexed entries see:

o-NITROANISOLE, $C_7H_7NO_3$: Hydrogen

IRIDIUM, Ir
NICKEL, Ni : Alone
: *p*-Dioxan
: Hydrogen
: Methanol
: Sulphur compounds
PALLADIUM, Pd: Carbon
PLATINUM, Pt: Ethanol
: Hydrogen, etc.
RHODIUM, Rh

HYDROXYLAMMONIUM SALTS

Anon., *Chem. Processing (Chicago)*, 1963, **26**(24), 30
Some decompositions of salts of hydroxylamine are discussed, including violent decomposition of crude hydrochloride solutions at 140°C, the exothermic decomposition of the pure chloride and explosive decomposition of the solid sulphate at 170°C. The phosphinate and nitrate decompose violently above 92 and 100°C, respectively.
See other OXOSALTS OF NITROGENOUS BASES

3-HYDROXYTRIAZENES

Houben-Weyl, 1965, Vol. 10(3), 717
Many of the 3-hydroxytriazene derivatives produced by diazo-coupling on to *N*-alkyl or *N*-aryl hydroxylamines decompose explosively above their m.p.s. The heavy-metal derivatives are, however, stable and used in analytical chemistry.
See other TRIAZENES

HYPOHALITES

1. Sidgwick, 1950, 1218; Chattaway, F.D., *J. Chem. Soc.*, 1923, **123**, 2999
2. Anbar, M. *et al.*, *Chem. Rev.*, 1954, **54**, 927

This class of oxidant compounds, all containing O–X bonds, includes widely differing types and many compounds of limited stability. Alkyl hypochlorites, readily formed from alcohols and chlorinating agents, will explode on ignition, irradiation or contact with copper powder [1]. Of the many alkyl hypohalites described, only ethyl, *tert*-butyl and *tert*-pentyl are stable enough to isolate, purify and handle [2], though care is needed. Individually indexed compounds are:

TRIFLUOROMETHYL HYPOFLUORITE, CF_4O
DIFLUOROMETHYLENE DIHYPOFLUORITE, CF_4O_2
METHYL HYPOCHLORITE, CH_3ClO
TRIFLUOROACETYL HYPOFLUORITE, $C_2F_4O_2$
ETHYL HYPOCHLORITE, C_2H_5ClO
PENTAFLUOROPROPIONYL HYPOFLUORITE, $C_3F_6O_2$
ISOPROPYL HYPOCHLORITE, C_3H_7ClO
HEPTAFLUOROBUTYRYL HYPOFLUORITE, $C_4F_8O_2$
tert-BUTYL HYPOCHLORITE, C_4H_9ClO
CHLORYL HYPOFLUORITE, $ClFO_3$
NITRYL HYPOFLUORITE, FNO_3
* FLUORINE FLUOROSULPHATE, F_2O_3S
PENTAFLUOROSULPHUR HYPOFLUORITE, F_6OS
PENTAFLUOROSELENIUM HYPOFLUORITE, F_6OSe

See also BLEACHING POWDER
METAL OXOHALOGENATES

INORGANIC AZIDES

1. Mellor, 1967, Vol. 8, Suppl. 2,42
2. Evans, B. L. *et al.*, *Chem. Rev.*, 1959, **59**, 515
3 Deb, S. K. *et al. Proc. 8th Combust. Symp.*, 1960, 829

Relationships existing between structure, stability and thermal, photochemical and explosive decomposition (sometimes spontaneous) of the inorganic azides have been extensively investigated and reviewed [1,2]. The ignition characteristics of explosive inorganic azides, with or without added impurities under initiation by heat or light have been discussed [3]. For individually indexed compounds, see:

HALOGEN AZIDES
METAL AZIDE HALIDES
METAL AZIDES
NON-METAL AZIDES

INORGANIC PEROXIDES

1. Castrantas, 1970
2. Castrantas, 1965

The guide to safe handling and storage of peroxides also contains a comprehensive bibliography of detailed information [1]. The earlier publication contains tabulated data on fire and explosion hazards of inorganic peroxides, with a comprehensive bibliography [2].

See HYDROGEN PEROXIDE, H_2O_2
METAL PEROXIDES
PEROXOACIDS
PEROXOACID SALTS

INTERHALOGENS

Kirk-Othmer, 1966, Vol. 9, 585–598

The fluorine-containing members of this class are oxidants at least as powerful as fluorine itself. Individually indexed compounds are:

BROMINE FLUORIDE, BrF
BROMINE TRIFLUORIDE, BrF_3
BROMINE PENTAFLUORIDE, BrF_5
CHLORINE FLUORIDE, ClF
CHLORINE TRIFLUORIDE, ClF_3
CHLORINE PENTAFLUORIDE, ClF_5
IODINE CHLORIDE, ClI
IODINE PENTAFLUORIDE, F_5I
IODINE HEPTAFLUORIDE, F_7I

IODINE COMPOUNDS

Several iodine compounds are explosively unstable, individually indexed compounds being:

IODINE ISOCYANATE, CINO
2-IODOSYLVINYL CHLORIDE, C_2H_2ClIO
2-IODYLVINYL CHLORIDE, $C_2H_2ClIO_2$
DIIODOACETYLENE, C_2I_2
1,6-DIIODO-2,4-HEXADIYNE, $C_6H_4I_2$
[(CHROMYLDIOXY)IODO] BENZENE, $C_6H_5CrIO_4$

IODOSYLBENZENE, C_6H_5IO
IODYLBENZENE, $C_6H_5IO_2$
IODYLBENZENE PERCHLORATE, $C_6H_6ClIO_6$
IODINE TRIACETATE, $C_6H_9IO_6$
1-IODO-3-PHENYL-2-PROPYNE, C_9H_7I
CALCIUM BIS-*p*-IODYLBENZOATE, $C_{14}H_8CaI_2O_8$
9-PHENYL-9-IODAFLUORENE, $C_{18}H_{13}I$
IODINE(III) PERCHLORATE, Cl_3IO_{12}
AMMONIUM PERIODATE, H_4INO_4
IODINE AZIDE, IN_3
NITROGEN TRIIODIDE-AMMONIA, $I_3N \cdot H_3N$

ION EXCHANGE RESINS

Anon., *Chem. Eng. News*, 1953, **31**, 5120; *Chem. Eng. News*, 1968, **26**, 1480

A three-year-old sample of ion exchange resin was soaked in dilute hydrochloric acid, and then charged into a 2.5 cm diameter glass column. After soaking in distilled water for 15 min, the tube exploded violently, presumably owing to swelling of the resin. Process resins as far as possible before charging into column. The earlier incident involved a column charged with dry resin which burst when wetted.

See NITRIC ACID, HNO_3: Ion exchange resins

ISOCYANIDES

Acids

Sidgwick, 1950, 673

Acid-catalysed hydrolysis of isocyanides ('carbylamines'), to primary amine and formic acid is very rapid, sometimes explosively so.

KETONE PEROXIDES

Davies, 1961, 72

Kirk-Othmer, 1967, Vol. 14, 777

The variety of peroxides (monomeric, dimeric and polymeric) which can be produced from interaction of a given ketone with hydrogen peroxide is very wide (see group-types below). The proportions of the products in the reaction mixture depend on the reaction conditions used, as well as the structure of the ketone. Many of the products appear to coexist in equilibrium, and several types of structure are explosive and sensitive in varying degrees to heat or shock. Extreme caution is therefore required in handling ketone peroxides in high concentrations, particularly those derived from ketones of low molecular weight. Acetone is thus entirely unsuitable as a reaction or cleaning solvent whenever hydrogen peroxide is used.

See CYCLIC PEROXIDES
1-OXYPEROXY COMPOUNDS
POLYPEROXIDES
3,3,6,6-TETRAMETHYL-1,2,4,5-TETRAOXANE, $C_6H_{12}O_4$
3,6-DIETHYL-3,6-DIMETHYL-1,2,4,5-TETRAOXANE, $C_8H_{16}O_4$
3,3,6,6,9,9-HEXAMETHYL-1,2,4,5,7,8-HEXAOXAONANE, $C_9H_{18}O_6$
3,6-DI(SPIROCYCLOHEXANE) TETRAOXANE, $C_{12}H_{20}O_4$
3,6,9-TRIETHYL-3,6,9-TRIMETHYL-1,2,4,5,7,8-HEXAOXAONANE, $C_{12}H_{24}O_6$
TRI(SPIROCYCLOPENTANE)-1,1,4,4,7,7-HEXAOXAONANE, $C_{15}H_{24}O_6$

KJELDAHL METHOD

Beet, A. E., *J.R. Inst. Chem.*, 1955, **79**, 163, 299
Possible hazards introduced by variations in experimental technique in Kjeldahl nitrogen determination are discussed.

LASSAIGNE TEST

See SODIUM, Na: Fluorinated compounds
: Halocarbons

LIGHT ALLOYS

Fire Prot. Assoc. J., 1959 (44), 28

Experiments to determine the probability of ignition of gas or vapour by incendive sparks arising from impact of aluminium-, magnesium- and zinc-containing alloys with rusty steel are described. The risk is greatest with magnesium alloys, where, the higher the magnesium content, the lower the impact energy necessary for incendive sparking. Wide ranges of ignitable gas concentrations also tend to promote ignition.

See DIIRON TRIOXIDE, Fe_2O_3: Aluminium
: Aluminium, propene
MAGNESIUM, Mg: Metal oxides

LINSEED OIL

Watts, B. G., private comm., 1965
Cloths used to apply linseed oil to laboratory benches were not burnt as directed, but dropped into a waste bin. A fire developed during a few hours and destroyed the laboratory. Tests showed that heating and ignition were rapid if a draught of warm air impinged on the oil-soaked cloth. Many other incidents involving ignition of peroxidisable materials dispersed on absorbent combustible fibrous materials have been recorded.

LIQUEFIED NATURAL GAS

Organic liquids,
or Water

1. Katz, D.L. *et al.*, *Hydrocarbon Proc.*, 1971, **50**, 240; Anon., *Chem. Eng. News,* 1972, **50**(8), 57
2. Yang, K., *Nature,* 1973, **243**, 221–222

The quite loud 'explosions' (either immediate or delayed) which occur when LNG (containing unusually high contents of heavier materials) is spilled on to water are non-combustive and harmless. Superheating and shock-wave phenomena are involved [1]. There is a similar effect when LNG of normal composition (90% methane) is spilled on to some C_5-C_8 hydrocarbons or methanol, acetone or 2-butanone [2].

LIQUEFIED PETROLEUM GASES

1. *Storage and Handling of LPG,* London, Fire Prot. Assoc., 1964
2. *LPG Safety, Model Code,* part 9, London, Inst. Petr., 1967

The considerable fire hazards of propane and butane, the most commonly used LPGs, reside in their high volatility, high vapour density and tendency to collect in drains, hollows and basements, and a lower explosive limit of only 2% by volume. The booklet gives guidance on precautions necessary for bulk storage sites and for filling and storage of cylinders, and refers to existing Codes of Practice and Standards [1], of which the most recent is reference 2.

LIQUID AIR

Liquid air, formerly used widely as a laboratory or industrial cryogenic liquid, has been involved in many violent incidents. Many of these have involved the increased content of residual liquid oxygen produced by fractional evaporation of liquid air during storage. However, liquid air is still a powerful oxidant in its own right. Liquid nitrogen, now widely available, is recommended as a safer coolant than liquid air.

Charcoal
Taylor, J., *J. Sci. Instrum.,* 1928, **5**, 24
Accidental contact via a cracked tube caused violent explosion. Nitrogen is a safer coolant.

Ether
Danckwort, P. W., *Angew. Chem.,* 1927, **40**, 1317
Addition of liquid air to ether in a dish caused a violent explosion after a short delay. Previous demonstrations had been uneventful, though it was known that such mixtures were impact- and friction-sensitive.

Hydrocarbons
McCartey, L. V. *et al.*, *Chem. Eng. News*, 1949, 27, 2612
All hydrocarbons (and most reducing agents) form explosive mixtures with liquid air.
See NITROGEN (Liquid), N_2
OXYGEN (Liquid), O_2

METAL ABIETATES

Fire Prot. Assoc. J., 1958, 255
Aluminium, zinc, calcium, sodium, cobalt, lead and manganese abietates ('resinates'), when finely divided, are subject to spontaneous heating and ignition. Store in sealed metal containers away from fire hazards.

METAL ACETYLIDES

1. Brameld, V.F. *et al.*, *J. Soc. Chem. Ind.*, 1947, **66**, 346
2. Houben-Weyl, 1970, Vol. 13(1), 739
3. Rutledge, 1968, 85–86
4. Miller, 1965, Vol. 1, 486

Previous literature on formation of various types of copper acetylides is discussed and the mechanism of their formation is examined, with experimental detail. Whenever a copper or a copper-rich alloy is likely to come into contact with atmospheres containing (1) ammonia, water vapour and acetylene or (2) lime-sludge, water vapour and acetylene, or a combination of these two, there is the probability of acetylide formation with danger of explosion.

The action is aided by the presence of air, or air with carbon dioxide, and hindered by the presence of nitrogen. Explosive acetylides may be formed on copper and brasses containing more than 50% copper when these are exposed to acetylene atmospheres. The acetylides produced by action of acetylene on ammoniacal or alkaline solutions of copper(II) salts are more explosive than those from the corresponding copper(I) salts [1]. The hydrated forms are less explosive than the anhydrous material [2].

Catalytic forms of copper, mercury and silver acetylides, supported on alumina, carbon or silica and used for polymerisation of alkanes, are relatively stable [3].

In contact with acetylene, silver and mercury salt solutions will also give explosive acetylides, the mercury derivatives being complex [4]. Many of the metal acetylides react violently with oxidants.

Individually indexed compounds in this often dangerously explosive class are:

DISILVER ACETYLIDE, C_2Ag_2
SILVER ACETYLIDE-SILVER NITRATE, $C_2Ag_2 \cdot AgNO_3$
DIGOLD(I) ACETYLIDE, C_2Au_2
BARIUM ACETYLIDE, C_2Ba
CALCIUM ACETYLIDE (CARBIDE), C_2Ca
DICAESIUM ACETYLIDE, C_2Cs_2
COPPER(II) ACETYLIDE, C_2Cu
DICOPPER(I) ACETYLIDE, C_2Cu_2
SILVER ACETYLIDE, C_2HAg
CAESIUM ACETYLIDE, C_2HCs
POTASSIUM ACETYLIDE, C_2HK
LITHIUM ACETYLIDE, C_2HLi
SODIUM ACETYLIDE, C_2HNa
RUBIDIUM ACETYLIDE, C_2HRb
LITHIUM ACETYLIDE-AMMONIA, $C_2HLi \cdot H_3N$
DIPOTASSIUM ACETYLIDE, C_2K_2
DILITHIUM ACETYLIDE, C_2Li_2
DISODIUM ACETYLIDE, C_2Na_2
DIRUBIDIUM ACETYLIDE, C_2Rb_2
STRONTIUM ACETYLIDE, C_2Sr
SILVER TRIFLUOROMETHYLACETYLIDE, C_3AgF_3
SODIUM METHOXYACETYLIDE, C_3H_3NaO
SODIUM ETHOXYACETYLIDE, C_4H_5NaO
1,3 PENTADIYN-1-YLSILVER, C_5H_3Ag
1,3-PENTADIYN-1-YLCOPPER, C_5H_3Cu
DIMETHYL-1-PROPYNYLTHALLIUM, C_5H_9Tl
TRIETHYNYLALUMINIUM, C_6H_3Al
TRIETHYNYLANTIMONY, C_6H_3Sb
BIS(DIMETHYLTHALLIUM) ACETYLIDE, $C_6H_{12}Tl_2$
TETRAETHYNYLGERMANIUM, C_8H_4Ge
TETRAETHYNYLTIN, C_8H_4Sn
SODIUM PHENYLACETYLIDE, C_8H_5Na
3-BUTEN-1-YNYLDIETHYLALUMINIUM, $C_8H_{13}Al$
DIMETHYL-PHENYLETHYNYLTHALLIUM, $C_{10}H_{11}Tl$
3-BUTEN-1-YNYLTRIETHYLLEAD, $C_{10}H_{18}Pb$
3-METHYL-3-BUTEN-1-YNYLTRIETHYLLEAD, $C_{11}H_{20}Pb$
3-BUTEN-1-YNYLDIISOBUTYLALUMINIUM, $C_{12}H_{21}Al$
BIS(TRIETHYLTIN)ACETYLENE, $C_{14}H_{30}Sn_2$

See other ACETYLENIC COMPOUNDS

METAL AMIDOSULPHATES

Metal nitrates,
or Nitrites

Heubel, J. *et al., Compt. Rend.* [3], 1963, **257**, 684
Heating mixtures of barium, potassium or sodium amidosulphates or amidosulphuric acid with sodium or potassium nitrates or nitrites, leads to reactions which may be explosive. Thermogravimetric plots are given.

METAL AZIDE HALIDES

Dehnicke, K., *Angew. Chem. (Intern. Ed.),* 1967, **6,** 243
Metal halides and halogen azides react to give a range of metal azide halides, many of which are explosive.

See SILVER AZIDE CHLORIDE, $AgClN_3$
TUNGSTEN AZIDE PENTABROMIDE, Br_5N_3W
CHROMYL AZIDE CHLORIDE, $ClCrN_3O_2$
VANADYL AZIDE DICHLORIDE, Cl_2N_3OV
TIN AZIDE TRICHLORIDE, Cl_3N_3Sn
TITANIUM AZIDE TRICHLORIDE, Cl_3N_3Ti
MOLYBDENUM DIAZIDE TETRACHLORIDE, Cl_4MoN_6
VANADIUM AZIDE TETRACHLORIDE, Cl_4N_3V
MOLYBDENUM AZIDE PENTACHLORIDE, Cl_5MoN_3
URANIUM AZIDE PENTACHLORIDE, Cl_5N_3U
TUNGSTEN AZIDE PENTACHLORIDE, Cl_5N_3W

METAL AZIDES

Mellor, 1940, Vol. 8, 344–355; Vol. 8, Suppl. 2, 16–54
This large and well-documented group of explosive compounds contains some which are widely used industrially. Related to this group are:

METAL AZIDE HALIDES

and, of the many known simple and complex explosive metal azides, those few selected for individual indexing here are:

SILVER AZIDE, AgN_3
LITHIUM TETRAAZIDOALUMINATE(1–), $AlLiN_{12}$

ALUMINIUM TRIAZIDE, AlN_9
TRIGOLD DISODIUM HEXAAZIDE, $Au_3N_{18}Na_2$
BARIUM DIAZIDE, BaN_6
DIMETHYLGOLD AZIDE: see the dimer, $C_4H_{12}Au_2N_6$
TETRAMETHYLDIGOLD DIAZIDE, $C_4H_{12}Au_2N_6$
CALCIUM DIAZIDE, CaN_6
CADMIUM DIAZIDE, CdN_6
POTASSIUM TRIAZIDOCOBALTATE(1–), $CoKN_9$
CHROMYL AZIDE, CrN_6O_2
TETRAAMMINECOPPER(II) AZIDE, $CuH_{12}N_{10}$
LITHIUM HEXAAZIDOCUPRATE(4–), $CuLi_4N_{18}$
COPPER(I) AZIDE, CuN_3
COPPER(II) AZIDE, CuN_6
* AZIDOGERMANE, GeH_3N_3
MERCURY(II) AZIDE, HgN_6
MERCURY(I) AZIDE, Hg_2N_6
POTASSIUM AZIDE, KN_3
LITHIUM AZIDE, LiN_3
SODIUM AZIDE, N_3Na
THALLIUM(I) AZIDE, N_3Tl
LEAD(II) AZIDE, N_6Pb
STRONTIUM AZIDE, N_6Sr
LEAD(IV) AZIDE, $N_{12}Pb$

See related AMMINECOBALT(III) AZIDES

METAL CHLORATES

Phosphorus,
Sugar,
Sulphur

1. Black, H. K., *School Sci. Rev.*, 1963, 44(53), 462
2. *59th Ann. Rep. HM Insp. Explosives* (Cmd. 4934), 5, London, HMSO, 1934
3. Berger, A., *Arbeits-Schutz.*, 1934, **2**, 20

The extremely hazardous nature of the mixtures of metal chlorates with phosphorus, sugar or sulphur, sometimes with addition of permanganates and metal powders, frequently prepared as amateur fireworks, is stressed. Apart from being powerfully explosive, such mixtures are dangerously sensitive to friction or shock, and spontaneous ignition sometimes occurs [1]. Chlorates containing 1–2% of bromates or sulphur as impurities are liable to spontaneous explosion [3]. The danger of mixtures of chlorates with sulphur or phosphorus

is such that their preparation without licence was prohibited by Orders in Council many years ago [2].
See other METAL HALOGENATES

Acids
1. Mellor, 1941, Vol. 2, 315
2. Stossel, E. *et al.*, US Pat. 2 338 268, 1944

Additionally to being oxidants in contact with strong acids, metal chlorates liberate explosive chlorine dioxide gas. With concentrated sulphuric acid, a violent explosion may occur unless effective cooling is used [1]. Heating a moist mixture of a metal chlorate and a dibasic organic acid (tartaric or citric acid) liberates chlorine dioxide mixed with carbon dioxide [2].
See other METAL HALOGENATES

METAL CYANIDES AND CYANOCOMPLEXES

1. von Schwartz, 1918, 399, 327; Pieters, 1957, 30
2. *Res. Rep. No. 2,* New York, Nat. Board Fire Underwriters, 1950

Several members of this class which contain heavy metals tend to explosive instability, and most are capable of violent oxidation under appropriate circumstances. Fusion of mixtures of metal cyanides with metal chlorates, perchlorates, nitrates or nitrites causes a violent explosion [1]. Addition of one solid component (even as a residue in small amount) to another molten component is also highly dangerous [2]. Individually indexed compounds are:

SILVER CYANIDE, CAgN
GOLD(I) CYANIDE, CAuN
POTASSIUM CYANIDE, CKN
SODIUM CYANIDE, CNNa
CADMIUM DICYANIDE, C_2CdN_2
COPPER(II) CYANIDE, C_2CuN_2
MERCURY(II) CYANIDE, C_2HgN_2
* DIMERCURY DICYANIDE OXIDE, $C_2Hg_2N_2O$
NICKEL(II) CYANIDE, C_2N_2Ni
LEAD(II) CYANIDE, C_2N_2Pb
ZINC DICYANIDE, C_2N_2Zn
POTASSIUM TETRACYANOMERCURATE (2–), $C_4HgK_2N_4$
SODIUM PENTACYANONITROSYLFERRATE (2–), $C_5FeN_6Na_2O$

POTASSIUM HEXACYANOFERRATE(3–) ('FERRICYANIDE'), $C_6FeK_3N_6$
POTASSIUM HEXACYANOFERRATE(4–) ('FERROCYANIDE'), $C_6FeK_4N_6$
IRON(3+) HEXACYANOFERRATE(4–), $C_{18}Fe_7N_{18}$

See also MOLTEN SALT BATHS

N-METAL DERIVATIVES

Many metal derivatives of nitrogenous systems containing one or more bonds linking nitrogen to a metal (usually but not exclusively a heavy metal) show explosive instability. Individually indexed compounds are:

SILVER PERCHLORYLAMIDE, $AgClHNO_3$
SILVER AMIDE, AgH_2N
TRISILVER NITRIDE, Ag_3N
TETRASILVER DIIMIDOTRIPHOSPHATE, $Ag_4H_3N_2O_8P_3$
SILVER DINITRIDODIOXOSULPHATE(4–), $Ag_4N_2O_2S$
PENTASILVER DIIMIDOTRIPHOSPHATE, $Ag_5H_2N_2O_8P_3$
GOLD NITRIDE-AMMONIA, $Au_3N \cdot H_3N$
BISMUTH AMIDE OXIDE, BiH_2NO
BISMUTH NITRIDE, BiN
DISILVER CYANAMIDE, CAg_2N_2
SILVER TETRAZOLIDE, $CHAgN_4$
SILVER NITROUREIDE, $CH_2AgN_3O_3$
SILVER 5-AMINOTETRAZOLIDE, CH_2AgN_5
SILVER NITROGUANIDIDE, CH_3AgN_4
POTASSIUM METHYLAMIDE, CH_4KN
5-HYDROXY-1(*N*-SODIO-5-TETRAZOLYLAZO)-TETRAZOLE, $C_2HN_{10}NaO$
N,N-DISODIUM *N,N'*-DIMETHOXYSULPHONYLDIAMIDE, $C_2H_6N_2Na_2O_2S$
COPPER(II)1,3-DI(5-TETRAZOLYL)TRIAZENIDE, $C_4H_4CuN_{22}$
TRICALCIUM DINITRIDE, Ca_3N_2
CADMIUM DIAMIDE, CdH_4N_2
TRICADMIUM DINITRIDE, Cd_3N_2
CERIUM NITRIDE, CeN
COBALT(III) AMIDE, CoH_6N_3
COBALT NITRIDE, CoN
CAESIUM AMIDE, CsH_2N
TRICAESIUM NITRIDE, Cs_3N
COPPER(I) NITRIDE, Cu_3N
GERMANIUM(II) IMIDE, GeHN

DIMERCURY IMIDE OXIDE, HHg_2NO
LEAD IMIDE, HNPb
POTASSIUM AMIDE, H_2KN
SODIUM AMIDE, H_2NNa
SODIUM HYDRAZIDE, H_3N_2Na
ZINC DIHYDRAZIDE, H_6N_4Zn
TRIMERCURY DINITRIDE, Hg_3N_2
NITROGEN TRIIODIDE-SILVER AMIDE, $I_3N \cdot AgH_2N$
POTASSIUM NITRIDOOSMATE (1–), KNO_3Os
POTASSIUM SULPHURDIIMIDE (2–), K_2N_2S
POTASSIUM THALLIUM(I) AMIDE (2–) AMMONIATE, $K_2NTl–H_3N$
POTASSIUM NITRIDE (3–), K_3N
LITHIUM SODIUM NITROXYLATE, $LiNNaO_2$
LITHIUM NITRIDE, Li_3N
SODIUM NITROXYLATE (2–), NNa_2O_2
SODIUM NITRIDE (3–), NNa_3
RUBIDIUM NITRIDE (3–), NRb_3
ANTIMONY(III)NITRIDE, NSb
THALLIUM(I) NITRIDE, NTl_3
TRILEAD DINITRIDE, N_2Pb_3
* TETRASELENIUM TETRANITRIDE, N_4Se_4
* TRITELLURIUM TETRANITRIDE, N_4Te_3
TRITHORIUM TETRANITRIDE, N_4Th_3

See PERCHLORYLAMIDE SALTS

METAL DUSTS

1. Jacobson, M. *et al.*, *Rep. Invest.*, 6516(9) Washington, US Bureau of Mines, 1961
2. Brown, H. R., *Chem. Eng. News,* 1956, **34**, 87

Of the 313 samples examined, the dust explosion hazards of finely divided aluminium, aluminium–magnesium alloys, magnesium, thorium, titanium and uranium, and the hydrides of thorium and uranium, are rated highest [1]. The need to exercise caution when handling dusts of some recently introduced reactive metals is briefly discussed. Some will form explosive mixtures, not only with air or oxygen, but also with nitrogen and carbon dioxide, reacting to give the carbonate or nitride. Beryllium, cerium, germanium, hafnium, lithium, niobium, potassium, sodium, thorium, titanium, uranium and zirconium are discussed [2].

See the individual metals

METAL FULMINATES

Urbanski, 1967, Vol. 3, 157

The metal salts are all powerfully explosive. Of several salts examined, those of cadmium, copper and silver were more powerful detonators than mercury fulminate, while thallium fulminate was much more sensitive to heating and impact. Formally related salts are also explosive. Individually indexed compounds are:

SODIUM FULMINATE, CNNaO
THALLIUM FULMINATE, CNOTl
SILVER FULMINATE, $C_2Ag_2N_2O_2$
CADMIUM FULMINATE, $C_2CdN_2O_2$
COPPER(II) FULMINATE, $C_2CuN_2O_2$
* MERCURY(II) METHYLNITROLATE, $C_2H_2HgN_4O_6$
* MERCURY(II) FORMHYDROXAMATE, $C_2H_4HgN_2O_4$
MERCURY(II) FULMINATE, $C_2HgN_2O_2$
DIMETHYLTHALLIUM FULMINATE, C_3H_6NOTl
DIPHENYLTHALLIUM FULMINATE, $C_{13}H_{10}NOTl$

METAL HALIDES

Members of this class have usually featured as secondary reagents in hazardous combination of chemicals.

Individually indexed compounds are:

SILVER CHLORIDE, AgCl
SILVER FLUORIDE, AgF
SILVER DIFLUORIDE, AgF_2
ALUMINIUM TRIBROMIDE, $AlBr_3$
ALUMINIUM TRICHLORIDE, $AlCl_3$
AMERICIUM CHLORIDE, $AmCl_3$
GOLD(III) CHLORIDE, $AuCl_3$
BERYLLIUM FLUORIDE, BeF_2
BISMUTH PENTAFLUORIDE, BiF_5
CALCIUM BROMIDE, Br_2Ca
COBALT(II) BROMIDE, Br_2Co
IRON(II) BROMIDE, Br_2Fe
MERCURY(II) BROMIDE, Br_2Hg
IRON(III) BROMIDE, Br_3Fe
CALCIUM CHLORIDE, $CaCl_2$
* AMMONIUM CHLORIDE, ClH_4N
SODIUM CHLORIDE, ClNa

THALLIUM(I) CHLORIDE, ClTl
COBALT(II) CHLORIDE, Cl_2Co
CHROMYL CHLORIDE, Cl_2CrO_2
MANGANESE(II) CHLORIDE, Cl_2Mn
LEAD DICHLORIDE, Cl_2Pb
TIN DICHLORIDE, Cl_2Sn
TITANIUM DICHLORIDE, Cl_2Ti
ZIRCONIUM DICHLORIDE, Cl_2Zr
COBALT(III) CHLORIDE, Cl_3Co
CHROMIUM(III) CHLORIDE, Cl_3Cr
IRON(III) CHLORIDE, Cl_3Fe
* HEXAAMMINETITANIUM(III) CHLORIDE, $Cl_3H_{18}N_6Ti$
* ANTIMONY TRICHLORIDE OXIDE, Cl_3OSb
RHODIUM(III) CHLORIDE, Cl_3Rh
RUTHENIUM(III) CHLORIDE, Cl_3Ru
ANTIMONY TRICHLORIDE, Cl_3Sb
TITANIUM TRICHLORIDE, Cl_3Ti
VANADIUM TRICHLORIDE, Cl_3V
GERMANIUM TETRACHLORIDE, Cl_4Ge
* RHENIUM TETRACHLORIDE OXIDE, Cl_4ORe
LEAD TETRACHLORIDE, Cl_4Pb
TIN TETRACHLORIDE, Cl_4Sn
TITANIUM TETRACHLORIDE, Cl_4Ti
ZIRCONIUM TETRACHLORIDE, Cl_4Zr
ANTIMONY PENTACHLORIDE, Cl_5Sb
COBALT TRIFLUORIDE, CoF_3
CAESIUM FLUORIDE, CsF
LEAD DIFLUORIDE, F_2Pb
MANGANESE TRIFLUORIDE, F_3Mn
PALLADIUM TRIFLUORIDE, F_3Pd
PLATINUM TETRAFLUORIDE, F_4Pt
URANIUM HEXAFLUORIDE, F_6U
IRON(II) IODIDE, FeI_2
* AMMONIUM IODIDE, H_4IN
MERCURY(II) IODIDE, HgI_2
POTASSIUM IODIDE, IK
SODIUM IODIDE, INa

METAL HALOGENATES

Metals and oxidisable derivatives,
or Non-metals,
or Oxidisable materials

Mellor, 1946, Vol. 2, 310; 1956, Vol. 2, Suppl. 1, 583–584; 1941, Vol. 3, 651
von Schwartz, 1918, 323
Intimate mixtures of chlorates, bromates or iodates of barium, cadmium, calcium, magnesium, potassium, sodium or zinc, with finely divided aluminium, arsenic, copper; carbon, phosphorus, sulphur; hydrides of alkali and alkaline earth metals; sulphides of antimony, arsenic, copper or tin, metal cyanides, thiocyanates or impure manganese dioxide may react violently or explosively, either spontaneously (especially in presence of moisture) or on initiation by heat, friction, impact, sparks or addition of sulphuric acid.

Mixtures of sodium or potassium chlorate with sulphur or phosphorus are rated as being exceptionally dangerous on frictional initiation.

See METAL CHLORATES
See other METAL OXOHALOGENATES

METAL HYDRIDES

1. Banus, M. D., *Chem. Eng. News,* 1954, **32**, 2424–2427
2. Mackay, 1966, 66

Precautions necessary for safe handling of three main groups of hydrides of commercial significance are discussed. The first group (sodium hydride; lithium or sodium tetrahydroaluminates) ignite or explode in contact with liquid water or high humidity, while the second group (lithium, calcium, strontium, barium hydrides; sodium or potassium tetrahydroborates) do not. Burning sodium hydride is reactive enough to explode with the combined water in concrete. The third group ('alloy' or non-stoicheiometric hydrides of titanium, zirconium, thorium, uranium, vanadium, tantalum and palladium) are produced commercially in very finely divided form. Though less pyrophoric than the corresponding powdered metals, once burning is established, they are difficult to extinguish and water-, carbon dioxide- or halocarbon-based extinguishers caused violent explosions. Powdered dolomite is usually effective in smothering such fires [1]. The trihydrides of the lanthanoids (rare earth metals) are pyrophoric in air and the dihydrides, though less reactive, must be handled under inert atmosphere [2].

Individually indexed members of this class of active reducants are:

ALUMINIUM HYDRIDE, AlH_3
ALUMINIUM HYDRIDE-TRIMETHYLAMINE, $AlH_3 \cdot C_3H_9N$
BARIUM HYDRIDE, BaH_2
BERYLLIUM HYDRIDE, BeH_2
CALCIUM DIHYDRIDE, CaH_2
CERIUM TRIHYDRIDE, CeH_3
* CHLOROGERMANE, $ClGeH_3$
COPPER(I) HYDRIDE, CuH
GERMANIUM MONOHYDRIDE, $(GeH)_n$
GERMANE, GeH_4
DIGERMANE, Ge_2H_6
TRIGERMANE, Ge_3H_8
POTASSIUM HYDRIDE, HK
LITHIUM HYDRIDE, HLi
SOLIUM HYDRIDE, HNa
RUBIDIUM HYDRIDE, HRb
MAGNESIUM HYDRIDE, H_2Mg
ZINC HYDRIDE, H_2Zn
'ZIRCONIUM HYDRIDE', H_2Zr
LANTHANUM TRIHYDRIDE, H_3La
PLUTONIUM(III) HYDRIDE, H_3Pu
URANIUM(III) HYDRIDE, H_3U
THORIUM HYDRIDE, H_4Th

See also COMPLEX HYDRIDES

METAL HYPOCHLORITES

A widely used group of industrial oxidants which has been involved in numerous incidents, some with nitrogenous materials leading to formation of nitrogen trichloride.

See BLEACHING POWDER
CALCIUM HYPOCHLORITE, $CaCl_2O_2$
SODIUM HYPOCHLORITE, ClNaO
See also CHLORINE, Cl_2 : Nitrogen compounds

Amines
Kirk-Othmer, 1963, Vol. 2, 104
Primary or secondary amines react with sodium or calcium hypochlorites to give *N*-chloroamines, some of which are explosive when isolated. Application of other chlorinating agents to amines or their

precursors may also produce the same result under appropriate conditions.
See related HYPOHALITES
See other METAL OXOHALOGENATES

METAL NITRATES

Aluminium
See ALUMINIUM, Al: Metal nitrates, etc.

Citric acid
Shannon, I. R., *Chem. & Ind.*, 1970, 149
During vacuum evaporation of an aqueous mixture of unspecified mixed metal nitrates and citric acid, the amorphous solid exploded when nearly dry. This was attributed to oxidation of the organic residue by the nitrates present, possibly catalysed by one of the oxides expected to be produced.

Esters,
or Phosphorus,
or Tin(II) chloride
Pieters, 1957, 30
Mixtures of metal nitrates with alkyl esters may explode, owing to formation of alkyl nitrates. Mixtures of a nitrate with phosphorus, tin(II) chloride or other reducing agents may react explosively.

Metal phosphinates
1. Mellor, 1940, Vol. 8, 881
2. Costa, R.L., *Chem. Eng. News*, 1947, **25**, 3177
Mixtures of metal nitrates and phosphinates, previously proposed as explosives [1], explode on heating.

Organic matter
Bowen, H. J. M., *Anal. Chem.*, 1968, **40**, 969; private comm.
When organic matter is destroyed for residue analysis by heating with equimolar potassium–sodium nitrate mixture to 390° C, a twentyfold

excess of nitrate must be used. If over 10% of organic matter is present, pyrotechnic reactions occur which could be explosive.

See MOLTEN SALT BATHS

See other METAL OXONON-METALLATES

Potassium hexanitrocobaltate (3–)

See POTASSIUM HEXANITROCOBALTATE (3–), $CoK_3N_6O_{12}$

METAL NITRITES

Metal cyanides

See SODIUM NITRITE, $NNaO_2$: Metal cyanides

Nitrogenous bases

Metal nitrites react with salts of nitrogenous bases to give the corresponding nitrite salts, many of which are unstable.

See NITRITE SALTS OF NITROGENOUS BASES

Potassium hexanitrocobaltate (3–)

See POTASSIUM HEXANITROCOBALTATE (3–), $CoK_3N_6O_{12}$

See MOLTEN SALT BATHS

See other METAL OXONON-METALLATES

METAL NON-METALLIDES

This class includes the products of combination of metals and non-metals except C (as acetylene), H, N, O and S, which are separately treated in the groups:

METAL ACETYLIDES
N-METAL DERIVATIVES
METAL HYDRIDES
METAL OXIDES
METAL SULPHIDES

Individually indexed compounds are:

DIALUMINIUM OCTAVANADIUM TRIDECASILICIDE, $Al_2Si_{13}V_8$
PLATINUM DIARSENIDE, As_2Pt

MAGNESIUM BORIDE, B_2Mg_3
IRON CARBIDE, CFe_3
TITANIUM CARBIDE, CTi
TUNGSTEN CARBIDE, CW
DITUNGSTEN CARBIDE, CW_2
THORIUM DICARBIDE, C_2Th
URANIUM DICARBIDE, C_2U
ZIRCONIUM DICARBIDE, C_2Zr
TETRAALUMINIUM TRICARBIDE, C_3Al_4
RUBIDIUM OCTACARBIDE, C_8Rb
CALCIUM SILICIDE, CaSi
CALCIUM DISILICIDE, $CaSi_2$
TRICALCIUM DIPHOSPHIDE, Ca_3P_2
CADMIUM SELENIDE, CdSe
TRICADMIUM DIPHOSPHIDE, Cd_3P_2
COPPER MONOPHOSPHIDE, CuP
COPPER DIPHOSPHIDE, Cu_2P
TRICOPPER DIPHOSPHIDE, Cu_3P_2
IRON–SILICON, Fe–Si
TRIS(IODOMERCURI)PHOSPHINE, Hg_3I_3P
TRIMERCURY TETRAPHOSPHIDE, Hg_3P_4
HEXALITHIUM DISILICIDE, Li_6Si_2
TRIMAGNESIUM DIPHOSPHIDE, Mg_3P_2
MANGANESE(II) TELLURIDE, MnTe
TRIMANGANESE DIPHOSPHIDE, Mn_3P_2
SODIUM SILICIDE, NaSi
SODIUM PHOSPHIDE (3—), Na_3P
TRIZINC DIPHOSPHIDE, P_2Zn_3

METAL OXALATES

Several heavy metal oxalates are unstable to heat.

See SILVER OXALATE, $C_2Ag_2O_4$
COPPER(I) OXALATE, $C_2Cu_2O_4$
MERCURY(II) OXALATE, C_2HgO_4
IRON(III) OXALATE, $C_6Fe_2O_{12}$
See other HEAVY METAL DERIVATIVES

METAL OXIDES

This large class covers a wide range of types of reactivity and there is a separate class entry for

METAL PEROXIDES

Individually indexed compounds are:

SILVER(II) OXIDE, AgO
SILVER(I) OXIDE, Ag_2O
DISILVER PENTATIN UNDECAOXIDE, $Ag_2Sn_5O_{11}$
DIALUMINIUM TRIOXIDE, Al_2O_3
GOLD(III) OXIDE, Au_2O_3
BARIUM OXIDE, BaO
BERYLLIUM OXIDE, BeO
DIBISMUTH TRIOXIDE, Bi_2O_3
CALCIUM OXIDE, CaO
CADMIUM OXIDE, CdO
COBALT(II) OXIDE, CoO
DICOBALT TRIOXIDE, Co_2O_3
DICHROMIUM TRIOXIDE, Cr_2O_3
CAESIUM TRIOXIDE ('OZONATE'), CsO_3
DICAESIUM OXIDE, Cs_2O
COPPER(II) OXIDE, CuO
COPPER(I) OXIDE, Cu_2O
IRON(II) OXIDE, FeO
DIIRON TRIOXIDE, Fe_2O_3
TRIIRON TETRAOXIDE, Fe_3O_4
MERCURY(II) OXIDE, HgO
'MERCURY(I) OXIDE', Hg_2O
IRIDIUM(IV) OXIDE, IrO_2
POTASSIUM DIOXIDE (SUPEROXIDE), KO_2
POTASSIUM TRIOXIDE, KO_3
DILANTHANUM TRIOXIDE, La_2O_3
MAGNESIUM OXIDE, MgO
MANGANESE(II) OXIDE, MnO
MANGANESE(IV) OXIDE, MnO_2
MANGANESE(VII) OXIDE, Mn_2O_7
MOLYBDENUM(VI) OXIDE, MoO_3
SODIUM OXIDE, Na_2O
NIOBIUM(V) OXIDE, Nb_2O_5
NICKEL(II) OXIDE, NiO
NICKEL(IV) OXIDE, NiO_2
DINICKEL TRIOXIDE, Ni_2O_3
LEAD(II) OXIDE, OPb
PALLADIUM(II) OXIDE, OPd
TIN(II) OXIDE, OSn
ZINC OXIDE, OZn
OSMIUM(IV) OXIDE, O_2Os
LEAD(IV) OXIDE, O_2Pb
PLATINUM(IV) OXIDE, O_2Pt
TIN(IV) OXIDE, O_2Sn
TITANIUM(IV) OXIDE, O_2Ti
URANIUM(IV) OXIDE, O_2U

TUNGSTEN(IV) OXIDE, O_2W
ZINC PEROXIDE, O_2Zn
DIPALLADIUM TRIOXIDE, O_3Pd_2
ANTIMONY(III) OXIDE, O_3Sb_2
THALLIUM(III) OXIDE, O_3Tl_2
VANADIUM(III) OXIDE, O_3V_2
TUNGSTEN(VI) OXIDE, O_3W
OSMIUM(VIII) OXIDE, O_4Os
TRILEAD TETRAOXIDE, O_4Pb_3
RUTHENIUM(VIII) OXIDE, O_4Ru
TANTALUM(V) OXIDE, O_5Ta_2
VANADIUM(V) OXIDE, O_5V_2
TRIURANIUM OCTAOXIDE, O_8U_3

METAL OXOHALOGENATES

This group covers the four levels of oxidation represented in the series hypochlorite, chlorite, chlorate and perchlorate, and, as expected, the oxidising power of the anion is roughly proportional to the oxygen content.

The group has been subdivided under the headings:

METAL HYPOCHLORITES
CHLORITE SALTS
METAL HALOGENATES
METAL PERCHLORATES

for which separate entries exist. Individually indexed compounds for all four sub-groups are:

SILVER CHLORITE, $AgClO_2$
SILVER PERCHLORATE, $AgClO_4$
SILVER IODATE, $AgIO_3$
ALUMINIUM CHLORATE, $AlCl_3O_9$
BARIUM BROMATE, $BaBr_2O_6$
BERYLLIUM PERCHLORATE, $BeCl_2O_8$
LEAD BROMATE, Br_2O_6Pb
ZINC BROMATE, Br_2O_6Zn
* METHYLMERCURY PERCHLORATE, CH_3ClHgO_4
* LEAD ACETATE BROMATE, $C_2H_3BrO_5Pb$
CALCIUM HYPOCHLORITE, $CaCl_2O_2$
CADMIUM CHLORATE, $CdCl_2O_6$
POTASSIUM CHLORATE, $ClKO_3$
POTASSIUM PERCHLORATE, $ClKO_4$
LITHIUM PERCHLORATE, $ClLiO_4$

SODIUM HYPOCHLORITE, $ClNaO$
SODIUM CHLORITE, $ClNaO_2$
SODIUM CHLORATE, $ClNaO_3$
SODIUM PERCHLORATE, $ClNaO_4$
MAGNESIUM CHLORATE, Cl_2MgO_6
MAGNESIUM PERCHLORATE, Cl_2MgO_8
MANGANESE DIPERCHLORATE, Cl_2MnO_8
NICKEL DIPERCHLORATE, Cl_2NiO_8
LEAD DICHLORITE, Cl_2O_4Pb
ZINC DICHLORATE, Cl_2O_6Zn
LEAD DIPERCHLORATE, Cl_2O_8Pb
URANYL DIPERCHLORATE, $Cl_2O_{10}U$
GALLIUM TRIPERCHLORATE, Cl_3GaO_{12}
VANADYL TRIPERCHLORATE, $Cl_3O_{13}V$
TITANIUM TETRAPERCHLORATE, $Cl_4O_{16}Ti$
* TETRAZIRCONIUM TETRAOXIDE HYDROGEN NONAPERCHLORATE, $Cl_9HO_{40}Zr_4$
* AMMONIUM IODATE, H_4INO_3
* AMMONIUM PERIODATE, H_4INO_4
POTASSIUM IODATE, IKO_3
POTASSIUM PERIODATE, IKO_4
SODIUM IODATE, $INaO_3$

METAL OXOMETALLATES

Salts with oxometallate anions function as oxidants, those with oxygen present as peroxogroups being naturally the more powerful and separately grouped under

PEROXOACID SALTS

Individually indexed oxometallate salts are:

CALCIUM PERMANGANATE, $CaMn_2O_8$
LEAD CHROMATE, CrO_4Pb
POTASSIUM DICHROMATE, $Cr_2K_2O_7$
SODIUM DICHROMATE, $Cr_2Na_2O_7$
* AMMONIUM PERMANGANATE, H_4MnNO_4
POTASSIUM PERMANGANATE, $KMnO_4$
* POTASSIUM NITRIDOOSMATE(1−), KNO_3Os
SODIUM PERMANGANATE, $MnNaO_4$

METAL OXONON-METALLATES

This large and commonly used class of salts covers a wide range of oxidising potential. Among the most powerful oxidants are

METAL OXOHALOGENATES and
PEROXOACID SALTS

which are dealt with separately. There is a rough gradation down the sub-groups

METAL NITRATES
METAL NITRITES
METAL SULPHATES
METAL AMIDOSULPHATES

for which separate entries emphasise the individual features. Less highly oxidised anions function as reducants. Individually indexed entries are:

SILVER NITRATE, $AgNO_3$
SILVER HYPONITRITE, $Ag_2N_2O_2$
* HEPTASILVER NITRATE OCTAOXIDE, Ag_7NO_{11}
BARIUM NITRATE, BaN_2O_6
BARIUM SULPHATE, BaO_4S
CALCIUM CARBONATE, $CCaO_3$
POTASSIUM CARBONATE, CK_2O_3
LITHIUM CARBONATE, CLi_2O_3
MAGNESIUM CARBONATE, $CMgO_3$
SODIUM CARBONATE, CNa_2O_3
LEAD CARBONATE, CO_3Pb
SODIUM ACETATE, $C_2H_3NaO_2$
CALCIUM SULPHATE, CaO_4S
COPPER(II) PHOSPHINATE, $CuH_2O_4P_2$
COPPER(II) NITRATE, CuN_2O_6
COPPER(II) SULPHATE, CuO_4S
POTASSIUM AMIDOSULPHATE, H_2KNO_3S
POTASSIUM PHOSPHINATE ('HYPOPHOSPHITE'), H_2KO_2P
SODIUM AMIDOSULPHATE, H_2NNaO_3S
SODIUM PHOSPHINATE ('HYPOPHOSPHATE'), H_2NaO_2P
* AMMONIUM NITRITE, $H_4N_2O_2$
* AMMONIUM NITRATE, $H_4N_2O_3$
* HYDROXYLAMMONIUM PHOSPHINATE, H_6NO_3P
* AMMONIUM AMIDOSULPHATE ('SULPHAMATE'), $H_6N_2O_3S$
* AMMONIUM SULPHATE, $H_8N_2O_4S$
* HYDROXYLAMMONIUM SULPHATE, $H_8N_2O_6S$
MERCURY(I) NITRATE, $HgNO_3$
MERCURY(II) NITRATE, HgN_2O_6
POTASSIUM NITRITE, KNO_2
POTASSIUM NITROSODISULPHATE (2–), $K_2NO_7S_2$
POTASSIUM DINITROSOSULPHITE, $K_2N_2O_5S$
LITHIUM SODIUM NITROXYLATE, $LiNNaO_2$
LITHIUM NITRATE, $LiNO_3$

MAGNESIUM NITRATE, MgN_2O_6
MAGNESIUM SULPHATE, MgO_4S
SODIUM NITRITE, $NNaO_2$
SODIUM NITRATE, $NNaO_3$
SODIUM HYPONITRITE, $N_2Na_2O_2$
SODIUM TRIOXODINITRATE (2–), $N_2Na_2O_3$
SODIUM TETRAOXODINITRATE (2–), $N_2Na_2O_4$
SODIUM PENTAOXODINITRATE (2–), $N_2Na_2O_5$
SODIUM HEXAOXODINITRATE (2–), $N_2Na_2O_6$
LEAD HYPONITRITE, N_2O_2Pb
LEAD(II) NITRATE, N_2O_6Pb
ZINC NITRATE, N_2O_6Zn
* TIN(II) NITRATE OXIDE, $N_2O_7Sn_2$
* URANYL NITRATE, N_2O_8U
* VANADIUM TRINITRATE OXIDE, $N_3O_{10}V$
PLUTONIUM(IV) NITRATE, N_4O_8Pu
SODIUM THIOSULPHATE, $Na_2O_3S_2$
SODIUM METASILICATE, Na_2O_3Si
SODIUM DITHIONITE ('HYDROSULPHITE'), $Na_2O_4S_2$
SODIUM DISULPHITE (METABISULPHITE), $Na_2O_5S_2$
LEAD SULPHATE, O_4PbS

METAL PERCHLORATES

Burton, M., *Chem. Eng. News,* 1970, **48**(51), 55
Though metal periodates and perbromates are known, the perchlorates have most frequently been involved in hazardous incidents over a long period. These highly stable salts are powerful oxidants and contact with combustible materials or reducants must be under controlled conditions. A severe restriction on the use of metal perchlorates in laboratory work has been recommended.

See SILVER PERCHLORATE, $AgClO_4$
DIMETHYL SULPHOXIDE, C_2H_6OS: Magnesium perchlorate
: Metal oxosalts
2,2-DIMETHOXYPROPANE, $C_5H_{12}O_2$: Metal perchlorates
POTASSIUM PERCHLORATE, $ClKO_4$
AMMONIUM PERCHLORATE, ClH_4NO_4
IRON(II) PERCHLORATE, Cl_2FeO_8
MAGNESIUM PERCHLORATE, Cl_2MgO_8
HYDRAZINIUM SALTS

Calcium hydride
Mellor, 1941, Vol. 3, 651
Rubbing a mixture of calcium (or strontium) hydride with a metal perchlorate in a mortar causes a violent explosion.

Sulphuric acid
Pieters, 1957, 30
Schumacher, 1960, 190
Metal perchlorates with highly concentrated or anhydrous acid form the explosively unstable anhydrous perchloric acid.

See PERCHLORIC ACID, $ClHO_4$: Dehydrating agents
See other METAL OXOHALOGENATES

METAL PEROXIDES

Castrantas, 1965, 1,4
This group contains many powerful oxidants, the most common being sodium peroxide. Undoubtedly the most hazardous is potassium 'superoxide', readily formed on the metal exposed to air.

See BARIUM PEROXIDE, BaO_2
* DIMETHYLETHEROXODIPEROXOCHROMIUM(VI), $C_2H_6CrO_6$
CALCIUM PEROXIDE, CaO_2
MERCURY PEROXIDE, HgO_2
* POTASSIUM DIOXIDE (SUPEROXIDE), KO_2
SODIUM PEROXIDE, Na_2O_2
ZINC PEROXIDE, O_2Zn

METAL PEROXOMOLYBDATES

Sidgwick, 1950, 1045
Many of the red metal peroxomolybdates are explosive.

See POTASSIUM TETRAPEROXOMOLYBDATE (2–), K_2MoO_8
SODIUM TETRAPEROXOMOLYBDATE (2–), $MoNa_2O_8$

METAL POLYHALOHALOGENATES

Organic solvents,
or Water
1. Whitney, E. D. *et al.*, *J. Amer. Chem. Soc.*, 1964, **86**, 2583
2. Sharpe, A. G. *et al.*, *J. Chem. Soc.*, 1948, 2135

Potassium, rubidium and caesium tetrafluorochlorates and hexafluorobromates react violently with water, and explosively with common organic solvents, analogously to the parent halogen fluorides [1]. Silver and barium tetrafluorobromates ignite in contact with ether, acetone, dioxan and petrol [2].

METALS

Individually indexed metals are:

SILVER, Ag
ALUMINIUM, Al
GOLD, Au
BARIUM, Ba
BERYLLIUM, Be
BISMUTH, Bi
CALCIUM, Ca
CADMIUM, Cd
CERIUM, Ce
COBALT, Co
CHROMIUM, Cr
CAESIUM, Cs
COPPER, Cu
IRON, Fe
GALLIUM, Ga
GERMANIUM, Ge
HAFNIUM, Hf
MERCURY, Hg
INDIUM, In
IRIDIUM, Ir
POTASSIUM, K
LANTHANUM, La
LITHIUM, Li
MAGNESIUM, Mg
MANGANESE, Mn
MOLYBDENUM, Mo
SODIUM, Na
NIOBIUM, Nb
NICKEL, Ni
OSMIUM, Os
LEAD, Pb
PALLADIUM, Pd
PLATINUM, Pt
PLUTONIUM, Pu
RUBIDIUM, Rb
RHENIUM, Re
RHODIUM, Rh
RUTHENIUM, Ru
ANTIMONY, Sb
TIN, Sn
TANTALUM, Ta
THORIUM, Th
TITANIUM, Ti
THALLIUM, Tl
URANIUM, U
VANADIUM, V
TUNGSTEN, W
ZINC, Zn
ZIRCONIUM, Zr

Halocarbons

Several of the more reactive metals are hazardous in combination with halocarbons.

See ALUMINIUM, Al: Halocarbons
BARIUM, Ba: Halocarbons
BERYLLIUM, Be: Halocarbons
POTASSIUM, K: Halocarbons
POTASSIUM–SODIUM ALLOY, K–Na: Halocarbons

LITHIUM, Li: Halocarbons
MAGNESIUM, Mg: Halocarbons
SODIUM, Na: Halocarbons
PLUTONIUM, Pu: Carbon tetrachloride
SAMARIUM, Sm: 1,1,2-Trichlorotrifluoroethane
TITANIUM, Ti: Halocarbons
ZINC, Zn: Halocarbons
ZIRCONIUM, Zr: Carbon tetrachloride

METAL SALICYLATES

Nitric acid
See NITRIC ACID, HNO_3: Metal salicylates

METAL SALTS

By far the largest class of compound in this Handbook, the metal (and ammonium) salts have been allocated into two sub-classes dependent on the presence or absence of oxygen in the anion.

The main groupings adopted for the non-oxygenated salts are:

METAL ACETYLIDES
METAL AZIDES
METAL CYANIDES AND CYANOCOMPLEXES
N-METAL DERIVATIVES
METAL FULMINATES
METAL HALIDES
METAL POLYHALOHALOGENATES

and for the oxosalts:

METAL OXALATES
METAL OXHALOGENATES (anion an oxoderivative of a halogen)
METAL OXOMETALLATES (anion an oxoderivative of a metal)
METAL OXONON-METALLATES (anion an oxoderivative of a non-metal)
PEROXOACID SALTS (anion a peroxoderivative of a metal or non-metal)

There is a separate entry for

OXOSALTS OF NITROGENOUS BASES

In some cases it has been convenient to subdivide the latter groups into smaller sub-groups and such sub-division is indicated under the appropriate group heading.

METAL SULPHATES

Aluminium
See ALUMINIUM, Al: Metal oxides, etc.

Magnesium
See MAGNESIUM, Mg: Metal oxosalts

METAL SULPHIDES

Some metal sulphides are so readily oxidised as to be pyrophoric in air. Individually indexed compounds are:

SILVER SULPHIDE, Ag_2S
DIGOLD TRISULPHIDE, Au_2S_3
BARIUM SULPHIDE, BaS
DIBISMUTH TRISULPHIDE, Bi_2S_3
CALCIUM SULPHIDE, CaS
DICERIUM TRISULPHIDE, Ce_2S_3
CHROMIUM(II) SULPHIDE, CrS
* DICAESIUM SELENIDE, Cs_2Se
COPPER(II) SULPHIDE, CuS
EUROPIUM(II) SULPHIDE, EuS
IRON(II) SULPHIDE, FeS
IRON DISULPHIDE, FeS_2
GERMANIUM(II) SULPHIDE, GeS
* AMMONIUM SULPHIDE, H_8N_2S
MERCURY(II) SULPHIDE, HgS
POTASSIUM SULPHIDE, K_2S
MANGANESE(II) SULPHIDE, MnS
MOLYBDENUM(IV) SULPHIDE, MoS_2
SODIUM SULPHIDE, Na_2S
SODIUM DISULPHIDE, Na_2S_2
SODIUM POLYSULPHIDE, Na_2S_x
* THORIUM OXIDE SULPHIDE, OSTh
RHENIUM(VII) SULPHIDE, Re_2S_7
TIN(II) SULPHIDE, SSn

TIN(IV) SULPHIDE, S_2Sn
TITANIUM(IV) SULPHIDE, S_2Ti
URANIUM(IV) SULPHIDE, S_2U
DIANTIMONY TRISULPHIDE, S_3Sb_2

METAL THIOCYANATES

Oxidants
1. von Schwartz, 1918, 299–300
2. *MCA Case History No. 853*

Metal thiocyanates are oxidised explosively by chlorates or nitrates when fused, or if intimately mixed, at 400°C or on spark or flame ignition [1]. Nitric acid violently oxidised an aqueous thiocyanate solution [2].

See NITRIC ACID, HNO_3: Metal thiocyanate

MILD STEEL

MCA Case History No. 947

A small mild steel cylinder suitable for high pressure was two-thirds filled with liquid ammonia by connecting it to a large ammonia cylinder and cooling it to −70°C by immersion in dry ice and acetone. Some hours after filling, the small cylinder burst, splitting cleanly along its length. This was attributed to cryogenic embrittlement and weakening of the mild steel cylinder.

MOLTEN SALT BATHS

1. 'Precautions in the Use of Nitrate Salt Baths', Min. of Labour, SHW booklet, London, HMSO, 1964
2. Pieters, 1957, 30
3. *Potential Hazards in the Use of Salt Baths for Heat Treatment of Metals,* NBFU Res. Rep. No. 2, New York, 1946

The booklet covers hazards attendant upon the use of molten nitrate salt baths for heat treatment of metals, including storage and disposal of salts, starting up, electrical heating, and emptying of salt baths. Readily oxidisable materials must be rigorously excluded from the vicinity of nitrate baths [1].

Earlier it had been reported that aluminium and its alloys if contaminated with organic matter may explode in nitrate–nitrite fused salt heating baths [2].

Uses, composition and precautions in the use of molten salt baths are discussed. Most common causes of accidents are: steam explosions, trapping of air, explosive reactions with metals (magnesium) and organic matter or cyanides from other heat-treatment processes [3].

NITRATING AGENTS

Dubar, J. *et al., Compt. Rend., Ser. C,* 1968, **266**, 1114
The potentially explosive character of various nitration mixtures (2-cyano-2-propyl nitrate in acetonitrile; solutions of dinitrogen tetraoxide in esters, ethers or hydrocarbons; dinitrogen pentaoxide in methylene chloride; nitronium tetrafluoroborate in sulpholane) are mentioned.

NITRITE SALTS OF NITROGENOUS BASES

1. Mellor, 1940, Vol. 8, 289, 470–472
2. Ray, P. C. *et al., J. Chem. Soc.,* 1911, **99**, 1470; *J. Chem. Soc.*, 1912 **101**, 141, 216

Ammonium and substituted-ammonium salts exhibit a range of instability, and reaction mixtures which may be expected to yield these products should be handled with care. Ammonium nitrite will decompose explosively either as the solid, or in concentrated aqueous solution when heated to 60–70° C. Presence of traces of acid lowers the decomposition temperature markedly. Hydroxylammonium nitrite appears to be so unstable that it decomposes immediately in solution. Hydrazinium (1+) nitrite is a solid which explodes violently on percussion, or less vigorously if heated rapidly, and hydrogen azide may be a product of decomposition [1]. Mono- and dialkylammonium nitrites decompose at temperatures below 60–70°C, but usually without violence [2].

NITRO-ACYL HALIDES

Aromatic acyl halides containing a nitro group adjacent to the halide function show a tendency towards violent thermal decomposition. The few individually indexed compounds are:

2-NITROTHIOPHENE-4-SULPHONYL CHLORIDE, $C_4H_2ClNO_4S_2$
2,4-DINITROBENZENESULPHENYL CHLORIDE, $C_6H_3ClN_2O_4S$
o-NITROBENZOYL CHLORIDE, $C_7H_4ClNO_3$
2,4-DINITROPHENYLACETYL CHLORIDE, $C_8H_5ClN_2O_5$
3-METHYL-2-NITROBENZOYL CHLORIDE, $C_8H_6ClNO_3$
o-NITROPHENYLACETYL CHLORIDE, $C_8H_6ClNO_3$

NITROALKANES

1. 'Nitroparaffins', TDS1, New York, Commercial Solvents Corp., 1968
2. Hass, H.B. *et al., Chem. Rev.*, 1943, **32**, 388
3. Noble, P. *et al., Chem. Rev.*, 1964, **64**, 20

The nitroalkanes are mild oxidants under ordinary conditions, but precautions should be taken when they are subjected to high temperatures and pressures, since violent reactions may occur [1]. The polynitroalkanes, being more in oxygen balance than the mono-derivatives, tend to explode more easily [2], and caution is urged, particularly during distillation [3].
See also POLYNITROALKYL COMPOUNDS

Alkali metals,
or Inorganic bases
Watts, C. E., *Chem. Eng. News*, 1952, **30**, 2344
Contact of nitroalkanes with inorganic bases must be effected under conditions which will avoid isolation in a dry state of the explosive metal salts of the isomeric *aci*-nitroparaffins (or nitronic acids).

Metal oxides

Hermoni, A. *et al., Chem. & Ind.,* 1960, 1265

Contact with metal oxides increases the sensitivity of nitromethane, nitroethane and 1-nitropropane to heat (and of nitromethane to detonation). Twenty-four oxides were examined in a simple quantitative test, and a mechanism is proposed. Cobalt, nickel, chromium, lead and silver oxides were most effective in lowering ignition temperatures.

See *aci*-NITRO SALTS

NITROALKYL PEROXONITRATES

See DINITROGEN TETRAOXIDE (NITROGEN DIOXIDE), N_2O_4: Cycloalkenes, etc.

C-NITRO COMPOUNDS

This group contains compounds with a single nitro group (attached to either an aliphatic or an aromatic nucleus) which have been involved in hazardous incidents. Poly-substitution is covered in the separate groups:

FLUORODINITROMETHYL COMPOUNDS
POLYNITROALKYL COMPOUNDS
POLYNITROARYL COMPOUNDS

and individually indexed mononitro compounds are:

TRICHLORONITROMETHANE (CHLOROPICRIN), CCl_3NO_2
5-NITROTETRAZOLE, CHN_5O_2
CHLORONITROMETHANE, CH_2ClNO_2
NITROMETHANE, CH_3NO_2
NITROETHANE, $C_2H_5NO_2$
2-NITROETHANOL, $C_2H_5NO_3$
1-NITROPROPANE, $C_3H_7NO_2$
1-NITRO-3-BUTENE, $C_4H_7NO_2$
tert-NITROBUTANE, $C_4H_9NO_2$
NITROBENZENE, $C_6H_5NO_2$
m-NITROBENZALDEHYDE, $C_7H_5NO_3$
2-CHLORO-4-NITROTOLUENE, $C_7H_6ClNO_2$
o- and *p*-NITROTOLUENE, $C_7H_7NO_2$

4-METHYL-2-NITROPHENOL, $C_7H_7NO_3$
o-NITROANISOLE, $C_7H_7NO_3$
NITROINDANE, $C_9H_9NO_2$
1-NITRONAPHTHALENE, $C_{10}H_7NO_2$
2,6-DI-tert-BUTYL-4-NITROPHENOL, $C_{14}H_{21}NO_3$

See also NITROALKANES

N-NITRO COMPOUNDS

Many *N*-nitro derivatives show explosive instability, individually indexed compounds being:

NITROUREA, $CH_3N_3O_3$
N-NITROMETHYLAMINE, $CH_4N_2O_2$
NITROGUANIDINE, $CH_4N_4O_2$
1-AMINO-3-NITROGUANIDINE, $CH_5N_5O_2$
1,2-BIS(DIFLUOROAMINO)*N*-NITROETHYLAMINE, $C_2H_4F_4N_4O_2$
AZO-*N*-NITROFORMAMIDINE, $C_2H_4N_8O_2$
1-METHYL-3-NITRO-1-NITROSOGUANIDINE, $C_2H_5N_5O_3$
N,N' DINITRO-1,2-DIAMINOETHANE, $C_2H_6N_4O_4$
N,N' DINITRO-*N*-METHYL-1,2-DIAMINOETHANE, $C_3H_8N_4O_4$
1,3,5,7-TETRANITROPERHYDRO-1,3,5,7-TETRAZOCINE, $C_4H_8N_8O_8$
N,N'-DIACETYL-*N,N'*-DINITRO-1,2-DIAMINOETHANE, $C_6H_{10}N_4O_6$
N,2,4,6-TETRANITRO-*N*-METHYLANILINE (TETRYL), $C_7H_5N_4O_8$
1-NITRO-3(2,4-DINITROPHENYL)UREA, $C_7H_5N_5O_7$
NITRIC AMIDE (NITRAMIDE), $H_2N_2O_2$

aci-NITRO SALTS

Many *aci*-nitro salts derived from action of bases on nitroalkanes are explosive in the dry state. Individually indexed compounds are:

SODIUM *aci*-NITROMETHANE, CH_2NNaO_2
AMMONIUM *aci*-NITROMETHANE, $CH_7N_2O_2$
DIPOTASSIUM *aci*-NITROACETATE, $C_2HK_2NO_4$
MERCURY(II) METHYLNITROLATE, $C_2H_2HgN_4O_6$
MONOPOTASSIUM *aci*-1,1-DINITROETHANE, $C_2H_3KN_2O_3$
DIAMMONIUM *N,N'*DINITRO-1,2-DIAMINOETHANE, $C_2H_{12}N_6O_4$
DIPOTASSIUM BIS-*aci*-TETRANITROETHANE, $C_2K_2N_4O_8$
DILITHIUM BIS-*aci*-TETRANITROETHANE, $C_2Li_2N_4O_8$

DISODIUM BIS-*aci*-TETRANITROETHANE, $C_2N_4Na_2O_8$
SODIUM NITROMALONALDEHYDE, $C_3H_2NNaO_4$
POTASSIUM *aci*-1,1-DINITROPROPANE, $C_3H_5KN_2O_4$
* POTASSIUM 4,6-DINITROBENZOFUROXAN HYDROXIDE COMPLEX, $C_6H_3KN_4O_7$
* POTASSIUM *p*-NITROPHENOLATE, $C_6H_4KNO_3$
* SODIUM *o*-NITROTHIOPHENOLATE, $C_6H_4NNaO_2S$
DISODIUM 1,3-DIHYDROXY-1,3-BIS(*aci*-NITROMETHYL)-2,2,4,4-TETRAMETHYLCYCLOBUTANE, $C_{10}H_{16}N_2Na_2O_6$

See NITROALKANES-: Alkali metals or inorganic bases

NITROSATED NYLON

ABCM Quart. Safety Summ., 1963, **34**, 20
Nylon, nitrosated with dinitrogen trioxide according to Belg. Patent 606 944 and stored cold, exploded on being allowed to warm to ambient temperature. The *N*-nitroso nylon would be similar to *N*-nitroso-*N*-alkylamides, some of which are unstable. Nylon components should therefore be excluded from contact with nitrosating agents.
See other NITROSO COMPOUNDS

NITROSO COMPOUNDS

A number of compounds containing nitroso or coordinated nitrosyl groups exhibit instability under appropriate conditions. Individually indexed compounds are:

N-METHYL-*N*-NITROSOUREA, $C_2H_5N_3O_2$
1-METHYL-3-NITRO-1-NITROSOGUANIDINE, $C_2H_5N_5O_3$
1,2-DIMETHYLNITROSOHYDRAZINE, $C_2H_7N_3O$
1,3,5-TRINITROSOHEXAHYDRO-1,3,5-TRIAZINE, $C_3H_6N_6O_3$
N,N'-DIMETHYL-*N,N'*-DINITROSOOXAMIDE, $C_4H_6N_4O_4$
ETHYL *N*-METHYL-*N*-NITROSOCARBAMATE, $C_4H_8N_2O_3$
3,7-DINITROSO-1,3,5,7-TETRAAZABICYCLO[3.3.1] NONANE, $C_5H_{10}N_6O_2$
TRINITROSOPHLOROGLUCINOL, $C_6H_3N_3O_6$
o-NITROSOPHENOL, $C_6H_5NO_2$
p-NITROSOPHENOL, $C_6H_5NO_2$
2-CHLORO-1-NITROSO-2-PHENYLPROPANE, $C_9H_{10}ClO$
LEAD(II) TRINITROSOPHLOROGLUCINOLATE, $C_{12}N_6O_{12}Pb_3$

NITROSYLRUTHENIUM TRICHLORIDE, Cl_3NORu
POTASSIUM NITROSODISULPHATE(2–), $K_2NO_7S_2$
POTASSIUM DINITROSOSULPHITE(2–), $K_2N_2O_5S$

See also NITROSATED NYLON
3-NITROSOTRIAZENES

3-NITROSOTRIAZENES

Müller, E. *et al.*, *Ber.*, 1962, **95**, 1255
A very unstable series of compounds, many decomposing at well below 0°C. The products formed from sodium triazenes and nitrosyl chloride explode violently on being disturbed with a wooden spatula, and are much more sensitive than those derived from silver triazenes. These exploded under a hammer blow, or on friction from a metal spatula.
See other TRIAZENES

NON-METAL AZIDES

This group contains compounds with azide groups linked to non-oxygenated non-metals, individually indexed compounds being:

ALUMINIUM TRIS(TETRAAZIDOBORATE), AlB_3N_{36}
BORON AZIDE DICHLORIDE, BCl_2N_3
TRIAZIDOBORANE, BN_9
BROMINE AZIDE, BrN_3
BIS(TRIFLUOROMETHYL)PHOSPHORUS(III) AZIDE, $C_2F_6N_3P$
AZIDODIMETHYLBORANE, $C_2H_6BN_3$
DIAZIDODIMETHYLSILANE, $C_2H_6N_6Si$
* TRIAZIDOCHLOROSILANE, ClN_9Si
PHOSPHORUS AZIDE DIFLUORIDE, F_2N_3P
PHOSPHORUS AZIDE DIFLUORIDE-BORANE, $F_2N_3P \cdot BH_3$
HYDROGEN AZIDE (HYDRAZOIC ACID), HN_3
AZIDOSILANE, H_3N_3Si
* AMMONIUM AZIDE, H_4N_4
HYDRAZINIUM AZIDE, H_5N_5
* NITROSYL AZIDE, N_4O
PHOSPHORUS TRIAZIDE, N_9P
SILICON TETRAAZIDE, $N_{12}Si$
1,1,3,3,5,5-HEXAAZIDO-2,4,6-TRIAZA-1,3,5-TRIPHOSPHORINE, $N_{21}P_3$

See related ACYL HALIDES
HALOGEN AZIDES

NON-METAL HALIDES AND THEIR OXIDES

This highly reactive class includes the separately treated groups

N-HALOGEN COMPOUNDS
HALOSILANES

as well as the individually indexed compounds:

ARSENIC TRICHLORIDE, $AsCl_3$
ARSINE-BORON TRIBROMIDE, $AsH_3 \cdot BBr_3$
BORON BROMIDE DIIODIDE, $BBrI_2$
BORON DIBROMIDE IODIDE, BBr_2I
BORON TRIBROMIDE, BBr_3
BORON TRICHLORIDE, BCl_3
BORON TRIFLUORIDE, BF_3
BORON DIIODOPHOSPHIDE, BI_2P
BORON TRIIODIDE, BI_3
DIBORON TETRACHLORIDE, B_2Cl_4
DIBORON TETRAFLUORIDE, $(Br_2\ Si)_n$
(*B*)1,3,5-TRICHLOROBORAZINF, $B_3Cl_3H_3N_3$
SELENINYL BROMIDE, Br_2OSe
SULPHUR DIBROMIDE, Br_2S
DISULPHUR DIBROMIDE, Br_2S_2
POLYDIBROMOSILANE, $(Br_2Si)_n$
PHOSPHORUS TRIBROMIDE, Br_3P
PHOSPHORUS CHLORIDE DIFLUORIDE, ClF_2P
SULPHINYL CHLORIDE (THIONYL CHLORIDE), Cl_2OS
SELENINYL CHLORIDE, Cl_2OSe
SULPHONYL DICHLORIDE (SULPHURYL CHLORIDE), Cl_2O_2S
DISULPHURYL DICHLORIDE, $Cl_2O_5S_2$
SULPHUR DICHLORIDE, Cl_2S
DISULPHUR DICHLORIDE (THIOSULPHINYL CHLORIDE), Cl_2S_2
* DISELENIUM DICHLORIDE, Cl_2Se_2
PHOSPHORYL CHLORIDE, Cl_3OP
PHOSPHORUS TRICHLORIDE, Cl_3P
* THIOPHOSPHORYL CHLORIDE, Cl_3PS
TETRACHLORODIPHOSPHANE, Cl_4P_2
TETRACHLOROSILANE, Cl_4Si
TELLURIUM TETRACHLORIDE, Cl_4Te
PHOSPHORUS PENTACHLORIDE, Cl_5P
DISULPHURYL DIFLUORIDE, $F_2O_5S_2$
XENON DIFLUORIDE, F_2Xe
NITROGEN TRIFLUORIDE, F_3N
PHOSPHORUS TRIFLUORIDE, F_3P
* THIOPHOSPHORYL FLUORIDE, F_3PS
SULPHUR TETRAFLUORIDE, F_4S

SELENIUM TETRAFLUORIDE, F_4Se
SULPHUR HEXAFLUORIDE, F_6S

NON-METAL HYDRIDES

Most members of this readily oxidised class ignite in air, individually indexed compounds being:

ARSINE, AsH_3
ARSINE-BORON TRIBROMIDE, $AsH_3 \cdot BBr_3$
HYDRAZINE-MONOBORANE, BH_7N_2
DIBORANE, B_2H_6
HYDRAZINE-BISBORANE, $B_2H_{10}N_2$
BORAZINE, $B_3H_6N_3$
TETRABORANE(10), B_4H_{10}
PENTABORANE(9), B_5H_9
PENTABORANE(11), B_5H_{11}
DECABORANE(14), $B_{10}H_{14}$
* *B*-CHLORO-*N*,*N*-DIMETHYL-AMINODIBORANE, $C_2H_{10}B_2ClN$
'SOLID PHOSPHORUS HYDRIDE', HP_2
SILICON MONOHYDRIDE, $(HSi)_n$
* SODIUM DIHYDROGENPHOSPHIDE, H_2NaP
* OXOSILANE, H_2OSi
HYDROGEN SULPHIDE, H_2S
HYDROGEN SELENIDE, H_2Se
POLYSILYLENE, $(H_2Si)_n$
AMMONIA, H_3N
PHOSPHINE, H_3P
STIBINE, H_3Sb
HYDRAZINE, H_4N_2
* OXODISILANE, H_4OSi_2
DIPHOSPHANE, H_4P_2
SILANE, H_4Si
2,4,6-TRISILATRIOXANE ('TRIPROSILOXANE'), $H_6O_3Si_3$
DISILANE, H_6Si_2
TRISILANE, H_8Si_3
* TRISILYLAMINE, H_9NSi_3
TETRASILANE, $H_{10}Si_4$
* TETRAAMMINELITHIUM DIHYDROGENPHOSPHIDE, $H_{14}LiN_4P$

See related ALKYLNON-METAL HYDRIDES

NON-METAL OXIDES

The generally acidic materials in this class may function as oxidants, some quite powerful, under appropriate conditions. Individually indexed compounds are:

DIARSENIC TRIOXIDE, As_2O_3
DIARSENIC PENTAOXIDE, As_2O_5
DIBORON DIOXIDE, B_2O_2
DIBORON TRIOXIDE, B_2O_3
CARBON MONOXIDE, CO
CARBON DIOXIDE, CO_2
NITROGEN OXIDE ('NITRIC OXIDE'), NO
DINITROGEN OXIDE ('NITROUS OXIDE'), N_2O
DINITROGEN TRIOXIDE, N_2O_3
DINITROGEN TETRAOXIDE (NITROGEN DIOXIDE), N_2O_4
DINITROGEN PENTAOXIDE, N_2O_5
SILICON OXIDE, OSi
SULPHUR DIOXIDE, O_2S
SELENIUM DIOXIDE, O_2Se
SILICON DIOXIDE, O_2Si
PHOSPHORUS(III) OXIDE, O_3P_2
SULPHUR TRIOXIDE, O_3S
ANTIMONY(III) OXIDE, O_3Sb_2
PHOSPHORUS(V) OXIDE, O_5P_2
DISULPHUR HEPTAOXIDE, O_7S_2

NON-METAL PERCHLORATES

Several perchlorate derivatives of non-metallic elements (including some non-nitrogenous organic compounds) are noted for explosive instability.

See TROPYLIUM PERCHLORATE, $C_7H_7ClO_4$
2,4,6-TRIMETHYLPYRILIUM PERCHLORATE, $C_8H_{11}ClO_5$
THIANTHRENIUM PERCHLORATE, $C_{12}H_8ClO_4S_2$
FLUORONIUM PERCHLORATE, $ClFH_2O_4$
PHOSPHONIUM PERCHLORATE, ClH_4O_4P
IODINE(III) PERCHLORATE, Cl_3IO_{12}
CAESIUM TETRAPERCHLORATOIODATE, Cl_4CsIO_{16}
BIS-TRIPERCHLORATOSILICON OXIDE, $Cl_6O_{25}Si_2$
See OXOSALTS OF NITROGENOUS BASES
See other PERCHLORATES

NON-METALS

Most members of this group of elements are readily oxidised with more or less violence dependent upon the oxidant and conditions involved. Individually indexed elements are:

ARSENIC, As
BORON, B
CARBON, C
HYDROGEN, H_2
NITROGEN, N_2
OXYGEN, O_2
PHOSPHORUS, P
SULPHUR, S
SELENIUM, Se
SILICON, Si
TELLURIUM, Te

NON-METAL SULPHIDES

In this group of readily oxidisable materials, individually indexed compounds are:

ARSENIC DISULPHIDE, AsS_2
DIARSENIC DISULPHIDE, As_2S_2
DIBORON TRISULPHIDE, B_2S_3
* CARBONYL SULPHIDE, COS
CARBON SULPHIDE, CS
CARBON DISULPHIDE, CS_2
HYDROGEN SULPHIDE, H_2S
HYDROGEN DISULPHIDE, H_2S_2
HYDROGEN TRISULPHIDE, H_2S_3
DISULPHUR DINITRIDE, N_2S_2
TETRASULPHUR DINITRIDE, N_2S_4
TETRASULPHUR TETRANITRIDE, N_4S_4
PHOSPHORUS(V) SULPHIDE, P_2S_5

ORGANIC AZIDES

Boyer, J. H. *et al., J. Chem. Eng. Data,* 1964, **9**, 480; *Chem. Eng. News,* 1964, **42**(31),6
The need for careful and small-scale handling of organic azides, which are usually heat- or shock-sensitive compounds of varying degrees of stability, has been discussed. The presence of more than one azido group, particularly if on the same atom (C or N) greatly reduces the stability.

This class contains the separately treated groups:

ACYL AZIDES

2-AZIDOCARBONYL COMPOUNDS

as well as the individually indexed compounds:

TRIAZIDOMETHYLIUM HEXACHLOROANTIMONATE, CCl_6N_9Sb
FLUORODINITROMETHYL AZIDE, CFN_5O_4
5-AZIDOTETRAZOLE, CHN_7
METHYL AZIDE, CH_3N_3
AZIDOACETONITRILE, $C_2H_2N_4$
3-AZIDO-1,2,4-TRIAZOLE, $C_2H_2N_6$
VINYL AZIDE, $C_2H_3N_3$
1,1-DIAZIDOETHANE, $C_2H_4N_6$
1,2-DIAZIDOETHANE, $C_2H_4N_6$
ETHYL AZIDE, $C_2H_5N_3$
* *N*-AZIDODOMETHYLAMINE, $C_2H_6N_4$
DIAZIDOMETHYLENECYANAMIDE, C_2N_8
DIAZIDOMETHYLENEAZINE, C_2N_{14}
1,3-DIAZIDOPROPENE, $C_3H_4N_6$
DIAZIDOMALONONITRILE, C_3N_8
2,4,6-TRIAZIDO-1,3,5-TRIAZINE, C_3N_{12}
PICRYL AZIDE, $C_6H_2N_6O_6$
PHENYL AZIDE, $C_6H_5N_3$

See related NON-METAL AZIDES

ORGANIC PEROXIDES

1. Castrantas, 1970
2. Varjarvandi, J. *et al., J. Chem. Educ.,* 1971, **48**, A451
3. 'Code of Practice for Storage of Organic Peroxides', London, Laporte Chemicals Ltd., 1970
4. Jackson, H. L. *et al., J. Chem. Educ.,* 1970, **47**, A175
5. Castrantas, 1965
6. Davies, 1961
7. Swern, 1970, Vol. 1, 1–104
8. Houben-Weyl, 1952, Vol. 8(3), 1
9. Swern, 1972, Vol. 3, 341–364

Of two general guides to the safe handling and use of peroxides, the second includes details of hazard evaluation tests, and the first has a comprehensive bibliography [1,2]. Storage aspects are rather specific [1,3]. Procedures for the safe handling of peroxidisable compounds have also been described [4]. Tabulated data on fire and explosion hazards of classes of organic peroxides with an extensive bibliography are available [5]. Theoretical aspects have been considered [5–7].

The hazards involved in synthesis of organic peroxides have been detailed [8], and a further review on the evaluation and management of peroxide hazards has appeared recently [9].

Index entries have been assigned to the structurally based sub-groups:

ALKYL HYDROPEROXIDES
CYCLIC PEROXIDES
DIACYL PEROXIDES
DIALKYL PEROXIDES
KETONE PEROXIDES
1-OXYPEROXY COMPOUNDS
OZONIDES
PEROXYACIDS
PEROXYCARBONATE ESTERS
PEROXYESTERS
POLYPEROXIDES

See also COMMERCIAL ORGANIC PEROXIDES
ORGANOMINERAL PEROXIDES

ORGANOLITHIUM REAGENTS

1. Bretherick, L., *Chem. & Ind.*, 1971, 1017
2. Gilman, H., private comm., 1971
3. 'Benzotrifluorides Catalogue 6/15', West Chester, Pa., Marshallton Res. Labs., 1971

Several halo-aryllithium compounds are explosive in the solid state in absence or near-absence of solvents or diluents, and operations with them should be designed to avoid their separation from solution. Such compounds include *m*- and *p*-bromo-, *m*-chloro-, *p*-fluoro-, *m*- and *p*-trifluoromethyl-phenyllithiums [1] and 3:4-dichloro-2,5-dilithiothiophene [2], but *m*-bromophenyl- and *o*-trifluoromethyl-phenyllithium appear to be explosive in presence of solvent also [1,3]. *m*- and *p*-Dilithiobenzene are also explosively unstable under appropriate conditions.

See PENTAFLUOROPHENYLLITHIUM, C_6F_5Li
See other HALO-ARYLMETALS

ORGANOMETALLICS

This miscellaneous group of organometallic compounds contains the individually indexed compounds:

BIS(η-CYCLOPENTADIENYLDINITROSYLCHROMIUM), $C_{10}H_{10}Cr_2N_2O_4$
BIS(η-CYCLOPENTADIENYL)MAGNESIUM, $C_{10}H_{10}Mg$
BIS(2,4-PENTANEDIONATO)CHROMIUM, $C_{10}H_{14}CrO_4$
η-BENZENE-η-CYCLOPENTADIENYLIRON(II) PERCHLORATE, $C_{11}H_{11}ClFeO_4$
POTASSIUM HEXAETHYNYLCOBALTATE(4–), $C_{12}H_6CoK_4$
BIS(η-BENZENE)CHROMIUM(0), $C_{12}H_{12}Cr$
BIS(η-BENZENE)IRON(0), $C_{12}H_{12}Fe$
TRIS(2,4-PENTANEDIONATO)MOLYBDENUM(III), $C_{15}H_{21}MoO_6$
BIS(η-CYCLOOCTATETRAENE)URANIUM(0), $C_{16}H_{16}U$
1,3-BIS(DI-η-CYCLOPENTADIENYLIRON)-2-PROPEN-1-ONE, $C_{23}H_{20}Fe_2O$
BIS[DI-η-BENZENECHROMIUM(VI)] DICHROMATE, $C_{24}H_{24}Cr_4O_7$

See related ALKYLMETAL HALIDES
ALKYLMETALS
HALO-ARYMETALS
ORGANOLITHIUM REAGENTS

ORGANOMINERAL PEROXIDES

Castrantas, 1965, 18
Swern, 1970, Vol. 1, 13
Sosnovsky, G. *et al., Chem. Rev.,* 1966, **66**, 529
Available information suggests that both hydroperoxides and peroxides in this extensive class are in many cases stable to heat at temperatures rather below 100°C, but may decompose explosively at higher temperatures. There are, however, exceptions.

See TRIMETHYLSILYL HYDROPEROXIDE, $C_3H_{10}O_2Si$
TRIETHYLTIN HYDROPEROXIDE, $C_6H_{16}O_2Sn$
TRIPHENYLTIN HYDROPEROXIDE, $C_{18}H_{16}O_2Sn$

OXIDANTS

Members of this class of materials have been involved in the majority of the two-component reactive systems included in this Handbook, and the whole class is very large. Most oxidants have been collectively treated in the structurally based entries:

ALKYL HYDROPEROXIDES
* BLEACHING POWDER
CHLORITE SALTS
* COMMERCIAL ORGANIC PEROXIDES
CYCLIC PEROXIDES
DIACYL PEROXIDES
DIALKYL PEROXIDES
(DIBENZOYLDIOXYIODO)BENZENES
DIFLUOROAMINO COMPOUNDS
N-HALOGEN COMPOUNDS
HALOGEN OXIDES
HALOGENS
N-HALOIMIDES
HYPOHALITES
INTERHALOGENS
KETONE PEROXIDES
* LIQUID AIR
METAL CHLORATES
METAL HALOGENATES
METAL HYPOCHLORITES
METAL NITRATES
METAL NITRITES
METAL OXOHALOGENATES
METAL OXOMETALLATES
METAL OXONON-METALLATES
METAL OXIDES
METAL PERCHLORATES
METAL PEROXIDES
METAL PEROXOMOLYBDATES
METAL POLYHALOHALOGENATES
* MOLTEN SALT BATHS
NITROALKANES
NON-METAL PERCHLORATES
ORGANOMINERAL PEROXIDES
OXIDES OF NITROGEN
OXOHALOGEN ACIDS
* OXYGEN ENRICHMENT
OXYGEN FLUORIDES
1-OXYPEROXY COMPOUNDS
OZONIDES
PERCHLORYL COMPOUNDS
PEROXOACIDS
PEROXOACID SALTS
PEROXYACIDS

PEROXYCARBONATE ESTERS
POLYPEROXIDES
XENON COMPOUNDS

Other individually indexed oxidants (not covered in the above) are:

DIOXYGENYL TETRAFLUOROBORATE, BF_4O_2
TRIFLUOROMETHYL HYPOFLUORITE, CF_4O
NITROMETHANE, CH_3NO_2
CHLOROSULPHURIC ACID, $ClHO_3S$
CHLORINE NITRATE, $ClNO_3$
NITROSYL PERCHLORATE, $ClNO_5$
NITRYL PERCHLORATE, $ClNO_6$
FLUORINE NITRATE, FNO_3
PEROXODISULPHURYL DIFLUORIDE, $F_2O_6S_2$
PERMANGANIC ACID, $HMnO_4$
NITROUS ACID, HNO_2
NITRIC ACID, HNO_3
HYDROGEN PEROXIDE, H_2O_2
SULPHURIC ACID, H_2O_4S
OXYGEN, O_2 (Gas or liquid)
OZONE, O_3

OXIDANTS AS HERBICIDES

Cook, W. H., *Can. J. Res.*, 1933, **8**, 509
The effect of humidity upon combustibility of various mixtures of organic matter and sodium chlorate was studied. Addition of a proportion of hygroscopic material (calcium or magnesium chlorides) effectively reduces the hazard. Similar effects were found for sodium dichromate and barium chlorate.
See SODIUM CHLORATE, $ClNaO_3$: Organic matter

N-OXIDES

Baumgarten, H. E. *et al., J. Amer. Chem. Soc.*, 1957, **79**, 3145
A procedure for preparing *N*-oxides is described which avoids formation of peracetic acid. After prolonged treatment of the amine at 35–40° C with excess 30% hydrogen peroxide, excess of the latter is catalytically decomposed with platinum oxide.

OXIDES OF NITROGEN

The oxides of nitrogen collectively are oxidants with power increasing with the level of oxygen content. Dinitrogen oxide will often support violent combustion, since its oxygen content (36.5%) approaches double that of atmospheric air.

See NITROGEN OXIDE ('NITRIC OXIDE'), NO
DINITROGEN OXIDE ('NITROUS OXIDE'), N_2O
DINITROGEN TRIOXIDE, N_2O_3
DINITROGEN TETRAOXIDE (NITROGEN DIOXIDE), N_2O_4
DINITROGEN PENTAOXIDE, N_2O_5

Glyptal resin
ABCM Quart. Safety Summ., 1937, **8**, 31
A new wooden fume cupboard was varnished with glyptal (glyceryl phthalate) resin. After a few weeks' use with 'nitrous fumes', the resin spontaneously and violently ignited. This was attributed to formation of glyceryl trinitrate.

OXOHALOGEN ACIDS

The oxidising power of the group of oxohalogen acids increases directly with oxygen content, though the high stability of the perchlorate ion at ambient temperature must be taken into account. The corresponding 'anhydrides' (halogen oxides) are also powerful oxidants, several being explosively unstable.

See BROMIC ACID, $BrHO_3$
HYPOCHLOROUS ACID, ClHO
CHLORIC ACID, $ClHO_3$
PERCHLORIC ACID, $ClHO_4$
PERIODIC ACID, HIO_4
ORTHOPERIODIC ACID, H_5IO_6
See also HALOGEN OXIDES

OXOSALTS OF NITROGENOUS BASES

Many of the salts of nitrogenous bases (particularly of high nitrogen content) with oxoacids are unstable or explosive.

There are separate group entries for:

HYDRAZINIUM SALTS
HYDROXYLAMMONIUM SALTS
NITRITE SALTS OF NITROGENOUS BASES

and individually indexed compounds are:

AMMONIUM BROMATE, BrH_4NO_3
METHYLAMMONIUM CHLORITE, CH_6ClNO_2
METHYLAMMONIUM PERCHLORATE, CH_6ClNO_4
GUANIDINIUM NITRATE, $CH_6N_4O_3$
AMINOGUANIDINIUM NITRATE, $CH_7N_5O_3$
DIAMINOGUANIDINIUM NITRATE, $CH_8N_6O_3$
TRIAMINOGUANIDINIUM PERCHLORATE, $CH_9ClN_6O_4$
TRIAMINOGUANIDINIUM NITRATE, $CH_9N_7O_3$
2-AZA-1,3-DIOXOLANIUM PERCHLORATE (ETHYLENE-DIOXYAMMONIUM PERCHLORATE), $C_2H_6ClNO_6$
1,2-ETHYLENEBIS-AMMONIUM PERCHLORATE, $C_2H_{10}Cl_2N_2O_8$
TRIMETHYLAMINE *N*-OXIDE PERCHLORATE, $C_3H_{10}ClNO_5$
4-CHLORO-1-METHYLIMIDAZOLIUM NITRATE, $C_4H_6ClN_3O_3$
TETRAMETHYLAMMONIUM CHLORITE, $C_4H_{13}ClNO_2$
PYRIDINIUM PERCHLORATE, $C_5H_6ClO_4$
2,4-DINITROPHENYLHYDRAZINIUM PERCHLORATE, $C_6H_7ClN_4O_8$
* 2-DIETHYLAMMONIOETHYLNITRATE NITRATE, $C_6H_{15}N_3O_6$
TRIETHYLAMMONIUM NITRATE, $C_6H_{16}NO_3$
1-*p*-CHLOROPHENYLBIGUANIDIUM HYDROGENDICHROMATE, $C_8H_{12}ClCr_2N_5O_7$
1-PHENYLBIGUANIDIUM HYDROGENDICHROMATE, $C_8H_{13}Cr_2N_5O_7$
3-AZONIABICYCLO[3.2.2]NONANE NITRATE, $C_8H_{16}N_2O_3$
1,3,6,8-TETRAAZONIATRICYCLO[6.2.1.1.^{3,6}]DODECANE TETRANITRATE, $C_8H_{16}N_8O_{12}$
TETRAMETHYLAMMONIUM PENTAPEROXODICHROMATE, $C_8H_{24}Cr_2N_2O_{12}$
1,2,3,4-TETRAHYDROISOQUINOLINIUM NITRATE, $C_9H_{12}N_2O_3$
1,2,3,4-TETRAHYDROQUINOLINIUM NITRATE, $C_9H_{12}N_2O_3$
p-TOLYLBIGUANIDIUM HYDROGENDICHROMATE, $C_9H_{15}Cr_2N_5O_7$
HEXAMETHYLENETETRAMMONIUM TETRAPEROXOCHROMATE(V) (3–) (?), $C_{18}H_{43}Cr_4N_{12}O_{32}$
TRI-*p*-TOLYLAMMONIUM PERCHLORATE, $C_{21}H_{22}ClNO_4$
AMMONIUM CHLORATE, ClH_4NO_3
AMMONIUM PERCHLORATE, ClH_4NO_4
PHOSPHONIUM PERCHLORATE, ClH_4O_4P
HYDRAZINIUM CHLORITE, $ClH_5N_2O_2$
HYDRAZINIUM CHLORATE, $ClH_5N_2O_3$
HYDRAZINIUM PERCHLORATE, $ClH_5N_2O_4$
AMMONIUM DICHROMATE, $Cr_2H_8N_2O_7$

HYDRAZINIUM DIPERCHLORATE, $Cl_2H_6N_2O_8$
AMMONIUM IODATE, H_4INO_3
AMMONIUM PERIODATE, H_4INO_4
AMMONIUM PERMANGANATE, H_4MnNO_4
AMMONIUM NITRITE, $H_4N_2O_2$
AMMONIUM NITRATE, $H_4N_2O_3$
HYDROXYLAMMONIUM NITRATE, $H_4N_2O_4$
HYDRAZINIUM NITRITE, $H_5N_3O_2$
HYDRAZINIUM NITRATE, $H_5N_3O_3$
HYDROXYLAMMONIUM PHOSPHINATE, H_6NO_3P
AMMONIUM AMIDOSULPHATE ('SULPHAMATE'), $H_6N_2O_3S$
HYDRAZINIUM HYDROGENSELENATE, $H_6N_2O_4Se$
HYDROXYLAMMONIUM SULPHATE, $H_8N_2O_6S$
AMMONIUM PEROXODISULPHATE, $H_8N_2O_8S_2$

See also CHLORITE SALTS
DICHROMATE SALTS OF NITROGENOUS BASES

OXYGEN BALANCE

1. Kirk-Othmer, 1965, Vol. 8, 581
2. Slack, R., private comm., 1957

Oxygen balance is the difference between the oxygen content of a chemical compound and that required fully to oxidise the carbon, hydrogen and other oxidisable elements present to carbon dioxide, water, etc. The concept is of particular importance in the design of explosive compounds or compositions, since the explosive power is maximal at equivalence, or zero oxygen balance. If there is a deficiency of oxygen present, the balance is negative, while an excess of oxygen gives a positive balance, and such compounds can function as oxidants. The balance is usually expressed as a percentage. The nitrogen content of a compound is not considered as oxidisable, as it is usually liberated as the gaseous element in explosive decomposition [1].

While it is, then, possible to recognise highly explosive materials by consideration of their oxygen balance (e.g. ETHYLENE DINITRATE, $C_2H_4N_2O_6$ is zero-balanced; 3,4,-BIS(1,2,3,4-THIATRIAZOL-5-YLTHIO)-MALEIMIDE, $C_4HN_7O_2S_4$, has a positive balance), the tendency to instability becomes apparent well below the zero-balance point. The empirical statement that the stability of any organic compound is doubtful when the oxygen or sulphur content approaches that necessary to convert the other elements present to their lowest state of oxidation (one sulphur atom equalling two oxygen atoms here) forms a useful guide [2].

OXYGEN ENRICHMENT

1. 'Oxygen Enrichment of Confined Areas', Information Sheet, London, Inst. of Welding, 1966
2. Wilk, I. J., *J. Chem. Educ.,* 1968, **45**, A547–551
3. Johnson, J. E. *et al., NRL Rep 6470,* Washington, Nav. Res. Lab., 1966
4. Woods, F. J. *et al., NRL Rep. 6606,* Washington, Nav. Res. Lab., 1967
5. Denison, D. M. *et al., Nature,* 1968, **218**, 1111–1113

With the widening industrial use of oxygen, accidents caused by atmospheric enrichment are increasing. Most materials, especially clothing, burn fiercely in an atmosphere containing more than the usual 21% of oxygen. In presence of petroleum products, fire and explosion can be spontaneous. Equipment which may emit or leak oxygen should be used sparingly, and never stored, in confined spaces [1].

Fourteen case histories of accidents caused by oxygen enrichment of the atmosphere are discussed and safety precautions described [2].

The flammability of textiles and other solids was studied under the unusual atmospheric conditions which occur in deep diving operations. The greatest effect upon ease of ignition and linear burning rate was caused by oxygen enrichment; increase in pressure had a similar effect [3].

Ignition and flame spread of fabrics and paper were measured at pressures from 21 bar down to the limiting pressure for ignition to occur. Increase in oxygen concentration above 21% in mixtures with nitrogen caused rapid decrease of minimum pressure for ignition.

In general, but not invariably, materials ignite less readily but burn faster in helium mixtures than in nitrogen mixtures. Nature of material has a marked influence on effect of variables on rate of burning. At oxygen concentrations of 41% all materials examined would burn except for glass and polytetrafluoroethylene, which resisted ignition attempts in pure oxygen. Flame retardants become ineffective on cotton in atmospheres containing above 32% oxygen [4].

A brief summary of known hazards and information in this general area is available [5].

OXYGEN FLUORIDES

Streng, A. G., *Chem. Rev.*, 1963, **63**, 607

In the series oxygen difluoride, dioxygen difluoride, trioxygen difluoride and tetraoxygen difluoride, as the oxygen content increases, the stability decreases and the oxidising power increases, tetraoxygen difluoride, even at −200°C, being one of the most potent oxidants known. Applications to both chemical reaction and rocket propulsion systems are covered in some detail.

See OXYGEN DIFLUORIDE, F_2O
DIOXYGEN DIFLUORIDE, F_2O_2
TRIOXYGEN DIFLUORIDE, F_2O_3

See other HALOGEN OXIDES

1-OXYPEROXY COMPOUNDS

Swern, 1970, Vol. 1, 29, 33

This group of compounds includes those monomers with one or more carbon atoms carrying a hydroperoxy or peroxy group and also singly bonded to an oxygen atom present as hydroxyl, ether or cyclic ether functions. While the group of compounds, in general, is moderately stable, the lower 1-hydroxy- and 1,1′-dihydroxy-alkyl peroxides or hydroperoxides are explosive.

See HYDROXYMETHYL HYDROPEROXIDE, CH_4O_3
HYDROXYMETHYL METHYL PEROXIDE, $C_2H_6O_3$
BIS-HYDROXYMETHYL PEROXIDE, $C_2H_6O_4$
1-HYDROXY-3-BUTYL HYDROPEROXIDE, $C_4H_{10}O_3$
BIS(1-HYDROXYCYCLOHEXYL) PEROXIDE, $C_{12}H_{22}O_4$
1(1′-HYDROPEROXY-1′-CYCLOHEXYLPEROXY)-CYCLOHEXANOL, $C_{12}H_{22}O_5$
BIS(1-HYDROPEROXYCYCLOHEXYL) PEROXIDE, $C_{12}H_{22}O_6$
1-ACETOXY-1-HYDROPEROXY-6-CYCLODODECANONE, $C_{14}H_{24}O_5$

OZONIDES

1. Rieche, A., *Angew. Chem.*, 1958, **70**, 251
2. Swern, 1970, Vol. 1, 39; Bailey, P. S., *Chem. Rev.*, 1958, **58**, 928
3. Greenwood, F. L. *et al.*, *J. Org. Chem.*, 1967, **32**, 3373
4. Rieche, A. *et al.*, *Ann.*, 1942, **553**, 187, 224

The preparation, properties and uses of ozonides have been comprehensively reviewed [1]. Many pure ozonides are generally stable to storage; some may be distilled under reduced pressure. The presence of peroxidic impurities is thought to cause the violently explosive decomposition often observed in this group. Use of ozone is not essential for their formation, as they are also produced by dehydration of α, α'-dihydroxy peroxides [2].

Polymeric alkene ozonides are shock-sensitive; that of *trans*-2-butene exploded when exposed to friction in a ground glass joint. The use of GLC to analyse crude ozonisation products is questionable because of the heat-sensitivity of some constituents [3]. Ozonides are decomposed, sometimes explosively, by finely divided palladium, platinum or silver, or by iron(II) salts [4].

Individually indexed compounds are:

ETHYLENE OZONIDE, $C_2H_4O_3$
PROPENE OZONIDE, $C_3H_6O_3$
MALEIC ANHYDRIDE OZONIDE, $C_4H_2O_6$
VINYL ACETATE OZONIDE, $C_4H_6O_5$
trans-2-BUTENE OZONIDE, $C_4H_8O_3$
ISOPRENE DIOZONIDE, $C_5H_8O_6$
trans-2-PENTENE OZONIDE, $C_5H_{10}O_3$
BENZENE TRIOZONIDE, $C_6H_6O_9$
trans-2-HEXENE OZONIDE, $C_6H_{12}O_3$
1,2-DIMETHYLCYCLOPENTENE OZONIDE, $C_7H_{12}O_3$
4-HYDROXY-4-METHYL-1,6-HEPTADIENE DIOZONIDE, $C_8H_{14}O_7$
2,6-DIMETHYL-2,5-HEPTADIEN-4-ONE DIOZONIDE, $C_9H_{14}O_7$
1,3-DIPHENYL-1,3-EPIDIOXY-1,3-DIHYDROISOBENZOFURAN, $C_{20}H_{14}O_3$

PAPER TOWELS

Unpublished observations, 1970

The increasing use of disposable paper towels in chemical laboratories

accentuates the fire hazard potentially created by disposal of solid oxidising agents or reactive residues into a bin containing such towels. The partially wet paper, necessarily of high surface area and absorbency, presents favourable conditions for fire to be initiated and spread. Separate bins for paper towels and chemical residues seem advisable.
See SODA-LIME

PERCHLORATES

1. Schumacher, 1960
2. Burton, H. *et al.*, *Analyst*, 1955, **80**, 4

All perchlorates have some potential for hazard when in contact with other reactive materials, while many are intrinsically hazardous, owing to the high oxygen content.

Existing knowledge on perchloric acid and its salts was extensively reviewed in 1960 in a monograph including the chapters: Perchloric Acid; Alkali Metal, Ammonium and Alkaline Earth Perchlorates; Metal Perchlorates; Miscellaneous Perchlorates; Manufacture of Perchloric Acid and Perchlorates; Analytical Chemistry of Perchlorates; Perchlorates in Explosives and Propellants; Miscellaneous Uses of Perchlorates; Safety Considerations in Handling Perchlorates [1].

There is a shorter earlier review, with a detailed treatment of the potentially catastrophic acetic anhydride–acetic acid–perchloric acid system. The violently explosive properties of methyl, ethyl and lower alkyl perchlorates, and the likelihood of their formation in alcohol–perchloric acid systems, are stressed. The instability of diazonium perchlorates, some when damp, is mentioned [2].

The group has been divided into the separately treated sub-groups:

ALKYL PERCHLORATES
AMMINEMETAL OXOSALTS
AMMONIUM PERCHLORATES
DIAZONIUM PERCHLORATES
METAL PERCHLORATES
NON-METAL PERCHLORATES

Glycol,
Polymer
MCA Case History No. 464

A mixture of an inorganic perchlorate, a glycol and a polymer exploded violently after heating at 265–270° C. It was stated that the glycol may have become oxidised, but formation of a perchlorate ester seems a more likely cause.

Organic matter
Schumacher, 1960, 188
Mixtures with finely divided or fibrous organic material are likely to be explosive. Porous or fibrous materials exposed to aqueous solutions and then dried are rendered explosively flammable and are easily ignited.

Reducants
Mellor, 1941, Vol. 2, 387; Vol. 3, 651
Schumacher, 1960, 188
Perchlorate salts react explosively when rubbed in a mortar with calcium hydride or with sulphur and charcoal; when melted with reducants; or on contact with glowing charcoal. Mixtures with finely divided aluminium, magnesium, zinc or other metals are explosives.

PERCHLORYLAMIDE SALTS

'Perchloryl Fluoride', Booklet DC-1819, Philadelphia, Pennsalt Chem. Corp., 1957
Ammonium perchlorylamide and the corresponding silver and barium salts are shock-sensitive when dry and may detonate. Extreme care is required when handling such salts.
See other *N*-METAL DERIVATIVES

PERCHLORYL COMPOUNDS

Organic compounds containing the perchloryl substituent are inherently shock-sensitive and powerful explosives. The few available examples are:

1-PERCHLORYLPIPERDINE, $C_5H_{10}ClNO_3$
2,6-DINITRO-4-PERCHLORYLPHENOL, $C_6H_3ClN_2O_8$
NITROPERCHLORYLBENZENE, $C_6H_4ClNO_5$

PERCHLORYLBENZENE, $C_6H_5ClO_3$
2,6-DIPERCHLORYL-4,4′-DIPHENOQUINONE, $C_{12}H_6Cl_2O_8$
PERCHLORYL FLUORIDE, $ClFO_3$

See also PERCHLORYLAMIDE SALTS

PEROXIDES

This group name probably covers the largest group of hazardous compounds and there are three main divisions:

INORGANIC PEROXIDES
ORGANIC PEROXIDES
ORGANOMINERAL PEROXIDES

PEROXIDES IN SOLVENTS

Many laboratory accidents have been ascribed to presence of peroxides in solvents, usually, but not exclusively, ethers.

See ETHERS
PEROXIDISABLE COMPOUNDS
4-METHYL-2-PENTANONE, $C_6H_{12}O$
HYDROGEN PEROXIDE, H_2O_2: Acetone, etc.
PEROXOMONOSULPHURIC ACID, H_2O_5S: Acetone

PEROXIDISABLE COMPOUNDS

1. Jackson, H.L. *et al., J. Chem. Educ.,* 1970, **47**, A175
2. Brandsma, 1971, 13

An account of a Du Pont safety study of the control of peroxidisable compounds covers structure examples, handling procedures, distillation of peroxidisable compounds, and detection and elimination of peroxides [1].

Essential organic structural features for a peroxidisable hydrogen atom are recognised as:

$>\underset{\mathrm{H}}{\overset{|}{\mathrm{C}}}-\mathrm{O}-$ as in acetals, ethers, oxygen heterocycles

Structure	
$\begin{matrix}-CH_2 \\ -CH_2\end{matrix}\!\!>\!\underset{\mid \atop H}{C}-$	as in isopropyl compounds, decahydronaphthalenes
$>C=C-\underset{\mid \atop H}{C}-$	as in allyl compounds
$>C=\underset{\mid \atop H}{C}\diagup X$	as in haloalkenes
$>C=\underset{\mid \atop H}{C}\diagup$	as in other vinyl compounds (monomeric esters, ethers, etc.)
$>C=\underset{\mid \atop H}{C}-\underset{\mid \atop H}{C}=C<$	as in dienes
$>C=\underset{\mid \atop H}{C}-C\equiv C-$	as in vinylacetylenes
$-C-\underset{\mid \atop H}{C}-Ar$	as in cumenes, tetrahydronaphthalenes, styrenes
$-\underset{\mid \atop H}{C}=O$	as in aldehydes
$-\underset{\parallel \atop O}{C}-\underset{\mid}{N}-\underset{\mid \atop H}{C}<$	as in *N*-alkyl-amides or -ureas, lactams

While the two latter types readily peroxidise, the products are readily degraded and do not accumulate to a hazardous level.

Inorganic compounds which readily peroxidise are listed as potassium and higher alkali metals, alkali metal alkoxides and amides, and organometallic compounds.

Three lists of specific compounds or compound types indicate different types of potential hazard, and appropriate storage, handling and disposal procedures are detailed.

List A, giving examples of compounds which form explosive peroxides in storage only, include diisopropyl ether, divinylacetylene, vinylidene chloride, potassium and sodium amide. Review of stocks and testing for peroxide content by given tested procedures at three-monthly intervals is recommended, together with safe disposal of any peroxidic samples.

List B, giving examples of liquids where a degree of concentration is necessary before hazardous levels of peroxide will develop includes several common solvents containing one ether function (diethyl ether, tetrahydrofuran, ethyl vinyl ether) or two ether functions (*p*-dioxan, 1,1-diethoxyethane, the dimethyl ethers of ethylene glycol or 'diethylene glycol') as well as the susceptible hydrocarbons propyne, butadiyne, dicyclopentadiene, cyclohexene, and tetra- and decahydronaphthalenes. Checking stocks at 12-monthly intervals, with peroxidic samples being discarded or repurified, is recommended here [1].

A simple method of effectively preventing accumulation of dangerously high concentrations of peroxidic species in distillation residues is that detailed in an outstanding practical textbook of preparative acetylene chemistry [2]. The material to be distilled is mixed with an equal volume of non-volatile mineral oil. This remains after distillation as an inert diluent for polymeric peroxidic materials.

List C contains peroxidisable monomers where the presence of peroxide may initiate exothermic polymerisation of the bulk of material. Precautions and procedures for storage and use of monomers with or without the presence of inhibitors are discussed in detail. Examples cited are acrylic acid, acrylonitirile, butadiene, 2-chlorobutadiene, chlorotrifluoroethylene, methyl methacrylate, styrene, tetrafluoroethylene, vinyl acetate, vinylacetylene, vinyl chloride, vinylidene chloride and vinylpyridine [1].

In general terms, the presence of two or more of the structural features indicated above in the same compound will tend to increase the likelihood of hazard. The selection of compound classes and of individually indexed compounds below includes compounds known to have been involved or those with a multiplicity of such structural features which would be expected to be especially susceptible to peroxide formation.

Separately treated groups are:

ACETYLENIC COMPOUNDS
ALLYL COMPOUNDS
DIENES
HALOALKENES

and individually indexed compounds:

2-CHLOROACRYLONITRILE, C_3H_2ClN
ACRYLALDEHYDE, C_3H_4O
2-PROPYNE-1-THIOL, C_3H_4S
BUTEN-3-YNE, C_4H_4
1,2- and 1,3-BUTADIENE, C_4H_6

CROTONALDEHYDE, C_4H_6O
DIVINYL ETHER, C_4H_6O
VINYL ACETATE, $C_4H_6O_2$
ETHYL VINYL ETHER, C_4H_8O
TETRAHYDROFURAN, C_4H_8O
m-DIOXAN, $C_4H_8O_2$
p-DIOXAN, $C_4H_8O_2$
DIETHYL ETHER, $C_4H_{10}O$
1,1- and 1,2-DIMETHOXYETHANE, $C_4H_{10}O_2$
2-PENTEN-4-YN-3-OL, C_5H_6O
ALLYL VINYL ETHER, C_5H_8O
2,3-DIHYDROPYRAN, C_5H_8O
1-ETHOXY-2-PROPYNE, C_5H_8O
ALLYL ETHYL ETHER, $C_5H_{10}O$
ISOPROPYL VINYL ETHER, $C_5H_{10}O$
TETRAHYDROPYRAN, $C_5H_{10}O$
3,3-DIMETHOXYPROPENE, $C_5H_{10}O_2$
2,2-DIMETHYL-1,3-DIOXOLAN, $C_5H_{10}O_2$
2-METHOXYETHYL VINYL ETHER, $C_5H_{10}O_2$
4-METHYL-1,3-DIOXAN, $C_5H_{10}O_2$
ETHYL ISOPROPYL ETHER, $C_5H_{12}O$
DIETHOXYMETHANE, $C_5H_{12}O_2$
1,1-DIMETHOXYPROPANE, $C_5H_{12}O_2$
4,5-HEXADIEN-2-YN-1-OL, C_6H_6O
DIALLYL ETHER, $C_6H_{10}O$
BUTYL VINYL ETHER, $C_6H_{12}O$
ISOBUTYL VINYL ETHER, $C_6H_{12}O$
2,6-DIMETHYL-1,4-DIOXAN, $C_6H_{12}O_2$
DIISOPROPYL ETHER, $C_6H_{14}O$
1,1-DIETHOXYETHANE (DIETHYLACETAL), $C_6H_{14}O_2$
1,2-DIETHOXYETHANE, $C_6H_{14}O_2$
3,3-DIETHOXYPROPENE, $C_7H_{14}O_2$
4-VINYLCYCLOHEXENE, C_8H_{12}
CINNAMALDEHYDE, C_9H_8O
2-ETHYLHEXYL VINYL ETHER, $C_{10}H_{20}O$
DIBENZYL ETHER, $C_{14}H_{14}O$
α-PENTYLCINNAMALDEHYDE, $C_{14}H_{18}O$
POTASSIUM AMIDE, H_2KN
SODIUM AMIDE, H_2NNa
POTASSIUM, K

PEROXOACIDS

Inorganic acids with a peroxide function are given the IUPAC group name above, which distinguishes them from the organic PEROXYACIDS. Collectively they are a group of very powerful oxidants.

See PEROXONITRIC ACID, HNO_4
PEROXOMONOSULPHURIC ACID, H_2O_5S
PEROXODISULPHURIC ACID, $H_2O_8S_2$
PEROXOMONOPHOSPHORIC ACID, H_3O_5P

PEROXOACID SALTS

Many of the salts of peroxoacids are unstable or explosive, are capable of initiation by heat, friction or impact, and all are powerful oxidants. Individually indexed compounds are:

SILVER PEROXOCHROMATE, $AgCrO_5$
AMMONIUM PEROXOBORATE, BH_4NO_3
SODIUM PEROXOBORATE, $BNaO_3$
SODIUM PEROXYACETATE, $C_2H_3NaO_3$
TETRAMETHYLAMMONIUM PENTAPEROXODICHROMATE(2–), $C_8H_{24}Cr_2N_2O_{12}$
MERCURY PEROXYBENZOATE, $C_{14}H_{10}HgO_6$
HEXAMETHYLENETETRAMMONIUM TETRAPEROXOCHROMATE(V) (3–) (?), $C_{18}H_{48}Cr_4N_{12}O_{32}$
CALCIUM PEROXODISULPHATE, CaO_8S_2
CALCIUM PEROXOCHROMATE (3–), $Ca_3Cr_2O_{12}$
AMMONIUM TETRAPEROXOCHROMATE (3–), $CrH_{12}N_2O_8$
POTASSIUM TETRAPEROXOCHROMATE (3–), CrK_3O_8
SODIUM TETRAPEROXOCHROMATE(3–), $CrNa_3O_8$
AMMONIUM PENTAPEROXODICHROMATE(2–), $Cr_2H_8N_2O_{12}$
POTASSIUM PENTAPEROXODICHROMATE(2–), $Cr_2K_2O_{12}$
POTASSIUM PEROXOFERRATE (2–), FeK_2O_5
POTASSIUM PEROXOMONOSULPHATE, HKO_5S
AMMONIUM PEROXODISULPHATE, $H_8N_2O_8S_2$
TETRAAMMINEZINC PEROXODISULPHATE, $H_{12}N_4O_8S_2Zn$
POTASSIUM TETRAPEROXOMOLYBDATE (2–), K_2MoO_8
POTASSIUM DIPEROXOORTHOVANADATE(2–), K_2O_6V
POTASSIUM PEROXODISULPHATE(2–), $K_2O_8S_2$
POTASSIUM TETRAPEROXOTUNGSTATE(2–), K_2O_8W
SODIUM TETRAPEROXOMOLYBDATE (2–), $MoNa_2O_8$
SODIUM TETRAPEROXOTUNGSTATE (2–), $Na_2O_8W_2$

See other OXIDANTS

PEROXYACIDS

Castrantas, 1965, 12

Swern, 1970, Vol. 1, 59, 337

The peroxyacids were until recently the most powerful oxidants of all organic peroxides, and it is often unnecessary to isolate them from the mixture of acid and hydrogen peroxide used to generate them. The pure lower aliphatic members are explosive (performic, particularly) at high, but not low, concentrations, being sensitive to heat but not usually to shock. Dipicolinic acid, or phosphates, have been used to stabilise these solutions. The detonable limits of peroxyacid solutions can be plotted by extrapolation from known data. Aromatic peroxyacids are generally more stable, particularly if ring substituents are present. Salts of peroxyacids are listed with peroxoacid salts above.

See PEROXYFORMIC ACID, CH_2O_3
PEROXYTRIFLUOROACETIC ACID, $C_2HF_3O_3$
PEROXYACETIC ACID, $C_2H_4O_3$
PEROXYPROPIONIC ACID, $C_3H_6O_3$
MONOPEROXYSUCCINIC ACID, $C_4H_6O_5$
PEROXYHEXANOIC ACID, $C_6H_{12}O_3$
PEROXYBENZOIC ACID, $C_7H_6O_3$

See (DIBENZOYLDIOXYIODO)BENZENES

PEROXYCARBONATE ESTERS

Strain, F. *et al., J. Amer. Chem. Soc.*, 1950, **72**, 1254

Of the three possible types of peroxycarbonate esters—dialkyl monoperoxycarbonates, dialkyl diperoxycarbonates and dialkyl peroxydicarbonates—the latter are by far the least stable class. Several of the 16 alkyl and substituted alkyl esters prepared decomposed violently or explosively at temperatures only slightly above the temperature (0–10°C) of preparation, owing to self-accelerating exothermic decomposition. Several were also explosive on exposure to heat, friction or shock.

See DIMETHYL PEROXYDICARBONATE, $C_4H_6O_6$
DIETHYL PEROXYDICARBONATE, $C_6H_{10}O_6$
DIALLYL PEROXYDICARBONATE, $C_8H_{10}O_6$
DIISOPROPYL PEROXYDICARBONATE, $C_8H_{14}O_6$
DI(2-METHOXYETHYL) PEROXYDICARBONATE, $C_8H_{14}O_8$
DI-*tert*-BUTYL DIPEROXYCARBONATE, $C_9H_{18}O_5$

See other PEROXIDES

PEROXY COMPOUNDS

Castrantas, 1965

Detonation theory is used to clarify the explosive characteristics of peroxy compounds. Some typical accidents are described. Hazards involved in use of a large number of peroxy compounds (including all those commercially available) are tabulated. 134 References.

PEROXYESTERS

Castrantas, 1965, 13

Swern, 1970, Vol. 1, 79

Though as a group they are noted for instability, there is a fairly wide variation in stability between particular sub-groups and compounds. See the group:

PEROXYCARBONATE ESTERS

and the individually indexed compounds:

1-HYDROXYETHYL PERACETATE, $C_4H_3O_4$
tert-BUTYL PERACETATE, $C_6H_{12}O_3$
ISOBUTYL PERACETATE, $C_6H_{12}O_3$
BIS-TRIMETHYLSILYL PEROXOMONOSULPHATE, $C_6H_{18}O_5SSi_2$
O,O-tert-BUTYL HYDROGEN MONOPEROXOMALEATE, $C_8H_{12}O_5$
Di-*tert*-BUTYL DIPEROXYOXALATE, $C_{10}H_{13}O_6$
tert-BUTYL *p*-NITROPEROXYBENZOATE, $C_{11}H_{13}NO_5$
tert-BUTYL PERBENZOATE, $C_{11}H_{14}O_3$
tert-BUTYL 1-ADAMANTANEPEROXYCARBOXYLATE, $C_{15}H_{24}O_3$
Di-*tert*-BUTYL DIPEROXYPHTHALATE, $C_{16}H_{22}O_6$
1,1-BIS(*p*-NITROBENZOYLPEROXY)CYCLOHEXANE, $C_{20}H_{18}N_2O_{10}$
1,1-BIS(BENZOYLPEROXY)CYCLOHEXANE, $C_{20}H_{20}O_6$
1,1,6,6-TETRAKIS(ACETYLPEROXY)CYCLODODECANE, $C_{20}H_{32}O_{12}$

See related (DIBENZOYLDIOXYIODO)BENZENES

PHOSPHINE DERIVATIVES

Halogens

Van Wazer, 1958, Vol. 1, 196

Organic derivatives of phosphine react very vigorously with halogens.

See ALKYLNON-METALS, etc.

PICRATES

Anon., *Angew. Chem. (Nachr.)*, 1954, 2, 21

While the melting point of a picrate was being determined in a silicone oil bath approaching 250°C, an explosion occurred, scattering hot oil. It is recommended that picrates, styphnates and similar derivatives should not be heated above 210°C in a liquid-containing m.p. apparatus.

See LEAD DIPICRATE, $C_{12}H_4N_6O_{14}Pb$
NITRIC ACID, HNO_3 : Metal salicylates

PLATINUM COMPOUNDS

Cotton, F. A., *Chem. Rev.*, 1955, **55**, 577

Several platinum compounds, including trimethylplatinum derivatives, are explosively unstable. Individually indexed compounds are:

POTASSIUM DINITROOXALATOPLATINATE(2–), $C_2K_2N_2O_8Pt$
TRIMETHYLPLATINUM HYDROXIDE, $C_3H_{10}OPt$
DIACETATOPLATINUM(II), $C_4H_6O_4Pt$
TETRAMETHYLPLATINUM, $C_4H_{12}Pt$
HEXAMETHYLDIPLATINUM, $C_6H_{18}Pt_2$
AMMONIUM HEXACHLOROPLATINATE(2–), $Cl_6H_8N_2Pt$
SODIUM HEXAHYDROXOPLATINATE(2–), $H_4Na_2O_6Pt$
cis-DIAMMINEDINITROPLATINUM(II), $H_6N_4O_4Pt$
AMMINEPENTAHYDROXOPLATINUM, H_8NO_5Pt
AMMONIUM TETRANITROPLATINATE(II), $H_8N_6O_8Pt$
TRIAMMINENITRATOPLATINUM(II) NITRATE, $H_9N_5O_6Pt$
TETRAAMMINEHYDROXONITRATOPLATINUM(IV) NITRATE, $H_{13}N_7O_{10}Pt$
* PLATINUM(IV) OXIDE, O_2Pt

See also GOLD COMPOUNDS
See other HEAVY METAL DERIVATIVES

POLY(DIMETHYLSILYL) CHROMATE

See BIS-TRIMETHYLSILYL CHROMATE, $C_6H_{18}CrO_4Si_2$

POLYNITROALKYL COMPOUNDS

Hammond, G. S. *et al.*, *Tetrahedron,* 1963, **19**, Suppl. 1, 177, 188
Trinitromethane ('nitroform'), dinitroacetonitrile, their salts, and polynitroalkanes are all potentially dangerous, and must be carefully handled as explosive compounds. Individually indexed compounds are:

TRINITROMETHANE ('NITROFORM'), CHN_3O_6
POTASSIUM TRINITROMETHANIDE ('NITROFORM' SALT), CKN_3O_6
DINITRODIAZOMETHANE, CN_4O_4
TETRANITROMETHANE, CN_4O_8
DINITROACETONITRILE, $C_2HN_3O_4$
SODIUM 5-DINITROMETHYLTETRAZOLIDE, $C_2HN_6NaO_4$
2(?)-FLUORO-1,1-DINITROETHANE, $C_2H_3FN_2O_4$
2,2,2-TRINITROETHANOL, $C_2H_3N_3O_7$
TRINITROACETONITRILE, $C_2N_4O_6$
1,1-DINITRO-3-BUTENE, $C_4H_6N_2O_4$
2,3-DINITRO-2-BUTENE, $C_4H_6N_2O_4$

See also FLUORODINITROMETHYL COMPOUNDS
NITROALKANES
See related *aci*-NITRO SALTS

POLYNITROARYL COMPOUNDS

1. Urbanski, 1964, Vol. 1
2. Shipp, K. G. *et al., J. Org. Chem.,* 1972, **37**, 1966

Polynitro derivatives of monocyclic aromatic systems (trinitrobenzene, trinitrotoluene, tetranitro-*N*-methylaniline, trinitrophenol, etc.) have long been used as explosives [1]. It has recently been found that a series of polynitro derivatives of biphenyl, diphenylmethane and 1,2-diphenylethylene (stilbene) are explosives liable to detonate on grinding or impact [2]. The same may be true of other polynitro derivatives of polycyclic systems not normally used as explosives, e.g. polynitro-fluorenones, -carbazoles, etc.

The presence of two or more nitro groups (each with two oxygen atoms) on an aromatic nucleus often increases the reactivity of other substituents and the tendency towards explosive instability as oxygen balance is approached.

Individually indexed compounds are:

2,4-DINITROBENZENESULPHENYL CHLORIDE, $C_8H_3ClN_2O_4S$

* 2,6-DINITRO-4-PERCHLORYLPHENOL, $C_6H_3ClN_2O_8$
1-FLUORO-2,4-DINITROBENZENE, $C_6H_3FN_2O_4$
POTASSIUM 4,6-DINITROBENZOFUROXAN HYDROXIDE COMPLEX, $C_6H_3KN_4O_7$
PICRIC ACID, $C_6H_3N_3O_7$
4-CHLORO-2,6-DINITROANILINE, $C_6H_4ClN_3O_4$
4-HYDROXY-3,5-DINITROBENZENE ARSONIC ACID, $C_6H_5AsN_2O_6$
POTASSIUM 3,5-DINITRO-2(1-TETRAZENYL)PHENOLATE, $C_6H_5KN_6O_5$
2,4-DINITROPHENYLHYDRAZINIUM PERCHLORATE, $C_6H_7ClN_4O_8$
2,4,6-TRINITROBENZOIC ACID, $C_7H_3N_3O_8$
2,6-DINITROBENZYL BROMIDE, $C_7H_5BrN_2O_4$
2,4,6-TRINITROTOLUENE, $C_7H_5N_3O_6$
N,2,4,6-TETRANITRO-*N*-METHYLANILINE (TETRYL), $C_7H_5N_5O_8$
1-NITRO-3(2,4-DINITROPHENYL)UREA, $C_7H_5N_5O_7$
2,4-DINITROTOLUENE, $C_7H_6N_2O_4$
2,4-DINITROPHENYLACETYL CHLORIDE, $C_8H_5ClN_2O_5$
mixo-DIMETHOXYDINITROANTHRAQUINONE, $C_{16}H_{10}N_2O_8$
2,2,4-TRIMETHYLDECAHYDROQUINOLINE PICRATE, $C_{18}H_{26}N_4O_7$
2,7-DINITRO-9-PHENYLPHEANTHRIDINE, $C_{19}H_{11}N_3O_4$

See also *C*-NITRO COMPOUNDS

POLYPEROXIDES

This group covers polymeric peroxides of indeterminate structure rather than polyfunctional molecules of known structure. Polymeric peroxide species described as hazardous include those derived from: butadiene (highly explosive); isoprene, dimethylbutadiene (both strongly explosive); 1,5-*p*-menthadiene, 1,3-cyclohexadiene (both explode at 110°C); methyl methacrylate, vinyl acetate, styrene (all explode above 40°C); diethyl ether (extremely explosive even below 100°C); and 1,1-diphenylethylene, cyclopentadiene (both explode on heating).

See TETRAFLUOROETHYLENE, C_2F_4: Air
1,1-DICHLOROETHYLENE, $C_2H_2Cl_2$: Air
CHLOROETHYLENE (VINYL CHLORIDE), C_2H_3Cl: Air
HEXAFLUOROPROPENE, C_3F_6: Air
1,3-BUTADIENE, C_4H_6: Air
DIMETHYLKETENE, C_4H_6O: Air
VINYL ACETATE, $C_4H_6O_2$: Oxygen
TETRAHYDROFURAN, C_4H_8O
DIETHYL ETHER, $C_4H_{10}O$: Air
CYCLOPENTADIENE, C_5H_6: Oxygen

2-METHYL-1,3-BUTADIENE (ISOPRENE), C_5H_8: Air
METHYL METHACRYLATE, $C_5H_8O_2$: Air
1,3-CYCLOHEXADIENE, C_6H_8: Air
2,3-DIMETHYL-1,3-BUTADIENE, C_6H_{10}: Air
STYRENE, C_8H_8: Oxygen
6,6-DIMETHYLFULVENE, C_8H_{10}: Air
1,5-*p*-MENTHADIENE, $C_{10}H_{16}$: Air
1,1-DIPHENYLETHYLENE, $C_{14}H_{12}$: Oxygen

See also HYDROGEN PEROXIDE, H_2O_2: Ketones, Nitric acid

PROPELLANTS

Gould, R. F., *Advanced Propellant Chemistry* (ACS 54), Washington, Amer. Chem. Soc., 1966

This deals, in 26 chapters in 5 sections, with theoretical and practical aspects of the use and safe handling of powerful oxidisers, and their complementary reactive fuels.

Materials include: nitrogen pentaoxide, perfluoroammonium ion and salts, nitronium tetrafluoroborate, hydrazinium, mono- and diperchlorates, nitronium perchlorate, tricyanomethyl compounds, difluoroamine and its alkyl derivatives, oxygen difluoride, chlorine trifluoride, dinitrogen tetraoxide, bromine trifluoride, nitrogen fluorides, liquid ozone–fluorine system.

See also ROCKET PROPELLANTS

PYROPHORIC CATALYSTS

'Laboratory Handling of Metal Catalysts', *Chem. Safety,* 1949, (2), 5

Proposed Code of Practice for laboratory handling of possibly pyrophoric catalysts includes: storage in tightly closed containers; extreme care in transfer operations, with provision for immediate cleaning up of spills and copious water flushing; avoidance of air-drying during filtration, and storage of residues under water; use of water-flush in event of ignition.

See HYDROGENATION CATALYSTS

PYROPHORIC METALS

1. Feitknecht, W., *Conference on Finely Divided Solids,* Commis à l'Énergie Atom. Saclay, 27–29 Sept., 1967
2. Peer, L. H. *et al., Mill & Factory,* 1959, **65**(2), 79
3. Anon., *Chem. Eng. News,* 1952, **30**, 3210

Finely divided metal powders develop pyrophoricity when a critical specific surface area is exceeded; this is ascribed to high heat of oxide formation on exposure to air. Safe handling is possible in relatively low concentrations of oxygen in an inert gas [1].

Safe handling, storage, disposal and fire fighting techniques for hafnium, titanium, uranium, thorium and hazards of machining the two latter metals are discussed [2].

Dry, finely divided tantalum, thorium, titanium, zirconium metals or titanium–nickel, zirconium–copper alloys are not normally shock sensitive. However, if they are enclosed in glass bottles which break on impact, ignition will occur. Storage of these materials moist and in metal containers is recommended [3].

Individually indexed pyrophoric metals are:

CALCIUM, Ca
CERIUM, Ce
COBALT, Co
CHROMIUM, Cr
CAESIUM, Cs
IRON, Fe
HAFNIUM, Hf
IRIDIUM, Ir
POTASSIUM, K
LITHIUM, Li
MANGANESE, Mn
SODIUM, Na
NICKEL, Ni
LEAD, Pb
PALLADIUM, Pd
PLATINUM, Pt
PLUTONIUM, Pu
RUBIDIUM, Rb
TANTALUM, Ta
THORIUM, Th
TITANIUM, Ti
URANIUM, U
ZIRCONIUM, Zr

See also ALUMINIUM AMALGAM, Al–Hg
BISMUTH PLUTONIDE, BiPu

QUALITATIVE ANALYSIS

See LEAD DIPICRATE, $C_{12}H_4N_6O_{14}Pb$
NITRIC ACID, HNO_3: Metal salicylates

REACTIVE CHEMICAL HAZARDS

In simple terms, reactive chemical hazards involve the release of energy in quantities too great to be dissipated by the environment of the reaction system.

The energy release may derive from the decomposition of a single unstable compound, or from exothermic interaction of two or more elements or compounds, possibly supplemented by subsequent decomposition of the product(s) of reaction.

In many of either type of case, careful control of the reaction environment will permit of safe isolation of single compounds and manipulation of reaction mixtures with, of course, appropriate operator-safeguards.

Attention to such details as the following may be required for safe working: proportions of reactants and concentration of reaction mixtures; presence of solvents or diluents; control of rates of addition, heating or cooling; degree of agitation; control of reaction or distillation atmosphere or pressure; shielding from actinic radiation; avoiding mechanical friction or shock on unstable solids. Examples of violent incidents which have involved lack of control of these variables will be found throughout this Handbook.

In some cases, however, the degree of instability and potential for energy release of particular compounds may be so high, and the threshold energy to initiate decomposition so low, as to necessitate extreme precautions when any attempt is made to prepare or use such materials. Relevant examples here are:

ETHYLENE DIPERCHLORATE, $C_2H_4Cl_2O_8$
DIAZONIUM SULPHIDES
CHLORINE DIOXIDE, ClO_2
OZONE (liquid), O_3

An interesting account of the reactive chemical hazards associated with the involvement of hazardous chemicals in fire situations is given in the book by Bahme, 1972.

REACTIVE METALS

Stout, E. L., *Los Alamos Scientific Lab. Rep.*, Washington, USAEC, 1957; *Chem. Eng. News*, 1958, **36**(8), 64–65

Safety considerations in handling plutonium, uranium, thorium, alkali metals, titanium, magnesium and calcium are discussed.

REDUCANTS

Most of the compounds showing powerful reducing action have been separately collected under the group headings

COMPLEX HYDRIDES
METAL ACETYLIDES
METAL HYDRIDES

The remaining individually indexed compounds are:

POTASSIUM PHOSPHINATE ('HYPOPHOSPHITE'), H_2KO_2P
HYPONITROUS ACID, $H_2N_2O_2$
SODIUM PHOSPHINATE ('HYPOPHOSPHITE'), H_2NaO_2P
HYDROXYLAMINE, H_3NO
PHOSPHINIC ('HYPOPHOSPHOROUS') ACID, H_3O_2P
PHOSPHONIUM IODIDE, H_4IP
HYDRAZINE, H_4N_2
HYDROXYLAMMONIUM PHOSPHINATE, H_6NO_3P
MAGNESIUM, Mg
POTASSIUM, K
SODIUM, Na
SODIUM THIOSULPHATE, $Na_2O_3S_2$
SODIUM DITHIONITE ('HYDROSULPHITE'), $Na_2O_4S_2$
SODIUM DISULPHITE ('METABISULPHITE'), $Na_2O_5S_2$

ROCKET PROPELLANTS

1. Kirk-Othmer, 1965, Vol. 8, 659
2. Urbanski, 1967, Vol. 3, 291
3. *ACS 88*, 1969

All of the theoretically possible high-energy (and potentially hazardous) oxidant–fuel systems have been considered for use, and many have been evaluated, in rocket propulsion systems (with apparently the exception of the most potent combination, liquid ozone–liquid acetylene). Some of the materials which have been examined are listed below, and it is apparent that any preparative reactions deliberately involving oxidant–fuel pairs must be conducted under controlled conditions with appropriate precautions.

OXIDANTS	FUELS
CHLORINE TRIFLUORIDE	ALCOHOLS
DINITROGEN TETRAOXIDE	AMINES
FLUORINE	AMMONIA
FLUORINE OXIDES	BERYLLIUM ALKYLS
HALOGEN FLUORIDES	BORANES
METHYL NITRATE	DICYANOGEN
NITRIC ACID	HYDRAZINES
NITROGEN TRIFLUORIDE	HYDROCARBONS
OXYGEN	HYDROGEN
OXYGEN FLUORIDES	NITROALKANES
OZONE	POWDERED METALS
PERCHLORIC ACID	SILANES
PERCHLORYL FLUORIDE	THIOLS
TETRAFLUOROHYDRAZINE	
TETRANITROMETHANE	

Many of the above combinations are hypergolic (ignite on contact) or can be made so with additives.

A few single compounds have been examined as monopropellants, (alkyl nitrates, ethylene oxide, hydrazine, hydrogen peroxide), the two latter being catalytically decomposed in this application.

Solid propellant mixtures, which are of necessity storage-stable, often contain ammonium or hydrazinium perchlorates as oxidants.

The hazardous aspects of rocket propellant technology has been surveyed [3].

See also PROPELLANTS

SHOCK-SENSITIVE MATERIALS

Recommended Safe Practices and Procedures: Storage and Handling of Shock- and Impact-sensitive Materials, Pamphlet SE-7, Washington, MCA, 1961

Materials classified as explosives are excluded.

SILANES

1. Stock, A. *et al., Ber.,* 1922, **55**, 3961

2. Kirk-Othmer, 1969, Vol. 18, 177
All the lower silanes are extremely sensitive to oxygen and ignite in air. The liberated hydrogen often ignites explosively [1]. Only under certain critical experimental conditions can they be mixed with oxygen without igniting [2].

Chloroform,
or Carbon tetrachloride,
or Oxygen
Stock, A. *et al., Ber.,* 1923, **56,** 1087
The chlorination of the lower silanes proceeds explosively in presence of oxygen, but catalytic presence of aluminium chloride controls the reaction.

Halogens
Stock, A. *et al., Ber.,* 1919, **52**, 695
Reaction of silanes with chlorine or bromine is violent.
See other NON-METAL HYDRIDES

SILICONE GREASE

Bromine trifluoride
See BROMINE TRIFLUORIDE, BrF_3: Silicone grease

SILVER-CONTAINING EXPLOSIVES

Luchs, J. A., *Photog. Sci. Eng.,* 1966, **10**, 334
Silver solutions used in photography can become explosive under a variety of conditions. Ammoniacal silver nitrate solutions, on storage, heating or evaporation, eventually deposit silver nitride ('fulminating silver'). Silver nitrate and ethanol may give silver fulminate, and, in contact with azides or hydrazine, silver azide. These are all dangerously sensitive explosives and detonators.
See also SILVERING SOLUTIONS
TOLLENS' REAGENT

SILVERING SOLUTIONS

1. Smith, I. C. P., *Chem. & Ind.*, 1965, 1070; *J. Brit. Soc. Glassblowers*, 1964, 45
2. Ermes, M., *Diamant*, 1929, **51**, 62, 587
3. Lohmann, E., ibid., 526; Mylius, W., ibid., 42

Brashear's silvering solution (alkaline ammoniacal silver oxide containing glucose) or residues therefrom should not be kept for more than two hours after preparation, since an explosive precipitate forms on standing [1]. The danger of explosion may be avoided by working with dilute silver solutions (0.35M) in the Brashear process, when formation of $Ag(NH_3)_2OH$ (and explosive $AgNH_2$ and Ag_3N therefrom) is minimised. The use of Rochelle salt, rather than caustic, and shielding of solutions from direct sunlight, are also recommended safeguards [2,3].

See TOLLENS' REAGENT

SODA-LIME

Hydrogen sulphide

Bretherick, L., *Chem. & Ind.*, 1971, 1042

Soda-lime, after absorbing hydrogen sulphide, exhibits a considerable exotherm (100°C) when exposed simultaneously to moisture and air, particularly with carbon dioxide enrichment, and has caused fires in laboratory waste bins containing moist paper wipes. Saturation with water and disposal in sealed containers is recommended.

SODIUM PRESS

Blau, K., private comm., 1965

The jet of a sodium press became blocked during use, and the ram was tightened to free it. It suddenly cleared and a piece of sodium wire was extruded, piercing a finger, which had to be amputated later. Sodium in a blocked die should be dissolved out in a dry alcohol.

See POTASSIUM, K: Alcohols

SOLVENTS

Davies, A. G., *J. R. Inst. Chem.*, 1956, **80**, 386
A short, detailed account, with references, of explosion hazards of autoxidised solvents, including: the autoxidation reaction; solvent and peroxide content; inhibition of peroxide formation; detection and estimation of peroxides; removal of peroxides.

SPILLAGES

'How to Deal with Spillages of Hazardous Chemicals', Poole, BDH Chemicals Ltd., 1970
A revised wall chart, with standardised disposal procedures for some 330 toxic and hazardous chemicals.

STARCH

Calcium hypochlorite,
Sodium hydrogensulphate
See CALCIUM HYPOCHLORITE, $CaCl_2O_2$: Sodium hydrogensulphate, etc.

STATIC INITIATION

Initiation of explosive decomposition by sparks derived from static electricity is thought to have been involved in the incidents involving the compounds:

METHYLMERCURY PERCHLORATE, CH_3ClHgO_4
TITANIUM CARBIDE, CTi
CHLOROETHYLENE (VINYL CHLORIDE), C_2H_3Cl
DICYANODIAZOMETHANE, C_3N_4
BIS-*o*-AZIDOBENZOYL PEROXIDE, $C_{14}H_8N_6O_4$
POTASSIUM PERCHLORATE, $ClKO_4$: Metal powders
SODIUM CHLORATE, $ClNaO_3$: Organic matter
: Paper

STEEL WOOL

Anon., *Fire Prot. Assoc. J.*, 1953, **21**, 53
Ignition can occur if steel wool short-circuits the contacts of even a small dry-cell torch battery.

SULPHUR BLACK

Anon., *Ind. Eng. Chem.*, 1919, **11**, 892
Twenty-four hours after several barrels of the dyestuff were bulked, blended and repacked, spontaneous heating occurred. This was attributed to the aerial oxidation of excess sodium polysulphide used during manufacture.

See SODIUM SULPHIDE, Na_2S

TETRAZOLES

Benson, F. R., *Chem. Rev.*, 1947, **41**, 4–5
There is a wide variation in thermal stability in derivatives of this high-nitrogen nucleus and several show explosive properties. Compounds indexed individually under this group heading are:

1-DICHLOROAMINOTETRAZOLE, $CHCl_2N_5$
5-NITROTETRAZOLE, CHN_5O_2
5-AZIDOTETRAZOLE, CHN_7
TETRAZOLE, CH_2N_4
5-*N*-NITROAMINOTETRAZOLE, $CH_2N_6O_2$
5-AMINOTETRAZOLE, CH_3N_5
5-DIAZONIOTETRAZOLIDE, CN_6
SODIUM 5-DINITROMETHYLTETRAZOLIDE, $C_2HN_6NaO_4$
5-HYDROXY-1(*N*-SODIO-5-TETRAZOLYLAZO)TETRAZOLE, $C_2HN_{10}NaO$
1,3-DI(5-TETRAZOLYL)TRIAZENE, $C_2H_3N_{11}$
1,2-DI(5-TETRAZOLYL)HYDRAZINE, $C_2H_4N_{10}$
1,6-BIS(5-TETRAZOLYL)HEXAZ-1,5-DIENE, $C_2H_4N_{14}$
1,2-DIHYDROPYRIDO[2,1,e]TETRAZOLE, $C_5H_4N_4$
3-PHENYL-1-TETRAZOLYL-1-TETRAZENE, $C_7H_3N_8$
5-PHENYLTETRAZOLE, $C_7H_6N_4$
5(4-DIMETHYLAMINOBENZENEAZO)TETRAZOLE, $C_9H_{11}N_7$
1(2-NAPHTHYL)-3(5-TETRAZOLYL)TRIAZENE, $C_{11}H_8N_7$

See other HIGH-NITROGEN COMPOUNDS

THERMITE

See ALUMINIUM, Al: Metal oxides, etc.
DIIRON TRIOXIDE, Fe_2O_3: Calcium disilicide
MAGNESIUM, Mg: Metal oxides

THIOPHENOLATES

Diazonium salts
See DIAZONIUM SULPHIDES AND DERIVATIVES

TOLLENS' REAGENT

Green, E., *Chem. & Ind.*, 1965, 943
This mixture of ammoniacal silver oxide and sodium hydroxide solution is potentially dangerous, since if kept for a few hours, it deposits a highly explosive precipitate. This danger was described by Tollens in 1882 but it is not generally known now. Prepare the reagent in small amounts just before use, in the tube to be used for the test, and discard immediately after use, NOT into a container for silver residues.
See SILVERING SOLUTIONS

TOXIC HAZARDS

While toxic hazards have been specifically excluded from consideration in this Handbook, such hazards are at least as important as reactive ones, particularly on a long-term basis. Due account of toxic hazards must therefore be taken in planning and executing laboratory work, particularly if unfamiliar materials are being brought into use.

It is perhaps appropriate to point out that many of the elements or compounds listed in this Handbook are here because of a high degree of reactivity towards other materials. It may therefore be broadly anticipated that under suitable circumstances of contact with animal organisms, a high degree of interaction will ensue, with possible subsequent onset of toxic or other deleterious effects.

See APROTIC SOLVENTS

TRIALKYLALUMINIUMS

A highly reactive group of compounds, of which the lower members are extremely pyrophoric, with very short ignition delays of use in rocket- or jet-fuel systems. Storage stability is generally high (decomposition with alkene and hydrogen evolution begins above about 170–180°C), but branched alkylaluminiums (notably triisobutylaluminium) decompose above 50°C.

See ALKYLALUMINIUM DERIVATIVES (references 1–6)

Individually indexed compounds of the TRIALKYLALUMINIUM group, many of which are commercially available in bulk, are:

TRIMETHYLALUMINIUM, C_3H_9Al
TRIETHYLALUMINIUM, $C_6H_{15}Al$
3-BUTEN-1-YNYLDIETHYLALUMINIUM, $C_8H_{13}Al$
TRIISOPROPYLALUMINIUM, $C_9H_{21}Al$
TRIPROPYLALUMINIUM, $C_9H_{21}Al$
3-BUTEN-1-YNYLDIISOBUTYLALUMINIUM, $C_{12}H_{21}Al$
TRIISOBUTYLALUMINIUM, $C_{12}H_{27}Al$

TRIALKYLBISMUTHS

Oxidants

Gilman, H. *et al.*, *Chem. Rev.*, 1942, **30**, 291

The lower alkylbismuths ignite in air, and explode in contact with oxygen, concentrated nitric or sulphuric acids.

See TRIMETHYLBISMUTH, C_3H_9Bi
TRIETHYLBISMUTH, $C_6H_{15}Bi$
TRIBUTYLBISMUTH, $C_{12}H_{27}Bi$

See other ALKYLMETALS

TRIAZENES

Houben-Weyl, 1965, Vol. 10(3), 700, 717, 722, 731

A number of triazene derivatives bearing –H, –CN, –OH or –NO on the terminal nitrogen of the chain are explosively unstable, mainly to heat.

See 3-CYANOTRIAZENES
3-HYDROXYTRIAZENES
3-NITROSOTRIAZENES
1,3-DI(5-TETRAZOLYL)TRIAZENE, $C_2H_3N_{11}$
3,3-DIMETHYL-1-PHENYLTRIAZENE, $C_8H_{11}N_3$
3,3-DIMETHYL-1-(3-QUINOLYL)TRIAZENE, $C_{11}H_{12}N_4$
1,3-DIPHENYLTRIAZENE, $C_{12}H_{11}N_3$
1,3-BIS(PHENYLTRIAZENO)BENZENE, $C_{18}H_{16}N_6$

See other HIGH-NITROGEN COMPOUNDS

N,N,N'-TRIFLUOROALKYLAMIDINES

Ross, D. L. *et al., J. Org. Chem.*, 1970, **35**, 3093

All the N–F compounds involved in the synthesis of a group of *N,N,N'*-trifluoroalkylamidines (C_3–C_7) were shock-sensitive, explosive compounds in varying degrees. Several were only stable in solution, and others exploded during analytical combustion.

See other *N*-HALOGEN COMPOUNDS

TURPENTINE

Diatomaceous earth

Anon., *Ind. Eng. Chem.*, 1950, **42**(7), 77A

A large quantity of discoloured (and peroxidised) turpentine was heated with fuller's earth to decolourise it, and subsequently exploded. Fuller's earth causes exothermic catalytic decomposition of peroxides and rearrangement of the terpene molecule.

Halogens,
or Oxidants,
or Tin(IV) chloride

Mellor, 1941, Vol. 2, 11, 90; 1941, Vol. 7, 446: 1943, Vol. 11, 395

Turpentine ignites in contact with fluorine (at $-210°C$), chlorine, iodine, chromium trioxide and chromyl chloride, and usually with tin(IV) chloride. Other highly unsaturated liquid hydrocarbons may be expected to react similarly.

UNSATURATED OILS

See LINSEED OIL
CARBON, C: Unsaturated oils

WAX FIRE

Carbon tetrachloride
Gilmont, R., *Chem. Eng. News,* 1947, **25**, 2853
Use of carbon tetrachloride to extinguish a wax fire caused an explosion. This was attributed to a violent reaction between unsaturated wax components and carbon tetrachloride initiated by radicals from decomposing peroxides.
See DIBENZOYL PEROXIDE, $C_{14}H_{10}O_4$: Carbon tetrachloride, etc.

'XANTHATES'

See POTASSIUM *O*-METHYLDITHIOCARBONATE, $C_2H_3KOS_2$
POTASSIUM *O*-ETHYLDITHIOCARBONATE (XANTHATE), $C_3H_5KOS_2$

XENON COMPOUNDS

Jha, N. K., *RIC Rev.,* 1971, **4**, 167–168
Several references to hazards associated with xenon compounds have been collected. Individually indexed compounds are:

XENON(II) FLUORIDE TRIFLUOROACETATE, CF_4O_2Xe

XENON(II) FLUORIDE TRIFLUOROMETHANESULPHONATE, CF_4O_3SXe

XENON(II) TRIFLUOROACETATE, $C_2F_6O_4Xe$

XENON DIFLUORIDE, F_2Xe

XENON TETRAFLUORIDE, F_4Xe

XENON HEXAFLUORIDE, F_6Xe

XENON(II) PENTAFLUOROORTHOTELLURATE, $F_{10}O_2Te_2Xe$

POTASSIUM HEXAOXOXENONATE (4–)-XENON TRIOXIDE, $K_4O_6Xe \cdot 2O_3Xe$

XENON TRIOXIDE,

SPECIFIC CHEMICAL SECTION

(Elements and Compounds arranged in formula order)

EXPLANATORY NOTES

The prefix of * to one member of a group-list of individual compounds means that it is similar to, but not identical with, the general class.

All temperatures are in degrees Celsius.

SILVER

Aziridine

See AZIRIDINE (ETHYLENEIMINE), C_2H_5N: Silver

Bromine azide

See BROMINE AZIDE, BrN_3

1-Bromo-2-propyne

See 1-BROMO-2-PROPYNE, C_3H_3Br: Metals

Carboxylic acids

Koffolt, J. H., private comm., 1965.

Silver is incompatible with oxalic or tartaric acids, since the silver salts decompose on heating. Silver oxalate explodes at 140°C, and silver tartarate loses carbon dioxide.

See also METAL OXALATES

Chlorine trifluoride

See CHLORINE TRIFLUORIDE, ClF_3 · Metals

Ethanol,
Nitric acid

Luchs, J. K., *Photog. Sci. Eng.*, 1966, **10**, 334

Action of silver on nitric acid in presence of ethanol may form the readily detonable silver fulminate.

See NITRIC ACID, HNO_3: Alcohols
See also SILVER-CONTAINING EXPLOSIVES

Ethylene oxide

See ETHYLENE OXIDE, C_2H_4O: Silver

Ethyl hydroperoxide

See ETHYL HYDROPEROXIDE, $C_2H_6O_2$: Silver

Hydrogen peroxide

See HYDROGEN PEROXIDE, H_2O_2 : Metals

Ozonides

See OZONIDES

Peroxomonosulphuric acid

See PEROXOMONOSULPHURIC ACID, H_2O_5S: Catalysts

Peroxyformic acid

See PEROXYFORMIC ACID, CH_2O_3 : Metals

See other METALS

SILVER TETRAFLUOROBROMATE $AgBrF_4$

See METAL POLYHALOHALOGENATES

SILVER CHLORIDE AgCl

Aluminium

See ALUMINIUM, Al: Silver chloride

Ammonia

Mellor, 1941, Vol. 3, 382

Exposure of ammoniacal silver chloride solutions to air or heat produces a black crystalline deposit of 'fulminating silver', mainly silver nitride, with disilver imide and silver amide also possibly present.

See TRISILVER NITRIDE, Ag_3N

See other METAL HALIDES

SILVER PERCHLORYLAMIDE $AgClHNO_3$

See PERCHLORYLAMIDE SALTS

SILVER AZIDE CHLORIDE $AgClN_3$

Frierson, W. J. *et al.*, *J. Amer. Chem. Soc.*, 1943, **65**, 1698
It is shock-sensitive when dry.
See other METAL AZIDE HALIDES

SILVER CHLORITE $AgClO_2$

Alkyl iodides
Levi, G. R., *Gazz. Chim. Ital [2]*, 1923, **53**, 40
Attempts to react the chlorite with methyl or ethyl iodides caused explosions, immediate in the absence of solvents, or delayed in presence of solvents. Silver chlorite itself is impact-sensitive, cannot be ground finely and explodes at 105°C.
See other CHLORITE SALTS

Hydrochloric acid,
or Sulphur
Mellor, 1941, Vol. 2, 284
It explodes in contact with hydrochloric acid or on rubbing with sulphur.
See other METAL OXOHALOGENATES

SILVER PERCHLORATE

Aromatic compounds
1. Sidgwick, 1950, 1234
2. Brinkley, S. R., *J. Amer. Chem. Soc.*, 1940, **62**, 3524
Silver perchlorate forms solid complexes with aniline, pyridine, toluene, benzene and many other aromatic hydrocarbons [1]. A sample of the benzene complex exploded violently on crushing in a mortar. The ethanol complex also exploded similarly, and unspecified perchlorates dissolved in organic solvents were observed to explode [2].

Diethyl ether
1. Heim, F., *Angew. Chem.*, 1957, **69**, 274

2. Anon., *Angew. Chem. (Nachr.)*, 1962, **10**, 2
After crystallisation from ether, the material exploded violently on crushing in a mortar. It has been considered stable previously, since it melts without decomposition [1]. However, there was a similar incident with silver perchlorate not previously in contact with organic materials [2].
See other METAL OXOHALOGENATES

SILVER PEROXOCHROMATE $AgCrO_5$

Sulphuric acid
Riesenfeld, E. H., *et al., Ber.*, 1914, **47**, 548
In attempts to prepare 'perchromic acid', a mixture of silver (or barium) peroxochromate and 50% sulphuric acid prepared at –80°C reacted explosively on slow warming to about –30°C.
See other PEROXOACID SALTS

SILVER FLUORIDE AgF

Calcium hydride
Mellor, 1941, Vol. 3, 389
A mixture becomes incandescent on grinding.

Non-metals
Mellor, 1941, Vol. 3, 389
Boron reacts explosively when ground with silver fluoride; silicon reacts violently.

Titanium
Mellor, 1941, Vol. 7, 20
Interaction at 320°C is incandescent.
See other METAL HALIDES

SILVER DIFLUORIDE AgF_2

Dimethyl sulphoxide
See IODINE PENTAFLUORIDE, F_5I : Dimethyl sulphoxide

Hydrocarbons,
or Water
Priest, H. F., *Inorg. Synth.*, 1950, **3**, 176
It reacts even more vigorously with most substances than does cobalt fluoride.
See other METAL HALIDES

SILVER AMIDE AgH_2N

Brauer, 1965, Vol. 2, 1045
Extraordinarily explosive when dry.
See NITROGEN TRIIODIDE-SILVER AMIDE, $I_3N \cdot AgH_2N$
See other *N*-METAL DERIVATIVES

SILVER IODATE $AgIO_3$

Potassium
See POTASSIUM, K : Oxidants

SILVER NITRATE $AgNO_3$

Acetylene and derivatives
Mellor, 1946, Vol. 5, 854
Silver nitrate (or other soluble salt) reacts with acetylene in presence of ammonia to form silver acetylide, a sensitive, powerful detonator when dry. In the absence of ammonia, or when calcium acetylide is added to silver nitrate solution, explosive double salts of silver acetylide and silver nitrate are produced. Mercurous acetylide precipitates silver acetylide from the aqueous nitrate.
See other METAL ACETYLIDES

Acrylonitrile
See ACRYLONITRILE, C_3H_3N : Silver nitrate

Ammonia,
Sodium hydroxide

1. Milligan, T. W. *et al., J. Org. Chem.*, 1962, **27**, 4663
2. *MCA Case History No. 1554*
3. Morse, J. R., *School Sci. Rev.*, 1955, **37**(131), 147
4. Baldwin, J., *School Sci. Rev.*, 1967, **48**(165), 586

During preparation of an oxidising agent on a larger scale than described [1], addition of warm sodium hydroxide solution to warm ammoniacal silver nitrate with stirring caused immediate precipitation of black silver nitride which exploded [2]. Similar incidents had been reported previously [3], including one where explosion appeared to be initiated by addition of Devarda's alloy (Al–Cu–Zn [4]).

See TRISILVER NITRIDE, Ag_3N

See also SILVERING SOLUTIONS
TOLLENS' REAGENT

Arsenic
Mellor, 1941, Vol. 3, 470
A finely divided mixture with excess nitrate ignited when shaken out on to paper.

Chlorine trifluoride
See CHLORINE TRIFLUORIDE, ClF_3: Metals, etc.

Chlorosulphuric acid
Mellor, 1941, Vol. 3, 470
Interaction is violent, nitrosulphuric acid being formed.

Ethanol
1. Tully, J. P., *Ind. Eng. Chem. (News Ed.)* 1941, **19**, 250
2. Luchs, J. K., *Photog. Sci. Eng.*, 1966, **10**, 334
3. Garin, D.L. *et al., J. Chem. Educ.*, 1970, **47**, 741

Reclaimed silver nitrate crystals, damp with the alcohol used for washing, exploded violently when touched with a spatula, generating a strong smell of ethyl nitrate [1]. The explosion was attributed to formation of silver fulminate, which is produced on addition of ethanol to silver nitrate solutions). Ethyl nitrate may also have been involved. Alternatives to avoid ethanol washing of recovered silver nitrate are discussed [2], including use of 2-propanol [3].

See SILVER FULMINATE, $C_2Ag_2N_2O_2$

Magnesium,
Water
Marsden, F., private comm., 1973
Lyness, D. J. *et al., School. Sci. Rev.,* 1953, **35**(125), 139
An intimate mixture of dry powdered magnesium and silver nitrate may ignite explosively on contact with a drop of water.

Non-metals
Mellor, 1941, Vol. 3, 469–473
Under a hammer blow, a mixture with charcoal ignites, while mixtures with phosphorus and sulphur explode, the former violently.

Phosphine
Mellor, 1941, Vol. 3, 471
Rapid passage of gas into the concentrated nitrate solution caused an explosion, or ignition with a slower gas stream. The explosion may have been due to rapid oxidation of the precipitated silver phosphide derivative by the co-produced nitric acid or dinitrogen tetraoxide.

Phosphonium iodide
See PHOSPHONIUM IODIDE, H_4IP : Oxidants
See other METAL OXONON-METALLATES

SILVER AZIDE AgN_3

1. Mellor, 1940, Vol. 8, 349; 1967, Vol. 8 Suppl. 2, 47
2. Gray, P. *et al.*, *Chem. & Ind.*, 1955, 1255

While pure silver azide explodes at 340°C [1], the presence of impurities may cause explosion at 270°C. It is also impact-sensitive and explosions are usually violent [2]. Its use as a detonator has been proposed.

Chlorine azide
See CHLORINE AZIDE, ClN_3 : Ammonia, etc.

Halogens
Mellor, 1940, Vol 8, 336
Silver azide, itself a sensitive compound, is converted by ethereal

iodine into the less stable and explosive compound, iodine azide. Similarly, contact with nitrogen-diluted bromine vapour gives bromine azide, often causing explosions.

See SILVER(II)AZIDE CHLORIDE, $AgClN_3$
See other METAL AZIDES

(UNKNOWN STRUCTURE) AgN_5S_3

See 1,3,5-TRICHLOROTRITHIA-1,3,5-TRIAZINE (THIAZYL CHLORIDE), $Cl_3N_3S_3$: Ammonia, etc.

SILVER(II) OXIDE AgO

Hydrogen sulphide
See HYDROGEN SULPHIDE, H_2S: Metal oxides

SILVER HYPONITRITE $Ag_2N_2O_2$

See HYPONITROUS ACID, $H_2N_2O_2$
See other HEAVY METAL DERIVATIVES

SILVER(I) OXIDE Ag_2O

Aluminium
See COPPER(II) OXIDE, CuO : Metals

Ammonia or hydrazine
Ethanol
Silver oxide and ammonia or hydrazine slowly form explosive silver nitride and, in presence of alcohol, silver fulminate may also be produced.
See SILVER-CONTAINING EXPLOSIVES
SILVERING SOLUTIONS

Carbon monoxide
Mellor, 1941, Vol. 3, 377
Carbon monoxide is exothermically oxidised over silver oxide, and the temperature may attain 300°C.

Chlorine,
Ethylene
See ETHYLENE, C_2H_4 : Chlorine

Hydrogen sulphide
See HYDROGEN SULPHIDE, H_2S : Metal oxides

Magnesium
Mellor, 1941, Vol. 3, 378
Oxidation of magnesium proceeds explosively when warmed with silver oxide in a sealed tube.

Metal sulphides
Mellor, 1941, Vol. 3, 376
Mixtures with auric, antimony or mercuric sulphides ignite on grinding.

Nitroalkanes
See NITROALKANES : Metal oxides

Non-metals
Mellor, 1941, Vol. 3, 376–377
Selenium, sulphur or phosphorus ignite on grinding with the oxide.

Potassium–sodium alloy
See POTASSIUM–SODIUM ALLOY, K–Na: Metal oxides

Seleninyl chloride
See SELENINYL CHLORIDE, Cl_2OSe : Metal oxides
See other METAL OXIDES

SILVER SULPHIDE Ag_2S

Potassium chlorate
See POTASSIUM CHLORATE, $ClKO_3$: Metal sulphides

DISILVER PENTATIN UNDECAOXIDE $Ag_2Sn_5O_{11}$

Mellor, 1941, Vol. 7, 418
The compound 'silver beta-stannate' is formed by long contact between solutions of silver and stannous nitrates, and loses water on heating and decomposes explosively.
See other METAL OXIDES

TRISILVER NITRIDE Ag_3N

Hahn, H., *et al., Z. Anorg. Chem.*, 1949, **258**, 77
Very sensitive to contact with hard objects, exploding even when moist. An extremely sensitive explosive when dry, initiable by friction, impact or heating. The impure product produced by allowing ammoniacal silver oxide solution to stand seems even more sensitive, often exploding spontaneously in suspension.

See SILVER CHLORIDE, AgCl : Ammonia
SILVER-CONTAINING EXPLOSIVES
SILVERING SOLUTIONS
TOLLENS' REAGENT
See other *N*-METAL DERIVATIVES

TETRASILVER DIIMIDOTRIPHOSPHATE $Ag_4H_3N_2O_8P_3$

Alone,
or Sulphuric acid
Mellor, 1940, Vol. 8, 705; 1971, Vol. 8, Suppl. 3, 787
The dry material explodes on heating, and ignites in contact with sulphuric acid. The molecule contains one N–Ag bond.
See other *N*-METAL DERIVATIVES

SILVER DINITRIDODIOXOSULPHATE (4–) $Ag_4N_2O_2S$

Nachbaur, E. *et al., Angew. Chem. (Intern. Ed.)*, 1973, **12**, 339

The dry salt explodes on friction or impact.

See other *N*-METAL DERIVATIVES

PENTASILVER DIIMIDOTRIPHOSPHATE $Ag_5H_2N_2O_8P_3$

Alone,
or Sulphuric acid
Mellor, 1940, Vol. 8, 705; 1971, Vol. 8, Suppl. 3, 787
The salt is explosive and may readily be initiated by friction, heat or contact with sulphuric acid. The molecule contains two N–Ag bonds.

See other *N*-METAL DERIVATIVES

HEPTASILVER NITRATE OCTAOXIDE Ag_7NO_{11}

Alone,
or Sulphides,
or Non-metals
Mellor, 1941, Vol. 3, 483–485
The crystalline product produced by electrolytic oxidation of silver nitrate (possibly $(Ag_3O_4)_2 \cdot AgNO_3$) detonates feebly at 110°C. Mixtures with phosphorus or sulphur explode on impact, hydrogen sulphide ignites on contact, and antimony trisulphide ignites when ground with the salt.

See other METAL OXONON-METALLATES

ALUMINIUM Al

Ammonium nitrate
Mellor, 1946, Vol. 5, 219
Mixtures with the powdered metal are used as an explosive, sometimes with addition of carbon, hydrocarbons and other oxidising agents.

See AMMONIUM NITRATE, $H_4N_2O_3$: Metals

Ammonium peroxodisulphate

See AMMONIUM PEROXODISULPHATE, $H_8N_2O_8S_2$: Aluminium

Bismuth

Mellor, 1947, Vol. 9, 626

The finely divided mixture of metals produced by hydrogen reduction of co-precipitated bismuth and aluminium hydroxides is pyrophoric.

Butanol

Luberoff, B. J., private comm., 1964

Butanol, used as a solvent in an autoclave preparation at *ca* 100°C, severely attacked the aluminium gasket, liberating hydrogen which caused a sharp rise in pressure. Other alcohols would behave similarly, forming the aluminium alkoxide.

Carbon,
Chlorine trifluoride

See CHLORINE TRIFLUORIDE, ClF_3 : Metals, etc.

Carbon dioxide,
Sodium peroxide

See SODIUM PEROXIDE, Na_2O_2 : Metals, etc.

Copper oxide

Anon., *Chem. Age* , 1932, **27**, 23

A mixture of aluminium powder and hot copper oxide exploded violently during mixing with a steel shovel on an iron plate. The frictional mixing initiated the thermite-like mixture.

See Metal Oxides, etc., below

Disodium acetylide

See DISODIUM ACETYLIDE, C_2Na_2 : Metals

Halocarbons

1. Anon., *Angew. Chem.*, 1950, **62**, 584
2. Anon., *Chem. Age,* 1950, **63**, 155
3. Anon., *Chem. Eng. News,* 1954, **32**, 258
4. *Pot. Incid. Rep., ASESB,* 1968, **39**
5. Anon., *Chem. Eng. News*, 1955, **33**, 942
6. Eiseman, B. J., *J. Amer. Soc. Htg. Refr. Air Condg. Eng.*, 1963, **5**, 63

7. Eiseman, B. J., *Chem. Eng. News,* 1961, **39**(27), 44
8. Laccabue, J. R., *Fluorolube-Aluminum Detonation Point* : Report 7E.1500, San Diego, Gen. Dynamics, 1958
9. Atwell, V. J., *Chem. Eng. News,* 1954, **32**, 1824
10. ICI Mond Div., private comm., 1968
11. Anon., *Ind. Acc. Prev. Bull. RoSPA*, 1953, **21**, 60
12. Wendon, W. G., private comm., 1973

Heating aluminium powder with carbon tetrachloride, chloromethane or carbon tetrachloride–chloroform mixtures in closed systems to 152°C may cause an explosion, particularly if traces of aluminium chloride are present [1]. A mixture of carbon tetrachloride and aluminium powder exploded during ball-milling [2], and it was later shown that heavy impact would detonate the mixture [3]. Mixtures with monofluorotrichlorethane and trichlorotrifluoroethane will flash or spark on heavy impact [4].

A virtually unvented aluminium tank containing a mixture of *o*-dichlorobenzene, 1, 2-dichloroethane and 1, 2-dichloropropane exploded violently seven days after filling. This was attributed to formation of aluminium chloride which catalysed further accelerating attack on the aluminium tank [5].

In a dichlorodifluoromethane system, frictional wear exposed fresh metal surfaces on an aluminium compressor impellor, causing an exothermic reaction with the halocarbon which melted much of the impellor. Later tests showed similar results decreasing in order of intensity, with: tetrafluoromethane; chlorodifluoromethane; bromotrifluoromethane; dichlorodifluoromethane; 1,2-dichlorotetrafluoroethane; 1,1,2-trichlorotrifluoroethane [6].

In similar tests molten aluminium dropped into liquid dichlorodifluoromethane burned incandescently below the liquid [6].

Aluminium bearing surfaces under load react explosively with polychlorotrifluoroethylene greases or oils. The inactive oxide film will be removed from the metal by friction, and hot spots will initiate reaction [8].

An attempt to scale up the methylation of 2-methylpropane with chloromethane in presence of aluminium chloride and aluminium went out of control and detonated, destroying the autoclave. The preparation had been done on a smaller scale on 20 previous occasions without incident [9].

See BROMOMETHANE, CH_3Br : Metals

Violent decomposition, with evolution of hydrogen chloride, may occur when 1,1,1-trichloroethane comes into contact with aluminium or its alloys with magnesium [10].

Aluminium-dusty, greasy overalls were cleaned by immersion in trichloroethylene. During subsequent drying, violent ignition occurred. This was attributed to presence of free hydrogen chloride in the solvent, which reacted to produce aluminium chloride [10]. This is known to catalyse polymerisation of trichloroethylene, producing more hydrogen chloride and heat. The reaction is self-accelerating and can develop a temperature of 1350°C [11].

Trichloroethylene cleaning baths must be kept neutral with sodium carbonate, and free of aluminium dust.

Halocarbons now available with added stabilisers (probably amines) show a reduced tendency to react with aluminium powder [12].

See other METALS: Halocarbons

Halogens

1. Mellor, 1946, Vol. 2, 92, 135; Vol. 5, 209
2. Azmathulla, S. *et al.*, *J. Chem. Educ.*, 1955, **32**, 447; *School Sci. Rev.*, 1956, **38** (134), 107
3. Hammerton, C. M., *School Sci. Rev.*, 1957, **38**(136), 459

Aluminium powder ignites in chlorine without heating, and foil reacts vigorously with liquid bromine at 15°C, and incandesces on warming in the vapour [1]. The metal and iodine react violently in the presence of water, either as liquid, vapour or that in hydrated salts [2]. Moistening a powdered mixture causes incandescence and will initiate a thermite mixture [3].

Hydrogen chloride

Batty, G. F., private comm., 1972

Erroneous use of aluminium instead of alumina in a hydrogen chloride purification reactor caused a vigorous exothermic reaction which distorted the steel reactor shell.

Interhalogens

See BROMINE PENTAFLUORIDE, BrF_5 : Acids, etc.
CHLORINE FLUORIDE, ClF : Aluminium
IODINE CHLORIDE, ClI : Metals
IODINE PENTAFLUORIDE, F_5I : Metals
IODINE HEPTAFLUORIDE, F_7I : Metals

Iron,
Water
Chen, W. Y. *et al.*, *Ind. Eng. Chem.*, 1955, **47**(7), 32A
A sludge of aluminium dust (containing iron and sand) removed from castings in water was found to undergo sudden exotherms (in summer weather) to 95°C with hydrogen evolution. Similar effects with aluminium-sprayed steel plates exposed to water were attributed to electrolytic action, as addition of iron filings to an aluminium dust slurry in water caused hydrogen evolution to occur without application of heat.
See Water, below

Mercury(II) salts
See ALUMINIUM AMALGAM, Al–Hg

Mercury(II) salts
Woelfel, W. C., *J. Chem. Educ.*, 1967, **44**, 484
Aluminium foil is unsuitable as a packing material in contact with mercury(II) salts in presence of moisture, when vigorous amalgamation ensues.

Metal nitrates,
Sulphur,
Water
Anon., *Chem. Eng. News*, 1954, **32**, 258
Berufsgenossenschaft, 1954, 184
Aluminium powder, barium and potassium nitrates, sulphur and vegetable adhesives, mixed to a paste with water, exploded on two occasions. Laboratory investigation showed initial interaction of water and aluminium to produce hydrogen. It is supposed that nascent hydrogen reduced the nitrates present, increasing the alkalinity and hence the rate of attack on aluminium, the reaction becoming self-accelerating. Cause of ignition was unknown. Other examples of interaction of aluminium with water are known.
See Iron, Water, above
Water, below

Metal oxides,
or Oxosalts,
or Sulphides
1. Mellor, 1946, Vol. 5, 217

2. Price, D. J. *et al.*, *Chem. Met. Eng.*, 1923, **29**, 878

Many metal oxo-compounds (nitrates, oxides and particularly sulphates) and sulphides are reduced violently or explosively (i.e. undergo 'thermite' reaction) on heating an intimate mixture with aluminium powder to a suitably high temperature. Contact of massive aluminium with molten salts may give explosions [1]. Application of sodium carbonate to molten (red-hot) aluminium caused an explosion [2].

See MOLTEN SALT BATHS
DIIRON TRIOXIDE, Fe_2O_3 : Aluminium

Oleic acid

de Ment, J., *J. Chem. Educ.*, 1959, **36**, 308

Shortly after mixing the two, an explosion occurred. This could not be repeated. The acid may have been peroxidised.

Oxidants

1. Kirshenbaum, 1956, 4, 13
2. Mellor, 1947, Vol. 2, 310

Mixtures of aluminium powder with liquid chlorine, dinitrogen tetraoxide or tetranitromethane are detonable explosives, but not as powerful as aluminium–liquid oxygen mixtures, some of which exceed TNT in effect by a factor of 3 to 4 [1]. Mixtures of the powdered metal and various bromates may explode on impact, heating or friction. Iodates and chlorates act similarly [2].

Other combinations are:

Halogens, above
POTASSIUM CHLORATE, $ClKO_3$: Metals
POTASSIUM PERCHLORATE, $ClKO_4$: Aluminium, etc.
: Metal powders
NITRYL FLUORIDE, FNO_2 : Metals
AMMONIUM PEROXODISULPHATE, $H_8N_2O_8S_2$: Aluminium, etc.
SODIUM NITRATE, $NNaO_3$: Aluminium
SODIUM PEROXIDE, Na_2O_2 : Aluminium, etc.
ZINC PEROXIDE, O_2Zn : Metals

Paint

See ZINC, Zn; Paint primer base

Phosphorus pentachloride

Berger, E. *Compt. Rend.*, 1920, **170**, 29

Aluminium powder ignites in contact with the pentachloride.

2-Propanol
Wilds, A. L., *Org. React.*, 1944, **2**, 198
Muir, G. D., private comm., 1968
Dissolution of aluminium in 2-propanol to give the isopropoxide is rather exothermic, but often subject to an induction period similar to that in preparation of Grignard reagents. Only small amounts of aluminium should be present until reaction begins.
See also MAGNESIUM, Mg: Methanol

Silver chloride
Anon., *Chem. Eng. News,* 1954, **32**, 258
An intimate mixture of the two powders may lead to reaction of explosive violence, unless excess aluminium is present.

Sulphur
Read, C. W. W., *School Sci. Rev.*, 1940, **21**(83), 977
The violent interaction of aluminium powder and sulphur on heating is considered too dangerous for a school experiment.
See SULPHUR, S : Metals

Water
MCA Case History No. 462
Cans of aluminium paint contaminated with water contained a considerable pressure of hydrogen from interaction of finely divided metal and moisture.
See other METALS

ALUMINIUM–COPPER–ZINC ALLOY — Al–Cu–Zn

See DEVARDA'S ALLOY
SILVER NITRATE, $AgNO_3$: Ammonia, etc.
See other ALLOYS (INTERMETALLIC COMPOUNDS)

ALUMINIUM AMALGAM — Al–Hg

1. Neely, T. A. *et al.*, *Org. Synth.*, 1965, **45**, 109
2. Calder, A. *et al.*, *Org. Synth.*, 1972, 52, 78

The amalgamated aluminium wool remaining from preparation of triphenylaluminium will rapidly oxidise and become hot on exposure to air. Careful disposal is necessary [1]. Amalgamated aluminium foil may be pyrophoric and should be kept moist and used immediately [2].

See other ALLOYS (INTERMETALLIC COMPOUNDS)

ALUMINIUM–MAGNESIUM ALLOY **Al–Mg**

Barium nitrate

See BARIUM NITRATE, BaN_2O_6 : Aluminium–magnesium alloy

ALUMINIUM–NICKEL ALLOYS **Al–Ni**

Water

Anon., *Angew. Chem. Nachr.* (1968), **16**, 2

Heating moist Raney nickel alloy containing 20% of aluminium in an autoclave under hydrogen caused the aluminium and water to interact explosively, generating 1000 bar pressure of hydrogen.

See also HYDROGENATION CATALYSTS

See other ALLOYS (INTERMETALLIC COMPOUNDS)

ALUMINIUM–TITANIUM ALLOYS **Al–Ti**

Oxidants

Mellor, 1941, Vol. 7, 20–21

Alloys ranging from Al_3Ti_2 to Al_4Ti have been described, which ignite or incandesce on heating in chlorine; or bromine or iodine vapour; or hydrogen chloride; or oxygen.

See other ALLOYS (INTERMETALLIC COMPOUNDS)

ALUMINIUM TETRAHYDROBORATE AlB_3H_{12}

1. Schlessinger, H. I. *et al.*, *J. Amer. Chem. Soc.*, 1940, **62**, 3421
2. Badin, F. J. *et al.*, *J. Amer. Chem. Soc.*, **71**, 2950

The vapour is spontaneously inflammable in air [1], and explodes in oxygen, but only in presence of traces of moisture [2].

Alkenes,
Oxygen
Gaylord, 1956, 26
The tetrahydroborate reacts with alkenes and, in presence of oxygen, combustion is initiated, even in absence of moisture. Butene explodes after an induction period, while butadiene explodes immediately.
See other COMPLEX HYDRIDES

ALUMINIUM TRIS(TETRAAZIDOBORATE) AlB_3N_{36}

Mellor, 1967, Vol. 8, Suppl. 2.2, 2
A very shock-sensitive explosive, containing nearly 90% wt. of nitrogen.
See other HIGH-NITROGEN COMPOUNDS
NON-METAL AZIDES

ALUMINIUM TRIBROMIDE $AlBr_3$

Water
Nicholson, D. G. *et al.*, *Inorg. Synth.*, 1950, **3**, 35
The anhydrous bromide should be destroyed by melting and pouring slowly into running water. Hydrolysis is very violent and may destroy the container if water is added to the container.
See other METAL HALIDES

ALUMINIUM DICHLORIDE HYDRIDE DIETHYL ETHERATE

$AlCl_2H \cdot C_4H_{10}O$

Dibenzyl ether

Marconi, W. *et al.*, *Ann. Chim.*, 1965, **55**, 897

During attempted reductive cleavage of the ether with the etherate an explosion occurred. Peroxides may have been present.

See related COMPLEX HYDRIDES

ALUMINIUM TRICHLORIDE

$AlCl_3$

1. Popov, P. V., *Savodskaya Lab.*, 1946, **13**, 127 *(Chem. Abs.*, 1947, **41**, 6723d); Kitching, A. F., *School Sci. Rev.*, 1930, **12**(45), 79
2. *MCA SD-62,* 1956

Long storage of the anydrous salt in closed containers caused spontaneous decomposition and occasional explosion on opening [1]. General handling precautions are detailed [2].

Aluminium,
Sodium peroxide

See SODIUM PEROXIDE, Na_2O_2 : Aluminium, etc.

Ethylene oxide

See ETHYLENE OXIDE, C_2H_4O : Contaminants

Nitrobenzene,
Phenol

Anon., *Chem. Eng. News*, 1953, **31**, 4915

Addition of aluminium chloride to a large volume of recovered nitrobenzene containing 5% of phenol caused a violent explosion. Experiment showed that mixtures containing all three components reacted violently at 120°C.

Nitromethane

See ALUMINIUM TRICHLORIDE-NITROMETHANE, $AlCl_3 \cdot CH_3NO_2$

Oxygen difluoride

See OXYGEN DIFLUORIDE, F_2O : Halogens, etc.

Phenyl azide

See PHENYL AZIDE, $C_6H_5N_3$: Lewis acids

Perchlorylbenzene

See PERCHLORYLBENZENE, $C_6H_5ClO_3$: Aluminium trichloride

Water

Anon., *Ind. Eng. Chem. (News Ed.)*, 1934, **12**, 194

An unopened bottle of anhydrous aluminium chloride erupted when the rubber bung with which it was sealed was removed. The accumulation of pressure was attributed to absorption of moisture by the anhydrous chloride before packing. The presence of an adsorbed layer of moisture in the bottle used for packing may have contributed. Reaction with liquid water is violently exothermic.

See other METAL HALIDES

ALUMINIUM TRICHLORIDE-NITROMETHANE $AlCl_3 \cdot CH_3NO_2$

An alkene

Cowen, F.M. *et.al.*, *Chem. Eng. News,* 1948, **26**, 2257

A gaseous alkene was passed into a cooled autoclave containing the complex, initially with agitation, and later without. Later, when the alkene was admitted to a pressure of 5.6 bar at 2°C, a slight exotherm occurred, followed by an explosion. The autoclave contents were completely carbonised. Mixtures of ethylene, aluminium chloride and nitromethane had exploded previously, but at 75°C.

See ETHYLENE, C_2H_4 : Aluminium trichloride

ALUMINIUM CHLORATE $AlCl_3O_9$

Sidgwick, 1950, 428

During evaporation, its aqueous solution evolves chlorine dioxide, and eventually explodes.

See other METAL OXOHALOGENATES

CAESIUM HEXAHYDROALUMINATE (3–) $AlCs_3H_6$

See POTASSIUM HEXAHYDROALUMINATE (3–), AlH_6K_3
See other COMPLEX HYDRIDES

COPPER(I) TETRAHYDROALUMINATE $AlCuH_4$

Aubry, J. *et al.*, *Compt. Rend.*, 1954, **238**, 2535
The unstable hydride decomposed at –70°C, and ignited on contact with air.
See other COMPLEX HYDRIDES

ALUMINIUM HYDRIDE AlH_3

Mirviss, S. B. *et al.*, *Ind. Eng. Chem.*, 1961, **53**(1), 54A
It is very unstable and has been known to decompose spontaneously at ambient temperature with explosive violence. Its complexes (particularly the diethyl etherate) are considerably more stable.
See also ALUMINIUM HYDRIDE-TRIMETHYLAMINE, $AlH_3 \cdot C_3H_9N$

Carbon dioxide,
Methyl ethers
Barbaras, G. *et al.*, *J. Amer. Chem. Soc.*, 1948, **70**, 877
Presence of carbon dioxide in solutions of the hydride in dimethyl or di(2-methoxyethyl) ethers can cause a violent decomposition on warming the residue from evaporation. Presence of aluminium chloride tends to increase the vigour of decomposition to explosion. Lithium tetrahydroaluminate may behave similarly, but is generally more stable.
See other METAL HYDRIDES

ALUMINIUM HYDRIDE-TRIMETHYLAMINE $AlH_3 \cdot C_3H_9N$

Water
Ruff, J. K., *Inorg. Synth.*, 1967, **9**, 34

It ignites in moist air and is explosively hydrolysed by water.

See other METAL HYDRIDES

ALUMINIUM TRIHYDROXIDE **AlH_3O_3**

Chlorinated rubber

See CHLORINATED RUBBER : Metal oxides or hydroxides

LITHIUM TETRAHYDROALUMINATE **AlH_4Li**

1. Augustine, 1968, 12
2. Gaylord, 1956, 37

Care is necessary in handling this powerful reducant, which may ignite if lumps are pulverised with a pestle and mortar, even in a dry box [1]. A rubber mallet is recommended for the purpose [2].

Bis(2-methoxyethyl) ether

1. Watson, A. R., *Chem. & Ind.*, 1964, 665
2. Adams, R. M., *Chem. Eng. News,* 1953, **31**, 2334
3. Barbaras, G. *et al., J. Amer. Chem. Soc.,* 1948, **70**, 877
4. *MCA Case History No. 1494*

The peroxide-free ether, being dried by distillation at 162°C under inert atmosphere at ambient pressure, exploded violently when the heating bath temperature had been raised to 200°C towards the end of distillation. This was attributed to local overheating of an insulating crust of hydride in contact with oxygen-containing organic material [1]. Two previous explosions were attributed to peroxides [2] and the high solubility of carbon dioxide in such ethers [3]. Stirring during distillation would probably prevent crust formation. Alternatively, drying could be effected with a column of molecular sieve or activated alumina. During distillation of the solvent from the aluminate at 100°C at atmospheric pressure, the flask broke and its contents ignited explosively. The aluminate decomposes at 125–135°C [4].

See ALUMINIUM TRIHYDRIDE, AlH_3 : Carbon dioxide

Boron trifluoride diethyl etherate
1. Scott, R. B., *Chem. Eng. News,* 1967, **45**(28), 7; ibid., **45**(21),51
2. Shapiro, I. *et al.*, *J. Amer. Chem. Soc.*, 1952, **74**, 90
Use of lumps of the solid aluminate, rather than its ethereal solution, and of peroxide-containing etherate, rather than the peroxide-free material specified [2], caused an explosion during the attempted preparation of diborane.

Dibenzoyl peroxide
See DIBENZOYL PEROXIDE, $C_{14}H_{10}O_4$: Lithium tetrahydroaluminate

3,5-Dibromocyclopentene
Johnson, C. R. *et al.*, *Tetrahedron Lett.*, 1964, **45**, 3327
Preparation of the 4-bromo compound by partial debromination of crude 3,5-dibromocyclopentene by addition of its ethereal solution to the aluminate in ice-cold ether is hazardous. Explosions have occurred on two occasions about an hour after addition of dibromide.

1,2-Dimethoxyethane
1. *MCA Case History No. 1182*
2. Hoffmann, K. A. *et al.*, *Org. Synth.*, 1968, **48**, 62
The finely powdered aluminate was charged through a funnel into a nitrogen-purged flask. When the solvent was added through the same funnel, ignition occùrred, possibly due to local absence of purge gas in the funnel caused by turbulence [1]. Distillation of the solvent from the solid must not be taken to dryness, to avoid explosive decomposition of the residual aluminate [2].

Ethyl acetate
1. Bessant, K. H. C., *Chem. & Ind.*, 1957, 432
2. Yardley, J. T., ibid., 433
Following a reductive dechlorination in ether, a violent explosion occurred when ethyl acetate was added to decompose excess aluminate [1]. Ignition was attributed to the strongly exothermic reaction occurring when undiluted (and reducible) ethyl acetate contacts the solid aluminate. Addition of a solution of ethyl acetate in inert solvent or of a moist unreactive solvent to destroy excess reagen is preferable [2].

Fluoroamides

1. Karo, W., *Chem. Eng. News*, 1955, **33**, 1368
2. Reid, T. S. *et al.*, *Chem. Eng. News*, 1951, **29**, 3042

The reduction of amides of fluorocarboxylic acids with the tetra-hydroaluminate appears generally hazardous at all stages. During reduction of *N*-ethylheptafluorobutyramide in ether, violent and prolonged gas evolution caused a fire. Towards the end of reduction of trifluoroacetamide in ether, solid separated and stopped the stirrer. Attempts to restart the stirrer by hand caused a violent explosion [1]. During decomposition by water of the reaction complex formed by interaction with tetrafluorosuccinamide in ether, a violent explosion occurred. Experiment showed this was due to the low stability of the complex, particularly in absence of ether, when detonation at room temperature occurred. Reaction complexes similarly obtained from trifluoroacetic acid, heptafluorobutyramide and octafluoroadipamide also showed instability, decomposing when heated. General barricading of all reductions of fluoro compounds with lithium tetrahydroaluminate is recommended [2].

Pyridine

Augustine, 1968, 22–23

Addition of the aluminate (0.5 g) to pyridine (50 ml) must be effected very slowly with cooling. Addition of 1 g portions may cause a highly exothermic reaction.

Tetrahydrofuran

Moffett, R. B., *Chem. Eng. News*, 1954, **32**, 4328

The solvent had been dried over the aluminate and then stored over calcium hydride for 2 years 'to prevent peroxide formation'. Subsequent addition of more aluminate caused a strong exotherm and ignition of liberated hydrogen. Calcium hydride does not prevent peroxide formation in solvents.

SODIUM TETRAHYDROALUMINATE AlH_4Na

Tetrahydrofuran

Del Giudia, F. P. *et al.*, *Chem. Eng. News*, 1961, **39**(40), 57

During synthesis from its elements in tetrahydrofuran, a violent

explosion occurred when absorption of hydrogen had stopped. This was attributed to deposition of solid above the liquid level, overheating and reaction with solvent to give butoxyaluminohydrides. Vigorous stirring and avoiding local overheating are essential.
See other COMPLEX HYDRIDES

POTASSIUM HEXAHYDROALUMINATE (3–) **AlH_6K_3**

Ashby, E. C., *Chem. Eng. News,* 1969, **47**(1), 9
A 20 g sample, prepared and stored dry in a dry box for several months, developed a thin crust of oxidation/hydrolysis products. When the crust was disturbed, a violent explosion occurred (later confirmed as equivalent to 230 g TNT). A weaker explosion was observed with potassium tetrahydroaluminate. The effect was attributed to superoxidation of traces of metallic potassium, and subsequent interaction of the hexahydroaluminate and superoxide after frictional initiation. Precautions advised include use of freshly prepared material, minimal storage in a dry diluent under an inert atmosphere and destruction of solid residues.

Potassium hydrides and caesium hexahydroaluminate may behave similarly, as caesium also superoxidises in air.
See other COMPLEX HYDRIDES

LITHIUM TETRAAZIDOALUMINATE (1–) **$AlLiN_{12}$**

Mellor, 1967, Vol. 8, Suppl. 2.2, 2
A shock-sensitive explosive.
See other METAL AZIDES

ALUMINIUM TRIAZIDE **AlN_9**

Brauer, 1963, Vol. 1, 829
May be detonated by shock.
See other METAL AZIDES

MAGNESIUM TETRAHYDROALUMINATE Al_2H_8Mg

Gaylord, 1956, 25
It is similar to the lithium salt.
See other COMPLEX HYDRIDES

MANGANESE(II) TETRAHYDROALUMINATE Al_2H_8Mn

Aubry, J. *et al., Compt. Rend.*, 1954, **238**, 2535
The unstable hydride decomposed at –80°C and ignited in contact with air.
See other COMPLEX HYDRIDES

DIALUMINIUM TRIOXIDE Al_2O_3

Chlorine trifluoride
See CHLORINE TRIFLUORIDE, ClF_3: Metals, etc.

Ethylene oxide
See ETHYLENE OXIDE, C_2H_4O: Contaminants

Oxygen difluoride
See OXYGEN DIFLUORIDE, F_2O: Adsorbents

Sodium nitrate
See SODIUM NITRATE, $NNaO_3$: Aluminium, etc.

Vinyl acetate
See VINYL ACETATE, $C_4H_6O_2$: Desiccants

DIALUMINIUM OCTAVANADIUM TRIDECASILICIDE $Al_2Si_{13}V_8$

Hydrofluoric acid
Sidgwick, 1950, 833
The silicide reacts violently with aqueous hydrofluoric acid.
See other METAL NON-METALLIDES

CERIUM(III) TETRAHYDROALUMINATE Al_3CeH_{12}

Aubry, J. *et al.*, *Compt. Rend.*, 1954, **238**, 2535
The unstable hydride decomposed at –80°C, and ignited in contact with air.
See other COMPLEX HYDRIDES

AMERICIUM TRICHLORIDE $AmCl_3$

MCA Case History No. 1105
A multi-wall shipping container, containing 400 cm^3 of a solution of americium chloride in a polythene bottle and sealed for 3½ months, exploded. The reason could have been a slow pressure build-up of radiolysis products. Venting and other precautions are recommended.
See other METAL HALIDES

ARSENIC As

Bromine azide
see BROMINE AZIDE, BrN_3

Dirubidium acetylide
See DIRUBIDIUM ACETYLIDE, C_2Rb_2 : Non-metals

Halogens or Interhalogens
1. Mellor, 1946, Vol. 2, 92

2. Mellor, 1956, Vol. 2, Suppl. 1, 379

The finely powdered element inflames in gaseous chlorine or liquid chlorine at –33°C [1]. The latter is doubtful [2].

See BROMINE TRIFLUORIDE, BrF_3: Halogens, etc.
BROMINE PENTAFLUORIDE, BrF_5: Acids, etc.
CHLORINE TRIFLUORIDE, ClF_3: Metals, etc.
IODINE PENTAFLUORIDE, F_5I: Metals, etc.

Metals

Mellor, 1940, Vol. 4, 485–486; 1942, Vol. 15, 629; 1937, Vol. 16, 161

Palladium or zinc and arsenic react on heating with evolution of light and heat, and platinum with vivid incandescence.

Nitrogen trichloride

See NITROGEN TRICHLORIDE, Cl_3N: Initiators

Oxidants

See SILVER NITRATE, $AgNO_3$: Arsenic
DICHLORINE OXIDE, Cl_2O: Oxidisable materials
CHROMIUM TRIOXIDE, CrO_3: Arsenic
NITROSYL FLUORIDE, FNO: Metals, etc.
POTASSIUM PERMANGANATE, $KMnO_4$: Antimony, etc.
POTASSIUM DIOXIDE (SUPEROXIDE), KO_2: Metals
SODIUM PEROXIDE, Na_2O_2: Non-metals

See other NON-METALS

ARSENIC TRICHLORIDE $AsCl_3$

Hexafluoroisopropylideneaminolithium

See HEXAFLUOROISOPROPYLIDENEAMINOLITHIUM, C_3F_6LiN: Non-metal halides

TRIFLUOROSELENIUM HEXAFLUOROARSENATE AsF_9Se

Water

Bartlett, N. *et al.*, *J. Chem. Soc.*, 1956, 3423

Violent interaction.

ARSINE AsH_3

Gas above −62°C; flammable.

Chlorine
See CHLORINE, Cl_2 : Non-metal hydrides

Nitric acid
See NITRIC ACID, HNO_3 : Non-metal hydrides
See other NON-METAL HYDRIDES

ARSINE-BORON TRIBROMIDE $AsH_3 \cdot BBr_3$

Oxidants

1. Stock, A., *Ber.*, 1901, **34**, 949
2. Mellor, 1939, Vol. 9, 57

Unlike arsine, the complex ignites on exposure to air or oxygen, even at below 0°C [1]. It is violently oxidized by nitric acid [2].
See NON-METAL HYDRIDES

ARSENIC DISULPHIDE AsS_2

Potassium nitrate
See POTASSIUM NITRATE, KNO_3 : Metal sulphides

DIARSENIC TRIOXIDE As_2O_3

MCA SD-60, 1956
See CHLORINE TRIFLUORIDE, ClF_3 : Metals, etc.
HYDROGEN FLUORIDE, FH: Oxides
ZINC, Zn: Arsenic trioxide

DIARSENIC PENTAOXIDE As_2O_5

Bromine pentafluoride
See BROMINE PENTAFLUORIDE, BrF_5 : Acids, etc.

PLATINUM DIARSENIDE As_2Pt

Preparative hazard.
See PLATINUM, Pt: Arsenic

DIARSENIC DISULPHIDE As_2S_2

Chlorine
See CHLORINE, Cl_2 : Sulphides

GOLD Au

Hydrogen peroxide
See HYDROGEN PEROXIDE, H_2O_2 : Metals

GOLD(III) CHLORIDE $AuCl_3$

Ammonia and derivatives
Mellor, 1941, Vol.3, 582–583
Sidgwick, 1950, 178
Action of ammonia or ammonium salts on gold chloride, oxide or other salts under a wide variety of conditions gives explosive or 'fulminating' gold. Of uncertain composition but containing Au–N bonds, this is a heat-, friction- and impact-sensitive explosive when dry, similar to the related mercury and silver compounds.
See other METAL HALIDES

TRIAMMINEGOLD TRIHYDROXIDE $AuH_{12}N_3O_3$

White, J. H., private comm., 1965
The potentially explosive properties have been mentioned in the literature.
See other GOLD COMPOUNDS

GOLD(III) OXIDE Au_2O_3

Ammonium salts
See GOLD(III) CHLORIDE, $AuCl_3$: Ammonia, etc.

GOLD(III) SULPHIDE Au_2S_3

Silver oxide
See SILVER OXIDE, Ag_2O: Metal sulphides

GOLD NITRIDE-AMMONIA $Au_3N \cdot H_3N$

Raschig, F., *Ann.*, 1886, **235**, 349
An explosive compound, probably present in explosive gold, produced from action of ammonia on gold(I) oxide.
See other *N*-METAL DERIVATIVES

TRIGOLD DISODIUM HEXAAZIDE $Au_3N_{18}Na_2$

Mellor, 1940, Vol. 8, 349; 1967, Vol. 8, Suppl. 2, 30
Of uncertain constitution, both the dry solid and its aqueous solutions may explode.
See other METAL AZIDES

BORON B

Dirubidium acetylide

See DIRUBIDIUM ACETYLIDE, C_2Rb_2: Non-metals

Halogens or Interhalogens

Mellor, 1946, Vol. 2, 92

Boron ignites in gaseous chlorine or fluorine at ambient temperature, attaining incandescence in fluorine.

See BROMINE TRIFLUORIDE, BrF_3: Halogens, etc.
BROMINE PENTAFLUORIDE, BrF_5: Acids, etc.
IODINE PENTAFLUORIDE, F_5I: Metals, etc.
See Oxidants, below

Metal fluorides

Mellor 1946, Vol. 3, 389; Vol. 5, 15

Explosive interaction when boron and lead or silver fluoride are ground together at ambient temperature.

Oxidants

Mellor, 1946, Vol. 5, 16

The presence of traces of nitrites in fused metal nitrates increases the violence of interaction.

See Halogens or Interhalogens, above
See also NITROSYL FLUORIDE, FNO: Metals, etc.
NITRYL FLUORIDE, FNO_2: Non-metals
OXYGEN DIFLUORIDE, F_2O : Non-metals
NITRIC ACID, HNO_3: Non-metals
POTASSIUM NITRITE, KNO_2 : Boron
POTASSIUM NITRATE, KNO_3 : Non-metals
NITROGEN OXIDE ('NITRIC OXIDE'), NO: Non-metals
DINITROGEN OXIDE ('NITROUS OXIDE'), N_2O : Boron
SODIUM PEROXIDE, Na_2O_2: Non-metals
LEAD(II) OXIDE, OPb: Non-metals
LEAD(IV) OXIDE, O_2Pb: Non-metals
See other NON-METALS

BORON BROMIDE DIIODIDE $BBrI_2$

Water

Mellor, 1946, Vol. 5, 136

Interaction is violent, as for the tribromide or triiodide.
See other NON-METAL HALIDES

BORON DIBROMIDE IODIDE BBr_2I

Water
Mellor, 1946, Vol, 5, 136
Interaction is violent, as for the tribromide or triiodide.
See other NON-METAL HALIDES

BORON TRIBROMIDE BBr_3

Sodium
See SODIUM, Na: Non-metal halides

Water
1. *BCISC Quart. Safety Summ.*, 1966, **37**, 22
2. Anon., *Lab. Pract.*, 1966, **15**, 797

Boron halides react violently with water, and, particularly if there is a deficiency of water, a violent explosion may result. It is therefore highly dangerous to wash glass ampoules of boron tribromide in running water, or to surround a container of boron tribromide with water under any circumstances. Experiment showed that an ampoule of boron tribromide, when deliberately broken under water, caused a violent explosion, possibly a detonation. Dry non-polar solvents should be used for cleaning or cooling purposes[1]. Small quantities of boron tribromide may be destroyed by cautious addition to a large volume of water, or water containing ice [2].
See other NON-METAL HALIDES

BORON AZIDE DICHLORIDE BCl_2N_3

1. Anon., *Angew. Chem. (Nachr.)*, 1970, **18**, 27

2. Paetzold, P. I., *Z. Anorg. Chem.*, 1963, **326**, 47

A hard crust of sublimed material exploded when crushed with a spatula [1]. Previous explosions on sublimation or during solvent removal were known [2].

See other NON-METAL AZIDES

BORON TRICHLORIDE BCl_3

Aniline

Jones, R. G., *J. Amer. Chem. Soc.*, 1939, **61**, 1378

In absence of cooling or a diluent, interaction is violent.

Hexafluorisopropylideneaminolithium

See HEXAFLUORISOPROPYLIDENEAMINOLITHIUM, C_3F_6LiN: Non-metal halides

See other NON-METAL HALIDES

BORON TRIFLUORIDE BF_3

Alkali metals,
or Alkaline earth metals (not magnesium)

Merck Index, 1968, 162

Interaction hot causes incandescence.

Alkyl nitrates

See ALKYL NITRATES: Lewis acids

See other NON-METAL HALIDES

NITRONIUM TETRAFLUOROBORATE BF_4NO_2

Tetrahydrothiophene-1,1-dioxide

See NITRATING AGENTS

DIOXYGENYL TETRAFLUOROBORATE BF_4O_2

Organic materials

Goetschel, C. T. *et al., J. Amer. Chem Soc.*, 1969, **91**, 4706

It is a very powerful oxidant, addition of a small particle to small samples of benzene or 2-propanol at ambient temperature causing ignition. A mixture prepared at –196°C with either methane or ethane exploded when the temperature was raised to –78°C.

See other OXIDANTS

TETRAFLUOROAMMONIUM TETRAFLUOROBORATE BF_8N

2-Propanol

Goetschel, C. T. *et al., Inorg. Chem.*, 1972, **11**, 1700

When the fluorine used during synthesis contained traces of oxygen, the solid behaved as a powerful oxidant (causing 2-propanol to ignite on contact) and it also exploded on impact. Material prepared from oxygen-free fluorine did not show these properties, which were ascribed to the presence of traces of dioxygenyl tetrafluoroborate.

See DIOXYGENYL TETRAFLUOROBORATE, BF_4O_2

See other *N*-HALOGEN COMPOUNDS

BORIC ACID BH_3O_3

Potassium

See POTASSIUM, K: Oxidants

LITHIUM TETRAHYDROBORATE BH_4Li

Water

Gaylord, 1965, 22

Contact with limited amounts of water, either as liquid or present as

moisture in cellulose fibres, may cause ignition after a delay.
See other COMPLEX HYDRIDES

AMMONIUM PEROXOBORATE **BH_4NO_3**

Menzel, H. *et al., Oesterr. Chem. Zeit.*, 1925, 28, 162
Explosive decomposition under vacuum.
See other PEROXOACID SALTS

SODIUM TETRAHYDROBORATE **BH_4Na**

Alkali
Anon., *Angew. Chem. (Nachr.)*, 1960, 8, 238
A large volume of alkaline tetrahydroborate solution spontaneously heated and decomposed, liberating large volumes of hydrogen which burst the container. Decomposition is rapid when pH is below 10.5.

Palladium
See PALLADIUM, Pd : Sodium tetrahydroborate
See other COMPLEX HYDRIDES

HYDRAZINE-MONOBORANE **BH_7N_2**

Gunderloy, F. C., *Inorg. Synth.*, 1967, 9, 13
It is shock-sensitive and highly flammable, like the bis-compound.
See COMPLEX HYDRIDES
NON-METAL HYDRIDES

BORON DIIODOPHOSPHIDE **BI_2P**

Chlorine
See CHLORINE, Cl_2 : Phosphorus compounds

Metals
Mellor, 1947, Vol. 8, 845
It ignites in contact with mercury vapour or magnesium powder.
See other NON-METAL HALIDES

BORON TRIIODIDE BI_3

Ammonia
Mellor, 1945, Vol. 5, 136
Strong exotherm on contact.

Phosphorus
Mellor, 1946, Vol. 5, 136
Warm red or white phosphorus reacts incandescently.

Water
1. Moissan, H., *Compt. Rend.*, 1892, **115**, 204
2. Unpublished information
Violent reaction [1], particularly with limited amounts of water [2].
See other NON-METAL HALIDES

BORON NITRIDE BN

Sodium peroxide
See SODIUM PEROXIDE, Na_2O_2: Boron nitride

TRIAZIDOBORANE BN_9

Anon., *Angew. Chem. (Nachr.)*, 1970, **18**, 27
A sample of the vacuum-distilled pyridine complex exploded in a heated capillary sampling tube.
See other NON-METAL AZIDES

SODIUM BORATE HYDROGEN PEROXIDATE $BNaO_2 \cdot H_2O_2$

See CRYSTALLINE HYDROGEN PEROXIDATES
See also SODIUM PEROXOBORATE, $BNaO_3$

SODIUM PEROXOBORATE $BNaO_3$

1. Anon., *Angew. Chem.*, 1963, **65**, 41
2. Castrantas, 1965, 5

The true peroxoborate has been reported to detonate on light friction [1]. The common 'tetrahydrate' is not a peroxoborate, but $NaBO_2 \cdot H_2O_2 \cdot 3H_2O$, and while subject to catalytic decomposition by heavy metals and their salts, or easily oxidisable foreign matter, it is relatively stable under mild grinding with other substances [2].

See CRYSTALLINE HYDROGEN PEROXIDATES
See other PEROXOACID SALTS

BORON PHOSPHIDE BP

Oxidants
See NITRIC ACID, HNO_3: Non-metals
SODIUM NITRATE, $NNaO_3$: Boron phosphide

BERYLLIUM TETRAHYDROBORATE B_2BeH_8

Mackay, 1966, 169
It vigorously ignites and often explodes in air.
See other COMPLEX HYDRIDES

DIBORON TETRACHLORIDE B_2Cl_4

Air
Wartik, T. *et al.*, *Inorg. Synth.*, 1967, **10**, 125
Sudden exposure to air may cause explosion.

Dimethylmercury
Wartik, T. *et al., Inorg. Chem.*, 1971, **10**, 650
The reaction, starting at –63°C under vacuum, exploded violently on two occasions, after 23 uneventful runs. Investigation pending.
See other NON-METAL HALIDES

DIBORON TETRAFLUORIDE B_2F_4

Oxygen
Trefonas, L. *et al., J. Chem. Phys.*, 1958, **28**, 54
The gas is extremely explosive in presence of oxygen.
See other NON-METAL HALIDES

DIBORANE B_2H_6

Fl.P., –90°C (gas above –93°C); E.L., 0.9–98%; A.I.T., 38–52°C (lowered by moisture)
MCA SD-84, 1961

Air,
Benzene,
or Moisture
1. Mellor, 1946, Vol. 5, 36
2. Schlessinger, H. I. *et al., Chem Rev.*, 1942, **31**, 8
3. Simons, H. P. *et al., Ind. Eng. Chem.*, 1958, **50**, 1665, 1669

Usually ignites in air unless dry and free of impurities [1]. Ignition delays of 3–5 days, followed by violent explosions, have been experienced [2]. Effects of presence of moisture or benzene vapour in air on spontaneously explosive reaction have been assessed quantitatively [3].

Ammonia
See AMMONIUM [AMINYLENIUMBIS (TRIHYDROBORATE)] (1–), $B_2H_{12}N_2$

Chlorine
See CHLORINE, Cl_2 : Non-metal hydrides

Dimethyl sulphoxide
See DIMETHYL SULPHOXIDE, C_2H_6OS: Boron compounds

Halocarbons
Haz. Chem. Data, 1971, 88
Diborane reacts violently with halocarbon liquids used as vaporising fire-extinguishants.

Octanal oxime,
Sodium hydroxide
Augustine, 1968, 78
Addition of sodium hydroxide solution during work-up of a reaction mixture of oxime and diborane in THF is very exothermic, a mild explosion being noted on one occasion.
See other NON-METAL HYDRIDES

HYDRAZINE-BISBORANE **$B_2H_{10}N_2$**

Gunderloy, F. C., *Inorg. Synth.,* 1967, 9, 14
Explodes on impact or at over 100°C and is extremely flammable.
See other COMPLEX HYDRIDES
NON-METAL HYDRIDES

AMMONIUM [AMINYLENIUMBIS(TRIHYDROBORATE)] (1−)
$B_2H_{12}N_2$

Sidgwick, 1950, 354
The diammine complex of diborane (formulated as above), though less reactive than diborane, ignites on heating in air.
See other COMPLEX HYDRIDES

MAGNESIUM BORIDE **B_2Mg_3**

Acids
Mellor, 1946, Vol. 5, 25

The crude product containing some silicide evolves, in contact with hydrochloric or sulphuric acid, boron and silicon hydrides, which may ignite.
See other METAL NON-METALLIDES

DIBORON DIOXIDE **B_2O_2**

Water
Halliday, A. K. *et al., Chem. Rev.*, 1962, **62**, 316
At 400°C, traces of water react causing a violent eruption and incandescence.
See other NON-METAL OXIDES

DIBORON TRIOXIDE **B_2O_3**

Bromine pentafluoride
See BROMINE PENTAFLUORIDE, BrF_5 : Acids, etc.

DIBORON TRISULPHIDE **B_2S_3**

Chlorine
See CHLORINE, Cl_2 : Sulphides

***B*-1,3,5-TRICHLOROBORAZINE** **$B_3Cl_3H_3N_3$**

Water
Niedenzu, K. *et al.*, *Inorg. Synth.*, 1967, **10**, 141
Violent interaction.

BORAZINE **$B_3H_6N_3$**

Niedenzu, K. *et al.*, *Inorg. Synth.*, 1967, **10**, 144

Samples sealed into ampoules exploded when stored in daylight, but not in the dark.

See other NON-METAL HYDRIDES

URANIUM(III) TETRAHYDROBORATE $B_3H_{12}U$

Sidgwick, 1950, 1085

It explodes violently.

See other COMPLEX HYDRIDES

TETRABORANE(10) B_4H_{10}

Oxidants

Mellor, 1946, Vol. 5, 36

Ignites in air or oxygen, and explodes with concentrated nitric acid.

See also NITRIC ACID, HNO_3 : Non-metal hydrides

See other NON-METAL HYDRIDES

HAFNIUM(IV) TETRAHYDROBORATE $B_4H_{16}Hf$

Gaylord, 1956, 58

Violent ignition on exposure to air.

See other COMPLEX HYDRIDES

ZIRCONIUM(IV) TETRAHYDROBORATE $B_4H_{16}Zr$

Gaylord, 1956, 58

Violent ignition on exposure to air.

See other COMPLEX HYDRIDES

SODIUM TETRABORATE (2–) $B_4Na_2O_7$

Zirconium

See ZIRCONIUM, Zr: Oxygen-containing compounds

PENTABORANE(9) B_5H_9

Fl.P., ~ 30°C; E.L., 0.42– %; A.I.T., 35°C (ignites spontaneously if impure)

MCA SD-84, 1961

Reactive solvents

Cloyd, 1965, 35

Pentaborane is stable in inert hydrocarbon solvents but forms shock-sensitive solutions in most other solvents containing carbonyl, ether or ester functional groups and/or halogen substitutents.

See DIMETHYL SULPHOXIDE, C_2H_6OS: Boron compounds

See other NON-METAL HYDRIDES

PENTABORANE(11) B_5H_{11}

Kit and Evered, 1960, 69

Ignites in air

See other NON-METAL HYDRIDES

DECABORANE(14) $B_{10}H_{14}$

Ethers,

or Halocarbons,

or Oxygen

MCA SD-84, 1961

Hawthorne, M. F., *Inorg. Synth.,* 1967, **10**, 93–94

It forms impact-sensitive mixtures with ethers (dioxan, etc.) and halocarbons (carbon tetrachloride) and ignites in oxygen at 100°C.

See PENTABORANE(9), B_5H_9: Reactive solvents
DIMETHYL SULPHOXIDE, C_2H_6OS: Boron compounds
See other NON-METAL HYDRIDES

BARIUM **Ba**

Halocarbons

1. *Serious Acc. Ser.*, 1952, **23** and Suppl., Washington, USAEC
2. Anon., *Ind. Res.*, 1968, (9), 15
3. *Pot. Incid. Rep.*, 1968, **39**

A violent reaction occurred when cleaning lump metal under carbon tetrachloride [1]. Finely divided barium, slurried with trichlorotrifluoroethane, exploded during transfer owing to frictional initiation [2]. Granular barium in contact with fluorotrichloromethane, carbon tetrachloride, 1,1,2-trichlorotrifluoroethane, tetrachloroethylene or trichloroethylene is susceptible to detonation [3].

See other METALS: Halocarbons

Interhalogens

See BROMINE PENTAFLUORIDE, BrF_5: Acids, etc.
IODINE HEPTAFLUORIDE, F_7I: Metals
See other METALS

BARIUM TETRAFLUOROBROMATE **$BaBr_2F_8$**

See METAL POLYHALOHALOGENATES

BARIUM BROMATE **$BaBr_2O_6$**

Hackspill, L. *et al.*, *Compt. Rend.*, 1930, **191**, 663

Thermal decomposition with evolution of oxygen is almost explosive at 300°C.

See other METAL OXOHALOGENATES

BARIUM PERCHLORYLAMIDE $BaCl_2H_2N_2O_6$

See PERCHLORYLAMIDE SALTS

BARIUM HYDRIDE BaH_2

Metal halogenates

See METAL HALOGENATES: Metals, etc.

BARIUM HYDROXIDE BaH_2O_2

Chlorinated rubber

See CHLORINATED RUBBER: Metal oxides or hydroxides

BARIUM AMIDOSULPHATE $BaH_4N_2O_6S_2$

Metal nitrates or nitrites

See METAL AMIDOSULPHATES

BARIUM NITRATE BaN_2O_6

Aluminium–magnesium alloy

Tomlinson, W. R. *et al.*, *J. Chem. Educ.*, 1950, **27**, 606

An intimate mixture of the finely divided components, widely used as a photoflash composition, is readily ignitable and extremely sensitive to friction or impact.

See other METAL OXONION-METALLATES

BARIUM DIAZIDE BaN_6

1. Fagan, C. P., *J. and Proc. R. Inst. Chem.*, 1947, 126
2. Ficheroulle, H. *et al.*, *Mem. Poudres*, 1956, **33**, 7

This material is impact-sensitive when dry and is supplied and stored damp with alcohol. It is used as a saturated solution and it is important to prevent total evaporation of such solution, or the slow growth of large crystals which may become dried and shock-sensitive. Lead drains must not be used, to avoid formation of the detonator, lead azide. Exposure to acid conditions may generate explosive hydrazoic acid [1]. More recently it has been stated that barium azide is relatively insensitive to impact but highly sensitive to friction [2]. Strontium and, particularly, calcium azides show much more marked explosive properties than barium azide.

See other METAL AZIDES

BARIUM OXIDE BaO

Dinitrogen tetraoxide

See DINITROGEN TETRAOXIDE (NITROGEN DIOXIDE), N_2O_4: Barium oxide

Hydroxylamine

See HYDROXYLAMINE, H_3NO: Oxidants

Sulphur trioxide

See SULPHUR TRIOXIDE, O_3S: Metal oxides

Triuranium octaoxide

See TRIURANIUM OCTAOXIDE, O_8U_3: Barium oxide

BARIUM PEROXIDE BaO_2

Hydrogen sulphide

See HYDROGEN SULPHIDE, H_2S: Metal oxides

Hydroxylamine
See HYDROXYLAMINE, H_3NO: Oxidants

Organic materials,
Water
Koffolt., J. H., private comm., 1966
Contact of barium peroxide and water will readily produce a temperature and a local oxygen concentration high enough to ignite many organic compounds.

Peroxyformic acid
See PEROXYFORMIC ACID, CH_2O_3: Metals, etc.
See other METAL PEROXIDES

BARIUM SULPHATE BaO_4S

Aluminium
See ALUMINIUM, Al: Metal oxosalts

Phosphorus
See PHOSPHORUS, P: Metal sulphates

BARIUM SULPHIDE BaS

Dichlorine oxide
See DICHLORINE OXIDE, Cl_2O: Oxidisable materials

Oxidants
Mellor, 1941, Vol. 3, 745
Barium sulphide explodes weakly on heating with lead dioxide or potassium chlorate, and strongly with potassium nitrate. Calcium and strontium are similar.
See other METAL SULPHIDES

BERYLLIUM Be

Halocarbons

Pot. Incid. Rep., 1968, **39**
Mixtures of powdered beryllium with carbon tetrachloride or trichloroethylene will flash on heavy impact.
See other METALS: Halocarbons

Phosphorus
See PHOSPHORUS, P: Metals
See other METALS

BERYLLIUM PERCHLORATE $BeCl_2O_8$

Laran, R. J., US Pat. 3 157 464, 1964
A powerful oxidant, insensitive to heat or shock and useful in propellant and igniter systems.
See other METAL OXOHALOGENATES

BERYLLIUM FLUORIDE BeF_2

Magnesium
See MAGNESIUM, Mg: Beryllium fluoride

BERYLLIUM HYDRIDE BeH_2

Methanol,
or Water
Barbaras, G. D., *J. Amer. Chem. Soc.*, 1951, **73**, 48
Reaction of the ether-containing hydride with methanol or water is violent, even at $-196°C$.
See other METAL HYDRIDES

BERYLLIUM OXIDE BeO

Magnesium
See MAGNESIUM, Mg: Metal oxides

BISMUTH Bi

Aluminium
See ALUMINIUM, Al: Bismuth

Oxidants
See BROMINE PENTAFLUORIDE, BrF_5: Acids, etc.
PERCHLORIC ACID, $ClHO_4$: Bismuth
NITROSYL FLUORIDE, FNO: Metals
IODINE PENTAFLUORIDE, F_5I: Metals
NITRIC ACID, HNO_3: Metals
AMMONIUM NITRATE, $H_4N_2O_3$: Metals

BISMUTH PENTAFLUORIDE BiF_5

von Wartenberg, H., *Z. Anorg. Chem.*, 1940, **224**, 344
It reacts vigorously with water, sometimes with ignition.
See other METAL HALIDES

BISMUTHIC ACID $BiHO_3$

Hydrofluoric acid
Mellor, 1939, Vol. 9, 657
Interaction of the solid acid with 40% hydrofluoric acid solution is violent, ozonised oxygen being evolved.

BISMUTH AMIDE OXIDE BiH_2NO

Watt, G. W. *et al.*, *J. Amer. Chem. Soc.*, 1939, **61**, 1693

The solid, prepared in liquid ammonia, explodes when free of ammonia and exposed to air.
See other *N*-METAL DERIVATIVES

BISMUTH NITRIDE

BiN

Alone,
or Water
1. Fischer F. *et al.*, *Ber.*, 1910, **43**, 1471
2. Franklin, E. C., *J. Amer. Chem. Soc.*, 1905, **27**, 847

Very unstable, exploded on shaking [1] or heating, or in contact with water or dilute acids [2].
See other *N*-METAL DERIVATIVES

BISMUTH PLUTONIDE

BiPu

Williamson, G. K., *Chem. & Ind.*, 1960, 1384
Extremely pyrophoric.
See other ALLOYS (INTERMETALLIC COMPOUNDS)

DIBISMUTH TRIOXIDE

$\mathbf{Bi_2O_3}$

Chlorine trifluoride
See CHLORINE TRIFLUORIDE, ClF_3: Metals, etc.

Sodium
See SODIUM, Na: Metal oxides

DIBISMUTH TRISULPHIDE

$\mathbf{Bi_2S_3}$

Glatz, A. C. *et al.*, *J. Electrochem. Soc.*, 1963, **110**, 1231

Possible causes of explosions in direct synthesis are discussed.
See also SULPHUR,S: Metals
See other METAL SULPHIDES

BROMINE PERCHLORATE $BrClO_4$

Schack, C. J. *et al.*, *Inorg. Chem.*, 1971, **10**, 1078
Shock-sensitive.
See other HALOGEN OXIDES

CAESIUM HEXAFLUOROBROMATE(1−) $BrCsF_6$

See METAL POLYHALOHALOGENATES

BROMINE FLUORIDE BrF

Sidgwick, 1950, 1149
Chemically it behaves like the other bromine fluorides, but is more reactive.
See other INTERHALOGENS

PERBROMYL FLUORIDE $BrFO_3$

Polymers
Johnson, G. K. *et al.*, *Inorg. Chem.*, 1972, **11**, 800
It is considerably more reactive than perchloryl fluoride, and attacks glass and the usually inert polytetrafluoroethylene and polychlorotrifluoroethylene.
See other HALOGEN OXIDES

Davis, R. A. *et al.*, *J. Org. Chem.*, 1967, **32**, 3478
Musgrove, W. K. R., *Advan. Fluorine Chem.*, 1964, **1**, 12
The hazards and precautions involved in the use of this very reactive fluorinating agent are outlined. Contact with rubber, plastics or other organic materials may be explosively violent, and reaction with water is very vigorous.

Ammonium halides
Sharpe, A. G. *et al.*, *J. Chem. Soc.*, 1948, 2137
Explosive reaction.

Antimony trichloride oxide
Mellor, 1956, Vol. 2, Suppl. 1, 166
Interaction is violent, even more so than with diantimony trioxide.

Carbon monoxide
Mellor, 1956, Vol. 2, Suppl. 1, 166
At temperatures rather above 30°C, explosions occurred.

Halogens
or Metals,
or Non-metals,
or Organic materials
Mellor, 1941, Vol. 2, 113; 1956, Vol. 2, Suppl. 1, 164–167
'Chlorine Trifluoride Technical Bull.', Morristown, Baker and Adamson, 1970
Incandescence is caused by contact with bromine, iodine, arsenic, antimony (even at −10°C); powdered molybdenum, niobium, tantalum, titanium, vanadium; boron, carbon, phosphorus or sulphur. Carbon tetraiodide, chloromethane, benzene, ether ignite or explode on contact, as do organic materials generally.
See Uranium, below

2-Pentanone
Stevens, T. E., *J. Org. Chem.*, 1961, **26**, 1629, footnote 11
During evaporation of solvent hydrogen fluoride, an exothermic reaction between residual ketone and bromine trifluoride set in and accelerated to explosion.

Pyridine
Kirk-Othmer, 1966, Vol. 9, 592
The solid, produced by action of bromine trifluoride on pyridine in carbon tetrachloride, ignites when dry. 2-Fluoropyridine reacts similarly.

Silicone grease
Sharpe, A. G. *et al.*, *J. Chem. Soc.*, 1948, 2136
As it reacts explosively in bulk, the amount of silicone grease used on ground joints must be minimal.

Solvents
1. Sharpe, A. G. *et al.*, *J. Chem. Soc.*, 1948, 2135
2. Simons, J. H., *Inorg. Synth.*, 1950, **3**, 185

Bromine trifluoride explodes on contact with acetone or ether [1], and the frozen solid at –80°C reacts violently with toluene at –80°C.
See Halogens, etc., above

Tin dichloride
Mellor, 1956, Vol. 2, Suppl. 1, 165
Contact causes ignition.

Uranium,
Uranium hexafluoride
Johnson, R. *et al.*, *6th Nucl. Eng. Sci. Conf., New York, 1960,* Reprint Paper No. 23
Uranium may ignite or explode during dissolution in bromine trifluoride, particularly when high concentrations of the hexafluoride are present. Causative factors are identified.
See Halogens, etc., above

Water
Mellor, 1941, Vol. 2, 113
Interaction is violent, oxygen being evolved.
See other INTERHALOGENS

BROMINE PENTAFLUORIDE BrF_5

Acids,
or Halogens,
or Metal halides,
or Metals,
or Non-metals,
or Oxides
Mellor, 1956, Vol. 2, Suppl. 1, 172
Sidgwick, 1950, 1158
Contact with the following at ambient or slightly elevated temperatures is violent, ignition often occurring: strong nitric and sulphuric acids; chlorine (explodes on heating), iodine; ammonium chloride, potassium iodide; antimony, arsenic, boron powder, selenium, tellurium; aluminium powder, barium, bismuth, cobalt powder, chromium, iridium powder, iron powder, lithium powder, manganese, molybdenum, nickel powder, rhodium powder, tungsten, zinc; charcoal, red phosphorus, sulphur; arsenic pentoxide, boron trioxide, calcium oxide, carbon monoxide, chromium trioxide, iodine pentoxide, magnesium oxide, molybdenum trioxide, phosphorus pentoxide, sulphur dioxide, tungsten trioxide.

Hydrogen-containing materials
1. Mellor, 1956, Vol. 2, Suppl. 1., 172
2. Braker, 1971, 49
Contact with the following materials, containing combined hydrogen, is likely to cause fire or explosion : acetic acid, ammonia, benzene, ethanol, hydrogen, hydrogen sulphide, methane; cork, grease, paper, wax, etc. The carbon content further contributes to the reactivity observed [1]. Chloromethane reacts with explosive violence [2].

Water
Sidgwick, 1950, 1158
Contact with water causes a violent reaction or explosion, oxygen being evolved.
See other INTERHALOGENS

POTASSIUM HEXAFLUOROBROMATE (1−) BrF_6K

See METAL POLYHALOHALOGENATES

RUBIDIUM HEXAFLUOROBROMATE (1–) BrF_6Rb

See METAL POLYHALOHALOGENATES

HYDROGEN BROMIDE BrH

Preparative hazard.
See BROMINE, Br_2 : Phosphorus

1,2-Diaminoethaneamminediperoxochromium(IV)
See 1,2-DIAMINOETHANEAMMINEDIPEROXOCHROMIUM(IV), $C_2H_{11}CrN_3O_4$: Hydrogen bromide

Fluorine
See FLUORINE, F_2 : Hydrogen halides

Ozone
See OZONE, O_3: Hydrogen bromide

BROMIC ACID $BrHO_3$

In contact with oxidisable materials, reactions are similar to those of the metal bromates.
See METAL HALOGENATES

BROMOAMINE BrH_2N

Jander, J. *et al.*, *Z. Anorg. Chem.*, 1958, **296**, 117
The isolated material decomposes violently at –70°C, while an ethereal solution is stable for a few hours at that temperature.
See other *N*-HALOGEN COMPOUNDS

BROMOSILANE BrH_3Si

Ward, L. G. L., *Inorg. Synth.*, 1968, **11**, 161
Ignites in air (gas above 2°C).
See other HALOSILANES

AMMONIUM BROMIDE BrH_4N

Bromine trifluoride
See BROMINE TRIFLUORIDE, BrF_3 : Ammonium halides

AMMONIUM BROMATE BrH_4NO_3

Sidgwick, 1950, 1227
An explosive salt.
See also LEAD BROMATE, Br_2O_6Pb
See other OXOSALTS OF NITROGENOUS BASES

BROMINE AZIDE BrN_3

Alone,
or Arsenic,
or Metals,
or Phosphorus
Mellor, 1940, Vol. 8, 336
The solid, liquid and vapour are all very shock-sensitive, and the liquid explodes on contact with arsenic; sodium, silver foil; or phosphorus. Concentrated solutions in organic solvents may explode on shaking.
See other HALOGEN AZIDES

SODIUM BROMATE $BrNaO_3$

Fluorine
See FLUORINE, F_2 : Sodium bromate

Grease
MCA Case History No. 874
A bearing assembly from a sodium bromate crusher had been degreased at 120°C, and, while still hot, the sleeve was hammered to free it. The assembly exploded violently probably because of the presence of a hot mixture of sodium bromate and a grease component (possibly a sulphurised derivative). It is known that mixtures of bromates and organic or sulphurous matter are heat- and friction-sensitive.
See other METAL OXOHALOGENATES

BROMINE DIOXIDE **BrO_2**

Brauer, 1963, Vol. 1, 306
Unstable unless stored at low temperatures, it may explode if heated rapidly.
See other HALOGEN OXIDES

BROMINE TRIOXIDE **BrO_3**

1. Lewis, B. *et al.*, *Z. Elektrochem.*, 1929, **35**, 648–652
2. Pflugmacher, A. *et al.*, *Z. Anorg. Chem.*, 1955, **279**, 313

The solid produced at –5°C by interaction of bromine and ozone is only stable at –80°C or in presence of ozone, and decomposition may be violently explosive in presence of trace impurities [1]. The structure may be the dimeric bromyl perbromate, analogous to Cl_2O_6 [2].
See other HALOGEN OXIDES

BROMINE **Br_2**

MCA SD-49, 1968

Acetone
Levene, P. A., *Org. Synth.,* 1943, Coll. Vol. 2, 89

During bromination of acetone to bromoacetone, presence of a large excess of bromine must be avoided to prevent a sudden very violent reaction.

Acrylonitrile

See ACRYLONITRILE, C_3H_3N:Bromine

Ammonia

Mellor, **1967**, Vol. 8, Suppl. 2, 417

Interaction at normal or elevated temperatures, followed by cooling to –95°C, gives an explosive red oil.

See NITROGEN TRIBROMIDE HEXAAMMONIATE, $Br_3N \cdot H_{18}N_6$

Bromine trifluoride

See BROMINE TRIFLUORIDE, BrF_3 : Halogens, etc.

Copper(I) hydride

See COPPER(I) HYDRIDE, CuH: Halogens

Diethyl ether

Tucker, H., private comm., 1972

Shortly after adding bromine to ether the solution erupted violently (or exploded softly).

See also CHLORINE, Cl_2 : Diethyl ether

N,N-Dimethylformamide

Tayim, H. A. *et al.*, *Chem. & Ind.*, 1973, 347

Interaction is extremely exothermic, and under confinement in an autoclave the internal temperature and pressure exceeded 100°C and 135 bar, causing failure of the bursting disc. The product of interaction is hydroxymethylenedimethylammonium bromide, and the explosive decomposition may have involved formation of *N*-bromodimethylamine, carbon monoxide and hydrogen bromide.

See other *N*-HALOGEN COMPOUNDS

Ethanol,

Phosphorus

Read, C. W. W., *School Sci. Rev.*, 1940, **21**(83), 967

Interaction of ethanol, phosphorus and bromine to give bromoethane is considered too dangerous for a school experiment.

Fluorine

See FLUORINE, F_2 : Halogens

Germane

See Non-metal hydrides below

Hydrogen

Mellor, 1956, Vol. 2, Suppl. 1, 707

Combination is explosive under appropriate pressure and temperature conditions.

Metal acetylides and carbides

Several of the mono- and di-alkali metal acetylides and copper acetylides ignite at ambient temperature or on slight warming, with either liquid or vapour. The alkaline earth, iron, uranium and zirconium carbides ignite in the vapour on heating.

See TRIIRON CARBIDE, CFe_3 : Halogens
DICAESIUM ACETYLIDE, C_2Cs_2 : Halogens
COPPER ACETYLIDE, C_2Cu: Halogens
DICOPPER(I) ACETYLIDE, C_2Cu_2 : Halogens
CAESIUM ACETYLIDE, C_2HCs
RUBIDIUM ACETYLIDE, C_2HRb
DILITHIUM ACETYLIDE, C_2Li_2 : Halogens
DIRUBIDIUM ACETYLIDE, C_2Rb_2: Halogens
STRONTIUM ACETYLIDE, C_2Sr: Halogens
URANIUM DICARBIDE, C_2U: Halogens
ZIRCONIUM DICARBIDE, C_2Zr: Halogens

Metal azides

Mellor, 1940, Vol. 8, 336

Nitrogen-diluted bromine vapour passed over silver or sodium azide formed bromine azide, and often caused explosions.

Metals

1. Staudinger, H., *Z. Elektrochem.*, 1925, **31**, 549
2. Mellor, 1941, Vol. 2, 469; 1963, Vol. 2, Suppl. 2.2, 1563, 2174
3. *MCA SD-49*, 1968

Lithium is stable in contact with dry bromine, but heavy impact will initiate explosion, while sodium in contact with bromine needs only moderate impact for initiation [1]. Potassium ignites in bromine vapour and explodes violently in contact with liquid bromine, and rubidium ignites in bromine vapour [2]. Aluminium, mercury and titanium react violently with dry bromine [3].

See GALLIUM, Ga: Halogens

Methanol

1. Muir, G. D., *Chem. Brit.*, 1972, **8**, 136
2. Bush, E. L., private comm., 1968

Reaction may be vigorously exothermic. A mixture of bromine (9 ml) and methanol (15 ml) boiled in 2 min and in a previous incident such a mixture had erupted from a measuring cylinder [1]. The exotherm with industrial ethanol (containing 5% methanol) is much greater, and addition of 10 ml of bromine to 40 ml of IMS rapidly causes violent boiling [2].

Ozone

See OZONE, O_3: Bromine

Non-metal hydrides

1. Stock, A. *et al.*, *Ber.*, 1917, **50**, 1739
2. Sujishi, S. *et al.*, *J. Amer. Chem. Soc.*, 1954, **76**, 4631
3. Geisler, T. C. *et al.*, *Inorg. Chem.*, 1972, **11**, 1710

Interaction of silane and its homologues with bromine at ambient temperature is explosively violent [1] and temperatures of below –30°C are necessary to avoid ignition of the reactants [2]. Ignition of disilane at –95°C and of germane at –112°C emphasises the need for good mixing to dissipate the large exotherm [3].

See also ETHYLPHOSPHINE, C_2H_7P: Halogens
PHOSPHINE, H_3P: Halogens

Oxygen difluoride

See OXYGEN DIFLUORIDE, F_2O: Halogens

Phosphorus

1. Bandar, L. S. *et al.*, *Zh. Prikl. Khim.*, 1966, **39**, 2304
2. 'Leaflet No. 2', Inst. of Chem., London, 1939

During preparation of hydrogen bromide by addition of bromine to a suspension of red phosphorus in water, the latter must be freshly prepared to avoid possibility of explosion. This is due to formation of peroxides in the suspension on standing and subsequent thermal decomposition [1]. In the earlier description of such an explosion,

action of bromine on boiling tetralin was preferred to generate hydrogen bromide [2], which is now also available in cylinders.
See also PHOSPHORUS, P: Halogens

Tetracarbonylnickel
See TETRACARBONYLNICKEL, C_4NiO_4 : Bromine

Tetraselenium tetranitride
See TETRASELENIUM TETRANITRIDE, N_4Se_4

Trialkylboranes
Coates, 1967, Vol. 1, 199
The lower homologues tend to ignite in bromine or chlorine.

Trioxygen difluoride
See TRIOXYGEN DIFLUORIDE, F_2O_3
See other HALOGENS

CALCIUM BROMIDE Br_2Ca

Potassium
See POTASSIUM, K: Metal halides

COBALT(II) BROMIDE Br_2Co

Sodium
See SODIUM, Na: Metal halides

IRON(II) BROMIDE Br_2Fe

Potassium
See POTASSIUM, K: Metal halides

Sodium
See SODIUM, Na: Metal halides

N,N-BIS(BROMOMERCURI)HYDRAZINE $Br_2H_2Hg_2N_2$

Hofmann, K. A. *et al., Ann.*, 1899, **305**, 217
An explosive compound.
See other *N*-METAL DERIVATIVES

MERCURY(II) BROMIDE Br_2Hg

Indium
Clark, R. J. *et al., Inorg. Synth.*, 1963, **7**, 19–20
Interaction at 350°C is so vigorous that it is unsafe to increase the scale of this preparation of indium bromide.
See other METAL HALIDES

SELENINYL BROMIDE Br_2OSe

Metals
Mellor, 1947, Vol. 10, 912
Sodium and potassium react explosively (the latter more violently), and zinc dust ignites, in contact with the liquid bromide.
See SODIUM, Na: Non-metal halides

Phosphorus
Mellor, 1947, Vol. 10, 912
Red phosphorus ignites, and white phosphorus explodes in contact with the liquid bromide.
See other NON-METAL HALIDES

LEAD BROMATE Br_2O_6Pb

Sidgwick, 1950, 1227
An explosive salt.
See also AMMONIUM BROMATE, BrH_4NO_3
LEAD ACETATE-BROMATE, $C_2H_3BrO_5Pb$
See other METAL OXOHALOGENATES

ZINC BROMATE Br_2O_6Zn

See METAL HALOGENATES

SULPHUR DIBROMIDE Br_2S

Nitric acid
See NITRIC ACID, HNO_3: Sulphur halides

Sodium
See SODIUM, Na: Non-metal halides

DISULPHUR DIBROMIDE Br_2S_2

Metals
Mellor, 1947, Vol. 10, 652
Thin sections of potassium or sodium usually ignite and incandesce in the liquid bromide. Iron at about 650°C ignites and incandesces in the vapour.
See other NON-METAL HALIDES

POLYDIBROMOSILANE $(Br_2Si)n$

Oxidants
Brauer, 1963, Vol. 1, 688
Ignites in air at 120°C and reacts explosively with oxidants (e.g. nitric acid).
See HALOSILANES
See other NON-METAL HALIDES

IRON(III) BROMIDE Br_3Fe

Potassium
See POTASSIUM, K: Metal halides

Sodium
See SODIUM, Na: Metal halides

TRIBROMOSILANE Br_3HSi

Schumb, W. C., *Inorg. Synth.*, 1939, 1, 42
It usually ignites when poured in air.
See HALOSILANES

NITROGEN TRIBROMIDE HEXAMMONIATE $Br_3N{\cdot}H_{18}N_6$

Mellor, 1967, Vol. 8, Suppl. 2, 417; 1940, Vol. 8, 605
The compound ($NBr_3{\cdot}6NH_3$) prepared by condensation of its vapour at –95°C explodes suddenly at –67°C. Prepared in another way, it is stable under water but explodes violently in contact with phosphorus or arsenic.
See other N-HALOGEN COMPOUNDS

NITROSYL TRIBROMIDE Br_3NO

Sodium–antimony alloy
Mellor, 1940, Vol. 8, 621
Powdered sodium antimonide ignites when dropped into the vapour.
See other N-HALOGEN COMPOUNDS

PHOSPHORUS TRIBROMIDE Br_3P

Mellor, 1940, Vol. 8, 1032
Interaction with warm water is very rapid and may be violent with limited quantities.
See SODIUM, Na: Non-metal halides

Potassium
See POTASSIUM, K: Non-metal halides

Sodium,
Water
See SODIUM, Na: Non-metal halides
See other NON-METAL HALIDES

TUNGSTEN AZIDE PENTABROMIDE **Br_5N_3W**

Extremely explosive.
See other METAL AZIDE HALIDES

CARBON **C**

Air
1. *Fire Prot. Assoc. J.*, 1964, 337
2. Cameron, A. *et al.*, *J. Appl. Chem.*, 1972, **22**, 1007

Activated carbon is a potential fire hazard because of its very high surface area and adsorptive capacity. Freshly prepared material may heat spontaneously in air, and presence of water accelerates this. Spontaneous heating and ignition may occur if contamination by drying oils or oxidising agents occurs [1]. The spontaneous heating effect has now been related to the composition and method of preparation of activated carbon, and the relative hazards may be readily assessed [2].

Metals
See POTASSIUM, K: Carbon
SODIUM, Na: Non-metals

Metal salts
MCA Case History No. 1094
Dry metal-impregnated charcoal catalyst was being added from a polythene bag to an aqueous solution under nitrogen. Static so generated ignited the charcoal dust and caused a flash fire. The risk

was eliminated by adding a slurry of catalyst from a metal container.
See COBALT(II) NITRATE, CoN_2O_6 : Carbon

Oxidants

Carbon has been frequently involved in hazardous reactions, particularly finely divided or high-porosity forms exhibiting a high ratio of surface area to mass (up to 2000 m^2/g). It then functions as an unusually active fuel which possesses adsorptive and catalytic properties to accelerate the rate of energy release involved in combustion reactions with virtually any oxidant.

Less active forms of carbon will ignite or explode on suitably intimate contact with oxygen, oxides, peroxides, oxosalts, halogens, interhalogens and other oxidising species. Individual combinations are:

SILVER NITRATE, $AgNO_3$: Non-metals
BROMINE TRIFLUORIDE, BrF_3 : Halogens, etc.
BROMINE PENTAFLUORIDE, BrF_5 : Acids, etc.
PEROXYFORMIC ACID, CH_2O_3 : Non-metals
CHLORINE TRIFLUORIDE, ClF_3 : Metals, etc.
AMMONIUM PERCHLORATE, ClH_4NO_4 : Carbon
DICHLORINE OXIDE, Cl_2O: Carbon
: Oxidisable materials
COBALT(II) NITRATE, CoN_2O_6 : Carbon
FLUORINE, F_2 : Non-metals
OXYGEN DIFLUORIDE, F_2O: Non-metals
TRIOXYGEN DIFLUORIDE, F_2O_3
NITROGEN TRIFLUORIDE, F_3N: Charcoal
IODINE HEPTAFLUORIDE, F_7I: Carbon
IODINE(V) OXIDE, I_2O_5 : Non-metals
POTASSIUM PERMANGANATE, $KMnO_4$: Non-metals
POTASSIUM NITRATE, KNO_3 : Non-metals
POTASSIUM DIOXIDE, KO_2 : Carbon
SODIUM NITRATE, $NNaO_3$: Non-metals
NITROGEN OXIDE ('NITRIC OXIDE'), NO: Non-metals
ZINC NITRATE, N_2O_6Zn: Carbon
SODIUM PEROXIDE, Na_2O_2 : Non-metals
SODIUM SULPHIDE, Na_2S: Carbon
OXYGEN (Liquid), O_2 : Carbon

Unsaturated oils

Bahme, C. W., *NFPA Quart.*, 1952, **45**, 431
von Schwartz, 1918, 326

Unsaturated (drying) oils, like linseed oil, etc., will rapidly heat and ignite when distributed on active carbon, owing to enormous increase

in surface area of the oil exposed to air. A similar, but slower, effect occurs on fibrous material such as cotton waste.
See other NON-METALS

SILVER CYANIDE **CAgN**

Phosphorus tricyanide
See PHOSPHORUS TRICYANIDE, C_3N_3P

Fluorine
See FLUORINE, F_2: Metal salts

SILVER TRINITROMETHANIDE **$CAgN_3O_6$**

Witucki, E. F. *et al.*, *J. Org. Chem.*, 1972, **37**, 152
The explosive silver salt may be replaced with advantage by the potassium salt in preparation of 1,1,1-trinitroalkanes.
See other HEAVY METAL DERIVATIVES

DISILVER CYANAMIDE **CAg_2N_2**

Chrétien, A. *et al.*, *Compt. Rend.*, 1951, **232**, 1114
During pyrolysis to silver (via silver dicyanamide), initial heating must be slow to avoid explosion.
See other *N*-METAL DERIVATIVES

GOLD(I) CYANIDE **CAuN**

Magnesium
See MAGNESIUM, Mg: Metal cyanides

BROMOTRICHLOROMETHANE $CBrCl_3$

Ethylene
See ETHYLENE, C_2H_4: Bromotrichloromethane

BROMOTRIFLUOROMETHANE $CBrF_3$

Aluminium
See ALUMINIUM, Al: Halocarbons

CARBON TETRABROMIDE CBr_4

Lithium
See LITHIUM, Li: Halocarbons

CALCIUM CARBONATE $CCaO_3$

Fluorine
See FLUORINE, F_2: Metal salts

TRIFLUOROMETHANESULPHENYL CHLORIDE $CClF_3S$

Hexafluoroisopropylideneaminolithium
See HEXAFLUOROISOPROPYLIDENEAMINOLITHIUM, C_3F_6LiN: Non-metal halides

CHLOROSULPHONYLISOCYANATE $CClNO_3S$

Water
Graf, R., *Org. Synth.*, 1966, **46**, 25
Interaction is violent.

DICHLORODIFLUOROMETHANE CCl_2F_2

Aluminium

See ALUMINIUM, Al: Halocarbons

CARBON DICHLORIDE (PHOSGENE) CCl_2O

MCA SD-95, 1967

Hexafluoroisopropylideneaminolithium

See HEXAFLUOROISOPROPYLIDENEAMINOLITHIUM, C_3F_6LiN: Non-metal halides

Potassium

See POTASSIUM, K: Non-metal halides

TRICHLOROFLUOROMETHANE CCl_3F

Metals

See ALUMINIUM, Al: Halocarbons
BARIUM, Ba: Halocarbons
LITHIUM, Li: Halocarbons
See other METALS: Halocarbons

TRICHLORONITROMETHANE (CHLOROPICRIN) CCl_3NO_2

Anon., *Chem. Eng. News.*, 1972, **50**(38), 13

Tests have shown that, above a critical volume, bulk containers of chloropicrin can be shocked into detonation. Containers below 700 kg content will now be the maximum size as against rail-tanks previously used.

Aniline

Jackson, K. E., *Chem. Rev.*, 1934, **14**, 269

Reaction at 145°C with excess aniline is violent.

Sodium hydroxide
Scholz, S., *Explosivstoffe,* 1963, **11**, 159, 181
During destruction of chemical warfare ammunition, pierced shells containing chloropicrin reacted violently with alcoholic sodium hydroxide.

Sodium methoxide
Ramsey, B. G. *et al.*, *J. Amer. Chem. Soc.*, 1966, **88**, 3059
During addition of the nitro compound in methanol to sodium methoxide solution, the temperature must not be allowed to fall much below 50°C. If this happens, excess nitro compound will accumulate and cause a violent and dangerous exotherm.
See other C-NITRO COMPOUNDS

CARBON TETRACHLORIDE CCl_4

MCA SD-3, 1963

Aluminium trichloride,
Triethylaluminium
See TRIETHYLDIALUMINIUM TRICHLORIDE, $C_6H_{15}Al_2Cl_3$: Carbon tetrachloride

Calcium disilicide
See CALCIUM DISILICIDE, $CaSi_2$: Carbon tetrachloride

Chlorine trifluoride
See CHLORINE TRIFLUORIDE, ClF_3 : Carbon tetrachloride

Decaborane(14)
See DECABORANE(14), $B_{10}H_{14}$: Ethers, etc.

Dibenzoyl peroxide,
Ethylene
See ETHYLENE, C_2H_4 : Carbon tetrachloride
DIBENZOYL PEROXIDE, $C_{14}H_{10}O_4$: Carbon tetrachloride, Ethylene
See also WAX FIRE

N,N-Dimethylformamide

'DMF Brochure', Billingham, ICI, 1965

There is a potentially dangerous reaction of carbon tetrachloride with dimethylformamide in presence of iron. The same occurs with 1,2,3,4,5,6-hexachlorocyclohexane, but not with dichloromethane or 1,2-dichloroethane under the same conditions.

Dinitrogen tetraoxide

See DINITROGEN TETRAOXIDE (NITROGEN DIOXIDE), N_2O_4: Halocarbons

Fluorine

See FLUORINE, F_2: Halocarbons

Metals

See ALUMINIUM, Al: Halocarbons
BARIUM, Ba: Halocarbons
BERYLLIUM, Be : Halocarbons
POTASSIUM, K: Halocarbons
POTASSIUM–SODIUM ALLOY, K–Na: Halocarbons
SODIUM, Na: Halocarbons
ZINC, Zn: Halocarbons

Potassium *tert*-butoxide

See POTASSIUM *tert*-BUTOXIDE, C_4H_9KO: Acids, etc.

1,11-Diamino-3,6,9-triazundecane ('tetraethylenepentamine')

1. Hudson, F. L., private comm., 1973
2. Collins, R. F., *Chem. & Ind.*, 1957, 704

A mixture erupted violently one hour after preparation [1]. Interaction (not vigorous) of amines and halocarbons at ambient temperature had been previously recorded [2]. The presence of five basic centres in the amine would be expected to exaggerate exothermic effects.

See other HALOCARBONS

TRICHLOROMETHYL PERCHLORATE CCl_4O_4

Sidgwick, 1950, 1236
An extremely explosive liquid, only capable of preparation in minute amounts.
See other ALKYL PERCHLORATES

TRIAZIDOMETHYLIUM HEXACHLOROANTIMONATE CCl_6N_9Sb

Müller, U. *et al., Angew. Chem.*, 1966, **78**, 825
The salt containing the $C(N_3)_3$ cation is sensitive to shock or rapid heating.
See other ORGANIC AZIDES

FLUORODINITROMETHYL AZIDE CFN_5O_4

Unstable at ambient temperature.
See other FLUORODINITROMETHYL COMPOUNDS
ORGANIC AZIDES

CARBONYL DIFLUORIDE CF_2O

Hexafluoroisopropylideneaminolithium
See HEXAFLUOROISOPROPYLIDENEAMINOLITHIUM, C_3F_6LiN: Non-metal halides

TRIFLUOROMETHANESULPHONYL AZIDE $CF_3N_3O_2S$

Cavender, C. J. *et al., J. Org. Chem.*, 1972, **37**, 3568
An explosion occurred when the azide separated during its preparation in absence of solvent.
See other ACYL AZIDES

CARBON TETRAFLUORIDE CF_4

Aluminium
See ALUMINIUM, Al: Halocarbons

TETRAFLUORODIAZIRIDINE CF_4N_2

Firth, W. C., *J. Org. Chem.*, 1968, **33**, 3489
Explodes on evaporation or condensation at low temperatures. Treat as a powerful explosive and handle on small scale.
See other N-HALOGEN COMPOUNDS

PERFLUOROUREA CF_4N_2O

Acetonitrile
Fraser, G. W. *et al., Chem. Comm.*, 1966, 532
The solution of the perfluoro compound in acetonitrile prepared at −40°C must not be kept at room temperature, since difluorodiazene is formed.
See ACETONITRILE, C_2H_3N: *N*-Fluoro compounds

TRIFLUOROMETHYL HYPOFLUORITE CF_4O

Hydrocarbons
Allison, J. A. C. *et al., J. Amer. Chem. Soc.*, 1959, **81**, 1089
In absence of nitrogen as diluent, interaction with acetylene, cyclopropane or ethylene is explosive on mixing.

Lithium
Porter, R. S. *et al., J. Amer. Chem. Soc.*, 1957, **79**, 5625
Interaction set in at about 170°C with a sufficient exotherm to melt the glass container.

Polymers

Barton, D. H. R. *et al., J. Org. Chem.*, 1972, **37**, 329

It is a powerful oxidant and only all-glass apparatus, free of polythene, PVC, rubber or similar elastomers should be used. Appreciable concentrations of the gas in oxidisable materials should be avoided.

See other HYPOHALITES

TRIFLUOROMETHANESULPHINYL FLUORIDE CF_4OS

Hexafluoroisopropylideneaminolithium

See HEXAFLUOROISOPROPYLIDENEAMINOLITHIUM, C_3F_6LiN: Non-metal halides

DIFLUOROMETHYLENE DIHYPOFLUORITE CF_4O_2

Haloalkenes

Hohorst, F. A. *et al., Inorg. Chem.*, 1968, **7**, 624

Attempts to react the fluoro compound with *trans*-dichloroethylene or tetrafluoroethylene at room temperature in absence of diluent caused violent explosions. The title compound should not be allowed to contact organic or easily oxidised material without adequate precautions.

See other HYPOHALITES

XENON(II) FLUORIDE TRIFLUOROACETATE CF_4O_2Xe

Jha, N. K., *RIC Rev.*, 1971, **4**, 157

Explodes on thermal or mechanical shock.

See other XENON COMPOUNDS

XENON(II) FLUORIDE TRIFLUOROMETHANESULPHONATE CF_4O_3SXe

Wechsberg, M. *et al., Inorg. Chem.*, 1972, **11**, 3066
Unless a deficiency of xenon difluoride was used in the preparation at 0°C or below, the product exploded violently on warming to ambient temperature.
See other XENON COMPOUNDS

3-DIFLUOROAMINO-1,2,3-TRIFLUORODIAZIRIDINE CF_5N_3

Firth, W. C., *J. Org. Chem.*, 1968, **33**, 3489
Explodes on evaporation or condensation at low temperatures. Treat as a powerful explosive and handle on small scale.
See other N-HALOGEN COMPOUNDS

PENTAFLUOROGUANIDINE CF_5N_3

Zollinger, J. L. *et al., J. Org. Chem.*, 1973, **38**, 1070–1071
This, and several of its adducts with alcohols, are shatteringly explosive compounds, frequently exploding during phase changes at low temperatures, or on friction or impact.
See other N-HALOGEN COMPOUNDS

BIS(DIFLUOROAMINO)DIFLUOROMETHANE CF_6N_2

Koshar, R. J. *et al., J. Org. Chem.*, 1966, **31**, 4233
Explosions occurred during handling of this material, especially during phase transitions. Use of protective equipment is recommended for preparation, handling and storage, even on microscale.
See other N-HALOGEN COMPOUNDS

IRON CARBIDE CFe_3

Halogens

Mellor, 1946, Vol. 5, 898

Incandesces in chlorine below 100°C, and in bromine at 100°C.

See other METAL NON-METALLIDES

SILVER TETRAZOLIDE $CHAgN_4$

Thiele, J., *Ann.*, 1892, **270**, 59

It explodes on heating.

See other *N*-METAL DERIVATIVES

BROMOFORM $CHBr_3$

Acetone,
Potassium hydroxide

1. Willgerodt, C., *Ber.*, 1881, **14**, 2451
2. Weizmann, C. *et al.*, *J. Amer. Chem. Soc.*, 1948, **70**, 1189

Interaction in presence of powdered potassium hydroxide (or other bases) is violently exothermic, even in presence of diluting solvents [1, 2].

See also CHLOROFORM, $CHCl_3$: Acetone, etc.

Metals

See POTASSIUM, K: Halocarbons
LITHIUM, Li: Halocarbons

CHLORODIFLUOROMETHANE $CHClF_2$

Aluminium

See ALUMINIUM, Al: Halocarbons

TETRAZOLE-5-DIAZONIUM CHLORIDE $CHClN_6$

See 5-AMINOTETRAZOLE, CH_3N_5
See other DIAZONIUM SALTS

1-DICHLOROAMINOTETRAZOLE $CHCl_2N_5$

Karrer, 1950, 804
1-Dichloroaminotetrazole and its 5-derivatives are extremely explosive, as expected in an *N,N*-dichloro derivative of a high-nitrogen nucleus.
See other *N*-HALOGEN COMPOUNDS

CHLOROFORM $CHCl_3$

MCA SD-89, 1962
Acetone,
Alkali

1. Willgerodt, C., *Ber.,* 1881, **14**, 258
2. King, H. K., *Chem. & Ind.,* 1970, 185
3. Ekely, J. B. *et al., J. Amer. Chem. Soc.,* 1924, **46**, 1253
4. Grant, D. H., *Chem. & Ind.,* 1970, 919

Chloroform and acetone interact vigorously and exothermally in presence of potassium hydroxide or calcium hydroxide to form 1,1,1-trichloro-2-hydroxy-2-methylpropane [1], and a laboratory solvent-residues explosion was attributed to this cause [2]. No reaction occurs in absence of base [1], and other haloforms and ketones also react similarly in presence of base [3]. A minor explosion (or sudden boiling) of a chloroform–acetone mixture in new glassware may have been caused by surface alkali [4].
See also BROMOFORM, $CHBr_3$: Acetone, etc.

Dinitrogen tetraoxide
See DINITROGEN TETRAOXIDE (NITROGEN DIOXIDE), N_2O_4: Halocarbons

Fluorine
See FLUORINE, F_2: Halocarbons

Metals

See ALUMINIUM, Al: Halocarbons
LITHIUM, Li: Halocarbons
SODIUM, Na: Halocarbons

Potassium *tert*-butoxide

See POTASSIUM *tert*-BUTOXIDE, C_4H_9KO: Acids, etc.
See also Sodium methoxide, below

Sodium,
Methanol
Unpublished information, 1948
During attempted preparation of trimethyl orthoformate, addition of sodium to an inadequately cooled chloroform–methanol mixture caused a violent explosion.
See Sodium methoxide, below

Sodium hydroxide,
Methanol
MCA Case History No. 498
A chloroform–methanol mixture was put into a drum contaminated with sodium hydroxide. A vigorous reaction set in, and the drum exploded. Chloroform normally reacts slowly with sodium hydroxide owing to the insolubility of the latter. The presence of methanol (or other solubiliser) increases the rate of reaction by increasing the degree of contact between chloroform and alkali.

Sodium methoxide
1. *MCA Case History No. 693*
2. Kaufmann, W. E. *et al., Org. Synth.,* 1944, Coll. Vol. 1, 258
For the preparation of methyl orthoformate, solid sodium methoxide, methanol and chloroform were mixed together. The mixture boiled violently and then exploded [1]. The analogous preparation of ethyl orthoformate [2] involves the slow addition of sodium or sodium ethoxide solution to a chloroform–ethanol mixture. The explosion was caused by the addition of the sodium methoxide in one portion.
See Sodium, above

Triisopropylphosphine
See TRIISOPROPYLPHOSPHINE, $C_9H_{21}P$: Chloroform
See other HALOCARBONS

FLUORODINITROMETHANE $CHFN_2O_4$

Potentially explosive.
See FLUORODINITROMETHYL COMPOUNDS

TRIFLUOROMETHANESULPHONIC ACID CHF_3O_3S

Acyl chlorides,
Aromatic hydrocarbons
Effenberger, F. *et al., Angew. Chem. (Intern. Ed.)*, 1972, **11**, 300
Addition of catalytic amounts (1%) of trifluoromethanesulphonic acid to mixtures of acyl chlorides and aromatic hydrocarbons causes more or less violent evolution of hydrogen chloride, depending on reactivity of the components.

IODOFORM CHI_3

Acetone
See CHLOROFORM, $CHCl_3$: Acetone, etc.

DIAZOMETHYLLITHIUM $CHLiN_2$

Müller, E. *et al., Chem. Ber.*, 1954, **87**, 1887
Alkali metal salts of diazomethane are very explosive when exposed to air in the dry state, and should be handled, preferably wet with solvent, under an inert atmosphere.
See other DIAZO COMPOUNDS

HYDROGEN CYANIDE CHN

Fl.P., –18°C (gas above 25.7°C); E.L., 6.0–41%

MCA SD-67, 1961
Gause, E. H. *et al., J. Chem. Eng. Data,* 1960, **5**, 351
Anhydrous hydrogen cyanide is stable at or below room temperature if inhibited with acid (e.g. 0.1% sulphuric acid). In absence of inhibitor, exothermic polymerisation occurs, and if temperature attains 184°C, explosively rapid polymerisation occurs.
See also MERCURY(II) CYANIDE, C_2HgN_2: Hydrogen cyanide

DIAZOMETHYLSODIUM CHN_2Na

See DIAZOMETHYLLITHIUM, $CHLiN_2$

TRINITROMETHANE ('NITROFORM') CHN_3O_6

Marans, N. S. *et al., J. Amer. Chem. Soc.,* 1950, **72**, 5329
Explosions occurred during distillation of this polynitro compound.

Divinyl ketone
Graff, M. *et al., J. Org. Chem.,* 1968, **33**, 1247
One attempted reaction of trinitromethane with impure ketone caused an explosion at refrigerator temperature.

2-Propanol
MCA Case History No. 1010
Frozen mixtures of trinitromethane–2-propanol (9:1) exploded during thawing. The former dissolves exothermally in the alcohol, the heat effect increasing directly with the concentration above 50% w/w. Traces of nitric acid may also have been present.
See other POLYNITROALKYL COMPOUNDS

AZIDODITHIOFORMIC ACID CHN_3S_2

1. Mellor, 1947, Vol. 8, 338
2. Smith, G. B. L., *Inorg. Synth.,* 1939, **1**, 81

The isolated solid or its salts are shock- and heat-sensitive explosives [1]. Safe preparative procedures have been detailed. The heavy metal salts, though powerful detonators, are too sensitive for practical use [2].

See CARBON DISULPHIDE, CS_2 : Metal azides
See also BIS(AZIDOTHIOCARBONYL)DISULPHIDE, $C_2N_6S_4$
See other ACYL AZIDES

5-NITROTETRAZOLE **CHN_5O_2**

Jenkins, J. M., *Chem. Brit.*, 1970, **6**, 401
An acidified solution of the sodium salt was allowed to evaporate during 3 days and spontaneously exploded 2 weeks later. Nature of the explosive species, possibly the tetrazolic acid, is being sought.
See other HIGH-NITROGEN COMPOUNDS
C-NITRO COMPOUNDS

5-AZIDOTETRAZOLE **CHN_7**

Alone,
or Acetic acid,
or Alkali
1. Thiele, J. *et al.*, *Ann.*, 1895, **287**, 238
2. Lieber, E. *et al.*, *J. Amer. Chem. Soc.*, 1951, **73**, 1313

Though explosive, it (and its ammonium salt) is much less sensitive to impact or friction than its sodium or potassium salts [1]. A small sample of the latter exploded violently during suction filtration. The parent compound explodes spontaneously even in acetone (but not in ethanol or aqueous) solution if traces of acetic acid are present [2]. The salts are readily formed from diaminoguanidine salts and alkali nitrites. The ammonium salt explodes on heating, and the silver salt is violently explosive, even when wet [1]. The sodium salt is also readily formed from cyanogen azide.
See other HIGH-NITROGEN COMPOUNDS
N-METAL DERIVATIVES
ORGANIC AZIDES

SILVER NITROUREIDE $CH_2AgN_3O_3$

See NITROUREA, $CH_3N_3O_3$

SILVER 5-AMINOTETRAZOLIDE CH_2AgN_5

Thiele, J., *Ann.*, 1892, **270**, 59
Explodes on heating, similarly to silver tetrazolide.
See other N-METAL DERIVATIVES

DIBROMOMETHANE CH_2Br_2

Potassium
See POTASSIUM, K: Halocarbons

CHLORONITROMETHANE CH_2ClNO_2

1. Seigle, L. W. *et al., J. Org. Chem.*, 1940, **5**, 100
2. Libman, D. D., private comm., 1968

Chlorination of nitromethane following the published general method [1] gave a product which decomposed explosively during distillation at 95 mbar [2]. A b.p. of 122°C/1 bar is quoted in the literature.
See other C-NITRO COMPOUNDS

DICHLOROMETHANE CH_2Cl_2

MCA SD-86, 1962
Normally non-flammable, but will form ignitable mixtures of wide limits in oxygen-enriched air or at elevated temperatures.

Dinitrogen pentaoxide
See NITRATING AGENTS

Dinitrogen tetraoxide

See DINITROGEN TETRAOXIDE, N_2O_4: Halocarbons

Metals

See LITHIUM, Li: Halocarbons
SODIUM, Na: Halocarbons

Nitric acid

See NITRIC ACID, HNO_3: Dichloromethane

Potassium *tert*-butoxide

See POTASSIUM *tert*-BUTOXIDE, C_4H_9KO: Acids, etc.
See other HALOCARBONS

DIIODOMETHANE CH_2I_2

Metals

See COPPER–ZINC COUPLE, Cu–Zn: Diiodomethane, etc.
POTASSIUM, K: Halocarbons
LITHIUM, K: Halocarbons

POLY(METHYLENEMAGNESIUM) $(CH_2Mg)_n$

Ziegler, K. *et al., Z. Anorg. Chem.*, 1955, **282**, 345
The polymer ignites in air.
See other ALKYLMETALS

SODIUM *aci*-NITROMETHANE CH_2NNaO_2

Zelinsky, N., *Ber.*, 1894, **27**, 3407
Salts of *aci*-nitromethane are easily detonated and powerful explosives.

Mercury(II) chloride,
Acids
Nef, J. U., *Ann.*, 1894, **280**, 263, 305

Interaction gives mercury nitromethane, which is converted by acids to mercury fulminate.
See MERCURY(II) FULMINATE, $C_2HgN_2O_2$

1,1,3,3-Tetramethyl-2,4-cyclobutanedione
See DISODIUM BIS-*aci*-1,3-DIHYDROXY-1,3-DI(NITROMETHYL)-2,2,4,4-TETRAMETHYLCYCLOBUTANE, $C_{10}H_{16}N_2Na_2O_6$

Water
Nef, J. U., *Ann.*, 1894, **280**, 273
The *aci*-sodium salt, normally crystallising with one molecule of ethanol and stable, will explode if moistened with water. This is due to liberation of heat and conversion to sodium fulminate.
See DISODIUM 1,3-DIHYDROXY-1,3-BIS(*aci*-NITROMETHYL)-2,2,4,4-TETRAMETHYLCYCLOBUTANE, $C_{10}H_{16}N_2Na_2O_6$
See other aci-NITRO SALTS

CYANAMIDE CH_2N_2

Anon., *Fire Prot. Assoc. J.*, 1966, 243
Cyanamide is thermally unstable and needs storage under controlled conditions. Contact with moisture, acids or alkalies accelerates the rate of decomposition, and at temperatures above 49°C thermal decomposition is rapid and may become violent. A maximum storage temperature of 27°C is recommended.

Water
Pinck, L. A. *et al.*, *Inorg. Synth*, 1950, **3**, 41
Evaporation of aqueous solutions to dryness is hazardous owing to possibility of explosive polymerisation in concentrated solution.
See other CYANO COMPOUNDS

DIAZIRINE CH_2N_2

1. Graham, W. H., *J. Org. Chem.*, 1965, **30**, 2108
2. Schmitz, E. *et al.*, *Chem. Ber.*, 1962, **95**, 800

This cyclic isomer of diazomethane is also a gas (b.p., –14°C) which explodes on heating. Several homologues are also thermally unstable [1,2].
See DIAZIRINES

DIAZOMETHANE **CH_2N_2**

1. Eistert, B., in *Newer Methods of Preparative Organic Chemistry,* 517–518, New York, Interscience, 1948
2. de Boer, H. J. *et al., Org. Synth.,* 1963, Coll. Vol. 4, 250
3. Gutsche, C. D., *Org. React.,* 1954, **8**, 392–393
4. Zollinger, H., *Azo and Diazo Chemistry,* 22, London, Interscience, 1961
5. Fieser, L. *et al., Reagents for Organic Synthesis,* Vol. 1, 191, New York, Wiley, 1967
6. Horàk, V. *et al., Chem. & Ind.,* 1961, 472

Diazomethane boils at –23°C and the undiluted liquid or concentrated solutions may explode if impurities or solids are present [1], including freshly crystallised products [2]. Gaseous diazomethane, even when diluted with nitrogen, may explode at elevated temperatures (100°C or above), or under high-intensity lighting, or if rough surfaces are present [1]. Ground glass apparatus or glass-sleeved stirrers are therefore undesirable when working with diazomethane. Explosive intermediates may also be formed during its use as a reagent, for which cold dilute solutions have been frequently used uneventfully [1]. Further safety precautions have been detailed [2–4]. Many precursors for diazomethane are available [5], including the stable water-soluble intermediate *N*-nitroso-3-methylaminosulpholane [6]. Many of the explosions observed are attributed to unsuitable conditions of contact between concentrated alkali and undiluted nitroso precursors [1].

Alkali metals
Eistert, B., in *Newer Methods of Preparative Organic Chemistry,* 518, New York, Interscience, 1948
Contact of diazomethane with alkali metals causes explosions.
See DIAZOMETHYLLITHIUM, $CHLiN_2$

Calcium sulphate
Gutsche, C. D., *Org. React.*, 1954, **8**, 392
Calcium sulphate is an unsuitable desiccant for drying-tubes in diazomethane systems. Contact of diazomethane vapour and the sulphate causes an exotherm which may lead to detonation. Potassium hydroxide is a suitable desiccant.
See other DIAZO COMPOUNDS

ISODIAZOMETHANE **CH_2N_2**

Müller, E. *et al.*, *Chem. Ber.*, 1954, **87**, 1887
This unstable liquid begins to decompose at 15°C and explodes exothermically at 35–40°C.
See other DIAZO COMPOUNDS

TETRAZOLE **CH_2N_4**

Benson, F.R., *Chem. Rev.*, 1947, **41**, 5
It explodes above its m.p., 155°C.
See other TETRAZOLES

5-AMINO-1,2,3,4-THIATRIAZOLE **CH_2N_4S**

Lieber, E. *et al.*, *Inorg. Synth.*, 1960, **6**, 44
It decomposes with a slight explosion in a capillary tube at 136°C.
See other HIGH-NITROGEN COMPOUNDS

5-*N*-NITROAMINOTETRAZOLE **$CH_2N_6O_2$**

1. Lieber, E. *et al.*, *J. Amer. Chem. Soc.*, 1951, **73**, 2328
2. O'Connor, T. E. *et al.*, *J. Soc. Chem. Ind.*, 1949, **68**, 309

It and its monopotassium salt explode at 140°C and the diammonium

salt explodes at 220°C after melting [1]. The disodium salt explodes at 207°C [2].
See other TETRAZOLES

FORMALDEHYDE CH_2O

Gas above –19°C; E.L., 7.0–73%
MCA SD-1, 1960

Hydrogen peroxide
See HYDROGEN PEROXIDE, H_2O_2: Oxygenated compounds

Magnesium carbonate
BCISC Quart. Safety Summ., 1965, (143), 44
During neutralisation of the formic acid present in formaldehyde solution by shaking with magnesium carbonate in a screw-capped bottle, the latter exploded owing to pressure of carbon dioxide. Periodical release of pressure should avoid this.

Nitromethane
See NITROMETHANE, CH_3NO_2: Formaldehyde

Peroxyformic acid
See PEROXYFORMIC ACID, CH_2O_3: Organic materials

FORMIC ACID CH_2O_2

Hydrogen peroxide
See HYDROGEN PEROXIDE, H_2O_2: Oxygenated compounds

Nitromethane
See NITROMETHANE, CH_3NO_2: Acids

Phosphorus pentaoxide
Muir, G. D., private comm., 1968
Attempted dehydration of 95% acid to anhydrous formic acid caused rapid evolution of carbon monoxide.

PEROXYFORMIC ACID CH_2O_3

1. Greenspan, F. P., *J. Amer. Chem. Soc.*, 1946, **68**, 907
2. D'Ans, J. *et al.*, *Ber.*, 1915, **48**, 1136
3. Weingartshofer, A. *et al.*, *Chem. Eng. News*, 1952, **30**, 3041
4. Swern, 1970, Vol. 1, 337

Peroxyformic acid solutions are unstable and undergo a self-accelerating exothermic decomposition at ambient temperature [1]. An 80% solution exploded at 80–85°C [2]. A small sample of the pure vacuum-distilled material cooled to below −10°C exploded violently when the flask was moved [3]. Though the acid has occasionally been distilled, this is an extremely dangerous operation [4].

Metals,
or Oxides
D'Ans, J. *et al.*, *Ber.*, 1915, **48**, 1136
Violence of reaction depends on concentration of acid and scale and proportion of reactants. The following observations were made with additions to 2–3 drops of *ca* 90% acid. Nickel powder becomes violent; mercury, colloidal silver and thallium powder readily cause explosions. Zinc powder causes a violent explosion immediately. Iron powder (and silicon) are ineffective alone, but a trace of maganese dioxide promotes deflagration. Barium peroxide, copper(I) oxide, impure chromium trioxide, iridium dioxide, lead dioxide, manganese dioxide and vanadium pentoxide all cause violent decomposition, sometimes accelerating to explosion. Lead oxide, trilead tetraoxide and sodium peroxide all cause an immediate violent explosion.

Non-metals
D'Ans, J. *et al.*, *Ber.*, 1915, **48**, 1136
Impure carbon and red phosphorus are oxidised violently, and silicon,

promoted by traces of manganese dioxide, is oxidised with ignition.

Organic materials

1. D'Ans, J. *et al., Ber.,* 1915, **48**, 1136
2. Anon., *Chem. Eng. News,* 1950, **28**, 418
3. Shanley, E. S., ibid., 3067

Formaldehyde, benzaldehyde and aniline react violently with 90% performic acid [1]. An unspecified organic compound was added to the acid (preformed from formic acid and 90% hydrogen peroxide), and soon after the initial vigorous reaction had subsided, the mixture exploded violently [2]. Reaction with alkenes is vigorously exothermic, and adequate cooling is necessary [3]. Reactions with performic acid can be more safely accomplished by slow addition of hydrogen peroxide to a solution of the compound in formic acid. Adequate safety screens should be used with all peracid preparations [3].
See other PEROXYACIDS

METHYLSILVER CH_3Ag

Thiele, H., *Z. Electrochem.*, 1943, **49**, 426
Prepared at −80°C, the addition compound with silver nitrate decomposes explosively on warming to −20°C.
See other ALKYLMETALS

SILVER NITROGUANIDIDE CH_3AgN_4

See NITROGUANIDINE, $CH_4N_4O_2$
See other *N*-METAL DERIVATIVES

METHYLALUMINIUM DIIODIDE CH_3AlI_2

Nitromethane
See NITROMETHANE, CH_3NO_2: Alkylmetal halides
See also ALKYLALUMINIUM HALIDES

METHYLBISMUTH OXIDE CH_3BiO

Marquardt, A., *Ber.*, 1887, **20**, 1522
Ignites on warming in air.
See related ALKYLMETALS

BROMOMETHANE CH_3Br

No Fl.P.; but E.L., 10–15% (5–25% at 10 bar)
1. *MCA SD-35,* 1968
2. *MCA Case History No. 746*
Though bromomethane is used as a fire extinguishant it does form explosive mixtures with air within narrow limits (13.5–14.5% also recorded) at atmospheric pressure, but considerably wider at higher pressure [1, 2].

Ethylene oxide
See ETHYLENE OXIDE, C_2H_4O: Air, Bromomethane

Metals
MCA Case History No. 746 and addendum
Metallic components of zinc, aluminium and magnesium are unsuitable for service with bromomethane because of the formation of pyrophoric Grignard-type compounds. The Case History attributes a severe explosion to ignition of a bromomethane–air mixture by pyrophoric methylaluminium bromides produced by corrosion of an aluminium component.
See other METALS: Halocarbons

CHLOROMETHANE CH_3Cl

Fl.P., below 0°C (gas above –24°C); E.L., 8.1–17%
MCA SD-40, 1970

Aluminium trichloride,
Ethylene
See ETHYLENE, C_2H_4: Aluminium trichloride

Interhalogens

See BROMINE TRIFLUORIDE, BrF_3: Halogens, etc.
BROMINE PENTAFLUORIDE, BrF_5: Hydrogen-containing materials

Metals

MCA SD-40, 1970

In presence of catalytic amounts of aluminium trichloride, powdered aluminium and chloromethane interact to form pyrophoric trimethylaluminium. Chloromethane may react explosively with magnesium, or potassium, sodium or their alloys. Zinc probably reacts similarly to magnesium.

See ALUMINIUM, Al: Halocarbons
SODIUM, Na: Halocarbons

See other HALOCARBONS

METHYLMERCURY PERCHLORATE CH_3ClHgO_4

Anon., *Angew. Chem. (Nachr.)*, 1970, **18**, 214

On rubbing with a glass rod, a sample exploded violently. As the explosion could not be reproduced using a metal rod, initiation by static electricity was suspected.

See STATIC INITIATION

METHYL HYPOCHLORITE CH_3ClO

Sandmeyer, T., *Ber.,* 1886, **19**, 859

The liquid could be very gently distilled (12°C) but the superheated vapour readily and violently explodes, as does the liquid on ignition.

See other HYPOHALITES

METHYL PERCHLORATE CH_3ClO_4

Sidgwick, 1950, 1236

The high explosive instability is due in part to the covalent character

of the alkyl perchlorates, and also to the excess of oxygen in the molecule over that required to completely combust the other elements present (i.e. negative oxygen balance).
See other ALKYL PERCHORATES

N,N-DICHLOROMETHYLAMINE CH_3Cl_2N

1. Bamberger, E. *et al., Ber.,* 1895, **28**, 1683
2. Okon, K. *et al., Chem. Abs.,* 1960, **54**, 17887b

A mixture with water exploded violently on warming [1]. Contact with solid sodium sulphide or distillation over calcium hypochlorite also caused explosions [2].
See other *N*-HALOGEN COMPOUNDS

TRICHLORO(METHYL)SILANE CH_3Cl_3Si

Fl.P., 8°C; E.L., 7.6– %
See other ALKYLNON-METAL HALIDES

METHYLCOPPER CH_3Cu

Coates, 1960, 348
The dry solid is very impact-sensitive and may explode spontaneously on being allowed to dry out at room temperature.
See other ALKYLMETALS

FLUOROMETHANE CH_3F

Gas above –78°C; flammable
See other HALOCARBONS

XENON(II) FLUORIDE METHANESULPHONATE CH_3FO_3SXe

Wechsberg, M. *et al., Inorg. Chem.*, 1972, **11**, 3066
The solid explodes on warming from 0°C to ambient temperature.
See other XENON COMPOUNDS

IODOMETHANE CH_3I

Silver chlorite
See SILVER CHLORITE, $AgClO_2$: Alkyl iodides

Sodium
1. Anon., *J. Chem. Educ.*, 1966, **43**, A236
2. Braidech, M. M., *J. Chem. Educ.*, 1967, **44**, A324

The first stage of a reaction involved the addition of sodium dispersed in toluene to a solution of adipic ester in toluene. The subsequent addition of iodomethane (b.p., 42°C) was too fast and vigorous boiling ejected some of the flask contents. Exposure of sodium particles to air caused ignition, and a violent toluene-vapour explosion followed [1]. When a reagent as volatile and reactive as iodomethane is added to a reaction mixture, controlled addition, and one or more wide-bore reflux condensers, are essential. A similar incident but involving benzene solvent was also reported [2].
See also SODIUM, Na: Halocarbons

METHYLMAGNESIUM IODIDE CH_3IMg

Vanadium trichloride
See VANADIUM TRICHLORIDE, Cl_3V: Methylmagnesium iodide

METHYLZINC IODIDE CH_3IZn

Nitromethane
See NITROMETHANE, CH_3NO_2 : Alkylmetal halides

METHYLPOTASSIUM CH_3K

Weiss E. *et al., Angew. Chem. (Intern. Ed.),* 1968, **7**, 133
The dry material is highly pyrophoric.
See other ALKYLMETALS

METHYLLITHIUM CH_3Li

Sidgwick, 1950, 71

Ignites and burns brilliantly in air.
See also ALKALI-METAL DERIVATIVES OF HYDROCARBONS
See other ALKYLMETALS

FORMAMIDE CH_3NO

Iodine,
Pyridine,
Sulphur trioxide
Anon., *J. Chem. Educ.*, 1973, **50**, A293
Bottles containing a modified Karl Fischer reagent with formamide replacing methanol developed gas pressure during several months and exploded. No reason is yet apparent.

METHYL NITRITE CH_3NO_2

Gas above $-12^\circ C$; flammable

NITROMETHANE CH_3NO_2

1. McKitterick, D. S. *et al., Ind. Eng. Chem. (Anal. Ed.)*, 1938, **10**, 630
2. Makovky, A. *et al., Chem. Rev.*, 1958, **58**, 627

Conditions under which it may explode by detonation, heat or shock were determined. It was concluded that it is potentially very explosive and precautions are necessary to prevent its exposure to severe shock or high temperatures in use [1]. Later work, following two rail tank explosions, showed that shock caused by sudden application of gas pressure, or by forced high-velocity flow through restrictions, could detonate the liquid. The stability and decomposition of nitromethane relevant to its use as a rocket fuel is also reviewed [2].

Acids,
or Bases
Makovky, A. *et al., Chem. Rev.*, 1958, **58**, 631
Addition of bases or acids to nitromethane renders it susceptible to initiation by a detonator. These include aniline, diaminoethane, morpholine, methylamine, ammonium hydroxide, potassium hydroxide, sodium carbonate, and formic, nitric, sulphuric and phosphoric acids.
See also 1,2-Diaminoethane, etc., below

Alkylmetal halides
Traverse, G., US Pat. 2 775 863, 1957
Contact with $R_m MX_n$ (R is methyl, ethyl; M is aluminium, zinc; X is Br, I) causes ignition. Diethylaluminium bromide, dimethylaluminium bromide, ethylaluminium bromide iodide, methylzinc iodide and methylaluminium diiodide are claimed as specially effective.

Aluminium trichloride
See ALUMINIUM TRICHLORIDE–NITROMETHANE, $AlCl_3 \cdot CH_3NO_2$

Aluminium trichloride,
Ethylene
See ETHYLENE, C_2H_4: Aluminium trichloride

Calcium hypochlorite
See CALCIUM HYPOCHLORITE, $CaCl_2O_2$: Nitromethane

1,2-Diaminoethane,
N,2,4,6-tetranitro-*N*-methylaniline
MCA Case History No. 1564
During preparations to initiate the explosion of nitromethane sensitised by addition of 20% of the diamine, accidental contact of the liquid mixture with the solid 'tetryl' detonator caused ignition of the latter.
See also Acids, or Bases, above

Formaldehyde
Noland, W. E., *Org. Synth.*, 1961, **41**, 69
Interaction of nitromethane and formaldehyde in presence of alkali gives not only 2-nitroethanol but also di- and tri-condensation products. After removal of the 2-nitroethanol by vacuum distillation the residue must be cooled before admitting air into the system to prevent a flash explosion or violent fume-off.

Hydrocarbons
1. Watts, C. E., *Chem. Eng. News*, 1952, **30**, 2344
2. Makovky, A. *et al., Chem. Rev.*, 1958, **58**, 631

Nitromethane may act as a mild oxidant and should not be heated with hydrocarbons or readily oxidisable materials under confinement [1]. Explosions may occur during cooling of such materials heated to high temperatures and pressures [2]. Mixtures of nitromethane and solvents which are to be heated above the boiling point of nitromethane should be first subjected to small-scale explosive tests [2].

Lithium perchlorate
1. Titus, J. A., *Chem. Eng. News*, 1971, **49**(23), 6
2. 'Nitroparaffin Data Sheet TDS 1', New York, Commercial Solvents Corp., 1965
3. Egly, R. S., *Chem. Eng. News*, 1973, **51**(6), 30

Explosions which occurred at the auxiliary electrode during electro-oxidation reactions in nitromethane-lithium perchlorate electrolytes, may have been caused by lithium fulminate. This could have been produced by formation of the lithium salt of nitromethane and subsequent dehydration to the fulminate [1], analogous to the known formation of mercuric fulminate [2]. This explanation is not considered tenable, however [3].

Metal oxides

See NITROALKANES: Metal oxides

Nitric acid

See NITRIC ACID, HNO_3: Nitromethane

See other C-NITRO COMPOUNDS

METHYL NITRATE CH_3NO_3

1. Kit and Evered, 1960, 268
2. Black, A. P. *et al., Org. Synth.,* 1943, Coll. Vol. 2, 412
3. Goodman, H. *et al., Comb. and Flame,* 1972, **19**, 157

It has high shock-sensitivity and thermal sensitivity, exploding at 65°C. It is too sensitive for use as a rocket mono-propellant [1]. Conditions during preparation of the ester from methanol and mixed nitric–sulphuric acids are fairly critical, and explosions may occur if the ester is suddenly heated, or distilled in presence of acid [2]. Spontaneous ignition or explosion of the vapour at 250–316°C in presence of gaseous diluents has been studied [3].

See other ALKYL NITRATES

METHYL AZIDE CH_3N_3

1. *MCA Case History No. 887*
2. Boyer, J. H. *et al., Chem. Rev.,* 1954, **54**, 32

The product, prepared by interaction of sodium azide with dimethyl sulphate and sodium hydroxide, exploded during concurrent vacuum distillation. The explosion was attributed to formation and co-distillation with the product, of hydrogen azide, due to variation in pH below 5 during the reaction. Free hydrogen azide itself is explosive, and it may also have reacted with mercury in a manometer to form the detonator, mercuric azide [1]. Methyl azide is stable at ambient temperature but may detonate on rapid heating [2].

Mercury

Currie, C. L. *et al., Can. J. Chem.,* 1963, **41**, 1048

Presence of mercury in methyl azide markedly reduces the stability towards shock or electric discharge.

Methanol

Grundmann, C. *et al., Angew. Chem.,* 1950, **62**, 410

In spite of extensive cooling and precautions, a mixture of methyl azide, methanol and dimethyl malonate exploded violently while being sealed into a Carius tube. The vapour of the azide is very easily initiated by heat, even at low concentrations.

See other ORGANIC AZIDES

NITROUREA $CH_3N_3O_3$

Urbanski, 1967, Vol. 3, 34

A rather unstable explosive material which gives mercuric and silver salts. These are much more impact-sensitive.

See other *N*-METAL DERIVATIVES
N-NITRO COMPOUNDS

5-AMINOTETRAZOLE CH_3N_5

1. Elmore, D. T., *Chem. Brit.,* 1966, **2**, 414
2. Thiele, J., *Ann.,* 1892, **270**, 59

Diazotised 5-aminotetrazole is unstable under the conditions recommended for its use as a biochemical reagent. While the pH of the diazotised material (the cation of which contains 87% of nitrogen) at 0°C was being reduced to 5 by addition of potassium hydroxide, a violent explosion occurred [1]. This may have been caused by a local excess of alkali causing the formation of 5-diazoniotetrazolide, which will explode in concentrated solution at 0°C [2]. The diazonium chloride is also very unstable in concentrated solution at 0°C.

See other DIAZONIUM SALTS
HIGH-NITROGEN COMPOUNDS
TETRAZOLES

METHYLSODIUM CH_3Na

Schlenk, W. *et al., Ber.*, 1917, **50**, 262
Ignites immediately in air. Tendency to ignition decreases with ascent of the homologous series of alkylsodiums.
See other ALKYLMETALS

p-Chloronitrobenzene
See *p*-CHLORONITROBENZENE, $C_6H_4ClNO_2$: Sodium methoxide

SODIUM METHOXIDE CH_3NaO

Chloroform,
Methanol
See CHLOROFORM, $CHCl_3$: Sodium methoxide

'SODIUM PERCARBONATE' $CH_3Na_2O_6$

See SODIUM CARBONATE HYDROGEN PEROXIDATE, $CNa_2O_3 \cdot 1.5H_2O_2$

METHANE CH_4

Fl.P., −187°C (gas above −161°C); E.L., 5.3–15%

Halogens or Interhalogens
See BROMINE PENTAFLUORIDE, BrF_5 : Hydrogen-containing materials
CHLORINE, Cl_2 : Hydrocarbons
FLUORINE, F_2 : Hydrocarbons
IODINE HEPTAFLUORIDE, F_7I: Carbon, etc.

Oxidants
See DIOXYGENYL TETRAFLUOROBORATE, BF_4O_2 : Organic materials
DIOXYGEN DIFLUORIDE, F_2O_2
TRIOXYGEN DIFLUORIDE, F_2O_3
OXYGEN (Liquid), O_2 : Hydrocarbons
: Liquefied gases

DICHLORO(METHYL)SILANE CH_4Cl_2Si

Fl.P., −32°C; ignites in air
See other ALKYLNON-METAL HALIDES

POTASSIUM METHYLAMIDE CH_4KN

Makhija, R. C. *et al., Can. J. Chem.*, 1971, **49**, 807
Extremely hygroscopic and pyrophoric; may explode on contact with air.
See other N-METAL DERIVATIVES

METHYLDIAZENE CH_4N_2

Oxygen
Ackermann, M. N. *et al., Inorg. Chem.*, 1972, **11**, 3077
Interaction on warming a mixture from −196°C rapidly to ambient temperature is explosive.
See other AZO COMPOUNDS

UREA CH_4N_2O

Nitrosyl perchlorate
See NITROSYL PERCHLORATE, $ClNO_5$: Organic materials

Phosphorus pentachloride
See PHOSPHORUS PENTACHLORIDE, Cl_5P: Urea

Sodium nitrite
See SODIUM NITRITE, $NNaO_2$: Urea

UREA HYDROGEN PEROXIDATE $CH_4N_2O \cdot H_2O_2$

MCA Case History No. 719

The contents of a screw-capped brown glass bottle spontaneously erupted after 4 years' storage at ambient temperature. All peroxides should be kept in special storage and periodically checked.

See other CRYSTALLINE HYDROGEN PEROXIDATES

N-NITROMETHYLAMINE $CH_4N_2O_2$

Sulphuric acid

Urbanski, 1967, Vol. 3, 16

The nitroamine is decomposed explosively by concentrated sulphuric acid.

See other *N*-NITRO COMPOUNDS

AMMONIUM THIOCYANATE CH_4N_2S

Potassium chlorate

See POTASSIUM CHLORATE, $ClKO_3$: Thiocyanates

THIOUREA

Acrylaldehyde

See ACRYLALDEHYDE, C_3H_4O: Acids, etc.

Hydrogen peroxide,
Nitric acid

See HYDROGEN PEROXIDE, H_2O_2: Nitric acid, etc.

NITROGUANIDINE $CH_4N_4O_2$

1. Urbanski, 1967, Vol. 3, 31
2. McKay, A. F., *Chem. Rev.*, 1952, **51**, 301

Nitroguanidine is difficult to detonate, but its mercury and silver complex salts are much more impact-sensitive [1]. Many nitroguanidine derivatives have been considered as explosives [2].
See other *N*-METAL DERIVATIVES
N-NITRO COMPOUNDS

METHANOL **CH_4O**

Fl.P., 10°C; E.L., 6.0–36.5% (latter at 60°C)
MCA SD-22, 1970

Beryllium dihydride
See BERYLLIUM DIHYDRIDE, BeH_2: Methanol

Chloroform,
Sodium
See CHLOROFORM, $CHCl_3$: Sodium, Methanol

Cyanuric chloride
See 2,4,6-TRICHLORO-1,3,5-TRIAZINE (CYANURIC CHLORIDE), $C_3Cl_3N_3$: Methanol

Metals
See POTASSIUM, K: Air (slow oxidation)
MAGNESIUM, Mg: Methanol

Oxidants
See *N*-HALOIMIDES: Alcohols
BROMINE, Br_2: Methanol
SODIUM HYPOCHLORITE, ClNaO: Methanol
LEAD DIPERCHLORATE, Cl_2O_8Pb: Methanol
CHROMIUM TRIOXIDE, CrO_3: Alcohols
NITRIC ACID, HNO_3: Alcohols
HYDROGEN PEROXIDE, H_2O_2: Oxygenated compounds

Potassium *tert*-butoxide
See POTASSIUM *tert*-BUTOXIDE, C_4H_9KO: Acids, etc.

METHYL HYDROPEROXIDE CH_4O_2

Rieche, A. *et al., Ber.*, 1929, **62**, 2458
Violently explosive, shock-sensitive, especially on warming; great care is necessary in handling. The barium salt is dangerously explosive when dry.
See other ALKYL HYDROPEROXIDES

HYDROXYMETHYL HYDROPEROXIDE CH_4O_3

Rieche, A., *Ber.*, 1931, **64**, 2328; ibid., 1935, **68**, 1465
Explodes on heating, but is friction-insensitive. Higher homologues are not explosive.
See other 1-OXYPEROXY COMPOUNDS

METHANESULPHONIC ACID CH_4O_3S

Ethyl vinyl ether
See ETHYL VINYL ETHER, C_4H_3O: Methanesulphonic acid

METHANETHIOL CH_4S

Fl.P., –18°C (gas above 6°C); E.L., 3.9–21.8%

Mercury(II) oxide
See MERCURY(II) OXIDE, HgO: Methanethiol

METHYLAMINE CH_5N

Fl.P., –18°C (gas above –6°C); E.L., 4.5–21%
Fl.P., 7.5°C for 35% w/v solution in water
MCA SD-57, 1955

Nitromethane

See NITROMETHANE, CH_3NO_2: Acids, etc.

1-AMINO-3-NITROGUANIDINE $CH_5N_5O_2$

Lieber, E. *et al., J. Amer. Chem. Soc.*, 1951, **73**, 2328

Explodes at the m.p., 190°C

See other HIGH-NITROGEN COMPOUNDS

METHYLPHOSPHINE CH_5P

Houben-Weyl, 1963, Vol. 12.1, 69

Primary lower-alkylphosphines readily ignite in air.

See other ALKYLNON-METALS

METHYLAMMONIUM CHLORITE CH_6ClNO_2

Levi, G. R., *Gazz. Chim. Ital.*, 1922, [1], **52**, 207

A concentrated solution caused a slight explosion when poured on to a cold iron plate.

See other CHLORITE SALTS

OXOSALTS OF NITROGENOUS BASES

METHYLAMMONIUM PERCHLORATE CH_6ClNO_4

Kasper, F., *Z. Chem.*, 1969, **9**, 34

The semi-crystalline mass exploded when stirred after standing overnight. The preparation was based on a published method used uneventfully for preparation of ammonium, dimethylammonium and piperidinium perchlorates.

See other AMMONIUM PERCHLORATES

OXOSALTS OF NITROGENOUS BASES

GUANIDINIUM PERCHLORATE $CH_6ClN_3O_4$

Davis, 1943, 121; Schumacher, 1960, 213
Unusually sensitive to initiation, and of high explosive power, it decomposes violently at 350°C.
See DIFFERENTIAL THERMAL ANALYSIS
See other OXOSALTS OF NITROGENOUS BASES

METHYLHYDRAZINE CH_6N_2

Fl.P., 23°C (17°C (o)); E.L., 2.5–97%; A.I.T., 196°C

Oxidants
Kirk-Othmer, 1966, Vol. 11, 186
A powerful reducing agent and fuel, hypergolic with many oxidants such as dinitrogen tetraoxide or hydrogen peroxide.
See ROCKET PROPELLANTS

'UREA PEROXIDE' $CH_6N_2O_3$

See UREA HYDROGEN PEROXIDATE, $CH_4N_2O \cdot H_2O_2$

AMINOGUANIDINE CH_6N_4

Kurzer, F. *et al., Chem. & Ind.*, 1962, 1585
All the oxoacid salts are potentially explosive, including the wet nitrate.
See AMINOGUANIDINIUM NITRATE, $CH_7N_5O_3$

CARBONOHYDRAZIDE CH_6N_4O

Nitrous acid
Curtius, J. *et al., Ber.*, 1894, **27**, 55
Interaction forms the highly explosive carbonyl diazide.
See CARBONYL DIAZIDE, CN_6O

GUANIDINIUM NITRATE

$CH_6N_4O_3$

1. Davis, T. L., *Org. Synth.*, 1941, Coll. Vol. 1, 302
2. Smith, G. B. L. and Schmidt, M. T., *Inorg. Synth.*, 1939, **1**, 96, 97

According to an *O. S.* amendment sheet, the procedure as described is dangerous because the reaction mixture [1] (dicyanodiamide and ammonium nitrate) is similar in composition to commercial blasting explosives. This probably also applies to similar earlier preparations [2].

See other OXOSALTS OF NITROGENOUS BASES

METHYLSILANE

CH_6Si

Mercury,
Oxygen

Stock, A. *et al.*, *Ber.*, 1919, **52**, 706

Does not ignite in air, but explodes if shaken with mercury in oxygen.

See other ALKYLNON-METAL HYDRIDES

AMMONIUM *aci*-NITROMETHANE

$CH_7N_2O_2$

Watts, C. E., *Chem. Eng. News*, 1952, **30**, 2344

The isolated salt is a friction-sensitive explosive.

*See other aci-*NITRO SALTS

AMINOGUANIDINIUM NITRATE

$CH_7N_5O_3$

Koopman, H., *Chem. Weekbl.*, 1957, **53**, 97

An aqueous solution exploded violently during evaporation on a steam bath. Nitrate salts of many organic bases are unstable and should be avoided.

See AMINOGUANIDINE, CH_6N_4
See other HIGH-NITROGEN COMPOUNDS
OXO-SALTS OF NITROGENOUS BASES

DIAMINOGUANIDINIUM NITRATE $CH_8N_6O_3$

Violent decomposition occurred at 260°C.
See DIFFERENTIAL THERMAL ANALYSIS
See other OXOSALTS OF NITROGENOUS BASES

TRIAMINOGUANIDINIUM PERCHLORATE $CH_9ClN_6O_4$

Violent decomposition occurred at 317°C.
See DIFFERENTIAL THERMAL ANALYSIS
See other OXOSALTS OF NITROGENOUS BASES

TRIAMINOGUANIDINIUM NITRATE $CH_9N_7O_3$

Violent decomposition occurred at 230°C.
See DIFFERENTIAL THERMAL ANALYSIS
See other OXOSALTS OF NITROGENOUS BASES

PENTAAMMINETHIOCYANATO COBALT(III) PERCHLORATE $CH_{15}Cl_2CoN_6O_8S$

Explodes at 325°C; medium impact-sensitivity.
See AMMINEMETAL OXOSALTS

PENTAAMMINETHIOCYANATORUTHENIUM(III) PERCHLORATE $CH_{15}Cl_2N_6O_8RuS$

Armor, J. N., private comm., 1969
After washing with ether, 0.1 g of the complex exploded violently when touched with a spatula.
See other AMMINEMETAL OXOSALTS

IODINE ISOCYANATE $CINO$

Rosen, S. *et al., Anal. Chem.*, 1966, **38**, 1394
On storage, solutions of iodine isocyanate gradually deposit a touch-sensitive, mildly explosive solid (possibly cyanogen peroxide).
See other IODINE COMPOUNDS

CARBON TETRAIODIDE CI_4

Bromine trifluoride
See BROMINE TRIFLUORIDE, BrF_3 : Halogens, etc.

Lithium
See LITHIUM, Li: Halocarbons

POTASSIUM CYANIDE CKN

Nitrogen trichloride
See NITROGEN TRICHLORIDE, Cl_3N: Initiators

Perchloryl fluoride
See PERCHLORYL FLUORIDE, $ClFO_3$: Calcium acetylide, etc.

Sodium nitrite
See SODIUM NITRITE, $NNaO_2$: Metal cyanides

POTASSIUM THIOCYANATE $CKNS$

Perchloryl fluoride
See PERCHLORYL FLUORIDE, $ClFO_3$: Calcium acetylide, etc.

POTASSIUM TRINITROMETHANIDE ('NITROFORM' SALT) CKN_3O_6

1. Sandler, S. R. *et al., Organic Functional Group Preparations,* 433, New York, Academic Press, 1968
2. Shulgin, A. T., private comm., 1968

This intermediate, produced by action of alkali on tetranitromethane, must be kept damp and used as soon as possible with great care, as it may be explosive [1]. Material produced as a by-product in a nitration reaction using tetranitromethane was washed with acetone. It exploded very violently after several months' storage [2].

See also TETRANITROMETHANE, CN_4O_8

See other POLYNITROALKYL COMPOUNDS

CARBONYLPOTASSIUM CKO

Oxygen,
or Water

Mellor, 1963, Vol. 2, Suppl. 2.2, 1567–1568; 1946, Vol. 5, 951

Carbonylpotassium reacts violently with oxygen, and explodes on heating in air or in contact with water.

See POTASSIUM, K: Non-metal oxides

See other CARBONYLMETALS

POTASSIUM CARBONATE CK_2O_3

Carbon

Druce, J. G. F., *School Sci. Rev.,* 1926, 7(28), 261

Potassium metal prepared by the old process of distilling an intimate mixture of the carbonate and carbon contained some carbonylpotassium, and several explosions with old samples of potassium may have involved this compound.

See POTASSIUM, K

Chlorine trifluoride

See CHLORINE TRIFLUORIDE, ClF_3: Metals, etc.

Magnesium

See MAGNESIUM, Mg: Potassium carbonate
See other METAL OXONON-METALLATES

CARBONYLLITHIUM **CLiO**

Mellor, 1961, Vol. 2, Suppl. 2.1, 88
Explodes with water.
See other CARBONYLMETALS

LITHIUM CARBONATE **CLi_2O_3**

Fluorine

See FLUORINE, F_2: Metal Salts

MAGNESIUM CARBONATE **$CMgO_3$**

Formaldehyde

See FORMALDEHYDE, CH_2O: Magnesium carbonate

***N*-FLUOROIMINODIFLUOROMETHANE** **CNF_3**

Ginsberg, V. A. *et al., Zh. Obsch. Khim.*, 1967, **37**, 1413
It boils at −60°C and explodes on warming.
See other N-HALOGEN COMPOUNDS

SODIUM CYANIDE **CNNa**

It is oxidised violently or explosively in contact with hot oxidants.
See METAL CYANIDES AND CYANOCOMPLEXES

SODIUM FULMINATE $CNNaO$

Smith, 1966, Vol. 2, 99
Even sodium fulminate detonates when touched lightly with a glass rod.
See other METAL FULMINATES

SODIUM THIOCYANATE $CNNaS$

Sodium nitrite
See SODIUM NITRITE, $NNaO_2$: Sodium thiocyanate

THALLIUM FULMINATE $CNOTl$

See METAL FULMINATES

CYANOGEN AZIDE CN_4

Marsh, F. D., *J. Amer. Chem. Soc.*, 1964, **86**, 4506; *J. Org. Chem.*, 1972, **37**, 2966
It detonates with great violence when subjected to mild mechanical, thermal or electrical shock. Special precautions are necessary to prevent separation of the azide from solution and to decontaminate equipment and materials.
See also DIAZIDOMETHYLENECYANAMIDE, C_2N_8

Sodium hydroxide
Marsh, F. D., *J. Org. Chem.*, 1972, **37**, 2966
Interaction with 10% alkali forms sodium 5-azidotetrazolide, which is violently explosive if isolated.
See 5-AZIDOTETRAZOLE, CHN_7
See related HALOGEN AZIDES

DINITRODIAZOMETHANE CN_4O_4

Schallkopf, V. *et al., Angew. Chem. (Intern. Ed.),* 1969, 8, 612
It explodes on impact, rapid heating or contact with conc. sulphuric acid.
See other DIAZO COMPOUNDS
POLYNITROALKYL COMPOUNDS

TETRANITROMETHANE CN_4O_8

Liang, P., *Org. Synth.*, 1955, Coll. Vol. 3, 804
During its preparation from fuming nitric acid and acetic anhydride, strict temperature control and rate of addition of anhydride are essential to prevent a runaway violent reaction. Tetranitromethane should not be distilled, since explosive decomposition may occur.
See NITRIC ACID, HNO_3: Acetic anhydride

Aluminium
See ALUMINIUM, Al: Oxidants

Aromatic nitrocompounds
Urbanski, 1964, Vol. 1, 592
Mixtures of nitrobenzene, *o*- or *p*-nitrotoluene, *m*-dinitrobenzene or 1-nitronaphthalene with tetranitromethane were found to be high explosives of high sensitivity and detonation velocities.

Hydrocarbons
1. Stettbacher, A., *Z. Ges. Schiess. und Sprengstoffw.*, 1930, **25**, 439
2. Hager, K. F., *Ind. Eng. Chem.*, 1949, **41**, 2168
3. Tschinkel, J. G., *Ind. Eng. Chem.* 1956, **48**, 732

When mixed with hydrocarbons in approximately stoicheiometric proportions, a sensitive, highly explosive mixture is produced, which needs careful handling [1,2]. The use of such mixtures as rocket propellants has also been investigated [3].

Toluene,
Cotton
Winderlich, R., *J. Chem. Educ.*, 1950, **27**, 669
A demonstration mixture, to show combustion of cotton by combined oxygen, exploded with great violence soon after ignition.
See Hydrocarbons, above
See other POLYNITROALKYL COMPOUNDS

5-DIAZONIOTETRAZOLIDE CN_6

Thiele, J., *Ann.*, 1892, **270**, 60
Concentrated aqueous solutions of this internal diazonium salt explode at 0°C.
See other DIAZONIUM SALTS
TETRAZOLES

CARBONYL DIAZIDE CN_6O

1. Chapman, L. E. *et al., Chem. & Ind.*, 1966, 1266
2. Kesting, W., *Ber.*, 1924, **57**, 1321

It is a violently explosive solid, which should be used only in solution, and on a small scale [1]. It exploded violently even under ice-water [2].

See other ACYL AZIDES

CARBONYLSODIUM $CNaO$

Mellor, 1961, Vol. 2, Suppl. 2.1, 467; 1946, Vol. 5, 951

It explodes when heated in air at 90°C, or in contact with water.

See other CARBONYLMETALS

SODIUM CARBONATE CNa_2O_3

Fluorine

See FLUORINE, F_2: Metal salts

Lithium

See LITHIUM, Li: Sodium carbonate

Phosphorus pentaoxide

See PHOSPHORUS(V) OXIDE, O_5P_2: Inorganic bases

Sulphuric acid

MCA Case History No. 888

Lack of any mixing arrangements caused stratification of strong sulphuric acid and (probably) sodium carbonate solutions in the same tank. When gas evolution caused intermixture of the layers, a violent eruption of the tank contents occurred.

2,4,6-Trinitrotoluene

See 2,4,6-TRINITROTOLUENE, $C_7H_3N_3O_6$: Added impurities

See other METAL OXONON-METALLATES

SODIUM CARBONATE HYDROGEN PEROXIDATE

$CNa_2O_3 \cdot 1.5H_2O_2$

See CRYSTALLINE HYDROGEN PEROXIDATES

CARBON MONOXIDE **CO**

Gas above −192°C; E.L., 12.5–74%

Fluorine,
Oxygen

See BISFLUOROFORMYL PEROXIDE, $C_2F_2O_4$

Interhalogens

See BROMINE TRIFLUORIDE, BrF_3: Carbon monoxide
BROMINE PENTAFLUORIDE, BrF_5: Acids, etc.
IODINE HEPTAFLUORIDE, F_7I: Carbon, etc.

Metal oxides

See SILVER(I) OXIDE, Ag_2O: Carbon monoxide
DICAESIUM OXIDE, Cs_2O: Halogens, etc.
DIIRON TRIOXIDE, Fe_2O_3: Carbon monoxide

Metals

See POTASSIUM, K: Non-metal oxides
SODIUM, Na: Non-metal oxides

Oxidants

See CHLORINE DIOXIDE, ClO_2: Carbon monoxide
OXYGEN (Liquid), O_2: Liquefied gases

CARBONYL SULPHIDE **COS**

Gas above −50°C; E.L., 12–28.2%

CARBON DIOXIDE CO_2

Acrylaldehyde

See ACRYLALDEHYDE, C_3H_4O: Acids

Aziridine

See AZIRIDINE (ETHYLENEIMINE), C_2H_5N: Acids

Dicaesium oxide

See DICAESIUM OXIDE, Cs_2O: Halogens, etc.

Metal hydrides

See ALUMINIUM TRIHYDRIDE, AlH_3: Carbon dioxide
LITHIUM TETRAHYDROALUMINATE, AlH_4Li: Bis(2-methoxyethyl) -ether

Metal acetylides

See POTASSIUM ACETYLIDE, C_2HK: Non-metal oxides
LITHIUM ACETYLIDE–AMMONIA, $C_2Li \cdot H_3N$: Gases
DISODIUM ACETYLIDE, C_2Na_2: Non-metal oxides
DIRUBIDIUM ACETYLIDE, C_2Rb_2: Non-metal oxides

Metals

See POTASSIUM, K: Non-metal oxides
POTASSIUM–SODIUM ALLOY, K–Na: Carbon dioxide
LITHIUM, Li: Atmospheric gases
: Non-metal oxides
SODIUM, Na: Non-metal oxides
TITANIUM, Ti: Carbon dioxide

Metals,
Sodium peroxide

See SODIUM PEROXIDE, Na_2O_2: Metals, Carbon dioxide

LEAD CARBONATE CO_3Pb

Fluorine

See FLUORINE, F_2: Metal salts

CARBON SULPHIDE CS

Pearson, T. G., *School. Sci. Rev.*, 1938, **20**(78), 189
The gaseous radical readily polymerises explosively.

CARBON DISULPHIDE CS_2

Fl.P., −30°C; E.L., 1.3–50%; A.I.T., 125°C (below 100°C if rust present)
MCA SD-12, 1967

Air,
Rust
1. Mee, A. J., *School Sci. Rev.*, 1940, **22**(86), 95
2. Dickens, G. A., *School Sci. Rev.*, 1950, **31**(114), 264
3. Anon., *J.R. Inst. Chem.*, 1956, **80**, 664

Disposal of 2 litres of the solvent into a rusted iron sewer caused an explosion. Initiation of the solvent–air mixture by the rust was suspected [1]. A hot gauze falling from a tripod into a laboratory sink containing some carbon disulphide initiated two explosions [2]. It is a hazardous solvent because of its high volatility and flammability. The vapour or liquid has been known to ignite on contact with steam pipes, particularly if rusted [3].

Halogens
See CHLORINE, Cl_2: Carbon disulphide
FLUORINE, F_2: Sulphides

Metal azides
Mellor, 1947, Vol. 8, 338
Carbon disulphide and aqueous solutions of metal azides interact to produce metal azidodithioformates most of which are explosive, with varying degrees of power and sensitivity to shock or heat.
See also AZIDODITHIOFORMIC ACID, CHN_3S_2

Metals
See POTASSIUM–SODIUM ALLOY, K–Na: Carbon dioxide, etc.
ZINC, Zn: Carbon disulphide

Oxidants

See Halogens, above
PERMANGANIC ACID, $HMnO_4$: Organic materials
NITROGEN OXIDE ('NITRIC OXIDE'), NO: Carbon disulphide
DINITROGEN TETRAOXIDE (NITROGEN DIOXIDE), N_2O_4: Carbon disulphide

TITANIUM CARBIDE CTi

MCA Case History No. 618

A violent, and apparently spontaneous, dust explosion occurred while the finely ground carbide was being removed from a ball-mill. Static initiation seems a likely possibility.

See also DUST EXPLOSIONS
See other METAL NON-METALLIDES

TUNGSTEN CARBIDE CW

Fluorine

See FLUORINE, F_2: Metal acetylides and carbides

Nitrogen oxides

Mellor, 1946, Vol. 5, 890

At about 600°C, the carbide ignites and incandesces in dinitrogen mono- or tetra-oxides.

DITUNGSTEN CARBIDE CW_2

Oxidants

Mellor, 1946, Vol. 5, 890

The carbide burns incandescently at red heat in contact with dinitrogen mono- or tetra-oxide.

See also FLUORINE, F_2: Metal acetylides and carbides
See other METAL NON-METALLIDES

DISILVER ACETYLIDE C_2Ag_2

Miller, 1965, Vol. 1, 486
Silver acetylide is a more powerful detonator than the copper derivative, but both will initiate explosive acetylene-containing gas mixtures.

DISILVER ACETYLIDE–SILVER NITRATE $C_2Ag_2 \cdot AgNO_3$

Baker, W. E., *et al. Chem. Eng. News,* 1965, **43**(49), 46
The dry complex is exploded by high-intensity light pulses, or by heat or sparks. As a slurry in acetone, it is stable for a week if kept dark.
See other METAL ACETYLIDES

SILVER FULMINATE $C_2Ag_2N_2O_2$

Urbanski, 1967, Vol. 3, 157
Dimeric silver fulminate, readily formed from silver or its salts, nitric acid and ethanol, is a much more sensitive and powerful detonator than mercuric fulminate.
See other METAL FULMINATES

Hydrogen sulphide
Boettger, A., *J. Prakt. Chem.,* 1868, **103,** 309
Contact with hydrogen sulphide at ambient temperature initiates violent explosion of the fulminate.
See METAL FULMINATES
See other HEAVY METAL DERIVATIVES

DISILVER KETENDIIDE C_2Ag_2O

Blues, E. T. *et al., Chem. Comm.,* 1970, 699
This and its pyridine complex explode violently if heated or struck.
See other HEAVY METAL DERIVATIVES

SILVER OXALATE $C_2Ag_2O_4$

Sidgwick, 1950, 126

Above 140°C its exothermic decomposition to metal and carbon dioxide readily becomes explosive.

See also SILVER, Ag: Carboxylic acids

See other METAL OXALATES

DIGOLD ACETYLIDE C_2Au_2

1. Mellor, 1946, Vol. 5, 855
2. Matthews, J. A. *et al., J. Amer. Chem. Soc.,* 1900, 22, 110

Produced by action of acetylene on sodium gold thiosulphate (or other gold salts), the acetylide is explosive and readily initiated by light impact, friction or rapid heating to 83°C [1]. This unstable detonator is noted for high brisance [2].

See other METAL ACETYLIDES

BARIUM ACETYLIDE C_2Ba

Halogens

Mellor, 1946, Vol. 5, 862

Barium acetylide incandesces with chlorine, bromine and iodine at 140, 130, 122°C, respectively.

Selenium

Mellor, 1946, Vol. 5, 862

A mixture incandesces when heated to 150°C.

See other METAL ACETYLIDES

BARIUM THIOCYANATE $C_2BaN_2S_2$

See POTASSIUM CHLORATE, $ClKO_3$: Metal thiocyanates

Sodium nitrate

See SODIUM NITRATE, $NNaO_3$: Barium thiocyanate

BROMOTRIFLUOROETHYLENE C_2BrF_3

Gas above −3°C; ignites in air

Oxygen
See OXYGEN (Gas), O_2: Halocarbons
See other HALOALKENES

LITHIUM BROMOACETYLIDE C_2BrLi

See LITHIUM CHLOROACETYLIDE, C_2ClLi

DIBROMOACETYLENE C_2Br_2

Rodd, 1951, Vol. 1A, 284
It ignites in air and explodes on heating.
See other HALOACETYLENE DERIVATIVES

OXALYL DIBROMIDE $C_2Br_2O_2$

Potassium
See POTASSIUM, K: Oxayl dihalides

CALCIUM ACETYLIDE (CARBIDE) C_2Ca

Anon., *Ind. Safety Bull.*, 1940, 8, 41
Use of a steel chisel to open a drum of carbide caused an incendive spark which ignited traces of acetylene in the drum. The non-ferrous tools normally used for this purpose should be kept free from embedded ferrous particles.

Halogens
Mellor, 1946, Vol. 5, 862
The acetylide incandesces with chlorine, bromine and iodine at 245, 350 and 305°C, respectively. Strontium and barium acetylides are more reactive.

Hydrogen chloride
Mellor, 1946, Vol. 5, 862
Incandescence on warming; strontium and barium acetylides are similar.

Iron(III) chloride,
Iron(III) oxide
Partington, 1967, 372
The carbide is an energetic reducant. A powdered mixture with iron oxide and chloride burns violently when ignited, producing molten iron.

Lead difluoride
Mellor, 1946, Vol. 5, 864
Incandescence on contact at ambient temperature.

Magnesium
Mellor, 1940, Vol. 4, 271
A mixture incandesces when heated in air.

Methanol
Unpublished observations, 1951
Interaction of calcium acetylide with boiling methanol to give calcium methoxide is very vigorous, but subject to an induction period of variable length. Once reaction starts, evolution of acetylene gas is very fast, and a wide-bore condenser and adequate ventilation are necessary.

Perchloryl fluoride
See PERCHLORYL FLUORIDE, $ClFO_3$: Calcium acetylide

Selenium
See SELENIUM, Se: Metal acetylides

Silver nitrate
Luchs, J. K., *Photog. Sci. Eng.*, 1966, **10**, 334
Addition of calcium acetylide to silver nitrate solution precipitates silver acetylide, a highly sensitive explosive. Copper salt solutions would behave similarly.
See other HEAVY METAL DERIVATIVES

Sodium peroxide
See SODIUM PEROXIDE, Na_2O_2 Calcium acetylide

Sulphur
See SULPHUR, S: Metal acetylides

Tin dichloride
Mellor, 1941, Vol. 7, 430
A mixture can be ignited with a match, and reduction to metallic tin proceeds with incandescence.

Water
Jones, G. W. *et al., Explosions in Med. Press. Acetylene Generators,* Invest. Report 3755, Washington, US Bureau of Mines, 1944
At the end of a generation run, maximum temperatures and high moisture content of acetylene may cause the finely divided acetylide to overheat and initiate explosion of compressed gas.
See other METAL ACETYLIDES

CADMIUM DICYANIDE C_2CdN_2

Magnesium
See MAGNESIUM, Mg: Metal cyanides

CADMIUM FULMINATE $C_2CdN_2O_2$

See METAL FULMINATES

CHLOROTRIFLUOROETHYLENE C_2ClF_3

Gas above –28°C; E.L., 8.4–38.7%

1,1-Dichloroethylene
See 1,1-DICHLOROETHYLENE, $C_2H_2Cl_2$: Chlorotrifluoroethylene

Oxygen
See OXYGEN (Gas), O_2 : Halocarbons

POLY(CHLOROTRIFLUOROETHYLENE) $(C_2ClF_3)_n$

See ALUMINIUM, Al: Halocarbons

LITHIUM CHLOROACETYLIDE C_2ClLi

Cadiot, P., *BCISC Quart. Safety Summ.*, 1965, (142), 27
During preparation of a chloroethynyl compound with lithium chloroacetylide in liquid ammonia, some of the salt separated as a crust due to evaporation of solvent, and exploded violently. Such salts are stable in solution, but dangerous in the solid state. Evaporation of ammonia must be prevented or made good until unreacted salt has been decomposed by addition of ammonium chloride. Lithium trifluoromethylacetylide behaves similarly.
See other HALOACETYLENE DERIVATIVES

SODIUM CHLOROACETYLIDE C_2ClNa

Viehe, H. G., *Chem. Ber.*, 1959, **92**, 1271
Though stable and usable in solution, the sodium salt, like the lithium and calcium salts, is dangerously explosive in the solid state.
See other HALOACETYLENE DERIVATIVES

DICHLOROACETYLENE C_2Cl_2

1. Wotiz, J. H. *et al., J. Org. Chem.*, 1961, **26**, 1626
2. Ott, E., *Ber.*, 1942, **75**, 1517
3. Kirk-Othmer, 1964, Vol. 5, 203–205
4. Siegel, J. *et al., J. Org. Chem.*, 1970, **35**, 3199
5. Riemschneider, R. *et al., Ann.*, 1961, **640**, 14

Dichloroacetylene is a heat-sensitive explosive gas which ignites in contact with air. However, its azeotrope with diethyl ether (55.4% dichloroacetylene) is not explosive and is stable to air [1, 2]. It is formed on catalysed contact between acetylene and chlorine, or sodium hypochlorite at low temperature; or by the action of alkali upon polychloro-ethane and -ethylene derivatives, notably trichloroethylene [3]. A safe synthesis has been described [4]. Ignition of 58% mol solution in ether on exposure to air of high humidity and violent explosion of a concentrated solution in carbon tetrachloride shortly after exposure to air have been reported. Stirring the ethereal solution with tap-water usually caused ignition and explosion [5].

See TETRACHLOROETHYLENE, C_2Cl_4 : Sodium hydroxide
TRICHLOROETHYLENE, C_2HCl_3 : Alkali
: Epoxides

See also CHLOROCYANOACETYLENE, C_3ClN
See other HALOACETYLENE DERIVATIVES

1,2-DICHLOROTETRAFLUOROETHANE $C_2Cl_2F_4$

Aluminium
See ALUMINIUM, Al: Halocarbons

OXALYL DICHLORIDE $C_2Cl_2O_2$

Potassium
See POTASSIUM, K: Oxalyl dihalides

1,1,2-TRICHLOROTRIFLUOROETHANE $C_2Cl_3F_3$

Metals

See ALUMINIUM, Al: Halocarbons
BARIUM, Ba: Halocarbons
LITHIUM, Li: Halocarbons
SAMARIUM, Sm: 1,1,2-Trichlorotrifluoroethane
TITANIUM, Ti: Halocarbons

TETRACHLOROETHYLENE C_2Cl_4

Dinitrogen tetraoxide

See DINITROGEN TETRAOXIDE (NITROGEN DIOXIDE), N_2O_4: Halocarbons

Metals

See BARIUM, Ba: Halocarbons
LITHIUM, Li: Halocarbons

Sodium hydroxide

Mitchell, P. R., private comm., 1973

The presence of 0.5% of trichloroethylene as impurity in tetrachloroethylene during unheated drying over solid sodium hydroxide caused the generation of dichloroacetylene. After subsequent fractional distillation, the volatile fore-run exploded.

See DICHLOROACETYLENE, C_2Cl_2
TRICHLOROETHYLENE, C_2HCl_3: Alkali

See other HALOALKENES

CHROMYL ISOTHIOCYANATE $C_2CrN_2O_2S_2$

Forbes, G. S. *et al., J. Amer. Chem. Soc.,* 1943, **65**, 2273

The salt, prepared in carbon tetrachloride solution, exploded feebly several times as the exothermic reaction proceeded. Oxidation of thiocyanate by chromium(VI) is postulated.

CHROMYL ISOCYANATE $C_2CrN_2O_4$

Forbes, G. S. *et al., J. Amer. Chem. Soc.*, 1943, **65**, 2273
The product is stable during unheated vacuum evaporation of its solutions in carbon tetrachloride. Evaporation with heat at atmospheric pressure led to a weak explosion of the salt.

DICAESIUM ACETYLIDE C_2Cs_2

Acids
Mellor, 1946, Vol. 5, 848
It ignites in contact with gaseous hydrogen chloride or its concentrated aqueous solution, and explodes with nitric acid.

Diiron trioxide
Mellor, 1946, Vol. 5, 849
Incandescence on warming.

Halogens
Mellor, 1946, Vol. 5, 848
It burns in all four halogens at or near ambient temperature.

Lead dioxide
See LEAD(IV) OXIDE, O_2Pb: Metal acetylides

Non-metals
Mellor, 1946, Vol. 5, 848
Mixtures of boron or silicon with dicaesium acetylide react vigorously on heating.
See other METAL ACETYLIDES

COPPER(II) ACETYLIDE C_2Cu

Urbanski, 1967, Vol. 3, 228
Copper(II) acetylide (black or brown) is much more sensitive to

impact, friction and heat than copper(I) acetylide (red-brown), which is used in electric fuses or detonators.
See other METAL ACETYLIDES

COPPER(II) CYANIDE C_2CuN_2

Magnesium
See MAGNESIUM, Mg: Metal cyanides

COPPER(II) FULMINATE $C_2CuN_2O_2$

See METAL FULMINATES

DICOPPER(I) ACETYLIDE C_2Cu_2

1. Mellor, 1946, Vol. 5, 851, 852
2. Rutledge, 1968, 84–85

Readily formed from copper or its compounds and acetylene, it detonates on impact or heating above 100°C. If warmed in air or oxygen, it explodes on subsequent contact with acetylene [1]. Explosivity of the precipitate increases with acidity of the salt solutions, while the stability increases in the presence of reducing agents (formaldehyde, hydrazine, or hydroxylamine). The form with a metallic lustre was the most explosive acetylide made. Catalysts with the acetylide supported on a porous solid are fairly stable [2].

Halogens
Mellor, 1946, Vol. 5, 852
Ignition occurs on contact with chlorine, bromine vapour or finely divided iodine.

Silver nitrate
Mellor, 1946, Vol. 5, 853
Contact with silver nitrate solution transforms copper(I) acetylide

into a sensitive and explosive mixture of silver acetylide and silver.
See also SILVER NITRATE, $AgNO_3$: Acetylene, etc.
See other METAL ACETYLIDES

COPPER(I) OXALATE $C_2Cu_2O_4$

Sidgwick, 1950, 126
Explodes feebly on heating.
See other METAL OXALATES

BISFLUOROFORMYL PEROXIDE $C_2F_2O_4$

1. Talbot, R. L., *J. Org. Chem.*, 1968, **33**, 2095
2. Czerepinski, R. *et al.*, *Inorg. Chem.*, 1968, **7**, 109

Several explosions occurred during the preparation, which involves charging carbon monoxide into a mixture of fluorine and oxygen [1]. It has been known to decompose or explode at elevated temperatures, and all samples should be maintained below 30°C and well shielded [2].
See other DIACYL PEROXIDES

TRIFLUOROACETYL NITRITE $C_2F_3NO_3$

1. Taylor, C. W. *et al.*, *J. Org. Chem.*, 1962, **27**, 1064
2. Gibbs, R. *et al.*, *J. Chem. Soc. Perkin II*, 1972, 1340

Though much more stable than acetyl nitrite even at 100°C, the vapour of trifluoroacetyl nitrite will explode at 160–200°C unless diluted with inert gas to below about 50% vol. concentration. Higher perfluoro-homologues are more stable [1]. A detailed examination of the explosion parameters has been made [2].
See other ACYL NITRITES

TETRAFLUOROETHYLENE C_2F_4

Gas above −76°C; E.L., 11−60%
Graham, D. P., *J. Org. Chem.*, 1966, **31**, 956
A terpene inhibitor is usually added to the monomer to prevent spontaneous polymerisation, and in its absence the monomer will spontaneously explode at pressures above 2.7 bar. The inhibited monomer will explode if ignited.

Air
Muller, R. *et al., Paste und Kaut.*, 1967, **14**(12), 903
Liquid tetrafluoroethylene, being collected in an open liquid nitrogen-cooled trap, formed a peroxidic polymer which exploded.

Air,
Hexafluoropropene
Dixon, G. D., private comm., 1968
A mixture of the two monomers and air sealed in an ampoule formed a gummy peroxide during several weeks. The residue left after opening the tube exploded violently on warming.
See other POLYPEROXIDES

Difluoromethylene dihypofluorite
See DIFLUOROMETHYLENE DIHYPOFLUORITE, CF_4O_2: Haloalkenes

Dioxygen difluoride
See DIOXYGEN DIFLUORIDE, F_2O_2

Iodine pentafluoride,
Limonene
MCA Case History No. 1520
Accidental contamination of a tetrafluoroethylene gas supply system with iodine pentafluoride caused a violent explosion in the cylinders. Exothermic reaction of the contaminant with the inhibitor (limonene) present in the gas cylinders may have initiated the explosion.

Oxygen
See OXYGEN, O_2: Tetrafluoroethylene
See other HALOALKENES

POLYTETRAFLUOROETHYLENE $(C_2F_4)_n$

Fluorine
See FLUORINE, F_2: Polymeric materials

TRIFLUOROACETYL HYPOFLUORITE $C_2F_4O_2$

Alone,
or Potassium iodide,
Water
Cady, G. H. *et al., J. Amer. Chem. Soc.,* 1953, **75**, 2501–2502
The gas explodes on sparking and often during preparation or distillation. Unless much diluted with nitrogen, it explodes on contact with aqueous potassium iodide.
See other HYPOHALITES

PERFLUORO-*N*-CYANODIAMINOMETHANE $C_2F_5N_3$

See FLUORINE, F_2: Sodium dicyanamide

BIS(TRIFLUOROMETHYL)PHOSPHORUS(III) AZIDE $C_2F_6N_3P$

1. Allcock, H. R., *Chem. Eng. News,* 1968, **46**(18), 70
2. Tesi, G. *et al., Proc. Chem. Soc.,* 1960, 219

Explosive, but stable if stored cold [1]. Previously it was found to be unpredictably unstable, violent explosions having occurred even at –196°C [2].
See other NON-METAL AZIDES

XENON(II) TRIFLUOROACETATE $C_2F_6O_4Xe$

Jha, N. K., *RIC Rev.,* 1971, **4**, 157

Explodes on thermal or mechanical shock.
See other XENON COMPOUNDS

PERFLUORO-1-AMINOMETHYLGUANIDINE $C_2F_8N_4$

See FLUORINE, F_2: Cyanoguanidine

PERFLUORO-*N*-AMINOMETHYLTRIAMINOMETHANE $C_2F_{10}N_4$

See FLUORINE, F_2: Cyanoguanidine

SILVER ACETYLIDE C_2HAg

Anon., *Chem. Trade J.*, 1966, **158**, 153
A poorly stoppered dropping bottle of silver nitrate solution absorbed sufficient acetylene from the atmosphere in an acetylene plant laboratory to block the dropping tube. A violent detonation occurred on moving the dropping tube.
See also METAL ACETYLIDES

BROMOACETYLENE C_2HBr

1. Tanaka, R. *et al., J. Org. Chem.*, 1971, **36**, 3856
2. Hucknall, D. J. *et al., Chem. & Ind.*, 1972, 116

It is dangerous and may burn or explode in contact with air, even when solid at −196°C [1]. Procedures for safe generation, transfer and storage, all under nitrogen, are described [2].
See other HALOACETYLENE DERIVATIVES

CHLOROACETYLENE C_2HCl

Tanaka, R. *et al., J. Org. Chem.*, 1971, **36**, 3856

Chloroacetylene may burn or explode in contact with air, and its greater volatility makes it more dangerous than bromoacetylene. Procedures for safe generation, handling and storage, all under nitrogen, are described.

See also TRICHLOROETHYLENE, C_2HCl_3 : Alkali
See other HALOACETYLENE DERIVATIVES

TRICHLOROETHYLENE C_2HCl_3

Fl.P., 32°C; E.L., 12.5–90% (at above 30°C)
MCA SD-14, 1956 considers it to be non-flammable at normal ambient temperatures, but note the very wide limits.

Alkali

1. Fabian, F., private comm., 1960
2. *ABCM Quart. Safety Summ.*, 1956, **27**, 17

An emulsion, formed during extraction of a strongly alkaline liquor with trichloroethylene, decomposed with the evolution of the spontaneously flammable gas, dichloroacetylene [1]. This reaction could also occur if alkaline metal-stripping preparations were used in conjunction with trichloroethylene de-greasing preparations, some of which also contain amine inhibitors which could cause the same reaction [2].

See also TETRACHLOROETHYLENE, C_2Cl_4 : Soium hydroxide

Epoxides

Dobinson, B. *et al., Chem. & Ind.*, 1972, 214

1-Chloro-2,3-epoxypropane, the mono- and di-2,3-expoypropyl ethers of 1,4-butanediol, and 2,2-bis [4(2′,3′-epoxpropoxy)-phenyl]-propane can, in presence of catalytic quantities of halide ions, cause dehydrochlorination of trichloroethylene to dichloro-acetylene, which causes minor explosions when the mixture is boiled under reflux. A mechanism is discussed.

See also Alkali, above

Metals

See ALUMINIUM, Al: Halocarbons
BARIUM, Ba: Halocarbons
BERYLLIUM, Be: Halocarbons

LITHIUM, Li: Halocarbons
MAGNESIUM, Mg: Halocarbons
TITANIUM, Ti: Halocarbons

Oxidants

See PERCHLORIC ACID, $ClHO_4$: Trichloroethylene
DINITROGEN TETRAOXIDE (NITROGEN DIOXIDE), N_2O_4: Halocarbons
OXYGEN (Gas), O_2: Halocarbons
OXYGEN (Liquid), O_2: Halocarbons
See other HALOALKENES

PENTACHLOROETHANE C_2HCl_5

Potassium

See POTASSIUM, K: Halocarbons

CAESIUM ACETYLIDE C_2HCs

Mellor, 1946, Vol. 5, 849–850
It reacts similarly to the dicaesium compound.
See DICAESIUM ACETYLIDE, C_2Cs_2
See other METAL ACETYLIDES

FLUOROACETYLENE C_2HF

Alone,
or Bromine

Middleton, W. J. *et al., J. Amer. Chem. Soc.,* 1959, **81**, 803–804
Liquid fluoroacetylene is treacherously explosive close to its b.p., –80°C. The gas does not ignite in air and is not explosive. Ignition occurred in contact with a solution of bromine in carbon tetrachloride. Mercury and silver salts were both stable to impact, but the latter exploded on heating, whereas the former decomposed violently.
See other HALOACETYLENE DERIVATIVES

TRIFLUOROACETIC ACID $C_2HF_3O_2$

Lithium tetrahydroaluminate

See LITHIUM TETRAHYDROALUMINATE, AlH_4Li: Fluoroamides

PEROXYTRIFLUOROACETIC ACID $C_2HF_3O_3$

Sundberg, R. J. *et al., J. Org. Chem.,* 1968, **33**, 4098

This extremely powerful oxidising agent must be handled and used with great care.

See other PEROXYACIDS

POTASSIUM ACETYLIDE C_2HK

Chlorine

Mellor, 1946, Vol. 5, 849

It ignites in chlorine.

Non-metal oxides

Mellor, 1946, Vol. 5, 849

Interaction with sulphur dioxide at ambient temperature, or with carbon dioxide on warming, causes incandescence.

See other METAL ACETYLIDES

DIPOTASSIUM *aci*-NITROACETATE $C_2HK_2NO_4$

Water

1. Lyttle, D. A., *Chem. Eng. News,* 1949, **27**, 1473
2. Whitmore, F. C., *Org. Synth.,* 1941, Coll. Vol. 1, 401

An inhomogeneous mixture of the dry salt with a little water exploded violently after 30 min [1]. This was probably due to exothermic decarboxylation generating free nitromethane in an alkaline system. The decomposition of sodium nitroacetate proceeds exothermically above 80°C [2].

See NITROMETHANE, CH_3NO_2: Acids, etc.

See other *aci*-NITRO SALTS

LITHIUM ACETYLIDE C_2HLi

Mellor, 1946, Vol. 5, 849
It reacts similarly to the dilithium compound.
See DILITHIUM ACETYLIDE, C_2Li_2

LITHIUM ACETYLIDE-AMMONIA $C_2HLi \cdot H_3N$

Gases,
or Water
Mellor, 1946, Vol. 5, 8
The complex ignites in contact with carbon dioxide, sulphur dioxide, chlorine or water.
See other METAL ACETYLIDES

DIAZOACETONITRILE C_2HN_3

Phillips, D. D. *et al., J. Amer. Chem. Soc.,* 1956, **78**, 5452
Removal of a rubber stopper from a flask of a concentrated solution in methylene chloride initiated an explosion, probably through friction on solvent-free material between the flask neck and bung. Handling only in dilute solution is recommended.
See other DIAZO COMPOUNDS
HIGH-NITROGEN COMPOUNDS

DINITROACETONITRILE $C_2HN_3O_4$

See POLYNITROALKYL COMPOUNDS

SODIUM 5-DINITROMETHYLTETRAZOLIDE $C_2HN_6NaO_4$

Einberg, F., *J. Org. Chem.,* 1964, **29**, 2021

It explodes violently at m.p., 160°C, and may be an *aci*-nitro salt.
See other POLYNITROALKYL COMPOUNDS
TETRAZOLES

5-HYDROXY-1(*N*-SODIO-5-TETRAZOLYLAZO)TETRAZOLE $C_2HN_{10}NaO$

Thiele, J. *et al., Ann.,* 1893, **273**, 150
The salt explodes very violently on heating.
See other *N*-METAL DERIVATIVES
TETRAZOLES

SODIUM ACETYLIDE C_2HNa

1. Mellor, 1946, Vol. 5, 849
2. Greenlee, K. W., *Inorg. Synth.,* 1947, **2**, 81

It reacts similarly to the disodium compound [1]. If heated to 150°C, it decomposes extensively, evolving gas which ignites in air owing to presence of pyrophoric carbon. The residual carbon is also highly reactive [2].
See DISODIUM ACETYLIDE, C_2Na_2
See other METAL ACETYLIDES

RUBIDIUM ACETYLIDE C_2HRb

Mellor, 1946, Vol. 5, 849
It reacts similarly to the di-rubidium compound.
See DIRUBIDIUM ACETYLIDE, C_2Rb_2
See other METAL ACETYLIDES

ACETYLENE C_2H_2

Gas above −83°C; E.L., 2.5–82% (but may explosively decompose without air)

MCA SD-7, 1957
1. Mayes, H. A., *Chem. Engr.*, 1965, (185), 25
2. Nedwick, J. J., *Ind. Eng. Chem. Proc. Des. Dev.*, 1962, **1**, 137; Tedeschi, R. J. *et al., ibid.*, 1968, **7**, 303
3. Miller, 1965, Vol. 1, 506
4. Kirk-Othmer, 1963, Vol. 1, 195–202
5. Rimarski, W., *Angew. Chem.*, 1929, **42**, 933
6. Foote, C. S., private comm., 1965
7. Sutherland, M. E. *et al., Chem. Eng. Progr.*, 1973, **69**(4), 48–51

Addition of up to 30% of miscible diluent did not sufficiently desensitise (explosive) liquid acetylene to permit of its transportation. The solid has similar properties [1]. However, safe techniques for the use of liquid acetylene in reaction systems at pressures up to 270 bar have been described [2].

Solid acetylene in presence of liquid nitrogen at −181°C is sensitised by presence of grit (carborundum) and may readily explode on impact [3]. The hazards of preparing and using acetylene have been adequately described [3–5]. A flask which had contained acetylene was left open to the air for a week, and then briefly purged with air. It still contained enough acetylene to form an explosive mixture [6].

The explosive decomposition of acetylene in absence of air readily escalates to detonation in tubular vessels. This type of explosive decomposition has been experienced in a 7 mile acetylene pipeline system [7].

Cobalt

Mellor, 1942, Vol. 14, 513

Finely divided (pyrophoric) cobalt decomposes and polymerises acetylene on contact, becoming incandescent.

Copper

Anon., *ABCM Quart. Safety Summ.*, 1946, **17**, 24

Rubber-covered electric cable, used as a makeshift handle in the effluent pit of an acetylene plant, formed copper acetylide with residual acetylene and the former detonated when disturbed and initiated explosion of the latter. All heavy metals must be rigorously excluded from locations where acetylene may be present.

See METAL ACETYLIDES

Copper (I) acetylide
See COPPER (I) ACETYLIDE, C_2Cu_2

Halogens
1. Muir, G. D., private comm., 1968
2. von Schwartz, 1918, 142, 321
3. *MCA SD-7,* 1957

Tetrabromoethane is made by passing acetylene into bromine in carbon tetrachloride at reflux. The rate of reaction falls off rapidly below reflux temperature, and if the rate of addition of acetylene is insufficient to maintain the temperature, high concentrations of unreacted acetylene build up, with the possibility of a violent delayed reaction [1]. In absence of a diluent, interaction may be explosive [2]. Mixtures of acetylene and chlorine may explode upon initiation by sunlight or other UV source, or high temperature, sometimes very violently [2]. Interaction with fluorine is very violent [3] and with iodine possibly explosive [2].

Heavy metals and salts
See METAL ACETYLIDES

Oxides of nitrogen
See NITROGEN OXIDE ('NITRIC OXIDE'), NO: Dienes, Oxygen

Oxygen
Fowles, G., *School Sci. Rev.,* 1940, **22**(85), 6
Holt, C., 1962, **44**(152), 161
Arculus, P. D., 1963, **44**(154), 706
Peacocke, T. A. H., *School Sci. Rev.,* 1964, **45**(156), 459

The explosion of acetylene–oxygen mixtures in open vessels is a very dangerous experiment (stoicheiometric mixtures detonate with great violence, completely shattering the container). Full precautions are essential for safety.

Potassium
Berthelot, M., *Bull. Soc. Chim. Fr.* [2], 1866, **5**, 188

Molten potassium ignites in acetylene, then explodes.

Trifluoromethyl hypofluorite
See TRIFLUOROMETHYL HYPOFLUORITE, CF_4O: Hydrocarbons
See other ACETYLENIC COMPOUNDS

SILVER 1,3-DI(5-TETRAZOLYL)TRIAZENE $C_2H_2AgN_{11}$

See 1,3-DI(5-TETRAZOLYL)TRIAZENE, $C_2H_3N_{11}$

2-IODOSYLVINYL CHLORIDE C_2H_2ClIO

Alone,
or Water
Thiele, J. *et al., Ann.*, 1909, **369**, 131
It explodes at 63°C and contact with water disproportionates it to the more explosive iodyl compound.

2-IODYLVINYL CHLORIDE $C_2H_2ClIO_2$

Thiele, J. *et al., Ann.*, 1909, **369**, 131
Explodes violently on impact, friction or heating to 135°C.
See other IODINE COMPOUNDS

1,1-DICHLOROETHYLENE $C_2H_2Cl_2$

Fl. P., –15°C (o); E. L., 7.3–16%

Air
1. Reinhardt, R. C., *Chem. Eng. News*, 1947, **25**, 2136
2. *MCA Case History No. 1172*

When stored between –40° and +25°C in the absence of inhibitor and presence of air, vinylidene chloride rapidly absorbs oxygen with formation of a violently explosive peroxide. The latter initiates polymerisation, producing an insoluble polymer which adsorbs the peroxide. Separation of this polymer in a dry state must be avoided, since if more than 15% of peroxide is present, the polymer may be detonable by slight shock or heat. Hindered phenols are suitable inhibitors to prevent peroxidation [1]. The Case History describes an

explosion during handling of a pipe used to transfer vinylidene chloride [2].
See also CHLOROETHYLENE (VINYL CHLORIDE), C_2H_3Cl: Air
See other POLYPEROXIDES

Chlorotrifluoroethylene
Raasch, M. S. *et al., Org. Synth.*, 1962, **42**, 46
Condensation of the reactants at 180°C under pressure to give 1,1,2-trichloro-2,3,3-trifluorocyclobutane was effected smoothly several times in a 1 litre autoclave. Scaling up to a 3 litre autoclave led to uncontrolled polymerisation which distorted the autoclave.

Ozone
'Vinylidene Chloride Monomer', 9, Midland (Mich.), Dow Chemical Co., 1966
The reaction products formed with ozone are particularly dangerous.

Perchloryl fluoride
See PERCHLORYL FLUORIDE, $ClFO_3$: Hydrocarbons, etc.
See other HALOALKENES

cis-1,2-DICHLOROETHYLENE $C_2H_2Cl_2$

Fl.P., 6°C; E. L., 3.3–15%

trans-1,2-DICHLOROETHYLENE $C_2H_2Cl_2$

Fl.P., 2°C; E. L., 9.7–12.8%
BCISC Quart. Safety Summ., 1964, **35**, 37
Sichere Chemiearbeit, 1964, **16**, 35
Under appropriate conditions, dichlorethylene, previously thought to be non-flammable, can cause a fire hazard. Addition of a hot liquid to the cold solvent caused sudden emission of sufficient vapour to cause a flame to flash back 12 m from a fire. Although the bulk of the solvent did not ignite, various items of paper and wood in the room were ignited by the transient flame.

Alkalies

Fire Accident Prev., 1956, **42**, 28

1,2-Dichloroethylene in contact with solid caustic alkalies or their concentrated solutions will form chloroacetylene which ignites in air.

See also TRICHLOROETHYLENE, C_2HCl_3: Alkalies
1,1,2,2-TETRACHLOROETHANE, $C_2H_2Cl_4$: Alkalies

Difluoromethylene dihypofluorite

See DIFLUOROMETHYLENE DIHYPOFLUORITE, CF_4O_2: Haloalkenes

Dinitrogen tetraoxide

See DINITROGEN TETRAOXIDE (NITROGEN DIOXIDE), N_2O_4: Halocarbons

See other HALOALKENES

1,1,1,2-TETRACHLOROETHANE $C_2H_2Cl_4$

Dinitrogen tetraoxide

See DINITROGEN TETRAOXIDE (NITROGEN DIOXIDE), N_2O_4: Halocarbons

1,1,2,2-TETRACHLOROETHANE $C_2H_2Cl_4$

Alkalies

MCA SD-34, 1949

It is not an inert solvent, and on heating with solid potassium hydroxide or other base, hydrogen chloride is eliminated and chloro- or dichloroacetylene, which ignite in air, are formed.

See also TRICHLOROETHYLENE, C_2HCl_3: Alkalies
1,2-DICHLOROETHYLENE, $C_2H_2Cl_2$: Alkalies

Metals

See POTASSIUM, K: Halocarbons
SODIUM, Na: Halocarbons

See other HALOCARBONS

1,1-DIFLUOROETHYLENE $C_2H_2F_2$

Gas above −86°C; E. L., 5.5–21.3%
See other HALOALKENES

TRIFLUOROACETAMIDE $C_2H_2F_3NO$

Lithium tetrahydroaluminate
See LITHIUM TETRAHYDROALUMINATE, AlH_4Li: Fluoroamides

MERCURY(II) METHYLNITROLATE $C_2H_2HgN_4O_6$

Urbanski, 1967, Vol. 3, 158
Formally a derivative of mercury fulminate, this salt of *aci*-dinitromethane also shows detonator properties
See related METAL FULMINATES

'ETHANE HEXAMERCARBIDE' $C_2H_2Hg_6O_4$

Hofmann, K. A., *Ber.*, 1898, **31**, 1904
This compound, formulated as 1,2-bis(hydroxymercuri)-1,1: 2,2-bis(oxydimercuri)ethane, explodes very violently at 230°C.
See other HEAVY METAL DERIVATIVES

DIAZOACETALDEHYDE $C_2H_2N_2O$

Arnold, Z., *Chem. Comm.*, 1967, 299
It may be distilled out continuously as formed at 40°C/0.013 bar, but readily detonates with great violence if overheated
See other DIAZOCOMPOUNDS

AZIDOACETONITRILE $C_2H_2N_4$

Freudenberg, K. *et al., Ber.,* 1932, **65**, 1188
With over 51% nitrogen content, it is, as expected, explosive, sensitive to impact or heating to 250°C.
See other HIGH-NITROGEN COMPOUNDS
ORGANIC AZIDES

3-AZIDO-1,2,4-TRIAZOLE $C_2H_2N_6$

Denault, G. C. *et al., J. Chem. Eng. Data,* 1968, **13**, 514
Samples exploded during analytical combustion.
See other ORGANIC AZIDES

1,2-DIAZIDOCARBONYLHYDRAZINE $C_2H_2N_8O_2$

Kesting, W., *Ber.,* 1924, **57**, 1321
Similar in properties to lead or silver azide, it explodes on heating or impact.
See other ACYL AZIDES
HIGH-NITROGEN COMPOUNDS

GLYOXAL $C_2H_2O_2$

Preparative hazard.
See NITRIC ACID, HNO_3 : 2,4,6-Trimethyltrioxane

OXALIC ACID $C_2H_2O_4$

Silver
See SILVER, Ag: Carboxylic acids

Sodium chlorite
See SODIUM CHLORITE, $ClNaO_2$: Oxalic acid

TRILEAD DICARBONATE DIHYDROXIDE $C_2H_2O_8Pb_2$

Fluorine
See FLUORINE, F_2 : Metal salts

BROMOETHYLENE (VINYL BROMIDE) C_2H_3Br

Non-flammable except at 6–15% with arc, hot flame or other high-energy source.
See other HALOALKENES

LEAD ACETATE BROMATE $C_2H_3BrO_5Pb$

Berger, A., *Arbeits-Schutz.*, 1934, 2, 20
The compound (possibly the double salt) may be formed during preparation of lead bromate from lead acetate and potassium bromate in acetic acid, and is explosive and very sensitive to friction. While pure lead bromate is stable up to 180°C, it is an explosive salt.
See LEAD BROMATE, Br_2O_6Pb
See other METAL OXOHALOGENATES

CHLOROETHYLENE (VINYL CHLORIDE) C_2H_3Cl

Fl. P., –8°C (gas above –14°C); E. L., 4–22% (33% also quoted)
MCA SD-56, 1972
MCA Case History No. 625
Discharge of a spray of vapour and liquid under pressure from a cylinder into a fume hood caused ignition of the vapour, due to static electricity. Discharge of the gas only did not cause static build-up.
See other HALOALKENES

Air
MCA Case History No. 1551
Accidental exposure of the recovered monomer to atmospheric oxygen for a long period caused formation of an unstable polyperoxide which initiated an explosion. Suitable precautions are discussed, including use of aqueous 20–30% sodium hydroxide solution to destroy the peroxide.
See also 1,1-DICHLOROETHYLENE, $C_2H_2Cl_2$: Air
See other POLYPEROXIDES

1-CHLORO-1,1-DIFLUOROETHANE $C_2H_3ClF_2$

Gas above –9°C; E. L., 9.0–14.8%

ACETYL CHLORIDE C_2H_3ClO

Fl. P., 4°C; E. L., 5.0– %
Preparative hazard.
See PHOSPHORUS TRICHLORIDE, Cl_3P: Acetic acid

Dimethyl sulphoxide
See DIMETHYL SULPHOXIDE, C_2H_6OS: Acyl halides

Water
Haz. Chem. Data, 1971, 24
Violent interaction

METHYL CHLOROFORMATE $C_2H_3ClO_2$

Fl. P., 12°C

1,1,1-TRICHLOROETHANE $C_2H_3Cl_3$

MCA SD-90, 1965

Dinitrogen tetraoxide

See DINITROGEN TETRAOXIDE (NITROGEN DIOXIDE), N_2O_4: Halocarbons

Metals

See ALUMINIUM, Al: Halocarbons
POTASSIUM, K: Halocarbons
POTASSIUM–SODIUM ALLOY, K–Na: Halocarbons
MAGNESIUM, Mg: Halocarbons

Oxygen

See OXYGEN (Gas), O_2: Halocarbons
OXYGEN (Liquid), O_2: Halocarbons

1,1,2-TRICHLOROETHANE $C_2H_3Cl_3$

Potassium

See POTASSIUM, K: Halocarbons

1,1,1-TRICHLOROETHANOL $C_2H_3Cl_3O$

Sodium hydroxide

MCA Case History No. 1574

Accidental contact of 50% sodium hydroxide solution with residual trichloroethanol in a pump caused an explosion. This was confirmed in laboratory experiments. Chlorohydroxyacetylene, the isomeric chloroketene or chlorooxirene, may have been formed by elimination of hydrogen chloride.

TRICHLORO(VINYL)SILANE $C_2H_3Cl_3Si$

Fl. P., below 10°C; may ignite in air

See other ALKYLNON-METAL HALIDES

FLUOROETHYLENE (VINYL FLUORIDE) C_2H_3F

Gas above –72°C; E. L., 2.6–22%
See other HALOALKENES

2(?)-FLUORO-1,1-DINITROETHANE $C_2H_3FN_2O_4$

MCA Case History No. 784
During a prolonged fractionation of the crude material at 75°C/0.05 bar, an exothermic decomposition began. As a remedial measure, air was admitted to the hot, decomposing residue, causing a violent explosion. Admission of nitrogen, or cooling of hot residues before admitting air, might have avoided the incident.
See other POLYNITROALKYL COMPOUNDS

2-FLUORO-2,2-DINITROETHANOL $C_2H_3FN_2O_5$

Cochoy, R. E. *et al., J. Org. Chem.*, 1972, **37**, 3041
It is a potentially explosive vesicant, from which a series of esters, expected to be explosive, was prepared.
See other FLUORODINTROMETHYL COMPOUNDS
POLYNITROALKYL COMPOUNDS

MONOPOTASSIUM *aci*-1,1-DINITROETHANE $C_2H_3KN_2O_3$

Rodd, 1965, Vol. 1B, 98
Isolated salts of nitrolic acids are explosive.
See other aci-NITRO SALTS

POTASSIUM *O*-METHYLDITHIOCARBONATE $C_2H_3KOS_2$

Benzenediazonium chloride
See DIAZONIUM SULPHIDES AND DERIVATIVES
BENZENETHIOL (THIOPHENOL), C_6H_6S

VINYLLITHIUM C_2H_3Li

Juenge, E. C. *et al., J. Org. Chem.,* 1961, **26**, 564
When freshly prepared, it is violently pyrophoric but on storage it becomes less reactive and slow to ignite in air, possibly owing to polymerisation.
See other ALKYLMETALS

ACETONITRILE C_2H_3N

Fl. P., 6°C (o); E. L., 4–16%

2-Cyano-2-propyl nitrate
See NITRATING AGENTS

Dinitrogen tetraoxide,
Indium
See DINITROGEN TETRAOXIDE (NITROGEN DIOXIDE), N_2O_4: Acetonitrile, Indium

N-Fluoro compounds
Fraser, G. W., *et al., Chem. Comm.,* 1966, 532
Nitrogen–fluorine compounds are potentially explosive in contact with acetonitrile.
See PERFLUOROUREA, CF_4N_2O: Acetonitrile

Nitric acid
See NITRIC ACID, HNO_3: Acetonitrile

Sulphuric acid,
Sulphur trioxide
Lee, S.A., private comm., 1972
A mixture of acetonitrile and sulphuric acid on heating (or self-heating) to 53°C underwent an uncontrollable exotherm to 160°C in a few seconds. The presence of 28 mol % of sulphur trioxide reduces the initiation temperature to about 15°C. Polymerisation of acetonitrile is suspected.
See other CYANO COMPOUNDS

METHYL ISOCYANIDE C_2H_3N

Lemoult, M. D., *Compt. Rend.*, 1906, **143**, 902
It exploded when heated sealed in an ampoule.
See ETHYL ISOCYANIDE, C_3H_5N
See related CYANO COMPOUNDS

GLYCOLONITRILE C_2H_3NO

1. *BCISC Quart. Safety Summ.*, 1964, **35**, 2
2. Gaudry, R., *Org. Synth.*, 1955, Coll. Vol. 3, 436
3. Anon., *Chem. Eng. News*, 1966, **44**, (49), 50

A year-old bottled sample, containing syrupy phosphoric acid as stabiliser, and which already showed signs of tar formation, exploded in storage. The pressure explosion appeared to be due to polymerisation, after occlusion of inhibitor in tar, in a container in which the stopper had become cemented by polymer [1]. A similar pressure explosion occurred when dry, redistilled nitrile, stabilised with ethanol [2], polymerised after 13 days [3].
See other CYANO COMPOUNDS

METHYL ISOCYANATE C_2H_3NO

Fl. p., below -15°C

ACETYL NITRITE $C_2H_3NO_3$

Francesconi, L. *et al.*, *Gazz. Chim. Ital.*, 1895, [1], **34**, 439
An unstable liquid, decomposed by light, of which the vapour is violently explosive on mild heating.
See other ACYL NITRITES

ACETYL NITRATE $C_2H_3NO_4$

1. Pictet, A. *et al., Ber.*, 1907, **40**, 1164
2. Bordwell, F. G. *et al., J. Amer. Chem. Soc.*, 1960, **82**, 3588
3. Konig, W., *Angew. Chem.*, 1955, **67**, 517

Acetyl nitrate, readily formed above 0°C from acetic anhydride and concentrated nitric acid, is thermally unstable, and its solutions may decompose violently above 60°C (forming tetranitromethane, a powerful oxidant). The pure nitrate explodes violently on rapid heating to above 100°C [1]. Isolation before use as a nitrating agent at −10°C is not necessary, but the mixture must be preformed at 20–25°C before cooling to −10°C to avoid violent reactions [2]. Spontaneous explosions of pure isolated material a few days old had been reported previously [3].

See also TETRANITROMETHANE, CN_4O_8

Mercury(II) oxide

Chretien, A. *et al., Compt. Rend.*, 1945, **220**, 823

Acetyl nitrate explodes when mixed with red mercury oxide, or other 'active' oxides.

See other ACYL NITRATES

VINYL AZIDE $C_2H_3N_3$

Wiley, R. H. *et al., J. Org. Chem.*, 1957, **22**, 995

A sample contained in a flask detonated when the ground-joint was rotated. Literature statements that it is surprisingly stable are erroneous.

See other ORGANIC AZIDES

ACETYL AZIDE $C_2H_3N_3O$

Smith, 1966, Vol. 2, 214

It is treacherously explosive.

See other ACYL AZIDES

AZIDOACETALDEHYDE $C_2H_3N_3O$

Forster, M. O., *et al., J. Chem. Soc.,* 1908, **93**, 1870
It decomposed with vigorous gas evolution below 80°C at 5 mbar. A reaction mixture of chloroacetaldehyde hydrate and sodium azide had previously exploded mildly on heating in absence of added water.
See other 2-AZIDOCARBONYL COMPOUNDS

5-METHOXY-1,2,3,4-THIATRIAZOLE $C_2H_3N_3OS$

Jensen, K. A., *et al., Acta. Chem. Scand.,* 1964, **18**, 825
It explodes at ambient temperature, and its higher homologues are unstable.
See other HIGH-NITROGEN COMPOUNDS

2,2,2-TRINITROETHANOL $C_2H_3N_3O_7$

1. Marans, N. S. *et al., J. Amer. Chem. Soc.,* 1950, **72**, 5329
2. Cochoy, R. E. *et al., J. Org. Chem.,* 1972, **37**, 3041

It is a moderately shock-sensitive explosive which has exploded during distillation [1]. A series of its esters, expected to be explosive, was prepared [2].
See other POLYNITROALKYL COMPOUNDS

1,3-DI(5-TETRAZOYL)TRIAZENE $C_2H_3N_{11}$

Hofmann, K. A. *et al., Ber.,* 1910, **43**, 1869–1870
The barium salt explodes weakly on heating, and the copper and silver salts strongly on heating or friction.
See other TETRAZOLES
TRIAZENES

SODIUM ACETATE $C_2H_3NaO_2$

Diketene
See DIKETENE, $C_4H_4O_2$: Acids, etc.

Potassium nitrate
See POTASSIUM NITRATE, KNO_3: Sodium acetate

SODIUM PEROXYACETATE $C_2H_3NaO_3$

Humber, L. G., *J. Org. Chem.*, 1959, **24**, 1789
A sample of the dry salt exploded at room temperature.
See other PEROXOACID SALTS

ETHYLENE C_2H_4

Gas above −104°C; E.L., 3.0–34%

Aluminium trichloride
Waterman, H. I. *et al.*, *J. Inst. Pet.*, 1947, **33**, 254
Mixtures of ethylene and aluminium chloride, initially at 30–60 bar, rapidly heat and explode in presence of supported nickel catalyst, methyl chloride or nitromethane.

Aluminium trichloride,
Nitromethane
See ALUMINIUM TRICHLORIDE-NITROMETHANE, $AlCl_3 \cdot CH_3NO_2$: Alkene

Bromotrichloromethane
Elsner, H. *et al.*, *Angew. Chem.*, 1962, **74**, 253
Following a literature method for preparation of 1-bromo-3,3,3-trichloropropane, the reagents were being heated at 120°C/51 bar. During the fourth preparation a violent explosion occurred.
See Carbon tetrachloride, below

Carbon tetrachloride
Zakaznov, V. F. *et al., Khim. Prom.*, 1968, 8, 584
Mixtures of ethylene and carbon tetrachloride can be initiated to explode at temperatures between 25 and 105°C and pressures of 30–80 bar, causing a sixfold pressure increase. At 100°C and 61 bar explosion initiated in the gas phase propagated into the liquid phase. Increase of carbon tetrachloride concentrations in the gas phase decreased the limiting decomposition pressure.
See Bromotrichloromethane, above
DIBENZOYL PEROXIDE, $C_{14}H_{10}O_4$: Carbon tetrachloride, etc.

Chlorine
See CHLORINE, Cl_2 : Hydrocarbons

Oxides of nitrogen
See NITROGEN OXIDE ('NITRIC OXIDE'), NO: Dienes, Oxygen

Tetrafluoroethylene
Coffman, D. D. *et al., J. Amer. Chem. Soc.*, 1949, **71**, 492
A violent explosion occurred when a mixture of tetrafluoroethylene and excess ethylene was heated at 160°C and 480 bar. Traces of oxygen must be rigorously excluded. Other olefins reacted smoothly.
See TETRAFLUOROETHYLENE, C_2F_4 : Air

Trifluoromethyl hypofluorite
See TRIFLUOROMETHYL HYPOFLUORITE, CF_4O: Hydrocarbons

1-BROMOAZIRIDINE C_2H_4BrN

Graefe, A. F., *J. Amer, Chem. Soc.*, 1958, **80**, 3940
The compound is very unstable and always decomposes, sometimes explosively during or shortly after distillation.
See other N-HALOGEN COMPOUNDS

N-BROMOACETAMIDE C_2H_4BrNO

'Organic Positive Bromine Compounds', Brochure, Boulder, Arapahoe Chemicals Inc., 1962

It tends to decompose rapidly at elevated temperatures in presence of moisture and light.

See also *N*-HALOIMIDES

See other *N*-HALOGEN COMPOUNDS

1-CHLOROAZIRIDINE C_2H_4ClN

1. Davies, C. S., *Chem. Eng. News,* 1964, **42**(8), 41
2. Graefe, A. F., *Chem. Eng. News,* 1958, **36**(43), 52

A sample of redistilled material exploded after keeping in an amber bottle at ambient temperature for 3 months [1]. A similar sample had exploded very violently when dropped [2].

See other *N*-HALOGEN COMPOUNDS

CHLOROACETALDEHYDE OXIME C_2H_4ClNO

Brintzinger, H. *et al., Chem. Ber.,* 1952, **85**, 345

Vacuum distillation of the product at 61°C/27 mbar must be interrupted when a solid separates from the residue to avoid an explosion.

See also BROMOACETONE OXIME, C_3H_6BrNO

***N*-CHLOROACETAMIDE** C_2H_4ClNO

Muir, G. D., private comm., 1968

It has exploded during desiccation of the solid or during concentration of its chloroform solution. It may be safely purified by pouring a solution in acetone into water and air-drying the product.

See also *N*-HALOIMIDES

See other *N*-HALOGEN COMPOUNDS

1,1-DICHLOROETHANE $C_2H_4Cl_2$

Fl.P., –6°C; E. L., 5.6–11.4%

1,2-DICHLOROETHANE $C_2H_4Cl_2$

Fl.P., 13°C; E. L., 6.2–15.9%
MCA SD-18, 1971

Dinitrogen tetraoxide
See DINITROGEN TETRAOXIDE (NITROGEN DIOXIDE), N_2O_4: Halocarbons

Metals
See ALUMINIUM, Al: Halocarbons
POTASSIUM, K: Halocarbons

AZO-*N*-CHLOROFORMAMIDINE $C_2H_4Cl_2N_6$

Braz, G. I. *et al., Zh. Prikl. Khim. USSR,* 1944, **17**, 565
Decomposes explosively at 155°C.
See other AZO COMPOUNDS
N-HALOGEN COMPOUNDS

BIS-CHLOROMETHYL ETHER $C_2H_4Cl_2O$

Fl.P., below 19°C

ETHYLENE DIPERCHLORATE $C_2H_4Cl_2O_8$

Schumacher, 1960, 214
A highly sensitive, violently explosive material, capable of initiation by addition of a few drops of water.
See other ALKYL PERCHLORATES

2-FLUORO-2,2-DINITROETHYLAMINE $C_2H_4FN_3O_4$

Adolph, H. G. *et al., J. Org. Chem.*, 1969, **34**, 47
Outstandingly explosive amongst fluorodinitromethyl compounds, samples stored neat at ambient temperature regularly exploded within a few hours. Occasionally concentrated solutions in dichloromethane have decomposed violently after long storage.
See other FLUORODINITROMETHYL COMPOUNDS

1,1-DIFLUOROETHANE $C_2H_4F_2$

Gas above −25°C; E.L., 3.7–18.0%

1,2-BIS(DIFLUOROAMINO)ETHANOL $C_2H_4F_4N_2O$

Reed, S. F., *J. Org. Chem.*, 1967, **32**, 2894
It is slightly more impact-sensitive than glyceryl trinitrate.
See DIFLUOROAMINO COMPOUNDS
See other *N*-HALOGEN COMPOUNDS

1,2-BIS(DIFLUOROAMINO)*N*-NITROETHYLAMINE $C_2H_4F_4N_4O_2$

Tyler, W. E., US Pat. 3 344 167, 1967
The crude product tends to explode spontaneously on storage, though the triple-distilled material appears stable on prolonged storage. Generally, such nitroamines are unstable and explode at 75°C or above.
See DIFLUOROAMINO COMPOUNDS
See other *N*-NITRO COMPOUNDS

MERCURY(III) FORMHYDROXAMATE $C_2H_4HgN_2O_4$

Urbanski, 1967, Vol. 3, 158
Formally a hydrate derivative of mercury fulminate, it also possesses detonator properties.
See related METAL FULMINATES

3-METHYLDIAZIRINE $C_2H_4N_2$

Schmitz, E. *et al., Chem. Ber.,* 1962, **95**, 795
The gas explodes on heating.
See DIAZIRINES

1,1-DIAZIDOETHANE $C_2H_4N_6$

Forster, M. O. *et al., J. Chem. Soc.,* 1908, **93**, 1070
The extreme instability and explosive behaviour of this diazide caused work on other *gem*-diazides to be abandoned.

1,2-DIAZIDOETHANE $C_2H_4N_6$

Alone,
or Sulphuric acid
Forster, M. O., *et al., J. Chem. Soc.,* 1908, **93**, 1070
Though less unstable than the 1,1-isomer (above), it explodes on heating, and unreproducibly in contact with sulphuric acid.
See other ORGANIC AZIDES

AZIDOCARBONYLGUANIDINE $C_2H_4N_6O$

Thiele, J. *et al., Ann.,* 1898, **303**, 93
It explodes violently on rapid heating.
See other ACYL AZIDES

AZO-*N*-NITROFORMAMIDINE $C_2H_4N_8O_4$

Wright, G. F., *Can. J. Chem.*, 1952, **30**, 64
Explosive decomposition at 165°C.
See other AZO COMPOUNDS
N-NITRO COMPOUNDS

1,2-DI(5-TETRAZOLYL)HYDRAZINE $C_2H_4N_{10}$

Thiele, J., *Ann.*, 1898, **303**, 66
Explodes without melting when heated.
See other TETRAZOLES

1,6-BIS(5-TETRAZOLYL)HEXAAZ-1,5-DIENE $C_2H_4N_{14}$

Hofmann, K. A. *et al.*, *Ber.*, 1911, **44**, 2953
This very high-nitrogen compound (87.5%) explodes violently on pressing with a glass rod, or on heating to 90°C.
See other HIGH-NITROGEN COMPOUNDS
TETRAZOLES

ACETALDEHYDE C_2H_4O

Fl.P., −38°C (gas above 20°C); E. L., 4.0–57%; A.I.T., 204°C (140°C by DIN 51794)
MCA SD-43, 1952
The MCA Data Sheet describes acetaldehyde as extremely or violently reactive with: acid anhydrides, alcohols, halogens, ketones, phenols, amines, ammonia, hydrogen cyanide or hydrogen sulphide.

Acetic acid
MCA Case History No. 1764
A drum contaminated with acetic acid was filled with acetaldehyde. The ensuing exothermic polymerisation reaction caused a mild eruption lasting for several hours.

Air
White, A. G. *et al., J. Soc. Chem. Ind.*, 1950, **69**, 206
Mixtures of 30–60% of acetaldehyde vapour with air or 60–80% with oxygen may ignite on surfaces at 176 and 105°C, respectively, owing to formation and subsequent violent decomposition of peracetic acid.

Cobalt acetate,
Oxygen
Phillips, B. *et al., J. Amer. Chem. Soc.*, 1957, **79**, 5982
Bloomfield, G. F. *et al., J. Soc. Chem. Ind.*, 1935, **54**, 129T
Oxygenation of acetaldehyde in presence of cobalt acetate at −20°C caused precipitation of 1-hydroxyethyl peracetate (acetaldehyde hemiperacetate), which exploded violently on stirring. Ozone or UV light also catalyses the autoxidation.

Hydrogen peroxide
See HYDROGEN PEROXIDE, H_2O_2: Oxygenated compounds

Oxygen
MCA Case History No. 117
Oxygen leaked into a free space in an acetaldehyde storage tank normally purged with nitrogen. Accelerating exothermic oxidation led to detonation.
See Air, above
See other PEROXIDISABLE COMPOUNDS

ETHYLENE OXIDE C_2H_4O

Fl.P., −20°C (gas above 11°C); E.L., 3.0–100%

1. Hess, L. G. *et al., Ind. Eng. Chem.*, 1950, **42**, 1251
2. *MCA SD-38*, 1971

Liquid ethylene oxide is not detonable, but the vapour may be readily initiated into explosive decomposition. Recommendations for storage and handling are discussed in detail [1]. Metal fittings containing copper, silver, mercury or magnesium should not be used in ethylene oxide service, since traces of acetylene could produce explosive acetylides capable of detonating ethylene oxide vapour [2].

See Contaminants, below

Air,
Bromomethane

Baratov, A. N. *et al.*, p.24 of British Lending Library translation (628. 74, issued 1966) of Russian book on *Fire Prevention and Firefighting Symposium*. The addition of bromomethane to ethylene oxide (used for germicidal sterilising) to reduce the risk of explosion is relatively ineffective, the inhibiting concentration being 31.2, as against 5.8 for hexane and 13.5% vol. for hydrogen.

Alkanethiols,
or An alcohol

Meigs, D. P., *Chem. Eng. News,* 1942, **20**, 1318

Autoclave reactions involving ethylene oxide with unspecified alkanethiols or an alcohol went out of control and exploded violently. Similar previous reactions had been uneventful.

Ammonia

MCA Case History No. 792

Accidental contamination of an ethylene oxide feed tank by ammonia caused violently explosive polymerisation.

See Contaminants, below
Trimethylamine, below

Contaminants

Gupta, A. K., *J. Soc. Chem. Ind.*, 1949, **68**, 179

Precautions designed to prevent explosive polymerisation of ethylene oxide are discussed, including rigid exclusion of acids, covalent halides such as aluminium, iron(III) and tin(IV) chlorides, basic materials like alkali hydroxides, ammonia, amines, metallic potassium, and catalytically active solids such as aluminium or iron oxides or rust.

See Ammonia, above
Trimethylamine, below

m-Nitroaniline

Anon., *Angew. Chem. (Nachr.)*, 1958, **70**, 150

Interaction of the two compounds in an autoclave at 150–160°C is described as safe in Swiss Pat. 171 721. During careful repetition of the reaction with stepwise heating, an autoclave exploded at 130°C.

Trimethylamine

1. *BCISC Quart. Safety Summ.*, 1966, **37**, 44
2. Anon., *Chem. Trade J.*, 1956, **138**, 1376

Accidental contamination of a large ethylene oxide feed-cylinder by reaction liquor containing trimethylamine caused the cylinder to explode 18 h later. Contamination was possible because of a faulty pressure gauge and suck-back of froth above the liquid level [1]. A similar incident occurred previously [2].

See Ammonia, above
Contaminants, above

See other 1,2-EPOXIDES

THIOACETIC *S*-ACID C_2H_4OS

Fl.P., below 23°C

ACETIC ACID $C_2H_4O_2$

MCA SD-41, 1951

Acetaldehyde

See ACETALDEHYDE, C_2H_4O: Acetic acid

5-Azidotetrazole

See 5-AZIDOTETRAZOLE, CHN_7: Acetic acid

Oxidants

See BROMINE PENTAFLUORIDE, BrF_5: Hydrogen-containing materials
CHROMIUM TRIOXIDE, CrO_3: Acetic acid
HYDROGEN PEROXIDE, H_2O_2: Acetic acid
POTASSIUM PERMANGANATE, $KMnO_4$: Acetic acid
SODIUM PEROXIDE, Na_2O_2: Acetic acid

Potassium *tert*-butoxide

See POTASSIUM *tert*-BUTOXIDE, C_4H_9KO: Acids

METHYL FORMATE $C_2H_4O_2$

Fl.P., −19°C; E. L., 5.9–20%

ETHYLENE OZONIDE $C_2H_4O_3$

1. Harries, C. *et al., Ber.*, 1909, **42**, 3305
2. Briner, E. *et al., Helv. Chim. Acta*, 1921, **12**, 154

It explodes very violently on heating, friction or shock [1]. Stable at 0°C but often decomposes explosively at ambient temperature [2].
See other OZONIDES

PEROXYACETIC ACID $C_2H_4O_3$

1. Swern, D., *Chem. Rev.*, 1949, **45**, 7
2. Davies, 1961, 56
3. Phillips, B. *et al., J. Org. Chem.*, 1959, **23**, 1823
4. Anon., *Angew. Chem. (Nachr.)*, 1957, **5**, 178
5. Smith, I.C.P., private comm., 1973

It is insensitive to impact but explodes violently at 110°C [1]. The solid acid has exploded at −20°C [2]. Safe procedures (on basis of detonability experiments) for preparation of anhydrous peracetic acid solutions in chloroform or ester solvents have been described [3]. However, a case of explosion on impact has been recorded [4]. During vacuum distillation, turning a ground glass stopcock in contact with the liquid initiated a violent explosion, but the grease may have been involved as well as friction [5].
See HYDROGEN PEROXIDE, H_2O_2: Acetic acid
See other PEROXYACIDS: Vinyl acetate

ETHYLALUMINIUM BROMIDE IODIDE C_2H_5AlBrI

Nitromethane
See NITROMETHANE, CH_3NO_2: Alkylmetal halides
See other ALKYLALUMINIUM HALIDES

ETHYLALUMINIUM DIBROMIDE $C_2H_5AlBr_2$

See ALKYLALUMINIUM HALIDES

ETHYLALUMINIUM DICHLORIDE $C_2H_5AlCl_2$

See ALKYLALUMINIUM HALIDES

BROMOETHANE C_2H_5Br

Fl.P., below −20°C; E. L., 6.7–11.3%
Preparative hazard.
See BROMINE, Br_2: Ethanol, etc.

CHLOROETHANE C_2H_5Cl

Fl.P., −50°C (gas above 12°C); E. L., 3.8–15.4%
MCA SD-50, 1953

Potassium
See POTASSIUM, K: Halocarbons

CHLOROMETHYL METHYL ETHER C_2H_5ClO

Fl.P., below 23°C

ETHYL HYPOCHLORITE C_2H_5ClO

Alone,
or Copper
Sandmeyer, T., *Ber.,* 1885, **18**, 1768

Though distillable slowly (36°C), ignition or rapid heating of the vapour causes explosion, as does contact of copper powder with the cold liquid.
See other HYPOHALITES

ETHYL PERCHLORATE $C_2H_5ClO_4$

Sidgwick, 1950, 1235
Reputedly the most explosive substance known, it is very sensitive to impact, friction and heat.
See other ALKYL PERCHLORATES

BIS(2-CHLOROETHYL)AMINE $C_2H_5Cl_2N$

Roedig, A. *et al., Chem. Ber.*, 1966, **99**, 121
A violent explosion occurred during evaporation of an ethereal solution at 260 mbar from a bath at 80–90°C. No explosion occurred when the bath temperature was limited to 40–45°, or during subsequent distillation at 60–64°/76 mbar. Aziridine derivatives may have been formed.

TRICHLORO(ETHYL)SILANE $C_2H_5Cl_3Si$

Fl.P., 14°C

FLUOROETHANE C_2H_5F

Gas above –38°C; flammable

IODOETHANE C_2H_5I

Preparative hazard.
See IODINE, I_2: Ethanol, etc.

Silver chlorite
See SILVER CHLORITE, $AgClO_2$: Alkyl iodides

ETHYLMAGNESIUM IODIDE C_2H_5IMg

Ethoxyacetylene
Jordan, C. F., *Chem, Eng. News,* 1966, **44**(8), 40
A stirred mixture of ethoxyacetylene with methylmagnesium iodide in ether exploded when the agitator was turned off. The corresponding bromide had been used, without incident, in earlier attempts on smaller scale.
See other ALKYLMETAL HALIDES

AZIRIDINE (ETHYLENEIMINE) C_2H_5N

Fl.P., −11°C; E. L., 3.3–54.8%
Preparative hazard.

1. Wystrach, V. P. *et al., J. Amer. Chem. Soc.*, 1955, **77**, 5915
2. Wystrach, V. P. *et al., Chem. Eng. News,* 1956, **34**, 1274

The procedure described previously [1] for the preparation is erroneous. 2-Chloroethylamine hydrochloride must be added with stirring as a 33% solution in water to strong sodium hydroxide solution. Addition of the solid hydrochloride to the alkali caused separation in bulk of 2-chloroethylamine, which polymerised explosively. Adequate dilution and stirring, and a temperature below 50°C are all essential [2].

Acids
'Ethylencimine', Brochure 125-521-65, Midland, Mich., Dow Chemical Co., 1965
It is very reactive chemically and subject to aqueous acid-catalysed exothermic polymerisation, which may be violent if uncontrolled by dilution, slow addition or cooling. Ethyleneimine is normally stored over solid caustic alkali, to minimise polymerisation catalysed by presence of carbon dioxide.

Chlorinating agents
Graefe, A. F. *et al., J. Amer. Chem. Soc.*, 1958, **80**, 3939
It gives the explosive 1-chloroaziridine on treatment with sodium hypochlorite solution.
See 1-CHLOROAZIRIDINE, C_2H_4ClN

Silver
'Ethyleneimine', Brochure 125-521-65, Midland, Mich., Dow Chemical Co., 1965
Explosive silver derivatives may be formed in contact with silver or its alloys, including silver solder, which is therefore unsuitable in handling equipment.
See other N-METAL DERIVATIVES

ACETALDEHYDE OXIME C_2H_5NO

Fl.P., below 22°C

***N*-METHYLFORMAMIDE** C_2H_5NO

Fl.P., below 22°C

ETHYL NITRITE $C_2H_5NO_2$

Fl.P., −35°C(gas above 17°C); E. L., 3.1–>50% explodes above 90°C

NITROETHANE $C_2H_5NO_2$

Metal oxides
See NITROALKANES: Metal oxides

ETHYL NITRATE $C_2H_5NO_3$

Fl.P., 10°C; E. L., 3.8– %; explodes at 85°C

Lewis acids
See ALKYL NITRATES

2-NITROETHANOL $C_2H_5NO_3$

ABCM Quart. Safety Summ., 1956, **27**, 24
An explosion occurred towards the end of vacuum distillation of a relatively small quantity of nitroethanol. This was attributed to the presence of peroxides, but the presence of traces of alkali seems a possible alternative cause.
See other C-NITRO COMPOUNDS

ETHYL AZIDE $C_2H_5N_3$

1. Boyer, J. H. *et al., Chem. Rev.*, 1954, **54**, 32
2. Koch, E., *Angew. Chem. (Nachr.)*, 1970, **18**, 26

Though stable at room temperature, it may detonate on rapid heating [1]. A sample stored at –55°C exploded after a few minutes' exposure at laboratory temperature, possibly owing to development of internal pressure in the stoppered vessel. It will also explode if dropped from 1 m in a small flask on to a stone floor [2].
See other ORGANIC AZIDES

BIURET $C_2H_5N_3O_2$

Chlorine
See CHLORINE, Cl_2: Nitrogen compounds

N-METHYL-*N*-NITROSOUREA $C_2H_5N_3O_2$

1. Arndt, F., *Org. Synth.*, 1943, Coll. Vol. 2, 462
2. Anon., *Angew. Chem. (Nachr.)*, 1957, **5**, 198

This must be stored under refrigeration to avoid sudden decomposition after storage at room temperature or slightly above (30°C) [1]. More stable materials for generation of diazomethane are now available. Material stored at 20°C exploded after 6 months [2].

See DIAZOMETHANE, CH_2N_2
See other NITROSO COMPOUNDS

1-METHYL-3-NITRO-1-NITROSOGUANIDINE $C_2H_5N_5O_3$

Eisendrath, J. N., *Chem. Eng. News*, 1953, **31**, 3016

Formerly used as a diazomethane precursor, this material will detonate under high impact, and a sample exploded when melted in a sealed capillary tube.

See other *N*-NITRO COMPOUNDS
NITROSO COMPOUNDS

SODIUM DIMETHYLSULPHINATE C_2H_5NaOS

p-Chlorotrifluoromethylbenzene

See *p*-CHLOROTRIFLUOROMETHYLBENZENE, $C_7H_4ClF_3$ Sodium dimethylsulphinate
See also DIMETHYL SULPHOXIDE, C_2H_6OS: Sodium hydride

ETHANE C_2H_6

Fl.P., −130°C (gas above −89°C); E. L., 3–12.5%

Chlorine

See CHLORINE, Cl_2 : Hydrocarbons

Dioxygenyl tetrafluoroborate

See DIOXYGENYL TETRAFLUOROBORATE, BF_4O_2 : Organic materials

DIMETHYLALUMINIUM BROMIDE C_2H_6AlBr

Nitromethane

See NITROMETHANE, CH_3NO_2: Alkylmetal halides

See other ALKYLALUMINIUM HALIDES

IODODIMETHYLARSINE C_2H_6AsI

Millar, I. T. *et al., Inorg. Synth.*, 1960, **6**, 117

It ignites when heated in air.

See other ALKYLNON-METAL HALIDES

DIMETHYLGOLD(III) AZIDE $C_2H_6AuN_3$

See the dimer, $C_4H_{12}Au_2N_6$

AZIDODIMETHYLBORANE $C_2H_6BN_3$

Anon., *Angew. Chem. (Nachr.)*, 1970, **18**, 27

A sample exploded on contact with a warm sampling capillary.

See other NONMETAL AZIDES

BARIUM METHYL PEROXIDE $C_2H_6BaO_4$

See METHYL HYDROPEROXIDE, CH_4O_2

DIMETHYLBERYLLIUM C_2H_6Be

Coates, 1967, Vol. 1, 106

Ignites in moist air or in carbon dioxide, and reacts explosively with water.

See other ALKYLMETALS

DIMETHYLBISMUTH CHLORIDE C_2H_6BiCl

Sidgwick, 1950, 781
Ignites when warm in air.
See other ALKYLMETAL HALIDES

DIMETHYLCADMIUM C_2H_6Cd

1. Egerton, A. *et al., Proc. R. Soc.,* 1954, **A225**, 429; Davies A. G., *Chem. & Ind.,* 1958, 1177
2. Sidgwick, 1950, 268

On exposure to air, dimethylcadmium peroxide is formed as a crust which explodes on friction [1]. Ignition of dimethylcadmium may occur if a large area : volume ratio is involved, as when it is dropped on to filter paper [2].
See other ALKYLMETALS

2-CHLOROETHYLAMINE C_2H_6ClN

May polymerise explosively.
See AZIRIDINE (ETHYLENEIMINE), C_2H_5N: Preparative hazard

2-AZA-1,3-DIOXOLANIUM PERCHLORATE (ETHYLENENEDIOXYAMMONIUM PERCHLORATE) $C_2H_6ClNO_6$

MCA Case History No. 1622
A batch of this sensitive compound exploded violently, probably during recrystallisation.
See other OXOSALTS OF NITROGENOUS BASES

DIMETHYLANTIMONY CHLORIDE C_2H_6ClSb

Sidgwick, 1950, 777
Ignites at 40°C in air.
See other ALKYLMETAL HALIDES

DIMETHYL *N,N*-DICHLOROPHOSPHORAMIDATE $C_2H_6Cl_2NO_3P$

Block, H. D. *et al., Angew. Chem. (Intern. Ed.),* 1971, **10**, 491
During preparation on 1.5g mol scale, a violent explosion occurred during stirring after reaction. This did not occur on 0.5g mol scale, and longer reaction time may have led to liberation of explosive nitrogen–chlorine compounds. Suggested precautions include a working scale below 0.2g mol and short reaction time.
See other *N*-HALOGEN COMPOUNDS

DICHLORODIMETHYLSILANE $C_2H_6Cl_2Si$

Fl.P., –9°C; E. L., 3.4–9.5%
See other ALKYLNON-METAL HALIDES

DICHLORO(ETHYL) SILANE $C_2H_6Cl_2Si$

Fl.P., below 23°C
See other ALKYLNON-METAL HALIDES

DIETHYLETHEROXODIPEROXOCHROMIUM(VI) $C_2H_6CrO_6$

Schwarz, R. *et al., Ber.,* 1936, **69**, 575
The blue solid explodes powerfully at –30°C.
See related AMMINECHROMIUM PEROXOCOMPLEXES

DIMETHYLMERCURY C_2H_6Hg

Diboron tetrachloride

See DIBORON TETRACHLORIDE, B_2Cl_4 : Dimethylmercury

DIMETHYLMAGNESIUM C_2H_6Mg

Gilman, H. *et al., J. Amer. Chem. Soc.,* 1930, **52**, 5049

Contact with moist air usually caused ignition of the dry powder, and water always ignited the solid or its ethereal solution.

See other ALKYLMETALS

N,N'-DISODIUM *N,N'*-DIMETHOXYSULPHONYLDIAMIDE $C_2H_6N_2Na_2O_2S$

Goehring, 1957, 87

The dry solid readily explodes, as do many *N*-metal hydroxylamides.

See other N-METAL DERIVATIVES

N-AZIDODIMETHYLAMINE $C_2H_6N_4$

Bock, H. *et al., Angew. Chem.,* 1962, **74**, 327

Rather explosive.

See ORGANIC AZIDES

N,N'-DINITRO-1,2-DIAMINOETHANE $C_2H_6N_4O_4$

Urbanski, 1967, Vol, 3, 20

This powerful but relatively insensitive explosive decomposes violently at 202°C, and gives lead and silver salts which are highly impact-sensitive.

See also DIFFERENTIAL THERMAL ANALYSIS

See other N-METAL DERIVATIVES

N-NITRO COMPOUNDS

DIAZIDODIMETHYLSILANE $C_2H_6N_6Si$

Anon., *Angew. Chem. (Nachr.)*, 1970, **18**, 26–27
A three-year-old sample exploded violently on removing the ground stopper.
See other NON-METAL AZIDES

DIMETHYL ETHER C_2H_6O

Fl.P., –41°C (gas above –24°C); E. L., 3.4–18% (27% also quoted)

ETHANOL C_2H_6O

Fl.P., 12°C; E. L., 3.3–19% (latter at 60°C)

Disulphuryl difluoride
See DISULPHURYL DIFLUORIDE, $F_2O_5S_2$: Ethanol

Nitric acid,
Silver
See SILVER, Ag : Ethanol, Nitric acid

Oxidants
See *N*-HALOMIDES: Alcohols
SILVER PERCHLORATE, $AgClO_4$: Aromatic compounds
BROMINE PENTAFLUORIDE, BrF_5: Hydrogen-containing materials
POTASSIUM PERCHLORATE, $ClKO_4$
NITROSYL PERCHLORATE, $ClNO_5$: Organic materials
CHROMYL CHLORIDE, Cl_2CrO_2: Organic solvents
CHLORYL PERCHLORATE, Cl_2O_{10} U: Ethanol
URANYL DIPERCHLORATE, $Cl_2O_{10}U$: Ethanol
CHROMIUM TRIOXIDE, CrO_3: Alcohols
FLUORINE NITRATE, FNO_3: Organic materials
DIOXYGEN DIFLUORIDE, F_2O_2
URANIUM HEXAFLUORIDE, F_6U: Aromatics, etc.
IODINE HEPTAFLUORIDE, F_7I: Organic solvents
PERMANGANIC ACID, $HMnO_4$: Organic materials
NITRIC ACID, HNO_3: Alcohols

HYDROGEN PEROXIDE, H_2O_2 : Alcohols
: Oxygenated compounds
PEROXODISULPHURIC ACID, $H_2O_8S_2$: Organic liquids
POTASSIUM DIOXIDE (SUPEROXIDE), KO_2 : Ethanol
SODIUM PEROXIDE, Na_2O_2 : Hydroxy compounds
RUTHENIUM(VIII) OXIDE, O_4Ru: Organic materials

Platinum
See PLATINUM, Pt: Ethanol

Potassium
See POTASSIUM, K : Air (slow oxidation)

Potassium *tert*-butoxide
See POTASSIUM *tert*-BUTOXIDE, C_4H_9KO : Acids, etc.

Silver nitrate
See SILVER NITRATE, $AgNO_3$: Ethanol

Silver oxide
See SILVER(I) OXIDE, Ag_2O: Ammonia, etc.

DIMETHYL SULPHOXIDE C_2H_6OS

A.I.T., 215°C

Acyl halides,
or Non-metal halides

1. Buckley, A., *J. Chem. Educ.*, 1965, **42**, 674
2. Allan, G. G. *et al., Chem. & Ind.*, 1967, 1706

In absence of diluent or other effective control of reaction rate, the sulphoxide reacts violently or explosively with the following: acetyl chloride, benzenesulphonyl chloride, cyanuric chloride, phosphorus trichloride, phosphoryl chloride, silicon tetrachloride, sulphur dichloride, disulphur dichloride, sulphuryl chloride or thionyl chloride [1]. These violent reactions are explained in terms of exothermic polymerisation of formaldehyde produced under a variety of conditions by interaction of the sulphoxide with reactive halides, acidic or basic reagents [2].

See Dinitrogen tetraoxide, below
Sodium Hydride, below
See also PERCHLORIC ACID, $ClHO_4$: Sulphoxides

Boron compounds
Shriver, 1969, 209
Dimethyl sulphoxide forms an explosive mixture with $B_9H_9{}^{2-}$ and with diborane. It is probable that other boron hydrides and hydroborates behave similarly.

Dinitrogen tetraoxide
Buckley, A., *J. Chem. Educ.*, 1965, **42**, 674
Interaction may be violent or explosive.
See Acyl halides, above

Iodine pentafluoride
Lawless, E. M., *Chem. Eng. News*, 1969, **47**(13), 8
Interaction is explosive, after a delay, in either tetrahydrothiophene-1,1-dioxide or trichlorofluoromethane as solvent, on 0.15g mol scale, though not on one-tenth this scale. Reaction of dimethyl sulphoxide with silver difluoride is also violent.

Magnesium perchlorate
See MAGNESIUM PERCHLORATE, Cl_2MgO_8 : Dimethyl sulphoxide

Metal oxosalts
Martin, 1971, 435
Mixtures of metal salts of oxoacids with the sulphoxide are powerful explosives. Examples are aluminium and sodium perchlorates and ferric nitrate.
See also PERCHLORIC ACID, $ClHO_4$: Sulphoxides
MAGNESIUM PERCHLORATE, Cl_2MgO_8 : Dimethyl sulphoxide

Non-metal halides
See Acyl halides, above

Perchloric acid
See PERCHLORIC ACID, $ClHO_4$: Sulphoxides

Periodic acid
See PERIODIC ACID, HIO_4 : Dimethyl sulphoxide

Silver difluoride

See Iodine Pentafluoride, above

Sodium hydride

1. French, F. A., *Chem. Eng. News*, 1966, **44**(15), 48
2. Olson, G. L., *Chem. Eng. News*, 1966, **44**(24), 7
3. Russell, G. A. *et al.*, *J. Org. Chem.*, 1966, **31**, 248

Two violent pressure-explosions occurred during preparations of dimethylsulphinyl anion on 3–4g mol scale by reaction of sodium hydride with excess solvent. In each case the explosion occurred soon after the separation of a solid. The first reaction involved addition of 4.5g mol of hydride to 18.4g mol of sulphoxide, heated to 70 °C [1], and the second 3.27 and 19.5g mol, respectively, heated to 50 °C [2]. A smaller-scale reaction, at the original lower hydride concentration [3], did not explode but methylation was incomplete. For explanation, see Acyl halides, above.

See also SODIUM DIMETHYLSULPHINATE, C_2H_5NaOS

Sulphur trioxide

See SULPHUR TRIOXIDE, O_3S: Dimethyl sulphoxide

DIMETHYL PEROXIDE $C_2H_6O_2$

1. Rieche, A. *et al.*, *Ber.*, 1928, **61**, 951
2. Baker, G. *et al.*, *Chem. & Ind.*, 1964, 1988

Extremely explosive, heat- and shock-sensitive as liquid or vapour [1]. During determination of the impact sensitivity of the confined material, rough handling of the container caused ignition. The material should only be handled in small quantity and with great care.

See other DIALKYL PEROXIDES

ETHYLENE GLYCOL $C_2H_6O_2$

Perchloric acid

See PERCHLORIC ACID, $ClHO_4$: Glycols and their ethers

Phosphorus pentasulphide
See PHOSPHORUS(V) SULPHIDE, P_2S_5: Alcohols

ETHYL HYDROPEROXIDE $C_2H_6O_2$

Baeyer, A. *et al.*, *Ber.*, 1901, **34**, 738
It explodes violently on superheating; the barium salt is heat- and impact-sensitive.

Hydriodic acid
Sidgwick, 1950, 873
The concentrated acid is oxidised explosively.

Silver
Sidgwick, 1950, 873
Finely divided 'molecular' silver decomposes the hydroperoxide, sometimes explosively.
See other ALKYL HYDROPEROXIDES

HYDROXYMETHYL METHYL PEROXIDE $C_2H_6O_3$

Rieche, A. *et al.*, *Ber.*, 1929, **62**, 2458
Violently explosive, impact-sensitive when heated.
See other 1-OXYPEROXY COMPOUNDS

BISHYDROXYMETHYL PEROXIDE $C_2H_6O_4$

Wieland, H. *et al.*, *Ber.*, 1930, **63**, 66
Highly explosive, very friction-sensitive. Higher homologues are more stable.
See other 1-OXYPEROXY COMPOUNDS

DIMETHYL SULPHATE $C_2H_6O_4S$

A.I.T., 188°C

Ammonia
1. Lindlar, H., *Angew. Chem.*, 1963, **75**, 297
2. Claesson, P. *et al.*, *Ber.*, 1880, **13**, 1700

A violent reaction occurred which shattered the flask when litre quantities of dimethyl sulphate and conc. aqueous ammonia were accidentally mixed. Use dilute ammonia in small quantities to destroy dimethyl sulphate [1]. Similar incidents had been noted previously with ammonia and other volatile bases [2].

DIMETHYL SELENATE $C_2H_6O_4Se$

Sidgwick, 1950, 977
Dimethyl selenate, like its ethyl and propyl homologues, can be distilled under reduced pressure, but explodes at *ca.* 150°C under ambient pressure, though less violently than the lower alkyl nitrates.

DIMETHYL SULPHIDE C_2H_6S

Fl.P., –34°C; E. L., 2.2–19.7%; A.I.T., 206°C

Dibenzoyl peroxide
See DIBENZOYL PEROXIDE, $C_{14}H_{10}O_4$: Dimethyl sulphide

ETHANETHIOL C_2H_6S

Fl.P., below –18°C; E. L., 2.8–18%

DIMETHYL DISULPHIDE $C_2H_6S_2$

Fl.P., 7°C

DIMETHYLZINC C_2H_6Zn

1. Egerton, A. *et al., Proc. R. Soc.,* 1954, **A225**, 429
2. Frankland, E., *Phil. Trans. R. Soc.,* 1852, 417

Ignites in air (owing to peroxide formation [1]), and explodes in oxygen [2].

See other ALKYLMETALS

DIMETHYLARSINE C_2H_7As

von Schwartz, 1918, 322; Sidgwick, 1950, 762

Inflames in air, even at 0°C.

See other ALKYLNON-METAL HYDRIDES

DIMETHYLAMINE C_2H_7N

Fl.P., –50°C (gas above 7°C); E. L., 2.8–14.4%
Fl.P., –16°C for 40% w/v solution in water
Fl.P., + 6°C for 25% w/v solution in water
MCA SD-57, 1955

Acrylaldehyde
See ACRYLALDEHYDE, C_3H_4O: Acids, etc.

Fluorine
See FLUORINE, F_2 : Nitrogenous bases

Maleic anhydride
See MALEIC ANHYDRIDE, $C_4H_2O_3$: Cations, etc.

ETHYLAMINE C_2H_7N

Fl.P., below –18°C (gas above 17°C); E. L., 3.5–14%
Fl.P., below 22°C for 70% w/v solution in water

Cellulose nitrate
See CELLULOSE NITRATE : Amines

2-HYDROXYETHYLAMINE (ETHANOLAMINE) **C_2H_7NO**

Cellulose nitrate
See CELLULOSE NITRATE: Amines

1,2-DIMETHYLNITROSOHYDRAZINE **$C_2H_7N_3O$**

Smith, 1966, Vol. 2, 459
The liquid deflagrates on heating.
See other NITROSO COMPOUNDS

DIMETHYLPHOSPHINE **C_2H_7P**

Houben-Weyl, 1963, Vol. 12(1), 69
Parshall, G. W., *Inorg. Synth., 1968,* **11**, 158
Secondary lower-alkylphosphines readily ignite in air.
See other ALKYLNON-METAL HYDRIDES

ETHYLPHOSPHINE **C_2H_7P**

Houben-Weyl, 1963, Vol. 12(1), 69
Primary lower-alkylphosphines readily ignite in air.

Halogens,
or Nitric acid
von Schwartz, 1918, 324–325
It explodes on contact with chlorine, bromine or fuming nitric acid, inflames with concentrated acid.
See other ALKYLNON-METAL HYDRIDES

1,2-DIAMINOETHANE $C_2H_8N_2$

Cellulose nitrate

See CELLULOSE NITRATE: Amines

Diisopropyl peroxydicarbonate

See DIISOPROPYL PEROXYDICARBONATE, $C_8H_{14}O_6$

Nitromethane

See NITROMETHANE, CH_3NO_2 : Acids, etc.
: 1,2-Diaminoethane, etc.

1,1-DIMETHYLHYDRAZINE $C_2H_8N_2$

Fl.P., –15°C; E.L., 2.1–95%

Oxidants

Kirk-Othmer, 1966, Vol. 11, 186

A powerful reducing agent and fuel, hypergolic with many oxidants, such as dinitrogen tetraoxide, hydrogen peroxide, nitric acid.

See ROCKET PROPELLANTS

1,2-DIMETHYLHYDRAZINE $C_2H_8N_2$

Fl.P., below 23°C

1,1-DIMETHYLDIBORANE $C_2H_{10}B_2$

Fl.P., below –10°C (gas above –3°C)

1,2-DIMETHYLDIBORANE $C_2H_{10}B_2$

Fl.P., below –55°C (gas above –49°C)

See other ALKYLNON-METAL HYDRIDES

B-CHLORO-*N*,*N*-DIMETHYLAMINODIBORANE $C_2H_{10}B_2ClN$

Burg, A. B. *et al., J. Amer. Chem. Soc.*, 1949, **71**, 3454
Ignites in air.
See other NON-METAL HYDRIDES

1,2-ETHYLENEBIS-AMMONIUM PERCHLORATE

$C_2H_{10}Cl_2N_2O_8$

Lothrop, W. C. *et al., Chem. Rev.*, 1949, **44**, 432
An explosive which appreciably exceeds the power and brisance of TNT.
See AMMONIUM PERCHLORATES
See other OXOSALTS OF NITROGENOUS BASES

1,2-DIAMINOETHANEAQUADIPEROXOCHROMIUM(IV)

$C_2H_{10}CrN_2O_5$

Childers, R. F. *et al., Inorg. Chem.*, 1968, **7**, 749
House, D. A. *et al., Inorg. Chem.*, 1966, **5**, 840
The monohydrate is light-sensitive and explodes at 96–97°C if heated at 2°C/min. It effervesces vigorously on dissolution in perchloric acid.
See other AMMINECHROMIUM PEROXOCOMPLEXES

1,2-DIAMINOETHANEAMMINEDIPEROXOCHROMIUM(IV)

$C_2H_{11}CrN_3O_4$

House, D. A. *et al., Inorg. Chem.*, 1967, **6**, 1077
The monohydrate is potentially explosive at 25°C and decomposes or explodes at 115°C during slow or moderate heating.

Hydrogen bromide
Hughes, R. G. *et al., Inorg. Chem.*, 1968, **7**, 74

Interaction must be slow with cooling to prevent explosion.
See other AMMINECHROMIUM PEROXOCOMPLEXES

TETRAAMMINEDITHIOCYANATOCOBALT(III) PERCHLORATE $C_2H_{12}ClCoN_6O_4S_2$

Tomlinson, W. R. *et al., J. Amer. Chem. Soc.*, 1949, **71**, 375
Explodes at 335°C, medium impact-sensitivity.
See other AMMINEMETAL OXOSALTS

DIAMMONIUM *N,N'*-DINITRO-1,2-DIAMINOETHANE $C_2H_{12}N_6O_4$

Violent decomposition occurred at 191°C.
See DIFFERENTIAL THERMAL ANALYSIS
See other aci-NITROSALTS

MERCURY(II) CYANIDE C_2HgN_2

Fluorine
See FLUORINE, F_2: Metal salts

Hydrogen cyanide
Wöhler, L. *et al., Chem. Ztg.*, 1926, **50**, 761
The cyanide is a friction- and impact-sensitive explosive and may initiate detonation of liquid hydrogen cyanide. Other heavy metal cyanides are similar.

Magnesium
See MAGNESIUM, Mg: Metal cyanides

Sodium nitrite
See SODIUM NITRITE, $NNaO_2$: Metal cyanides
See other METAL CYANIDES

MERCURY(II) FULMINATE $C_2HgN_2O_2$

Alone,
or Sulphuric acid
Urbanski, 1967, Vol. 3, 135, 140
Mercury fulminate, readily formed by interaction of mercury nitrite, nitric acid and ethanol, is a very widely used detonator. It may be initiated when dry by flame, heat, impact, friction or intense radiation. Contact with sulphuric acid causes explosion.
See other METAL FULMINATES

MERCURY(II) OXALATE C_2HgO_4

1. *ABCM Quart. Safety Summ.*, 1953, **24**, 30, 45
2. Muir, G. D., private comm., 1968

When dry, it explodes readily on percussion, grinding or heating to 105°C. This instability is attributed to presence of impurities (nitrate, oxide or basic oxalate) in the product [1]. It is so thermally unstable that storage is inadvisable [2].
See other METAL OXALATES

DIMERCURY DICYANIDE OXIDE $C_2Hg_2N_2O$

Merck Index, 1968, 660
This addition compound, $Hg(CN)_2 \cdot HgO$, when pure is explosive, sensitive to impact or heat.
See related METAL CYANIDES

DIIODOACETYLENE C_2I_2

1. Anon., *Chemiearbeit,* 1955, **7**, 55
2. Vaughn, T. H. *et al., J. Amer. Chem. Soc.,* 1932, **54**, 789

Pure, recrystallised material exploded while being crushed manually in a mortar. The decomposition temperature is 125°C, and this may

have been reached locally during crushing [1]. Explosion on impact, on heating to 84°C, and during attempted distillation at 98°C/5 mbar had been reported previously [2].
See other HALOACETYLENE DERIVATIVES

DIPOTASSIUM ACETYLIDE C_2K_2

Water
Bahme, 1972, 80
Contact with limited amounts of water may cause ignition and explosion of evolved acetylene.
See other METAL ACETYLIDES

POTASSIUM DINITROOXALATOPLATINATE (2−) $C_2K_2N_2O_8Pt$

Vèzes, M., *Compt. Rend.*, 1897, **125**, 525
The salt decomposes violently at 240°C.
See other PLATINUM COMPOUNDS

DIPOTASSIUM BIS-*aci*-TETRANITROETHANE $C_2K_2N_4O_8$

Borgardt, F. G. *et al., J. Org. Chem.*, 1966, **31**, 2806
This anhydrous salt, and the mono- and dihydrates of the analogous lithium and sodium salts, are all very impact-sensitive.
See other *aci*-NITRO SALTS
POLYNITROALKYL COMPOUNDS

DILITHIUM ACETYLIDE C_2Li_2

Halogens
Mellor, 1946, Vol. 5, 848
It burns brilliantly when cold in fluorine or chlorine but must be warm before ignition occurs in bromine or iodine vapours.

Lead oxide

See LEAD(II) OXIDE, OPb: Metal acetylides

Non-metals

Mellor, 1946, Vol. 5, 848

It burns vigorously in phosphorus, sulphur or selenium vapours.

See other METAL ACETYLIDES

DILITHIUM BIS-*aci*-TETRANITROETHANE $C_2Li_2N_4O_8$

See DIPOTASSIUM BIS-*aci*-TETRANITROETHANE, $C_2K_2N_4O_8$

DICYANOGEN C_2N_2

Gas above $-21°C$; E.L., 6–32%

Oxidants

The potential energy of mixtures of (endothermic) dicyanogen and powerful oxidants may be released explosively under appropriate circumstances.

See ROCKET PROPELLANTS
DICHLORINE OXIDE, Cl_2O: Dicyanogen
FLUORINE, F_2 : Halogens
OXYGEN (Liquid), O_2 : Liquefied gases
OZONE, O_3: Dicyanogen

NICKEL(II) CYANIDE C_2N_2Ni

Magnesium

See MAGNESIUM, Mg: Metal cyanides

DICYANOGEN *N,N*-DIOXIDE $C_2N_2O_2$

Grundmann, C., *Angew. Chem.*, 1963, **75**, 450

Ann., 1965, **687**, 194

Solid decomposes at –45°C under vacuum, emitting a brilliant light before exploding.
See related HALOGEN OXIDES

LEAD(II) CYANIDE C_2N_2Pb

Magnesium
See MAGNESIUM, Mg : Metal cyanides

LEAD DITHIOCYANATE $C_2N_2PbS_2$

Urbanski, 1967, Vol. 3, 230
The explosive properties of lead dithiocyanate have found limited use.
See other HEAVY METAL DERIVATIVES

THIOCYANOGEN $C_2N_2S_2$

Söderbäck, E., *Ann.,* 1919, **419**, 217

Low-temperature storage is necessary, as it polymerises explosively above its m.p., 16°C.

ZINC DICYANIDE C_2N_2Zn

Magnesium
See MAGNESIUM, Mg: Metal cyanides

DISODIUM BIS-*aci*-TETRANITROETHANE $C_2N_4Na_2O_8$

See DIPOTASSIUM BIS-*aci*-TETRANITROETHANE, $C_2K_2N_4O_8$

TRINITROACETONITRILE $C_2N_4O_6$

Schischkow, A., *Ann. Chim.* [3], 1857, **49**, 310
It explodes if heated quickly to 220°C.
See other POLYNITROALKYL COMPOUNDS

BIS(AZIDOTHIOCARBONYL) DISULPHIDE $C_2N_6S_4$

Smith, G. B. L., *Inorg. Synth.*, 1939, **1**, 81
This compound, readily formed by iodine oxidation of solutions of azidodithioformic acid or its salts, is a powerful explosive. It is sensitive to mechanical impact or heating to 40°C, and slow decomposition during storage increases the sensitivity. Preparative precautions are detailed.
See AZIDODITHIOFORMIC ACID, CHN_3S_2
See other ACYL AZIDES

DIAZIDOMETHYLENECYANAMIDE C_2N_8

Marsh, F. D., *J. Org. Chem.*, 1972, **37**, 2967
This explosive solid may be produced during preparation of cyanogen azide, CN_4.

See other HIGH-NITROGEN COMPOUNDS
ORGANIC AZIDES

DIAZIDOMETHYLENEAZINE C_2N_{14}

Houben-Weyl, 1965, Vol. 10(3), 793
This very explosive bis-*gem*-diazide contains over 89% of nitrogen.
See other HIGH-NITROGEN COMPOUNDS
ORGANIC AZIDES

DISODIUM ACETYLIDE C_2Na_2

Opolsky, S., *Bull. Acad. Cracow*, 1905, 548
A brown explosive form is produced if excess sodium is used in preparation of thiophene homologues – possibly because of sulphur compounds.

Halogens
Mellor, 1946, Vol. 5, 848
Disodium acetylide burns in chlorine and (though not stated) probably also in fluorine, and in contact with bromine and iodine on warming.

Metals
Mellor, 1946, Vol. 5, 848
Trituration in a mortar with finely divided lead, aluminium, iron or mercury may be violent, carbon being liberated.

Metal salts
Mellor, 1946, Vol. 5, 848
Rubbing in a mortar with some chlorides or iodides may cause incandescence or explosion. Sulphates are reduced, and nitrates would be expected to behave similarly.

Non-metal oxides
von Schwartz, 1918, 328
Disodium acetylide incandesces in carbon dioxide or sulphur dioxide.

Oxidants
Mellor, 1946, Vol. 5, 848
Ignites on warming in oxygen, and incandesces at 150°C in dinitrogen pentaoxide.
See Halogens above

Phosphorus
See PHOSPHORUS, P: Metal acetylides

Water
1. Mellor, 1946, Vol. 5, 848
2. Davidsohn, W. E., *Chem. Rev.*, 1967, **67**, 74

Excess water is necessary to avoid explosion [1] ; the need for care in handling is stressed [2].
See other METAL ACETYLIDES

DIRUBIDIUM ACETYLIDE **C_2Rb_2**

Acids
Mellor, 1946, Vol. 5, 848
With concentrated hydrochloric acid ignition occurs, and contact with nitric acid causes explosion.

Halogens
Mellor, 1946, Vol. 5, 848
It burns in all four halogens.

Metal oxides
Mellor, 1946, Vol. 5, 848–850
Iron(III) and chromium(III) oxides react exothermically, and lead oxide explosively. Copper oxide and manganese dioxide react at 350°C incandescently.

Non-metal oxides
Mellor, 1946, Vol. 5, 848
Warming in carbon dioxide, nitrogen oxide or sulphur dioxide causes ignition.

Non-metals
Mellor, 1946, Vol. 5, 848
It reacts vigorously with boron and silicon on warming, ignites with arsenic, and burns in sulphur or selenium vapours.
See other METAL ACETYLIDES

STRONTIUM ACETYLIDE **C_2Sr**

Halogens
Mellor, 1946, Vol. 5, 862

Strontium acetylide incandesces with chlorine, bromine and iodine at 197, 174 and 182°C, respectively.
See other METAL ACETYLIDES

THORIUM DICARBIDE C_2Th

Non-metals,
or Oxidants
Mellor, 1946, Vol. 5, 885–886
Contact with selenium or sulphur vapour causes the heated carbide to incandesce. Contact of the carbide with molten potassium chlorate, potassium nitrate or even potassium hydroxide causes incandescence.
See other METAL NON-METALLIDES

URANIUM DICARBIDE C_2U

Air
Mellor, 1946, Vol. 5, 890
Sidgwick, 1950, 1071
Uranium dicarbide emits brilliant sparks on impact and ignites on grinding in a mortar or on heating in air to 400°C.

Halogens
Mellor, 1946, Vol. 5, 891
Incandescence occurs in warm fluorine, in chlorine at 300°C and weakly in bromine at 390°C.

Hydrogen chloride
See HYDROGEN CHLORIDE, ClH: Metal acetylides or carbides

Nitrogen oxide
See NITROGEN OXIDE ('NITRIC OXIDE'), NO: Metal acetylides or carbides

Water
Sidgwick, 1950, 1071

Mellor, 1946, Vol. 5, 890–891
Interaction with hot water is violent, and the carbide ignites in steam at dull red heat.
See other METAL NON-METALLIDES

ZIRCONIUM DICARBIDE C_2Zr

Halogens
Mellor, 1946, Vol. 5, 885
Ignites in cold fluorine, in chlorine at 250°C, bromine at 300°C and iodine at 400°C.
See other METAL NON-METALLIDES

SILVER TRIFLUOROMETHYLACETYLIDE C_3AgF_3

Henne, A. L. *et al., J. Amer. Chem. Soc.,* 1951, **73**, 1042
Explosive decomposition on heating.
See HALOACETYLENE DERIVATIVES
See other HEAVY METAL DERIVATIVES

TETRAALUMINIUM TRICARBIDE C_3Al_4

Oxidants
Mellor, 1946, Vol. 5, 872
Incandescence on warming with lead peroxide or potassium permanganate.
See other METAL NON-METALLIDES

POTASSIUM 1,3-DIBROMO-2,4-DIKETO-1,3,5-TRIAZINE-6-OLATE $C_3Br_2KN_3O_3$

It may be expected to show similar properties to the chloro-analogue.
See SODIUM 1,3-DICHLORO-2,4-DIKETO-1,3,5-TRIAZINE-6-OLATE, $C_3Cl_2N_3NaO_3$
See other *N*-HALOIMIDES

CHLOROCYANOACETYLENE C_3ClN

Hashimoto, N. *et al., J. Org. Chem.,* 1970, **35**, 675
Avoid contact with air at elevated temperature because of its low ignition temperature. Burns moderately in the open, but may explode in a nearly closed vessel. Presence of mono- and dichloroacetylenes as impurities increases flammability hazard, which may be reduced by addition of 1% of ethyl ether.
See other HALOACETYLENE DERIVATIVES

SODIUM 1,3-DICHLORO-2,4-DIKETO-1,3,5-TRIAZINE-6-OLATE $C_3Cl_2N_3NaO_3$

'FI-CLOR 60S', Brochure NH/FS/67.4, Loughborough, Fisons, 1967
This compound, used in chlorination of swimming pools, is a powerful oxidant, and indiscriminate contact with combustible materials must be avoided. Ammonium salts and other nitrogenous materials are incompatible in formulated products.
See other N-HALOIMIDES

2,4,6-TRICHLORO-1,3,5-TRIAZINE (CYANURIC CHLORIDE) $C_3Cl_3N_3$

N,N-Dimethylformamide
BCISC Quart. Safety Summ., 1964, **35**, 24
Cyanuric chloride reacts vigorously and exothermically with dimethylformamide after a deceptively long induction period. The 1:1 adduct initially formed decomposes above 60°C with evolution of carbon dioxide and formation of a dimeric unsaturated quaternary ammonium salt. Dimethylformamide is appreciably basic and is not a suitable solvent for acyl halides.

Dimethyl sulphoxide
See DIMETHYL SULPHOXIDE, C_2H_6OS : Acyl halides

Methanol
ABCM Quart. Safety Summ., 1960, **31**, 40

Cyanuric chloride dissolved in methanol reacted violently and uncontrollably with the solvent. This was attributed to the absence of an acid acceptor to prevent the initially acid-catalysed (and later auto-catalysed) exothermic reaction of all three chlorine atoms simultaneously.

1,3,5-TRICHLORO-1,3,5-TRIAZINETRIONE $C_3Cl_3N_3O_3$

'FI-CLOR 91' Brochure, Loughborough, Fisons, 1967

This compound, used in chlorination of swimming pools, is a powerful oxidant, and indiscriminate contact with combustible materials must be avoided.

See other N-HALOIMIDES

2,4,6-TRIS(DICHLOROAMINO)-1,3,5-TRIAZINE (HEXACHLOROMELAMINE) $C_3Cl_6N_6$

As a trifunctional *N, N*-dichloro-compound, it is probably more reactive and less stable than the trichloro- derivative.

See 2,4,6-TRIS(CHLOROAMINO)1,3,5-TRIAZINE, $C_3H_3Cl_3N_6$

LITHIUM TRIFLUOROMETHYLACETYLIDE C_3F_3Li

See LITHIUM CHLOROACETYLIDE, C_2ClLi

See other HALOACETYLENE DERIVATIVES

O-TRIFLUOROACETYL-*S*-FLUOROFORMYL THIOPEROXIDE $C_3F_4O_3S$

Anon., *Angew. Chem. (Nachr.)*, 1970, **18**, 378

It exploded spontaneously in a glass bomb closed with a PTFE-lined

valve. No previous indications of instability had been noted during distillation, pyrolysis or irradiation.
See related DIACYL PEROXIDES

HEXAFLUOROPROPENE C_3F_6

Air,
Tetrafluoroethylene
See TETRAFLUOROETHYLENE, C_2F_4 : Air, Hexafluoropropene

HEXAFLUOROISOPROPYLIDENEAMINOLITHIUM C_3F_6LiN

Non-metal halides
Swindell, R.F., *Inorg. Chem.*, 1972, **11**, 242
Interaction of the lithium derivative with a range of chloro- and fluoro- derivatives of arsenic, boron, phosphorus, silicon and sulphur during warming to 25°C tended to be violently exothermic in absence of solvent. Thionyl chloride reacted with explosion.
See HEXAFLUOROISOPROPYLIDENEAMINE, C_3HF_6N : Butyllithium

PENTAFLUOROPROPIONYL HYPOFLUORITE $C_3F_6O_2$

Menefee, A. *et al.*, *J. Amer. Chem. Soc.*, 1954, **76**, 2020
Less stable than its lower homologue, the hypofluorite explodes on sparking, or on distillation at atmospheric pressure (b.p., 2°C), though not at below 0.13 bar.
See FLUORINE, F_2 : Caesium heptafluoropropoxide
See other HYPOHALITES

3,3,3-TRIFLUOROPROPYNE C_3HF_3

Haszeldine, R. N., *J. Chem. Soc.*, 1951, 590

It tends to explode during analytical combustion and the cuprous and silver derivatives decomposed violently (with occasional explosion) on rapid heating.
See other HALOACETYLENE DERIVATIVES

HEXAFLUOROISOPROPYLIDENEAMINE C_3HF_6N

Butyllithium
Swindell, R. F. *et al., Inorg. Chem.,* 1972, **11**, 242
The exothermic reaction which set in on warming the reagents in hexane to 0°C sometimes exploded if concentrated solutions of butyllithium (above 2.5M) were used, but not if diluted (to about 1.2M) with pentane.

4-AZIDOCARBONYL-1,2,3-THIADIAZOLE C_3HN_5OS

Pain, D. L. *et al., J. Chem. Soc.,* 1965, 5167
The azide is extremely explosive in the dry state.
See other ACYL AZIDES

2-CHLOROACRYLONITRILE C_3H_2ClN

Fl.P., 8°C
See other PEROXIDISABLE COMPOUNDS

SODIUM NITROMALONALDEHYDE $C_3H_2NNaO_4$

Fanta, P. E., *Org. Synth.,* 1962, Coll. Vol. 4, 844
The monohydrate, possibly an *aci*-nitro salt, is an impact-sensitive solid and must be carefully handled with precautions.
See other aci-NITRO SALTS

MALONONITRILE $C_3H_2N_2$

Alone,
or Bases
Personal experience
It may polymerise violently on prolonged heating at 130°C, or in contact with strong bases at lower temperatures.

BIS(1,2,3,4-THIATRIAZOL-5-YLTHIO)METHANE $C_3H_2N_6S_4$

Pilgram, K. *et al., Angew. Chem.*, 1965, 77, 348
This compound, and its three longer-chain homologues, explodes loudly with a flash on impact or on heating to the m.p.
See other HIGH-NITROGEN COMPOUNDS

PROPIOLALDEHYDE C_3H_2O

Bases
Sauer, J. C., *Org. Synth.*, 1963, Coll. Vol. 4, 814
This acetylenic aldehyde undergoes vigorous polymerisation in presence of alkalies and, with pyridine, the reaction is almost explosive.
See also ACRYLALDEHYDE, C_3H_4O
See other ACETYLENIC COMPOUNDS

ALUMINIUM TRIFORMATE $C_3H_3AlO_6$

ABCM Quart. Safety Summ., 1939, **10**, 1
An aqueous solution of aluminium formate was being evaporated over a low flame. When the surface crust was disturbed, an explosion occurred. This seems likely to have been due to decomposition, liberation of carbon monoxide and ignition of the latter admixed with air.

1-BROMO-2-PROPYNE C_3H_3Br

Fl.P., 10°C; E.L., 3.0 – %

1. Coffee, R. D. *et al.*, *Loss Prevention,* 1967, **1**, 6–9
2. Driedger, P. E. *et al.*, *Chem. Eng. News,* 1972, **50**(12), 51

This liquid acetylenic compound may be decomposed by mild shock, and when heated under confinement, it decomposes with explosive violence and may detonate. Addition of 20–30% wt. of toluene makes the bromide insensitive to laboratory impact and confinement tests [1]. More recently, it was classed as extremely shock-sensitive [2].

Metals

Dangerous Substances, 1972, Sect. 1, 27

There is a danger of explosion in contact with copper, high-copper alloys, mercury or silver.

See METAL ACETYLIDES

See other HALOACETYLENE DERIVATIVES

2,4,6-TRIS(BROMOAMINO)-1,3,5-TRIAZINE $C_3H_3Br_3N_6$

Vona, J.A., *et al.*, *Chem. Eng. News,* 1952, **30**, 1916

Bromination with this and similar *N*-halo-compounds may become violent or explosive after an induction period as long as 15 minutes. Small scale preliminary experiments, designed to avoid the initial presence of excess brominating agent, are recommended.

Allyl alcohol

Vona, J. A. *et al.*, *Chem. Eng. News,* 1952, **30**, 1916

The components reacted violently 15 min after mixing at ambient temperature. This seems likely to have been a radical-initiated polymerisation of the alcohol (possibly peroxidised) in absence of diluent.

See other *N*-HALOGEN COMPOUNDS

1-CHLORO-2-PROPYNE C_3H_3Cl

Fl.P., below 15°C

Ammonia

Anon., *Chemiearbeit*, 1956, 8(6), 45

Interaction of 1-chloro-2-propyne and liquid ammonia under pressure in a steel bomb had been used several times to prepare the amine. On one occasion the usual slow exothermic reaction did not occur, and the bomb was shaken mechanically. A rapid exothermic reaction, followed by an explosion, occurred. Other cases of instability in propyne derivatives are known.

See 2-PROPYNE-1-OL (PROPARGYL ALCOHOL), C_3H_4O
2-PROPYNE-1-THIOL, C_3H_4S

See other HALOACETYLENE DERIVATIVES

2,4,6-TRIS(CHLOROAMINO)-1,3,5-TRIAZINE $H_3Cl_3N_6$

See 2,4,6-TRIS(BROMOAMINO)–1,3,5-TRIAZINE, $C_3H_3Br_3N_6$

See other *N*-HALOGEN COMPOUNDS

1,1,1,-TRIFLUOROACETONE $C_3H_3F_3O$

Fl.P., below 10°C (gas above 22°C)

ACRYLONITRILE C_3H_3N

Fl.P., –1°C; E.L., 3.0–17%

Fl.P. of 5% solution in water, below 9°C

MCA SD-31, 1964

Acids

1. 'Acrylonitrile', London, British Hydrocarbon Chemicals Ltd., 1965
2, Kaszuba, F. J., *J. Amer. Chem. Soc.,* 1945, **67**, 1227

3. Shirley, D. A., *Preparation of Organic Intermediates,* 3, New York, Wiley, 1951
4. Kaszuba, F. J., *Chem. Eng. News,* 1952, **30**, 824

Contact of strong acids (sulphuric or nitric) with acrylonitrile may lead to vigorous reactions. Even small amounts of acid are potentially dangerous, as these may neutralise the aqueous ammonia present as polymerisation inhibitor and leave the nitrile unstabilised [1]. Precautions necessary in the hydrolysis of acrylonitrile [2] are omitted in the later reference [3]. It is essential to use well-chilled ingredients (acrylonitrile, diluted sulphuric acid, hydroquinone, copper powder) to avoid eruption and carbonisation. A really wide bore condenser is necessary to cope with vigorous boiling of unhydrolysed acrylonitrile.

Bases

MCA SD-31, 1964

Acrylonitrile polymerises violently in contact with bases. In the absence of inhibitor, the pure nitrile will also polymerise.

Bromine

MCA Case History No. 1214

Bromine was being added in portions to acrylonitrile with ice cooling, with intermediate warming to 20°C between portions. After half the bromine was added, the temperature increased to 70°C; then the flask exploded. This was attributed either to an accumulation of unreacted bromine (which would be obvious) or to violent polymerisation. The latter seems more likely, catalysed by hydrogen bromide formed by substitutive bromination.

See Acids, above

Silver nitrate

ABCM Quart. Safety Summ., 1962, **33**, 24

Acrylonitrile containing undissolved solid silver nitrate is liable, on long standing, to polymerise explosively and ignite. This is attributed to the slow deposition of a thermally insulating layer of polymer on the solid nitrate, which gradually gets hotter and catalyses rapid polymerisation.

Tetrahydrocarbazole,
Benzyltrimethylammonium hydroxide

BCISC Quart. Safety Summ., 1968, **39**, 36

Cyanoethylation of 1,2,3,4-tetrahydrocarbazole initiated by the hydroxide had been effected smoothly on twice a published scale of working. During a further fourfold increase in scale, the initiator was added at 0°C, and shortly after cooling had been stopped and heating begun, the mixture exploded. A smaller proportion of initiator and very slow warming to effect reaction are recommended (to avoid rapid polymerisation of the nitrile by the base).

See Bases, above

See other CYANO COMPOUNDS

CYANOACETIC ACID **$C_3H_3NO_2$**

See FURFURYL ALCOHOL, $C_5H_6O_2$: Acids

2-AMINO-5-NITROTHIAZOLE **$C_3H_3N_3O_2S$**

Preparative hazard.

See NITRIC ACID, HNO_3 : 2-Aminothiazole, Sulphuric acid

2,4,6-TRIHYDROXY-1,3,5-TRIAZINE (CYANURIC ACID) **$C_3H_3N_3O_3$**

Chlorine

See CHLORINE, Cl_2 : Nitrogen compounds

SODIUM METHOXYACETYLIDE **C_3H_3NaO**

Brine

Jones, E. R. H. *et al.*, *Org. Synth.*, 1963, Coll. Vol. 4, 406

During addition of saturated brine at −20°C to the sodium derivative at −70°C, minor explosions occur. These may have been due to particles of sodium igniting the liberated methoxyacetylene.

See other METAL ACETYLIDES

PROPADIENE C_3H_4

Gas above −32°C; E.L., 1.7–12%

Bondor, A. M. *et al., Khim. Prom.*, 1965, **41**, 923

The pure diene can decompose explosively under a pressure of 2 bar.

Oxides of nitrogen

See NITROGEN OXIDE ('NITRIC OXIDE'), NO: Dienes, Oxygen

See other DIENES

PROPYNE C_3H_4

Gas above −23°C; E.L., 1.7–11.7%

MCA Case History No. 632

The liquid material (which contains ca. 30% of propadiene) in cylinders is not shock-sensitive, but a temperature of 95°C (even very localised) accompanied by pressures of *ca.* 3.5 bar, will cause a detonation to propagate.

See other ACETYLENIC COMPOUNDS

3-BROMO-1,1,1-TRICHLOROPROPANE $C_3H_4BrCl_3$

Preparative hazard.

See ETHYLENE, C_2H_4 : Bromotrichloromethane

2-CHLORO-1-CYANOETHANOL C_3H_4ClNO

Scotti, F. *et al., J. Org. Chem.*, 1964, **29**, 1800

Distillation (at 110°C/4 mbar) is hazardous, since slight overheating may cause explosive decomposition to 2-chloroacetaldehyde and hydrogen cyanide.

See other CYANO COMPOUNDS

1,3-DICHLOROPROPENE $C_3H_4Cl_2$

Fl.P., 21°C

2,3-DICHLOROPROPENE $C_3H_4Cl_2$

Fl.P., 10°C
See other HALOALKENES

2,2,3,3-TETRAFLUOROPROPANOL $C_3H_4F_4O$

Potassium hydroxide,
or Sodium
Bagnall, R. D., private comm., 1972
Attempted formation of sodium tetrafluoropropoxide by adding the alcohol to sodium (40 g) caused ignition and a fierce fire which melted the flask. This was attributed to alkoxide-induced elimination of hydrogen fluoride, and subsequent exothermic polymerisation. In an alternative preparation of the potassium alkoxide by adding alcohol to solid potassium hydroxide a vigorous exotherm occurred. This was not observed when the base was added slowly to the alcohol.

METHYL DIAZOACETATE $C_3H_4N_2O_2$

Searle, N. E., *Org. Synth.*, 1963, Coll. Vol. 4, 426
This ester must be handled with particular caution as it explodes with extreme violence on heating.
See other DIAZO COMPOUNDS

2-AMINOTHIAZOLE $C_3H_4N_2S$

MCA Case History No. 1587
Drying 2-aminothiazole in an oven without forced air circulation caused development of hot spots and eventual ignition. It has a low auto-ignition temperature and will ignite after 3.5 h at 100°C.

Nitric acid,
Sulphuric acid
See NITRIC ACID, HNO_3: 2-Aminothiazole, Sulphuric acid

1,3-DINITRO-2-IMIDAZOLIDONE $C_3H_4N_4O_5$

Violent decomposition occurred at 238°C.
See DIFFERENTIAL THERMAL ANALYSIS

1,3-DIAZIDOPROPENE $C_3H_4N_6$

Forster, M. O. *et al., J. Chem. Soc.*, 1912, **101**, 489
A sample exploded while being weighed.
See other ORGANIC AZIDES

ACRYLALDEHYDE C_3H_4O

Fl.P., −26°C; E.L., 2.8–31%
MCA SD-85, 1961

Acids,
or Bases
1. *MCA SD-85*, 1961
2. Hearsey, C. J., private comm., 1973
3. Catalogue note, Hopkin and Williams, 1973

Acrylaldehyde is very reactive and will polymerise rapidly, accelerating to violence, in contact with strong acid or basic catalysts. Normally an induction period, shortened by increase in contamination, water content or initial temperature, precedes the onset of polymerisation. Uncatalysed polymerisation sets in at 200°C in the pure material [1]. Exposure to weakly acid condition (nitrous fumes, sulphur dioxide, carbon dioxide), some hydrolysable salts, or thiourea will also cause exothermic and violent polymerisation. A two-year-old sample stored in a refrigerator close to a bottle of dimethylamine exploded violently, presumably after absorbing enough volatile amine (which penetrates plastics closures) to initiate polymerisation [2]. The stabilising effect of the added hydroquinone may cease after a comparatively short storage time. Such unstabilised material could polymerise explosively [3].
See other PEROXIDISABLE COMPOUNDS

METHOXYACETYLENE C_3H_4O

Fl.P., below $-20°C$ (gas above $23°C$)
See SODIUM METHOXYACETYLIDE, C_3H_3NaO: Brine
See other ACETYLENIC COMPOUNDS

2-PROPYN-1-OL (PROPARGYL ALCOHOL) C_3H_4O

Alkalies
Anon., *Angew. Chem.(Nachr.)*, 1954, **2**, 209
If propargyl alcohol and similar acetylenic compounds are dried with alkali before distillation, the residue may explode (probably due to salt formation). Sodium sulphate is recommended as a suitable desiccant.

Mercury(II) sulphate,
Sulphuric acid
Nettleton, J., private comm., 1972
Reppe, W. *et al., Ann.*, 1955, **596**, 38
Following the published procedure, hydroxyacetone was being prepared on half the scale by treating propargyl alcohol as a 30% wt. aqueous solution with mercury sulphate and sulphuric acid (6 g and 0.6 g per mol of alcohol, respectively). On stirring and warming the mixture to 70°C a violent exothermic eruption occurred. Quartering the scale of operations to 1g mol and reducing the amount of acid to 0.37g/mol gave a controllable reaction at 70°C. Adding the alcohol to the other reactants maintained at 70°C is an alternative possibility to avoid the suspected protonation and polymerisation of the propargyl alcohol.
See other ACETYLENIC COMPOUNDS

VINYL FORMATE $C_3H_4O_2$

Fl.P., below 0°C

2-PROPYN-1-THIOL C_3H_4S

1. Sato, K. *et al., Chem. Abs.*, 1956, **53**, 5112b

2. Brandsma, 1971, 179

When distilled at atmospheric pressure, it polymerised explosively. It distils smoothly under reduced pressure at 33–35°C/127 mbar [1]. The polymer produced on exposure to air may explode on heating. Presence of a stabiliser is essential during handling or storage under nitrogen at −20°C [2].

See other ACETYLENIC COMPOUNDS

1-BROMO-2-PROPENE (ALLYL BROMIDE) **C_3H_5Br**

Fl.P., −1°C; E.L., 4.4–7.3%

See other ALLYL COMPOUNDS
HALOALKENES

1-BROMO-2,3-EXPOXYPROPANE **C_3H_5BrO**

Fl.P., below 22°C

See other 1,2-EPOXIDES

1-CHLORO-1-PROPENE **C_3H_5Cl**

Fl.P., below −6°C; E.L., 4.5–16%

See other HALOALKENES

2-CHLOROPROPENE **C_3H_5Cl**

Fl.P., below −20°C (gas above 22°C); E.L., 4.5–16%

See other HALOALKENES

1-CHLORO-2-PROPENE (ALLYL CHLORIDE) **C_3H_5Cl**

Fl.P., −32°C; E.L., 3.3–11.2%

Lewis acids,
Metals
MCA SD-99, 1973
Contact with aluminium chloride, boron trifluoride, sulphuric acid, etc., may cause violently exothermic polymerisation. Products of contact with aluminium, magnesium, zinc (or galvanised metal) may produce similar results.
See other ALLYL COMPOUNDS
HALOALKENES

CHLOROACETONE **C_3H_5ClO**

1. Allen, C. F. H. *et al., Ind. Eng. Chem. (News Ed.)*, 1931, 9, 184
2. Ewe, G. E., ibid., 229

Two separate incidents involved explosive polymerisation of chloroacetone stored in glass bottles under ambient conditions for extended periods.
See also BROMOACETONE OXIME, C_3H_6BrNO

1-CHLORO-2,3-EPOXYPROPANE **C_3H_5ClO**

Fl.P., 21°C

Isopropylamine
Barton, N. *et al., Chem. & Ind.*, 1971, 994
With slow mixing and adequate cooling, smooth condensation to 1-chloro-3-isopropylamino-2-propanol occurs. With rapid mixing and poor cooling, a variable induction period with slow warming precedes a rapid, violent exotherm (to 350°C in 6s). Other primary and secondary amines behave similarly. Moderating effect of water and nature of the products are discussed.

Potassium *tert*-butoxide
See POTASSIUM *tert*-BUTOXIDE, C_4H_9KO: Acids, etc.

Trichloroethylene
See TRICHLOROETHYLENE, C_2HCl_3: Epoxides
See other 1,2-EPOXIDES

PROPIONYL CHLORIDE C_3H_5ClO

Fl.P., 12°C

ETHYL CHLOROFORMATE $C_3H_5ClO_2$

Fl.P., 2°C

METHYL TRICHLOROACETATE $C_3H_5Cl_3O_2$

Trimethylamine

Anon., *Angew. Chem. (Nachr.)*, 1962, **10**, 197

A stirred, uncooled mixture in an autoclave reacted violently, the pressure developed exceeding 400 bar (6000 p.s.i.). Polymerisation of a reactive species formed by dehydrochlorination of the ester seems a possibility.

1-IODO-2-PROPENE (ALLYL IODIDE) C_3H_5I

Fl.P., below 21°C

See other ALLYL COMPOUNDS
HALOALKENES

POTASSIUM *aci*-1,1-DINITROPROPANE $C_3H_5KN_2O_4$

Bisgrove, D. E. *et al.*, *Org. Synth.*, 1963, Coll. Vol. 4, 373

The potassium salt of 1,1-dinitropropane, isolated as a by-product during preparation of 3,4-dinitro-3-hexene, is a hazardous explosive.

See NITROALKANES: Alkali metals
See other *aci*-NITRO SALTS

POTASSIUM *O*-ETHYL DITHIOCARBONATE (XANTHATE) $C_3H_5KOS_2$

Diazonium salts
See DIAZONIUM SULPHIDES AND DERIVATIVES

ETHYL ISOCYANIDE C_3H_5N

Lemoult, M. P., *Compt. Rend.*, 1906, **143**, 903
This showed a strong tendency to explode while being sealed into glass ampoules.
See METHYL ISOCYANIDE, C_2H_3N
See related CYANO COMPOUNDS

PROPIONITRILE C_3H_5N

Fl.P., 2°C; E.L., 3.1– %
See other CYANO COMPOUNDS

PROPIONYL NITRITE $C_3H_5NO_3$

See ACYL NITRITES

AZIDOACETONE $C_3H_5N_3O$

1. Spauschus, H. O. *et al., J. Amer. Chem. Soc.*, 1951, **73**, 209
2. Forster, M. O. *et al., J. Chem. Soc.*, 1908, **93**, 72

A small sample exploded after 6 months' storage in the dark [1]. The freshly prepared material explodes when dropped on to a hotplate and burns brilliantly [2].
See other 2-AZIDOCARBONYL COMPOUNDS

ETHYL AZIDOFORMATE $C_3H_5N_3O_2$

Forster, M. O. *et al., J. Chem. Soc.*, 1908, **93**, 81
It is liable to explode if boiled at atmospheric pressure (at 114°C).
See other ACYL AZIDES

CYCLOPROPANE C_3H_6

Gas above −33°C; E.L., 2.4–10.4%

PROPENE C_3H_6

Fl.P., −108°C (gas above −48°C); E.L., 2.4–11.1%

MCA SD-59, 1956
Russell, F. R. *et al., Chem. Eng. News,* 1952, **30**, 1239
Propene at 955 bar and 327°C was being subjected to further rapid compression. At 4860 bar explosive decomposition occurred, causing a pressure surge to 10 000 bar or above. Decomposition to carbon, hydrogen and methane must have occurred to account for this pressure. Ethylene behaves similarly at much lower pressure and cyclopentadiene, cyclohexadiene, acetylene and a few aromatic hydrocarbons have been explosively decomposed.

Lithium nitrate,
Sulphur dioxide
Pitkethly, R.C., private comm., 1973
A mixture under confinement in a glass pressure bottle at 20°C polymerised explosively, the polymerisation probably being initiated by access of light through the clear glass container. Such alkene–sulphur dioxide copolymerisations will not occur above a ceiling temperature, different for each alkene.

Oxides of nitrogen
See NITROGEN OXIDE ('NITRIC OXIDE'), NO: Dienes, Oxygen

Trifluoromethyl hypofluorite
See TRIFLUOROMETHYL HYPOFLUORITE, CF_4O: Hydrocarbons

DIMETHYLGOLD SELENOCYANATE C_3H_6AuNSe

Stocco, F. *et al.*, *Inorg. Chem.*, 1971, **10**, 2640
Very shock-sensitive and explodes readily when precipitated from aqueous solution. Crystals obtained by slow evaporation of a carbon tetrachloride extract were less sensitive.
See other GOLD COMPOUNDS

BROMOACETONE OXIME C_3H_6BrNO

Forster, M. O. *et al.*, *J. Chem. Soc.*, 1908, **93**, 84
It decomposes explosively during distillation.
See also CHLOROACETALDEHYDE OXIME, C_2H_4ClNO
CHLOROACETONE, C_3H_5ClO

1,1-DICHLOROPROPANE $C_3H_6Cl_2$

Fl.P., 21°C; E.L., 3.1– %

1,2-DICHLOROPROPANE $C_3H_6Cl_2$

Fl.P., 15°C; E.L., 3.4–14.5%

Aluminium,
o-Dichlorobenzene,
1,2-Dichloroethane
See ALUMINIUM, Al: Halocarbons
See other HALOCARBONS

3-CHLORO-2-HYDROXYPROPYL PERCHLORATE $C_3H_6Cl_2O_5$

Hofmann, K. A. *et al.*, *Ber.*, 1909, **42**, 4390
Explodes violently on heating in a capillary.
See other ALKYL PERCHLORATES

DICHLOROMETHYLENEDIMETHYLAMMONIUM CHLORIDE $C_3H_6Cl_3N$

Bader, A. R., private comm., 1971
Senning, A., *Chem. Rev.*, 1965, **65**, 388
The compound does not explode on heating; the published reference is in error.

DIMETHYLTHALLIUM FULMINATE C_3H_6NOTl

Beck, W. *et al., J. Organomet. Chem.*, 1965, **3**, 55
Highly explosive, like the diphenyl analogue.
See related METAL FULMINATES

2-AMINOPROPIONITRILE $C_3H_6N_2$

Shuer, V. F., *Zh. Prikl. Khim.*, 1966, **39**, 2386
Vacuum-distilled material kept dark for 6 months was found to have largely polymerised to a yellow solid, and it exploded 15 days later.
See other CYANO COMPOUNDS

AZIDOACETONE OXIME $C_3H_6N_4O$

Forster, M. O. *et al., J. Chem. Soc.*, 1908, **93**, 83
During distillation at 84°C/2.6 mbar, the large residue darkened and finally exploded violently.
See also BROMOACETONE OXIME, C_3H_6BrNO
See other 2-AZIDOCARBONYL COMPOUNDS

1,3,5-TRINITROSOHEXAHYDRO-1,3,5-TRIAZINE $C_3H_6N_6O_3$

Sulphuric acid
Urbanski, 1967, Vol. 3, 122
Concentrated sulphuric acid causes explosive decomposition.
See other NITROSO COMPOUNDS

ACETONE C_3H_6O

Fl.P., −17°C; E.L., 2.6–12.8%
MCA SD-87, 1962

Bromoform
See BROMOFORM, $CHBr_3$: Acetone, etc.

Carbon,
Air
Boiston, D. A., *Brit. Chem. Eng.*, 1968, **13**, 85
Fires in plant to recover acetone from air with active carbon are due to the bulk surface effect of oxidative heating when air flow is too low to cool effectively.

Chloroform
See CHLOROFORM, $CHCl_3$: Acetones, etc.

Nitric acid,
Sulphuric acid,
Fawcett, H.H., *Ind. Eng. Chem.*, 1959, **51**(4), 89A
Acetone will decompose with explosive violence if brought into contact with mixed nitric and sulphuric acids, particularly under confinement.
See Oxidants, below

Oxidants
See BROMINE TRIFLUORIDE, BrF_3 : Solvents
BROMINE, Br_2 : Acetone
NITROSYL CHLORIDE, ClNO: Acetone, etc.
NITROSYL PERCHLORATE, $ClNO_5$: Organic materials
NITRYL PERCHLORATE, $ClNO_6$: Organic solvents
CHROMYL CHLORIDE, Cl_2CrO_2 : Organic solvents
CHROMIUM TRIOXIDE, CrO_3 : Acetone
DIOXYGEN DIFLUORIDE, F_2O_2
NITRIC ACID, HNO_3 : Acetone, etc.
HYDROGEN PEROXIDE, H_2O_2 : Acetones, etc.
: Ketones
: Oxygenated compounds
PEROXOMONOSULPHURIC ACID, H_2O_5S : Acetone
See Nitric-acid, Sulphuric acid, above

Potassium *tert*-butoxide

See POTASSIUM *tert*-BUTOXIDE, C_4H_9KO: Acids, etc.

Sulphur dichloride

See SULPHUR DICHLORIDE, Cl_2S: Acetone

Thiotrithiazyl perchlorate

See THIOTRITHIAZYL PERCHLORATE, $ClN_3O_4S_4$: Organic solvents

2-PROPEN-1-OL (ALLYL ALCOHOL) C_3H_6O

Fl.P., 21°C; E.L., 3.0–18.0%

See other ALLYL COMPOUNDS

METHYL VINYL ETHER C_3H_6O

Fl.P., –56°C (o) (gas above 5.5°C); E.L., 2.6–39.0%

Acids

1. Braker, 1971, 382
2. MVE Brochure, Billingham, ICI, 1962

Methyl vinyl ether is rapidly hydrolysed by contact with dilute acids to produce acetaldehyde, which is more reactive and has wider flammability limits than the ether [1]. Presense of base is essential during storage or distillation of the ether to prevent rapid acid-catalysed homopolymerisation, which is not prevented by antoxidants. Even mildly acidic solids (calcium chloride or some ceramics) will initiate exothermic polymerisation [2].

See ACETALDEHYDE, C_2H_4O

Halogens,

or Hydrogen halides

Braker, 1971, 382

Addition reactions with bromine, chlorine, hydrogen bromide or chloride are very vigorous and may be explosive if uncontrolled.

See other PEROXIDISABLE COMPOUNDS

PROPIONALDEHYDE C_3H_6O

Fl.P., –9°C; E.L., 2.9–17%; A.I.T., 207°C

PROPYLENE OXIDE C_3H_6O

Fl.P., –37°C; E.L., 3.1–27.5%

1. Kulik, M. K., *Science,* 1967, **155**, 400
2. Smith, R. S., *Science,* 1967, **56**, 12

Use of propylene oxide as a biological sterilant is hazardous because of ready formation of explosive mixtures with air (2.8–37%). Commercially available mixtures with carbon dioxide, though non-explosive, may be asphyxiant and vesicant [1]. Such mixtures may be ineffective, but neat propylene oxide may be safely used, provided that it is removed by evacuation using a water-jet pump [2].

Sodium hydroxide

MCA Case History No. 31

A drum of a crude product containing unreacted propylene oxide and sodium hydroxide catalyst exploded and ignited, probably owing to base-catalysed polymerisation of the oxide.

See also ETHYLENE OXIDE, C_2H_4O: Contaminants
See other 1,2-EPOXIDES

ALLYL HYDROPEROXIDE $C_3H_6O_2$

Dykstra, S. *et al., J. Amer. Chem. Soc.,* 1957, **79**, 3474

When impure, the material is unstable towards heat and light and decomposes to give an explosive residue. The pure material is more stable to light, but detonates on heating or in contact with solid alkalies.

See other ALKYL HYDROPEROXIDES

1,3-DIOXOLANE $C_3H_6O_2$

Fl.P., 2°C

See other PEROXIDISABLE COMPOUNDS

ETHYL FORMATE $C_3H_6O_2$

Fl.P., −20°C; E.L., 2.7–13.5%

HYDROXYACETONE $C_3H_6O_2$

Preparative hazard.

See 2-PROPYNE-1-OL, C_3H_4O : Mercury(II) sulphate, etc.

METHYL ACETATE $C_3H_6O_2$

Fl.P., −9°C; E.L., 3.1–16%

DIMETHYL CARBONATE $C_3H_6O_3$

Fl.P., 18°C

Potassium *tert*-butoxide

See POTASSIUM *tert*-BUTOXIDE, C_4H_9KO: Acids, etc.

LACTIC ACID $C_3H_6O_3$

Hydrofluoric acid,

Nitric acid

See NITRIC ACID, HNO_3 : Lactic acid, etc.

PEROXYPROPIONIC ACID $C_3H_6O_3$

1. Swern, D., *Chem. Rev.*, 1949, **45**, 9
2. Phillips, B. *et al.*, *J. Org. Chem.*, 1959, **23**, 1823

More stable than its lower homologues, it merely deflagrates on heating. Higher homologues appear to be more stable [1]. Safe procedures (on the basis of detonability experiments) for preparation of anhydrous solutions of peroxypropionic acid in chloroform or ethyl propionate have been described [2].

See other PEROXYACIDS

PROPENE OZONIDE $C_3H_6O_3$

Briner, E. *et al.*, *Helv. Chim. Acta*, 1929, **12**, 181

Stable at 0°C but often explosively decomposes at ambient temperature.

See other OZONIDES

1,3,5-TRIOXANE $C_3H_6O_3$

MCA Case History No. 1129

Use of an axe to break up a large lump of trioxane caused ignition and a vigorous fire. Peroxides may have been involved.

Oxygen (liquid)

See OXYGEN (Liquid), O_2: 1,3,5-Trioxane ('paraformaldehyde')

See other PEROXIDISABLE COMPOUNDS

3-HYDROXYTHIETANE-1,1-DIOXIDE $C_3H_6O_3S$

Dittmer, D. C. *et al.*, *J. Org. Chem.*, 1971, **36**, 1324

Evaporation of the residue after treatment of the thietane with hydrogen peroxide is liable to explode, and must be done in an open dish. This is probably because of formation of a 2- or 3-hydroperoxide derivative.

2-PROPEN-1-THIOL C_3H_6S

Fl.P., −10°C
See other ALLYL COMPOUNDS

1-BROMOPROPANE C_3H_7Br

Fl.P., below 22°C; E.L., 4.6– %

2-BROMOPROPANE C_3H_7Br

Fl.P., −14°C

1-CHLOROPROPANE C_3H_7Cl

Fl.P., below −18°C; E.L., 2.6–11.1%

2-CHLOROPROPANE C_3H_7Cl

Fl.P., −32°C; E.L., 2.8–10.7%

CHLOROMETHYL ETHYL ETHER C_3H_7ClO

Fl.P., below 19°C

ISOPROPYL HYPOCHLORITE C_3H_7ClO

Chattaway, F. D. *et al., J. Chem. Soc.,* 1923, **123**, 3001

Of extremely low stability; explosions occurred during its preparation if cooling was inadequate.
See other HYPOHALITES

1-CHLORO-2,3-PROPANEDIOL $C_3H_7ClO_2$

Perchloric acid
See PERCHLORIC ACID, $ClHO_4$, Glycols, etc.

PROPYL PERCHLORATE $C_3H_7ClO_4$

See ALKYL PERCHLORATES

PROPYLCOPPER(I) C_3H_7Cu

Houben-Weyl, 1970, **13**(1), 737
Small quantities only should be handled because of the danger of explosion.
See other ALKYLMETALS

1-IODOPROPANE C_3H_7I

Fl.P., below 22°C

2-IODOPROPANE C_3H_7I

Fl.P., 20°C

ALLYLAMINE C_3H_7N

Fl.P., −29°C; E.L., 2.2–22%
See other ALLYL COMPOUNDS

CYCLOPROPYLAMINE C_3H_7N

Fl.P., 1°C

2-METHYLAZIRIDINE (PROPYLENEIMINE) C_3H_7N

Fl.P., −10°C

***N,N*-DIMETHYLFORMAMIDE** C_3H_7NO

This powerful solvent is not inert and reacts vigorously or violently with a range of materials.

See BROMINE, Br_2 : *N,N*-Dimethylformamide
CARBON TETRACHLORIDE, CCl_4 : *N,N*-Dimethylformamide
2,4,6-TRICHLORO-1,3,5-TRIAZINE, $C_3Cl_3N_3$: *N,N*-Dimethylformamide
1,2,3,4,5,6-HEXACHLOROCYCLOHEXANE, $C_6H_6Cl_6$: *N,N*-Dimethylformamide
TRIETHYLALUMINIUM, $C_6H_{15}Al$: *N,N*-Dimethylformamide
CHROMIUM TRIOXIDE, CrO_3 : *N,N*-Dimethylformamide
MAGNESIUM NITRATE, MgN_2O_6 : *N,N*-Dimethylformamide
SODIUM, Na: *N,N*-Dimethylformamide

ISOPROPYL NITRITE $C_3H_7NO_2$

Fl.P., below 10°C

1-NITROPROPANE $C_3H_7NO_2$

Metal oxides
See NITROALKANES: Metal oxides

ISOPROPYL NITRATE $C_3H_7NO_3$

Fl.P., 11°C; E.L., –100%
Tech. Bull. No. 2, London, ICI Ltd., 1964
It covers storage and handling precautions for this fuel of low oxygen balance. The self-sustaining exothermic decomposition renders it suitable as a rocket monopropellant.

Lewis acids
See ALKYL NITRATES

PROPYL NITRATE $C_3H_7NO_3$

Fl.P., 20°C; E.L., 2–100%
See ALKYL NITRATES

PROPANE C_3H_8

Fl.P., –104°C (gas above –45°C); E.L., 2.2–9.5%

ISOPROPYLDIAZENE $C_3H_8N_2$

Chlorine
Abendroth, H. J., *Angew. Chem.*, 1959, **71**, 340
The crude chlorination product 'isoacetone hydrazone' (presumably an *N*-chloro- compound) exploded violently during drying at 0°C.
See *N*-HALOGEN COMPOUNDS
See other AZO COMPOUNDS

N,N'-DINITRO-*N*-METHYL-1,2-DIAMINOETHANE $C_3H_8N_4O_4$

Violent decomposition occurred at 210°C.
See DIFFERENTIAL THERMAL ANALYSIS
See other *N*-NITRO COMPOUNDS

ETHYL METHYL ETHER C_3H_8O

Fl.P., −37°C (gas above 11°C); E.L., 2–10.1%; A.I.T., 190°C

2-PROPANOL C_3H_8O

Fl.P., 12°C; E.L., 2.3–12.7%
MCA SD-98, 1972

Aluminium
See ALUMINIUM, Al: 2-Propanol

Aluminium triisopropoxide,
Crotonaldehyde
Wagner-Jauregg, T., *Angew. Chem.*, 1939, **52**, 710
During distillation of 2-propanol recovered from the reduction of crotonaldehyde with aluminium isopropoxide, a violent explosion occurred. This was attributed either to peroxidised diisopropyl ether (a possible by-product) or to peroxidised crotonaldehyde.
See DIISOPROPYL ETHER, $C_6H_{14}O$

Oxidants
See DIOXYGENYL TETRAFLUOROBORATE, BF_4O_2: Organic materials
TRINITROMETHANE, CHN_3O_6: 2-Propanol
CHROMIUM TRIOXIDE, CrO_3: Alcohols
HYDROGEN PEROXIDE, H_2O_2: Oxygenated compounds
OXYGEN (Gas), O_2: Alcohols

Potassium *tert*-butoxide
See POTASSIUM *tert*-BUTOXIDE, C_4H_9KO: Acids, etc.

PROPANOL C_3H_8O

Fl.P., 15°C; E.L., 2.5–13.5%

Potassium *tert*-butoxide
See POTASSIUM *tert*-BUTOXIDE, C_4H_9KO: Acids, etc.

DIMETHOXYMETHANE $C_3H_8O_2$

Fl.P., –18°C (o)

ETHYL METHYL PEROXIDE $C_3H_8O_2$

Rieche, A., *Ber.*, 1929, **62**, 218
Shock-sensitive as liquid or vapour, it explodes violently on superheating.
See other DIALKYL PEROXIDES

ISOPROPYL HYDROPEROXIDE $C_3H_8O_2$

Medvedev, S., *et al.*, *Ber.*, 1932, **65**, 133
Explodes just above b.p., 107– 109°C.
See other ALKYL HYDROPEROXIDES

GLYCEROL $C_3H_8O_3$

Oxidants
The violent or explosive reactions exhibited by glycerol in contact with many solid oxidants are due to its unique properties of having three hydroxylic points of attack in the same molecule, since it is a liquid which ensures good contact, and of high boiling point and viscosity, which prevents dissipation of oxidative heat. The

difunctional, less viscous liquid glycols show similar but less violent behaviour.

See CALCIUM HYPOCHLORITE, $CaCl_2O_2$: Hydroxy compounds
CHLORINE, Cl_2 : Glycerol
CHROMIUM TRIOXIDE, CrO_3 : Glycerol
POTASSIUM PERMANGANATE, $KMnO_4$: Glycerol
SODIUM PEROXIDE, Na_2O_2 : Hydroxy compounds

Sodium hydride

See SODIUM HYDRIDE, HNa : Glycerol

ETHYL METHYL SULPHIDE C_3H_8S

Fl.P., below 21°C

PROPANE-1-THIOL C_3H_8S

Fl.P., −20°C

PROPANE-2-THIOL C_3H_8S

Fl.P., −35°C

TRIMETHYLALUMINIUM C_3H_9Al

Kirk-Othmer, 1963, Vol. 2, 40

Extremely pyrophoric, ignition delay is 13 ms in air at 232°C/0.16 bar.

See TRIALKYLALUMINIUMS

TRIMETHYLDIALUMINIUM TRICHLORIDE (ALUMINIUM TRICHLORIDE-TRIETHYLALUMINIUM COMPLEX $C_3H_9Al_2Cl_3$

See ALKYLALUMINIUM HALIDES

TRIMETHYLARSINE C_3H_9As

Air,
or Halogens
Sidgwick, 1950, 762, 769
Inflames in air, and interaction with halogens is violent.
See other ALKYLNON-METALS

TRIMETHYLBORANE C_3H_9B

Stock, A. *et al.*, *Ber.*, 1921, **54**, 535

The gas ignites in air.
See also CHLORINE, Cl_2: Trialkylboranes
See other ALKYLNON-METALS

TRIMETHYL BORATE $C_3H_9BO_3$

Fl.P., below 23°C

TRIMETHYLBISMUTH C_3H_9Bi

Coates, 1967, Vol. 1, 536
It ignites in air, but is unreactive towards water.
See TRIALKYLBISMUTHS

CHLOROTRIMETHYLSILANE C_3H_9ClSi

Fl.P., –20°C

Hexafluoroisopropylideneaminolithium
See HEXAFLUOROISOPROPYLIDENEAMINOLITHIUM, C_3F_6LiN: Non-metal halides
See other HALOSILANES

TRIMETHYLGALLIUM C_3H_9Ga

Sidgwick, 1950, 461
It ignites in air, even at –76°C, and reacts violently with water.
See other ALKYLMETALS

ISOPROPYLAMINE C_3H_9N

Fl.P., –37°C (o); E.L., 2.3–10.4%

1-Chloro-2,3-epoxypropane
See 1-CHLORO-2,3-EPOXYPROPANE, C_3H_5ClO: Isopropylamine

Perchloryl fluoride
See PERCHLORYL FLUORIDE, $ClFO_3$: Nitrogenous bases

PROPYLAMINE C_3H_9N

Fl.P., –10°C; E.L., 2.0–10.4%

Triethynylaluminium
See TRIETHYNYLALUMINIUM, C_6H_3Al : Dioxan, etc.

TRIMETHYLAMINE C_3H_9N

Fl.P., –5°C (gas above 3°C); E.L., 2.0–11.6%; A.I.T., 190°C
Fl.P., 5°C for 25% W/V solution in water
MCA SD-57, 1955

Ethylene oxide
See ETHYLENE OXIDE, C_2H_4O : Trimethylamine

1-AMINO-2-PROPANOL C_3H_9NO

Cellulose nitrate
See CELLULOSE NITRATE : Amines

TRIMETHYLAMINE OXIDE C_3H_9NO

Ringel, C., *Z. Chem.*, 1969, **9**, 188
A preparation, allowed to stand for a week rather than the specified day, exploded during concentration.

TRIMETHYL PHOSPHITE $C_3H_9O_3P$

Magnesium diperchlorate
See MAGNESIUM DIPERCHLORATE, Cl_2MgO_8: Trimethyl phosphite

TRIMETHYL PHOSPHATE $C_3H_9O_4P$

ABCM Quart. Safety Summ., 1953, **25**, 3
The residue from a large atmospheric distillation of trimethyl phosphate exploded violently. This was attributed to rapid decomposition of the ester, catalysed by acidic degradation products, with evolution of gaseous hydrocarbons. It is recommended that only small batches of alkylphosphate should be vacuum distilled and in presence of magnesium oxide to neutralise any acid by-products.

TRIMETHYLPHOSPHINE C_3H_9P

Personal experience
May ignite in air.
See other ALKYLNON-METALS

TRIMETHYLANTIMONY C_3H_9Sb

Air,
or Halogens
1. von Schwartz, 1918, 322

2. Sidgwick, 1950, 777
Inflames in air [1] and reacts violently with halogens [2].
See other ALKYLMETALS

TRIMETHYLTHALLIUM C_3H_9Tl

Sidgwick, 1950, 463

Liable to explode violently above 90°C, and ignites in air.
See other ALKYLMETALS

TRIMETHYLAMINE *N*-OXIDE PERCHLORATE $C_3H_{10}ClNO_5$

Hofmann, K. A. *et al., Ber.,* 1910, **43**, 2624
Explodes on heating, or under a hammer blow.
See related OXOSALTS OF NITROGENOUS BASES

1,2-DIAMINOPROPANE $C_3H_{10}N_2$

Fl.P., 24°C

1,3-DIAMINOPROPANE $C_3H_{10}N_2$

Fl.P., 24°C (o)

TRIMETHYLPLATINUM HYDROXIDE $C_3H_{10}OPt$

Hoechstetter, M. N. *et al., Inorg. Chem.,* 1969, 8, 400
The compound (originally reported as tetramethyl platinum), and

usually obtained in admixture with an alkoxy derivative, detonates on heating.
See other PLATINUM COMPOUNDS

TRIMETHYLSILYL HYDROPEROXIDE $C_3H_{10}O_2Si$

Buncel, E. *et al., J. Chem. Soc.,* 1958, 1551
It decomposes rapidly above 35°C and may have been involved in an explosion during distillation of the corresponding peroxide, which is stable to 135°C.
See ORGANOMINERAL PEROXIDES

ALUMINIUM HYDRIDE-TRIMETHYLAMINE $C_3H_{12}AlN$

See $AlH_3 \cdot C_3H_9N$

1,2-DIAMINOPROPANEAQUADIPEROXO-CHROMIUM(IV) $C_3H_{12}CrN_2O_5$

House, D. A. *et al., Inorg. Chem.,* 1966, **5**, 840
Several preparations of the dihydrate exploded spontaneously at 20–25°C, and it explodes at 88–90°C during slow heating.
See other AMMINECHROMIUM PEROXOCOMPLEXES

BERYLLIUM TETRAHYDROBORATE-TRIMETHYLAMINE $C_3H_{17}B_2BeN$

Burg, A. B. *et al., J. Amer. Chem. Soc.,* 1940, **62**, 3427
The complex ignites in contact with air or water.
See other COMPLEX HYDRIDES

MESOXALONITRILE C_3N_2O

Water
Martin, E. L., *Org. Synth.*, 1971, **51**, 70
The nitrile reacts explosively with water.
See other CYANO COMPOUNDS

PHOSPHORUS TRICYANIDE C_3N_3P

1. Absalom, R. *et al., Chem. & Ind.*, 1967, 1593
2. Smith, T. D. *et al.*, ibid., 1909

During vacuum sublimation at *ca.* 100°C, explosions occurred on three occasions. These were attributed to slight leakage of air into the sublimer. The use of joints remote from the heating bath, a high melting grease, and a cooling liquid other than water (to avoid rapid evolution of hydrogen cyanide in the event of breakage) are recommended, as well as working in a fume cupboard [1].

However, the later publication suggests that the silver cyanide used in the preparation could be the cause of the explosions. On long keeping, it discolours and may then contain the nitride (fulminating silver). Silver cyanide precipitated from slightly acidic solution and stored in a dark bottle is suitable.
See other CYANO COMPOUNDS

DICYANODIAZOMETHANE C_3N_4

Ciganek, E., *J. Org. Chem.*, 1965, **30**, 4200
Small (mg) quantities melt at 75°C, but larger amounts explode. It also has borderline sensitivity towards static electricity and must be handled with full precautions.
See other CYANO COMPOUNDS
DIAZO COMPOUNDS

DIAZIDOMALONONITRILE C_3N_8

MCA Case History No. 820

An ethereal solution of *ca.* 100 g of the crude nitrile was allowed to spontaneously evaporate and crystallise. The crystalline slurry so produced exploded violently without warning. Previously such material had been found not to be shock-sensitive to hammer blows, but dry, recrystallised material was very shock-sensitive. Traces of free hydrogen azide could have been present, and a metal spatula had been used to stir the slurry.

See other CYANO COMPOUNDS
ORGANIC AZIDES

2,4,6-TRIAZIDO-1,3,5-TRIAZINE C_3N_{12}

Ott, E. *et al., Ber.*, 1921, **54**, 183

Explodes on impact, shock or rapid heating to 170–180°C.

See other HIGH-NITROGEN COMPOUNDS
ORGANIC AZIDES

3,4-DICHLORO-2,5-DILITHIOTHIOPHENE $C_4Cl_2Li_2S$

Gilman, H., private comm., 1971

The dry solid is explosive, though relatively insensitive to shock.

See other ORGANOLITHIUM REAGENTS

1,1,4,4-TETRACHLOROBUTATRIENE C_4Cl_4

Heinrich, B. *et al., Angew. Chem. (Intern. Ed.)*, 1968, **7**, 375

One of the by-products formed by heating this compound is an explosive polymer.

See other HALOALKENES

BISTRICHLOROACETYL PEROXIDE $C_4Cl_6O_4$

Swern, 1971, Vol. 2, 815
Pure material explodes on standing at room temperature.
See other DIACYL PEROXIDES

1,1,4,4-TETRAFLUOROBUTATRIENE C_4F_4

Martin, E. L. *et al., J. Amer. Chem. Soc.,* 1959, **81**, 5256
In the liquid state, it explodes above $-5°C$.
See other HALOALKENES

BISTRIFLUOROACETYL PEROXIDE $C_4F_6O_4$

Swern, 1971, Vol. 2, 815
Pure material explodes on standing at room temperature.
See other DIACYL PEROXIDES

HEPTAFLUOROBUTYRYL HYPOFLUORITE $C_4F_8O_2$

Cady, G. H. *et al., J. Amer. Chem. Soc.,* 1954, **76**, 2020
The vapour slowly decomposes at ambient temperature, but will decompose explosively if initiated by a spark.
See other HYPOHALITES

DI-(BISTRIFLUOROMETHYLPHOSPHIDO)-MERCURY $C_4F_{12}HgP_2$

Grobe, J. *et al., Angew. Chem. (Intern. Ed.),* 1972, **11**, 1098
It ignites in air.

GERMANIUM ISOCYANATE $C_4GeN_4O_4$

Water

Johnson, D. H., *Chem. Rev.*, 1951, **48**, 287

The rate of (exothermic) hydrolysis becomes dangerously fast above 80°C.

1-IODO-1,3-BUTADIYNE C_4HI

1. Schluhbach, H. H. *et al., Ann.*, 1951, **573**, 118, 120
2. Kloster-Jensen, E., *Tetrahedron,* 1966, **22**, 969

Crude material exploded violently at 35°C during attempted vacuum distillation, and temperatures below 30° are essential for safe handling [1]. A sample of pure material exploded on scratching under illumination [2].

See other HALOACETYLENE DERIVATIVES

3,4-BIS(1,2,3,4-THIATRIAZOL-5-YLTHIO)MALEIMIDE $C_4HN_7O_2S_4$

Pilgram, K. *et al., Angew. Chem.*, 1965, **77**, 348

It explodes loudly on impact or at the m.p.

See other HIGH-NITROGEN COMPOUNDS

1,3-BUTADIYNE C_4H_2

Gas above 10°C; E.L., 1.4–100%; polymerises rapidly above 0°C

1. Armitage, J. B., *et al. J. Chem. Soc.*, 1951, 44
2. Mushii, R. Y. *et al. Khim, Prom.*, 1963, 109

Potentially very explosive, it may be handled and transferred by low-temperature distillation. It should be stored at –25°C to prevent decomposition and formation of explosive polymers [1]. The critical pressure for explosion is 0.04 bar, but presence of 15–40% of diluents (acetylene, ammonia, carbon dioxide or nitrogen) will raise the critical pressure to 0.92 bar.

See other ACETYLENIC COMPOUNDS
PEROXIDISABLE COMPOUNDS

2-NITROTHIOPHENE-4-SULPHONYL CHLORIDE $C_4H_2ClNO_4S_2$

Libman, D. D., private comm., 1968
After distilling the chloride up to 147°C/6 mbar, the residue decomposed vigorously.
See other NITRO-ACYL HALIDES

HEPTAFLUOROBUTYRAMIDE $C_4H_2F_7NO$

See LITHIUM TETRAHYDROALUMINATE, AlH_4Li:Fluoroamides

IRON(II) MALEATE $C_4H_2FeO_4$

Schwab, R. F. *et al., Chem. Eng. Prog.,* 1970, **66**(9), 53
The finely divided maleate, a by-product of phthalic anhydride manufacture, is subject to rapid aerial oxidation above 150°C, and has been involved in plant fires.

MALEIC ANHYDRIDE $C_4H_2O_3$

Cations,
or Bases

1. Vogler, C. E. *et al., J. Chem. Eng. Data,* 1963, **8**, 620
2. Davie, W. R., *Chem. Eng. News.,* 1964, **42**(8), 41
3. *MCA Case History No. 622*
4. *MCA SD-88,* 1962

Maleic anhydride decomposes exothermically, evolving carbon dioxide, in the presence of alkali- or alkaline earth-metal or ammonium ions, dimethylamine, triethylamine, pyridine or quinoline, at temperatures above 150°C [1]. Sodium ions and pyridine are particularly effective, even at concentrations below 0.1%, and decomposition is rapid [2]. The Case History describes gas-rupture of a large insulated tank of semi-solid maleic anhydride which had been contaminated by sodium hydroxide. Use of additives to reduce sensitivity of the anhydride has been described [3].

MALEIC ANHYDRIDE OZONIDE $C_4H_2O_6$

Briner, E. *et al., Helv. Chim. Acta,* 1937, **20**, 1211
Explodes on warming to –40°C.
See other OZONIDES

BUTEN-3-YNE C_4H_4

Fl.P., below –5°C (gas above 5°C); E.L., 2–100% (1.7–42% also quoted)

1. Hennion, G. F. *et al., Org. Synth.,* 1963, Coll. Vol. 4, 684
2. Rutledge, 1968, 28

It must be stored out of contact with air to avoid formation of explosive compounds [1]. The sodium salt is a safe source of butenyne for synthetic use [2].

See other ACETYLENIC COMPOUNDS
PEROXIDISABLE COMPOUNDS

BARIUM 1,3-DI(5-TETRAZOLYL)TRIAZENE $C_4H_4BaN_{22}$

See 1,3-DI(5-TETRAZOLYL)TRIAZENE, $C_2H_3N_{11}$
See other HEAVY METAL DERIVATIVES

N-BROMOSUCCINIMIDE $C_4H_4BrNO_2$

Aniline,
or Diallyl sulphide,
or Hydrazine hydrate
See *N*-HALOIMIDES: Alcohols, etc.

N-CHLOROSUCCINIMIDE $C_4H_4ClNO_2$

Alcohols,
or Benzylamine
See *N*-HALOIMIDES: Alcohols, etc.

Dust
Boscott, R. J., private comm., 1968
Smouldering in a stored drum of the chloroimide was attributed to dust contamination.

1,4-DICHLORO-2-BUTYNE $C_4H_4Cl_2$

Ford, M. *et al.*, *Chem. Eng. News*, 1972, **50**(3), 67
Preparation by conversion of the diol in pyridine with neat thionyl chloride is difficult to control and hazardous on a large scale. Use of dichloromethane as diluent and operation at −30°C renders the preparation reproducible and safer.
See other HALOACETYLENE DERIVATIVES

COPPER(II)1,3-DI(5-TETRAZOLYL)TRIAZENIDE $C_4H_4CuN_{22}$

See 1,3-DI(5-TETRAZOLYL)TRIAZENE, $C_2H_3N_{11}$
See other HEAVY METAL DERIVATIVES
N-METAL DERIVATIVES

TETRAFLUOROSUCCINAMIDE $C_4H_4F_4N_2O_2$

See LITHIUM TETRAHYDROALUMINATE, AlH_4Li:Fluoroamides

2-CYANO-1,2,3-TRIS(DIFLUOROAMINO)PROPANE $C_4H_4F_6N_4$

Reed, S. F., *J. Org. Chem.*, 1968, **33**, 1861
The 95% pure product is shock-sensitive.
See other DIFLUOROAMINO COMPOUNDS

SUCCINOYL DIAZIDE $C_4H_4N_6O_2$

1. Maclaren, J. A., *Chem. & Ind.*, 1971, 395
2. France, A. D. G. *et al.*, *Chem. & Ind.*, 1962, 2065

This intermediate for preparation of ethylene diisocyanate exploded violently during isolation [1]. Previously the almost dry solid had exploded violently on stirring with a spatula [2].

See other ACYL AZIDES

FURAN C_4H_4O

Fl.P., –36°C; E.L., 2.3–14.3%

DIKETENE $C_4H_4O_2$

Acids,
or Bases,
or Sodium acetate

1. *Haz. Chem. Data,* 1969, 99
2. Laboratory Chemical Disposal Co. Ltd., confidential information, 1968

Presence of mineral, or Lewis acids, or bases including amines, will catalyse violent polymerisation of this very reactive dimer, accompanied by gas evolution [1]. Sodium acetate is sufficiently basic to cause violent polymerisation at 0.1% concentration when added to diketene at 60°C [2].

DIHYDROXYMALEIC ACID $C_4H_4O_6$

Axelrod, B., *Chem. Eng. News,* 1955, **33**, 3024

A glass ampoule of the acid exploded during storage at ambient temperature. It is unstable and slowly loses carbon dioxide, even at 4°C.

THIOPHENE C_4H_4S

Fl.P., $-6^\circ C$

Nitric acid

See NITRIC ACID, HNO_3: Thiophene

ETHYNYL VINYL SELENIDE C_4H_4Se

Brandsma, L. *et al., Rec. Trav. Chim.*, 1962, **81**, 539

Distillable at 30–40°C/80 mbar, it explosively decomposes on heating at 1 bar.

See other ACETYLENIC COMPOUNDS

2-CHLORO-1,3-BUTADIENE C_4H_5Cl

Fl.P., $-20^\circ C$; E.L., 4–20%

See other HALOALKENES

1-CHLORO-1-BUTEN-3-ONE C_4H_5ClO

Pohland, A. E. *et al., Chem. Rev.*, 1966, **66**, 164

If prepared from vinyl chloride, this unstable ketone will decompose almost explosively within one day, apparently owing to the presence of some *cis*-isomer.

BIS(2-FLUORO-2,2-DINITROETHYL)AMINE $C_4H_5F_2N_5O_8$

Gilligan, W. H., *J. Org. Chem.*, 1972, **37**, 3947

The amine and several derived amides are explosives and need appropriate care in handling.

See other FLUORODINITROMETHYL COMPOUNDS

ETHYL TRIFLUOROACETATE $C_4H_5F_3O_2$

Fl.P., –17°C

1-CYANOPROPENE C_4H_5N

Fl.P., 16°C

3-CYANO-PROPENE C_4H_5N

Fl.P., 19°C
See other ALLYL COMPOUNDS
CYANO COMPOUNDS

1-CYANO-2-PROPEN-1-OL C_4H_5NO

Unpublished information
It shows a strong tendency to exothermic polymerisation of explosive violence in presence of light and air at temperatures above 25–30°C. Presence of an inhibitor and inert atmosphere are essential for stable storage.
See other CYANO COMPOUNDS

ALLYL ISOTHIOCYANATE C_4H_5NS

Anon., *Ind. Eng. Chem.(News Ed.)*, 1941, **19**, 1408
A routine preparation by interaction of allyl chloride and sodium thiocyanate in an autoclave at 5.5 bar exploded violently at the end of the reaction. Peroxides were not present or involved and no other cause could be found, but extensive decomposition occurred when allyl isothiocyanate was heated to 250°C in glass ampoules.
See other ALLYL COMPOUNDS

SODIUM ETHOXYACETYLIDE C_4H_5NaO

1. Jones, E. R. H. *et al., Org. Synth,* 1963, Coll. Vol. 4, 405
2. Brandsma. 1971, 120

This sodium derivative is extremely pyrophoric, apparently even at $-70^\circ C$ [1]. The solid may explode after prolonged contact with air [2].

See other METAL ACETYLIDES

1,2-BUTADIENE C_4H_6

Fl.P., below $0^\circ C$ (gas above $19^\circ C$)

See other DIENES

1,3-BUTADIENE C_4H_6

Fl.R., below $-17^\circ C$; E.L., 2.0–11.5%

MCA SD-55, 1954

Alone
or Air

1. Scott, D. A., *Chem. Eng. News,* 1940, **18**, 404
2. Hendry, D. G. *et al., Ind. Eng. Chem. Prod. Res. Dev.,* 1950, 7, 136, 145
3. Mayo, F. R. *et al., Prog. Rep.* No. 40, August 1971, Stamford Res. Inst. Project PRC-6217

When heated under pressure, butadiene may violently decompose thermally.

Solid butadiene at below $-113^\circ C$ will absorb enough oxygen at subatmospheric pressure to make it explode violently when allowed to melt. The peroxides formed on long contact with air are explosives sensitive to heat or shock, but may also initiate polymerisation [1].

The hazards associated with peroxidation of butadiene are closely related to the fact that the polyperoxide is insoluble in butadiene and progressively separates. If local concentrations build up, self-heating from the initially slow, spontaneous decomposition, and when a large enough mass of peroxide accumulates, explosion occurs. Critical mass at 27°C is a 9 cm sphere; size decreases rapidly with increasing temperature [2].

Although isoprene and styrene also readily peroxidise, the peroxides are soluble in the monomers, and the degree of hazard is correspondingly less [3].

See also BUTADIENE PEROXIDE, $C_4H_6O_2$: Butadiene
ISOPRENE, C_5H_8: Air

Aluminium tris-tetrahydroborate

See ALUMINIUM TRIS-TETRAHYDROBORATE, AlB_3H_{12}: Alkenes, etc.

Boron trifluoride etherate,
Phenol

MCA Case History No. 790

The hydrocarbon–phenol reaction, catalysed by the etherate, was being run in petroleum ether solution in a sealed pressure bottle. The bottle burst, possibly owing to exothermic polymerisation of the diene.

See BUTADIENE PEROXIDE, $C_4H_6O_2$: Butadiene

Crotonaldehyde

See CROTONALDEHYDE, C_4H_6O: Butadiene

Ethanol,
Iodine,
Mercury oxide

See 2-ETHOXY-1-IODO-3-BUTENE, $C_6H_{11}IO$

Oxides of nitrogen

See NITROGEN OXIDE, NO: Dienes, Oxygen

Sodium nitrite

See SODIUM NITRITE, $NNaO_2$: Butadiene
See other DIENES

1-BUTYNE C_4H_6

Fl.P., below −7°C (gas above 8°C)

2-BUTYNE C_4H_6

Fl.P., below −7°C; E.L., 1.4– %
See other ACETYLENIC COMPOUNDS

CYCLOBUTENE C_4H_6

Fl.P., below −10°C (gas above 2°C)

4-CHLORO-1-METHYLIMIDAZOLIUM NITRATE $C_4H_6ClN_3O_3$

Personal experience
Towards the end of evaporation of a dilute aqueous solution, the residue decomposed violently. Nitrate salts of many organic bases are unstable and should be avoided.
See other OXOSALTS OF NITROGENOUS BASES

CHROMIUM DIACETATE $C_4H_6CrO_4$

Ocone, L. R. *et al., Inorg. Synth,* 1966, 8, 129–131
The anhydrous salt is pyrophoric in air, or slowly chars in lower oxygen concentrations.
See also BIS(2,4-PENTANEDIONATO) CHROMIUM, $C_{10}H_{14}CrO_4$

CHROMYL ACETATE $C_4H_6CrO_6$

Preparative hazard.
See CHROMIUM TRIOXIDE, CrO_3: Acetic anhydride

1,2-BIS(DIFLUOROAMINO)ETHYL VINYL ETHER $C_4H_6F_4N_2O$

Reed, S. F., *J. Org. Chem.*, 1967, **32**, 2894
It is slightly less impact-sensitive than glyceryl nitrate.
See other DIFLUOROAMINO COMPOUNDS

DI-1,2-BIS(DIFLUOROAMINO)ETHYL ETHER $C_4H_6F_8N_4O$

Reed, S. F., *J. Org. Chem.*, 1967, **32**, 2894
It is slightly more impact-sensitive than glyceryl nitrate.
See other N-HALOGEN COMPOUNDS

ETHYL DIAZOACETATE $C_4H_6N_2O_2$

Searle, N. E., *Org. Synth.*, 1963, Coll. Vol. 4, 426
It is explosive, and distillation, even under reduced pressure as described, may be dangerous.
See other DIAZO COMPOUNDS

3-ETHYL-4-HYDROXY-1,2,5-OXADIAZOLE $C_4H_6N_2O_2$

Sodium hydroxide
Barker, M. D., *Chem. & Ind.*, 1971, 1234
The sodium salt obtained by vacuum evaporation at 50°C of an aqueous alcoholic mixture of the above ingredients exploded violently when disturbed.

2-CYANO-2-PROPYL NITRATE $C_4H_6N_2O_3$

Freeman, J. P., *Org. Synth.*, 1963, **43**, 85

It is a moderately explosive material of moderate impact-sensitivity.

See NITRATING AGENTS

See other CYANO COMPOUNDS

DIMETHYL AZODIFORMATE $C_4H_6N_2O_4$

1. US Pat. 3 347 845, 1967
2. Kauer, J. C., *Org. Synth.*, 1963, Coll. Vol. 4, 412

Shock-sensitive, burns explosively.

See DIETHYL AZODIFORMATE, $C_6H_{10}N_2O_4$

See other AZO COMPOUNDS

1,1-DINITRO-3-BUTENE $C_4H_6N_2O_4$

Witucki, E. F. *et al.*, *J. Org. Chem.*, 1972, **37**, 152

A fume-off during distillation of this compound (b.p., 63–65°C/2 mbar) illustrates the inherent instability of this type of compound.

See also 2,3-DINITRO-2-BUTENE, $C_4H_6N_2O_4$
1-NITRO-3-BUTENE, $C_4H_7NO_2$

2,3-DINITRO-2-BUTENE $C_4H_6N_2O_4$

Bisgrove, D. E. *et al.*, *Org. Synth*, 1963, Coll. Vol. 4, 374

Only one explosion has been recorded during vacuum distillation at 135°C/14 mbar.

See other POLYNITROALKYL COMPOUNDS

DIACETATOPLATINUM(II) NITRATE $C_4H_6N_2O_{10}Pt$

Preparative hazard.

See NITRIC ACID, HNO_3: Acetic acid, etc.

N,N'-DIMETHYL-*N,N'*-DINITROSOOXAMIDE $C_4H_6N_4O_4$

Preussmann, R., *Angew. Chem.*, 1963, **75**, 642
Removal of the solvent carbon tetrachloride (in which nitrosation had been effected) at atmospheric, rather than reduced, pressure caused a violent explosion at the end of distillation. Lowest possible temperatures should be maintained in the preparation. *N*-Nitroso-*p*-toluenesulphonmethylamide seems preferable as a source of diazomethane.

See DIAZOMETHANE, CH_2N_2
See other NITROSO COMPOUNDS

1-BUTEN-3-ONE C_4H_6O

Fl.P., –7°C

CROTONALDEHYDE C_4H_6O

Fl.P., 13°C; E.L. 2.1–15.5% (latter at 60°C); A.I.T., 207°C

Butadiene

1. Greenlee, K. W., *Chem. Eng. News,* 1948, **26**, 1985
2. Hanson, E. S., ibid., 2551
3. Watson, K. M., *Ind. Eng. Chem.,* 1943, **35**, 398

An autoclave without a bursting disc and containing the two poorly mixed reactants was wrecked by a violent explosion which occurred on heating the autoclave to 180°C [1]. This was attributed to not allowing sufficient free space for liquid expansion to occur [2]. The need to calculate separate reactant volumes under reaction conditions for autoclave preparations is stressed [3].

See 1,3-BUTADIENE, C_4H_6: Alone, etc.

Nitric acid

See NITRIC ACID, HNO_3: Crotonaldehyde
See other PEROXIDISABLE COMPOUNDS

DIMETHYLKETENE C_4H_6O

Air

1. Smith, G. W. *et al., Org. Synth.,* 1964, Coll. Vol. 4, 348
2. Staudinger, H., *Ber.,* 1925, **58**, 1079

Dimethylketene rapidly forms an extremely explosive peroxide when exposed to air at ambient temperatures. Drops of solution allowed to evaporate may explode. Inert atmosphere should be maintained above the monomer [1]. The peroxide is polymeric and very sensitive, exploding on friction at –80°C. Higher homologues are very unstable and unisolable [2].

See other POLYPEROXIDES

DIVINYL ETHER C_4H_6O

Fl.P., below –30°C; E.L., 1.7–27% (36.5% also quoted)

Anon., *Chemist & Druggist,* 1947, **157**, 258

The presence of *N*-phenyl-1-naphthylamine as inhibitor considerably reduces the development of peroxide in the ether.

Nitric acid

See NITRIC ACID, HNO_3: Divinyl ether

See other PEROXIDISABLE COMPOUNDS

3,4-EPOXYBUTENE C_4H_6O

Fl.P., below –50°C

See other 1,2-EPOXIDES

ETHOXYACETYLENE C_4H_6O

Fl.P., below –7°C

Jacobs, T. L. *et al., J. Amer. Chem. Soc.,* 1942, **64**, 223

Small samples rapidly heated in sealed tubes to around 100°C exploded.

Ethylmagnesium iodide
See ETHYLMAGNESIUM IODIDE, C_2H_5IMg: Ethoxyacetylene
See other ACETYLENIC COMPOUNDS

METHACRYLALDEHYDE **C_4H_6O**

Fl.P., below 23°C

3-METHOXYPROPYNE **C_4H_6O**

Brandsma, 1971, 13
It explodes on distillation at its atmospheric b.p., 61°C.
See other ACETYLENIC COMPOUNDS

ALLYL FORMATE **$C_4H_6O_2$**

Fl.P., below 22°C

BUTADIENE PEROXIDE **$C_4H_6O_2$**

Butadiene
Alexander, D. S., *Ind. Eng. Chem.*, 1959, **51**, 733
A violent explosion in a partially full butadiene storage sphere was traced to butadiene peroxide. The latter had been formed by contact with air over a long period, and had eventually initiated the exothermic polymerisation of the sphere contents. A tank monitoring and purging system was introduced to prevent recurrence.
See 1,3-BUTADIENE, C_4H_6: Air

BUTANE-2,3-DIONE $C_4H_6O_2$

Fl.P., below 21°C

2-BUTYNE-1,4-DIOL $C_4H_6O_2$

Alkalies,
or Halide salts,
or Mercury salts
Kirk-Othmer, 1960, Suppl. Vol. 2 (to 1st Ed.), 45
The pure diol may be distilled unchanged, but traces of alkali or alkaline earth hydroxides or halides may cause explosive decomposition during distillation. In presence of strong acids, mercury salts may cause violent decomposition of the diol.
See other ACETYLENIC COMPOUNDS

METHYL ACRYLATE $C_4H_6O_2$

Fl.P., –3°C (O); E.L., 2.8–25%
MCA SD-79, 1960

VINYL ACETATE $C_4H_6O_2$

Fl.P., –8°C, E.L., 2.6–13.4%
MCA SD-75, 1970

Air,
Water
MCA SD-75, 1970
Vinyl acetate is normally inhibited with hydroquinone to prevent polymerisation. A combination of too low a level of inhibitor and warm, moist storage conditions may lead to spontaneous polymerisation. This process involves autoxidation of acetaldehyde (a normal impurity produced by hydrolysis of the monomer) to a peroxide, which initiates

exothermic polymerisation as it decomposes. In bulk, this may accelerate to a dangerous extent. Other peroxides or radical sources will initiate the exothermic polymerisation.

Desiccants
MCA SD-75, 1970
Vinyl acetate vapour may react vigorously in contact with silica gel or alumina.

Hydrogen peroxide
See HYDROGEN PEROXIDE, H_2O_2: Vinyl acetate

Oxygen
Barnes, C. E. *et al.*, *J. Amer. Chem. Soc.*, 1950, **72**, 210
The unstabilised polymer exposed to oxygen at 50°C generated an ester–oxygen interpolymeric peroxide which, when isolated, exploded vigorously on gentle heating.
See other POLYPEROXIDES

Ozone
See VINYL ACETATE OZONIDE, $C_4H_6O_5$
See other PEROXIDISABLE COMPOUNDS

ACETIC ANHYDRIDE **$C_4H_6O_3$**

MCA SD-15, 1962

Chromic acid
1. Dawkins, A. E., *Chem. & Ind.*, 1956, 196
2. Baker, W., ibid., 280
Addition of acetic anhydride to a solution of chromium trioxide in water caused violent boiling [1], due to the exothermic acid-catalysed hydrolysis of the anhydride [2].

Chromium trioxide
See CHROMIUM TRIOXIDE, CrO_3: Acetic anhydride

1,3-Diphenyltriazene
See 1,3-DIPHENYLTRIAZENE, $C_{12}H_{11}N_3$: Acetic anhydride

Glycerol,
Phosphoryl chloride
Bellis, M. P., *Hexagon Alpha Chi Sigma (Indianapolis),* 1949, **40**(10), 40
Violent acylation occurs in catalytic presence of phosphoryl chloride. The high viscosity of the mixture in absence of solvent prevents mixing and dissipation of heat of reaction.

Metal nitrates
Davey, W. *et al., Chem. & Ind.,* 1948, 814
Use of mixtures of nitrates with acetic anhydride as a nitrating agent may be hazardous depending on the proportion of reactants and the cation; copper and sodium nitrates usually cause violent reaction.

Nitric acid
See NITRIC ACID, HNO_3: Acetic anhydride

Perchloric acid,
Water
1. *ABCM Quart. Safety Summ.,* 1954, **25**, 3
2. Turner, H. S. *et al., Chem. & Ind.,* 1965, 1933
3. McG. Tegart, W. J., *The Electrolytic and Chemical Polishing of Metals in Research and Industry,* London, Pergamon, 2nd Ed., 1959

Anhydrous solutions of perchloric acid in acetic acid are prepared by using acetic anhydride to remove the water introduced as diluent of perchloric acid. It is essential that the acetic anhydride be added slowly to the aqueous perchloric–acetic acid mixture under conditions where it will readily react with the water, i.e. at about 10°C. Use of anhydride cooled in a freezing mixture caused delayed and violent boiling to occur. Full directions for the preparation are given [1]. A violent explosion occurred during the preparation of an electro-polishing solution by addition of perchloric acid solution to a mixture of water and acetic anhydride. The cause of the explosion, the vigorously exothermic acid-catalysed hydrolysis of acetic anhydride, is avoided if water is added last to the mixture produced by adding

perchloric acid solution to acetic anhydride [2]. The published directions [3] are erroneous and insufficiently detailed for safe working.
See also PERCHLORIC ACID, $ClHO_4$: Acetic anhydride
: Dehydrating agents

Potassium permanganate
See POTASSIUM PERMANGANATE, $KMnO_4$: Acetic acid, etc.

Water
1. Leigh, W. R. D. *et al., Chem. & Ind.*, 1962, 778
2. Benson, G., *Chem. Eng. News,* 1947, **25**, 3458

Accidental slow addition of water to a mixture of the anhydride (85%) and acetic acid (15%) led to a violent, large-scale explosion. This was simulated closely in the laboratory again in absence of mineral-acid catalyst [1]. If unmoderated, the rate of acid-catalysed hydrolysis of (water-insoluble) acetic anhydride can accelerate to explosive boiling [2].

DIACETYL PEROXIDE $C_4H_6O_4$

1. Shanley, E. S., *Chem. Eng. News,* 1949, **27**, 175
2. Kuhn, L. P., *Chem. Eng. News,* 1948, **26**, 3197

Acetyl peroxide may be readily prepared and used in ethereal solution. It is essential to prevent separation of the crystalline peroxide even in traces, since, when dry, it is shock-sensitive and a high explosion risk [1]. Crystalline material, separated and dried deliberately, detonated violently [2]. The commercial material, supplied as a 30% solution in dimethyl phthalate, is free of the tendency to crystallise and is relatively safe. It is, however, a powerful oxidant [1].
See FLUORINE, F_2 : Sodium acetate
See othe. DIACYL PEROXIDES

DIACETATOPLATINUM(II) $C_4H_6O_4Pt$

Preparative hazard.
See NITRIC ACID, HNO_3 : Acetic acid, etc.

MONOPEROXYSUCCINIC ACID $C_4H_6O_5$

1. Lombard, R. *et al., Bull. Soc. Chim. Fr.*, 1963, **12**, 2800
2. Castrantas, 1965, 16

It explodes in contact with flame [1] and is weakly shock-sensitive [2].

See other PEROXYACIDS

VINYL ACETATE OZONIDE $C_4H_6O_5$

Kirk-Othmer, 1970, Vol. 21, 320

The ozonide formed by vinyl acetate is explosive when dry.

See other OZONIDES

DIMETHYL PEROXYDICARBONATE $C_4H_6O_6$

Explodes on heating to 55–60°C, or readily under a hammer blow, and more powerfully than dibenzoyl peroxide.

See PEROXYCARBONATE ESTERS

TARTARIC ACID $C_4H_6O_6$

Silver

See SILVER, Ag: Carboxylic acids

2-BUTYNE-1-THIOL C_4H_6S

Brandsma, 1971, 180

Presence of a stabiliser is essential during handling or storage at –20°C under nitrogen. Exposure to air leads to formation of a polymer which may explode on heating.

See other ACETYLENIC COMPOUNDS

1-BROMO-2-BUTENE C_4H_7Br

Fl.P., 13°C

4-BROMO-1-BUTENE C_4H_7Br

Fl.P., 1°C

2-CHLORO-2-BUTENE C_4H_7Cl

Fl.P., –25°C; E.L., 2.3–9.3%

3-CHLORO-1-BUTENE C_4H_7Cl

Fl.P., –27°C

3-CHLORO-2-METHYL-1-PROPENE C_4H_7Cl

Fl.P., –19°C; E.L., 2.3–9.3%
See other HALOALKENES

BUTYRYL CHLORIDE C_4H_7ClO

Fl.P., below 21°C

1-CHLORO-2-BUTANONE C_4H_7ClO

Tilford, C. H., private comm., 1965
An amber bottle of stabilised material spontaneously exploded.
See CHLOROACETONE, C_3H_5ClO

2-CHLOROETHYL VINYL ETHER C_4H_7ClO

Fl.P., 25°C

ISOBUTYRYL CHLORIDE C_4H_7ClO

Fl.P., below 21°C

ISOPROPYL CHLOROFORMATE $C_4H_7ClO_2$

Fl.P., below 23°C

BUTYRONITRILE C_4H_7N

Fl.P., 21°C
See other CYANO COMPOUNDS

DIMETHYLAMINOACETYLENE C_4H_7N

Brandsma, 1971, 139
The amine reacts extremely vigorously with neutral water.
See other ACETYLENIC COMPOUNDS

ISOBUTYRONITRILE C_4H_7N

Fl.P., 8°C
See other CYANO COMPOUNDS

2-CYANO-2-PROPANOL C_4H_7NO

Sulphuric acid
See SULPHURIC ACID, H_2O_4S : 2-Cyano-2-propanol

2,3-BUTANEDIONE MONOXIME $C_4H_7NO_2$

Muir, G. D., private comm., 1968
Distillation from a heating mantle at 6.5 mbar has caused explosions on several occasions. Distillation at below 1.3 mbar from a steam bath seems safe.

1-NITRO-3-BUTENE $C_4H_7NO_2$

Witucki, E. F. *et al.*, *J. Org. Chem.*, 1972, **37**, 152
A fume-off during distillation of one sample of this compound (b.p., 55°C/25 mbar) illustrates the inherent instability of this type of compound.
See also 1,1-DINITRO-3-BUTENE, $C_4H_6N_2O_4$
2,3-DINITRO-2-BUTENE, $C_4H_6N_2O_4$

BUTYRYL NITRATE $C_4H_7NO_4$

Francis, F. E., *Ber.*, 1906, **39**, 3798
It detonates on heating.
See other ACYL NITRATES

1,3-BIS(5-AMINO-1,3,4-TRIAZOL-2-YL)TRIAZENE $C_4H_7N_{11}$

Hauser, M., *J. Org. Chem.*, 1964, **29**, 3449
It explodes at the m.p., 187°C
See other HIGH-NITROGEN COMPOUNDS

1-BUTENE C_4H_8

Fl.P., −80°C (gas above −6°C); E.L., 1.6–9.3%

Aluminium tris-tetrahydroborate

See ALUMINIUM TRIS-TETRAHYDROBORATE, AlB_3H_{12}: Alkenes, etc.

cis-**2-BUTENE** C_4H_8

Fl.P., below −6°C (gas above 1°C); E.L., 1.7–9.0%

trans-**2-BUTENE** C_4H_8

Fl.P., below −6°C (gas above 3°C); E.L., 1.8–9.7%

CYCLOBUTANE C_4H_8

Fl.P., below 10°C (gas above 13°C); E.L., 1.8– %

2-METHYLCYCLOPROPANE C_4H_8

Fl.P., below 0°C (gas above 4°C)

2-METHYLPROPENE C_4H_8

Fl.P., below −10°C (gas above −7°C); E.L., 1.8–8.8%

mixo-DICHLOROBUTANE $C_4H_8Cl_2$

Fl.P., 21°C for unspecified mixture of 1,2- and 1,3-isomers

BIS(2-CHLOROETHYL) SULPHIDE $C_4H_8Cl_2S$

Bleaching powder
See BLEACHING POWDER: Bis(2-chloroethyl) sulphide

BIS (1-CHLOROETHYLTHALLIUM CHLORIDE) OXIDE $C_4H_8Cl_4OTl_2$

Yakubovich, A. Ya. *et al., Dokl. Akad. Nauk SSSR,* 1950, **73**, 957–958
An explosive solid of low stability.

DIMETHYLAMINO ACETONITRILE $C_4H_8N_2$

Fl.P., below 23°C
See other CYANO COMPOUNDS

3-PROPYLDIAZIRINE $C_4H_8N_2$

Schmitz, E., *et al., Ber.,* 1962, **95**, 800
It exploded on attempted distillation from calcium chloride at about 75°C.
See other DIAZIRINES

ETHYL *N*-METHYL-*N*-NITROSOCARBAMATE $C_4H_8N_2O_3$

1. Hartman, W. W., *Org. Synth.,* 1943, Coll. Vol. 2, 465
2. Druckrey, H. *et al., Nature,* 1962, 195, 1111

The material is unstable and explodes if distilled at atmospheric pressure [1] or may become explosive if stored above 15°C [2]. Its use for preparation of diazomethane has been superseded by stable intermediates.

See DIAZOMETHANE, CH_2N_2
See other NITROSO COMPOUNDS

2,2'-OXYDI(ETHYL NITRATE) **$C_4H_8N_2O_7$**

Kit and Evered, 1960, 268
A powerful explosive, sensitive to vibration and mechanical shock, too heat-sensitive for a rocket propellant.
See other ALKYL NITRATES

1,3,5,7-TETRANITRO PERHYDRO-1,3,5,7-TETRAZOCINE **$C_4H_8N_8O_8$**

Violent decomposition occurred at 279°C.

See DIFFERENTIAL THERMAL ANALYSIS
See other *N*-NITRO COMPOUNDS

2-BUTANONE **C_4H_8O**

Fl.P., –7°C; E.L., 1.8–11.5%
MCA SD-83, 1961

Chloroform
See CHLOROFORM, $CHCl_3$: Acetone, etc.

Hydrogen peroxide,
Nitric acid
See HYDROGEN PEROXIDE, H_2O_2: Ketones, etc.

Potassium *tert*-butoxide
See POTASSIUM *tert*-BUTOXIDE, C_4H_9KO; Acids, etc.

BUTYRALDEHYDE C_4H_8O

Fl.P., –6°C; E.L., 2.5– %
MCA SD-78, 1960

CYCLOPROPYL METHYL ETHER C_4H_8O

Fl.P., below 10°C

1,2-EPOXYBUTANE C_4H_8O

Fl.P., –15°C; E.L., 1.5–18.3%

ETHYL VINYL ETHER C_4H_8O

Fl.P., –46°C; E.L., 1.7–28%; A.I.T., 202°C

Methanesulphonic acid
Eaton, P. E. *et al., J. Org. Chem.*, 1972, **37**, 1947
Methanesulphonic acid is too powerful a catalyst for *O*-alkylation with the vinyl ether, causing explosive polymerisation of the latter on the multimol scale. Dichloroacetic acid is a satisfactory catalyst on the 3g mol scale.
See other PEROXIDISABLE COMPOUNDS

ISOBUTYRALDEHYDE C_4H_8O

Fl.P., –40°C; E.L., 1.6–10.6%; A.I.T., 223°C
MCA SD-78, 1960

TETRAHYDROFURAN C_4H_8O

Fl.P., –17°C; E.L., 1.8–11.8%

1. 'Recommended Handling Procedures', THF Brochure FC3-664, Wilmington, Du Pont, 1964
2. Anon., *Org. Synth.*, 1966, **46**, 105
3. Schurz, J. *et al.*, *Angew. Chem.*, 1956, **68**, 182
4. Davies, A. G., *J. R. Inst. Chem.*, 1956, **80**, 386

Like many other ethers and cyclic ethers, in absence of inhibitors tetrahydrofuran is subject to autoxidation on exposure to air, when initially the 2-hydroperoxide forms. This tends to decompose smoothly when heated, but if allowed to accumulate for a considerable period, it becomes transformed to other peroxidic species which will decompose violently [1]

Commercial material is supplied stabilised with a phenolic anti-oxidant which is effective under normal closed storage conditions in preventing the formation and accumulation of peroxide. Procedures for testing for the presence of peroxides and also for their removal are detailed [1,2,4]. The latest publication recommends the use of cuprous chloride for removal of trace quantities of peroxides. If more than trace quantities are present, the peroxidised solvent should be discarded by dilution and flushing away with water [2]. An attempt to remove peroxides by shaking with solid ferrous sulphate before distillation did not prevent explosion of the distillation residue [3]. Alkali treatment to destroy peroxides [1] appears not to be safe [2]. (See Caustic alkalies, below). Distillation or alkali treatment of stabilised tetrahydrofuran removes the involatile anti-oxidant and the solvent must be restabilised or stored under nitrogen to prevent peroxide formation during storage, which should not exceed a few days' duration in the absence of stabiliser [2]. The use of lithium tetrahydroaluminate is only recommended for drying tetrahydrofuran which is peroxide-free and not grossly wet [2].

See LITHIUM TETRAHYDROALUMINATE, AlH_4Li : Tetrahydrofuran
TETRAHYDRO-2-FURYL HYDROPEROXIDE, $C_4H_8O_3$
See other PEROXIDISABLE COMPOUNDS

Caustic alkalies

1. *NSC Newsletter, Chem. Section*, 1964(10); 1967(3)
2. Anon., *Org. Synth.*, 1966, **46**, 105

It is not safe to store quantities of tetrahydrofuran which have been purified (i.e. freed of the normally added inhibitors), since dangerous concentrations of peroxide may build up during storage. Peroxidised material should not be dried with sodium or potassium hydroxide, as explosions may occur [1,2].

Lithium tetrahydroaluminate
See LITHIUM TETRAHYDROALUMINATE, AlH_4Li: Tetrahydrofuran

Sodium tetrahydroaluminate
See SODIUM TETRAHYDROALUMINATE, AlH_4Na: Tetrahydrofuran

BUTYRIC ACID $C_4H_8O_2$

Chromium trioxide
See CHROMIUM TRIOXIDE, CrO_3: Butyric acid

m-DIOXAN $C_4H_8O_2$

Fl.P., 2°C; (O); E.L., 2–22%

p-DIOXAN $C_4H_8O_2$

Fl.P., 12°C; E.L., 2.0–22%; A.I.T., 180°C

Air
Dasler, W. *et al., Ind. Eng. Chem. (Anal. Ed.),* 1946, **18**, 52
Like all other ethers, dioxan is susceptible to autoxidation with formation of peroxides which may be hazardous if distillation (causing concentration) is attempted. Because it is water-miscible, treatment by shaking with aqueous reducants (iron(II) sulphate, sodium sulphide, etc.) is impracticable. Peroxides may be removed, however, under anhydrous conditions by passing dioxan (or any other ether) down a column of activated alumina. The peroxides (and any water) are removed by adsorption on to the alumina, which must then be washed

with water or methanol to remove them before being discarded.
See other PEROXIDISABLE COMPOUNDS

Decaborane(14)
See DECABORANE(14), $B_{10}H_{14}$: Ethers

Nickel
Mozingo, R., *Org. Synth.*, 1955, Coll. Vol. 3, 182
Dioxan reacts with Raney nickel catalyst almost explosively above 210°C.

Sulphide trioxide
See SULPHUR TRIOXIDE, O_3S: Dioxan

Triethynylaluminium
See TRIETHYNYLALUMINIUM, C_6H_3Al: Dioxan

ETHYL ACETATE $C_4H_8O_2$

Fl.P., –4°C; E.L., 2.2–11.5%
MCA SD-51, 1953

Lithium tetrahydroaluminate
See LITHIUM TETRAHYDROALUMINATE, AlH_4Li : Ethyl acetate

Potassium *tert*-butoxide
See POTASSIUM *tert*-BUTOXIDE, C_4H_9KO: Acids, etc.

ISOPROPYL FORMATE $C_4H_8O_2$

Fl.P., –6°C

METHYL PROPIONATE $C_4H_8O_2$

Fl.P., –2°C; E.L., 2.5–13%

PROPYL FORMATE $C_4H_8O_2$

Fl.P., –3°C; E.L., 2.3–   %

Potassium *tert*-butoxide
See POTASSIUM *tert*-BUTOXIDE, C_4H_9KO; Acids, etc.

TETRAHYDROTHIOPHENE-1,1-DIOXIDE ('SULPHOLANE') $C_4H_8O_2S$

Nitronium tetrafluoroborate
See NITRATING AGENTS

trans-2-BUTENE OZONIDE $C_4H_8O_3$

See 2-HEXENE OZONIDE, $C_6H_{12}O_3$
See other OZONIDES

2-TETRAHYDROFURYL HYDROPEROXIDE $C_4H_8O_3$

Rein, H. and Criegee, R., *Angew. Chem.*, 1950, **62**, 120
This occurs as the first product of the ready autoxidation of tetrahydrofuran, and is relatively stable. It readily changes, however, to a highly explosive polyalkylidene peroxide, which is responsible for the numerous explosions observed on distillation of peroxidised tetrahydrofuran.
See TETRAHYDROFURAN, C_4H_8O
See other ALKYL HYDROPEROXIDES

3,6-DIMETHYL-1,2,4,5-TETRAOXANE $C_4H_8O_4$

Rieche, A. *et al.*, *Ber.*, 1939, **72**, 1933
This dimeric 'ethylidene peroxide' is an extremely shock-sensitive

solid which explodes violently on being touched. Extreme caution in handling is required.

See other CYCLIC PEROXIDES

1-HYDROXYETHYL PERACETATE $C_4H_8O_4$

An explosive low-melting solid, readily formed during autoxidation of acetaldehyde.

See ACETALDEHYDE, C_2H_4O : Cobalt acetate, oxygen

See other PEROXYESTERS

TETRAHYDROTHIOPHENE C_4H_8S

Fl.P., 13°C

BUTYLDICHLOROBORANE $C_4H_9BCl_2$

Air,

or Water

Niedenzu, K. *et al.*, *Inorg. Synth.*, 1967, **10**, 126

Ignites on prolonged exposure to air; hydrolysis may be explosive.

See other ALKYLNON-METAL HALIDES

1-BROMOBUTANE C_4H_9Br

Fl.P., 18°C; E.L., 2.8–6.6% (latter at 100°C)

2-BROMOBUTANE C_4H_9Br

Fl.P., 21°C

1-BROMO-2-METHYLPROPANE C_4H_9Br

Fl.P., 22°C

2-BROMO-2-METHYLPROPANE C_4H_9Br

Fl.P., –18°C

2-BROMOETHYL ETHYL ETHER C_4H_9BrO

Fl.P., 5°C

1-CHLOROBUTANE C_4H_9Cl

Fl.P., –12°C; E.L., 1.9–10.1%

2-CHLOROBUTANE C_4H_9Cl

Fl.P., –10°C

1-CHLORO-2-METHYLPROPANE C_4H_9Cl

Fl.P., –6°C; E.L., 2.0–8.7%

2-CHLORO-2-METHYLPROPANE C_4H_9Cl

Fl.P., 0°C

tert-BUTYL HYPOCHLORITE C_4H_9ClO

1. Lewis, J. C., *Chem. Eng. News*, 1962, **40**(43), 62; Teeter, H.M. *et al.*, *Org. Synth.*, 1962, Coll. Vol. 4, 125
2. Mintz, M. J. *et al.*, *Org. Synth.*, 1969, **49**, 9

This material, normally in sealed ampoules and used as a paper chromatography spray reagent, is photo-sensitive. Exposure to UV light causes exothermic decomposition to acetone and chloromethane. Ampoules have burst because of pressure build-up after exposure to fluorescent or direct daylight. Store cool and dark, and open ampoules with personal protection. The material also reacts violently with rubber [1]. It should not be heated to above its boiling point [2]. There is also a preparative hazard.

See CHLORINE, Cl_2 : *tert*-Butanol
See other HYPOHALITES

2(2-HYDROXYETHOXY)ETHYL PERCHLORATE $C_4H_9ClO_6$

Hofman, K. A. *et al.*, *Ber.*, 1909, **42**, 4390
Explodes violently on heating in a capillary.
See other ALKYL PERCHLORATES

N-FLUORO-*N*-NITROBUTYLAMINE $C_4H_9FN_2O_2$

Grakauskas, V. *et al.*, *J. Org. Chem.*, 1972, **37**, 334
A sample exploded on vacuum distillation at 60°C, though not at 40°C/33 mbar.

2-IODOBUTANE C_4H_9I

Fl.P., –10°C

1-IODO-2-METHYLPROPANE C_4H_9I

Fl.P., 0°C

2-IODO-2-METHYLPROPANE C_4H_9I

Fl.P., −10°C

POTASSIUM *tert*-BUTOXIDE C_4H_9KO

Acids,
or Reactive solvents
Manwaring, R., *et al., Chem. & Ind.,* 1973, 172
Contact of 1.5 g portions of the solid butoxide with drops of the liquid (1) or with vapours (v) of the reagents below caused ignition after the indicated period (min).

Acetic acid,v,3; sulphuric acid, 1,0.5
Methanol, 1,2; ethanol, v,7; propanol, 1,1; isopropanol, 1, 1
Ethyl acetate, v,2; butyl acetate, v,2; propyl formate, v,4; dimethyl carbonate, 1,1; diethyl sulphate, 1,1
Acetone, v,4, 1,2; 2-butanone, v,1, 1,0.5; 4-methyl-2-butanone, v,3
Dichloromethane, 1,2; chloroform, v,2, 1,0; carbon tetrachloride, 1,1; 1-chloro-2,3-epoxypropane, 1,1

The potentially dangerous reactivity with water, acids or halocarbons was already known, but that arising from contact with alcohols, esters and ketones was unexpected. Under normal reaction conditions, little significant danger should exist where excess of solvent will dissipate the heat, but accidental spillage of the solid butoxide could be hazardous.

BUTYLLITHIUM C_4H_9Li

Air,
or Carbon dioxide,
or Water
MCA SD-91, 1966
Dezmelyk, E. W. *et al., Ind. Eng. Chem.,* 1961, **53**(6), 56A
This reactive liquid is now normally supplied commercially as a solution

(up to 25% wt.) in pentane, hexane or heptane, because it reacts with ether and such solutions must be stored under refrigeration. Reaction with atmospheric oxygen or water vapour is highly exothermic and solutions of 20% concentration will ignite on exposure to air at ambient temperature and about 70% relative humidity. Solutions of above 25% concentration ignite in air at any humidity. Contact with liquid water will cause immediate ignition of solutions in these highly flammable solvents.

Though contact with carbon dioxide causes an exothermic reaction, it is less than with atmospheric oxygen, so carbon dioxide extinguishers are effective on butyllithium fires.

Full handling, operating and disposal procedures are detailed.

See other ALKYLMETALS

PYRROLIDINE C_4H_9N

Fl.P., 3°C

BUTYRALDOXIME C_4H_9NO

Anon., *Chemiearbeit*, 1966, (3), 20

A large batch exploded violently (without flame) during vacuum distillation at 90–120°C/20–25 mbar. Since the distilled product contained up to 12% of butyronitrile, it was assumed that the oxime had rearranged to butyramide, and subsequently dehydrated to butyronitrile. The release of water into a system at 120°C would generate excessive pressure which the vessel could not withstand. The rearrangement may have been catalysed by metallic impurities.

See also ETHYL 2-FORMYLPROPIONATE OXIME, $C_6H_{11}NO_3$: Hydrogen chloride

SULPHURIC ACID, H_2O_4S : Cyclopentanone oxime

MORPHOLINE C_4H_9NO

Cellulose nitrate

See CELLULOSE NITRATE: Amines

Nitromethane

See NITROMETHANE, CH_3NO_2: Acids, etc.

BUTYL NITRITE $C_4H_9NO_2$

Fl.P., 10°C

***tert*-NITROBUTANE** $C_4H_9NO_2$

Mee, A. J., *School Sci. Rev.*, 1940, 22(85), 96

A sample exploded during distillation.

See other C-NITRO COMPOUNDS

BUTANE C_4H_{10}

Fl.P., –60°C (gas above 1°C); E.L., 1.9–8.5%

ISOBUTANE C_4H_{10}

Fl.P., –81°C (gas above –12°C); E.L., 1.9–8.5%

DIETHYLALUMINIUM BROMIDE $C_4H_{10}AlBr$

Nitromethane

See NITROMETHANE, CH_3NO_2: Alkylmetal halides

See ALKYLALUMINIUM HALIDES

DIETHYLALUMINIUM CHLORIDE $C_4H_{10}AlCl$

See ALKYLALUMINIUM HALIDES

BORON TRIFLUORIDE DIETHYL ETHERATE $C_4H_{10}BF_3O$

Fl.P., below 22°C

Lithium tetrahydroaluminate
See LITHIUM TETRAHYDROALUMINATE, AlH_4Li: Boron trifluoride diethyl etherate

DIETHYLBERYLLIUM $C_4H_{10}Be$

Coates, 1967, Vol. 1, 106
Ignites in air, even when containing ether, and reacts explosively with water.
See other ALKYLMETALS

DIETHYLBISMUTH CHLORIDE $C_4H_{10}BiCl$

Gilman, H. *et al., J. Org. Chem.*, 1939, **4**, 167
Ignites in air.
See other ALKYLMETAL HALIDES

DIETHYLCADMIUM $C_4H_{10}Cd$

Krause, E., *Ber.*, 1918, **50**, 1813
The vapour decomposes explosively at 180°C. Exposure to ambient air produces white fumes which turn brown and then explode violently.
See other ALKYLMETALS

DIETHYLPHOSPHOROCHLORIDATE $C_4H_{10}ClOP$

Silbert, L. A., *J. Org. Chem.*, 1971, **36**, 2162
Presence of hydrogen chloride as impurity causes an uncontrollable

exothermic reaction during preparation of diethyl phosphate from the title compound.

DICHLORODIETHYLSILANE $C_4H_{10}Cl_2Si$

Fl.P., 24°C
See other ALKYLNON-METAL HALIDES

DIETHYLMAGNESIUM $C_4H_{10}Mg$

Gilman, H. *et al., J. Amer. Chem. Soc.*, 1930, **52**, 5048
Contact with moist air usually caused ignition of the dry powder, and water always ignited the solid or its ethereal solution.
See other ALKYLMETALS

DIETHYL HYPONITRITE $C_4H_{10}N_2O_2$

See DIALKYL HYPONITRITES

***N*-BUTYLAMIDOSULPHURYL AZIDE** $C_4H_{10}N_4O_2S$

Shozda, R. J. *et al., J. Org. Chem.*, 1967, **32**, 2876
It exploded during analytical combustion.
See other ACYL AZIDES

BUTANOL $C_4H_{10}O$

Aluminium
See ALUMINIUM, Al: Butanol

Chromium trioxide
See CHROMIUM TRIOXIDE, CrO_3 : Alcohols

sec-BUTANOL $C_4H_{10}O$

Fl.P., 14°C; E.L., 1.7–9.8% (both at 100°C)

tert-BUTANOL $C_4H_{10}O$

Fl.P., 10°C; E.L., 2.4–8%

DIETHYL ETHER $C_4H_{10}O$

Fl.P., −45°C; E.L., 1.8–36.5%; A.I.T., 180°C
MCA SD-29, 1965

Air

1. Criegee, R. *et al., Angew. Chem.,* 1936, **49**, 101; ibid., 1958, **70**, 261
2. Mallinckrodt, E., *et al. Chem. Eng. News,* 1955, **33**, 3194
3. Ray, A., *J. Appl. Chem.,* 1955, 188
4. Anon., *Chemist & Druggist,* 1947, **157**, 258
5. Feinstein, R. N., *J. Org. Chem.,* 1959, **24**, 1172
6. Davies, A. G., *J.R. Inst. Chem.,* 1956, **80**, 386

The hydroperoxide initially formed by autoxidation of ether is not particularly explosive, but on standing and evaporation, polymeric 1-oxyperoxides are formed which are dangerously explosive, even below 100°C. Numerous laboratory explosions have been caused by evaporation of peroxidised ether [1]. Formation of peroxides in stored ether may be prevented by presence of sodium diethyldithio-carbamate (0.05 p.p.m; reference 2), which probably deactivates traces of metals which catalyse peroxidation [1]; of pyrogallol (1 p.p.m.; reference 3); or by larger proportions (0.005–0.02% of other inhibitors; reference 4). Once present in ether, peroxides may be detected by the iodine–starch test and removed by percolation through anion exchange resin [5], or activated alumina [6], which leaves the ether dry, or by shaking with ferrous sulphate or sodium sulphate solution. Many other methods have been described [6].

See other 1-OXYPEROXY COMPOUNDS
PEROXIDISABLE COMPOUNDS
POLYPEROXIDES

Halogens or Interhalogens

See BROMINE TRIFLUORIDE, BrF_3 : Halogens, etc.
: Solvents
BROMINE PENTAFLUORIDE, BrF_5 : Hydrogen-containing materials
BROMINE, Br_2 : Diethyl ether
CHLORINE, Cl_2 : Diethyl ether
IODINE HEPTAFLUORIDE, F_7I

Oxidants

See Halogens and Interhalogens, above
LIQUID AIR: Diethyl ether
SILVER PERCHLORATE, $AgClO_4$: Diethyl ether
PERCHLORIC ACID, $ClHO_4$: Diethyl ether
NITROSYL PERCHLORATE, $ClNO_5$: Organic materials
NITRYL PERCHLORATE, $ClNO_6$: Organic solvents
CHROMYL CHLORIDE, Cl_2CrO_2 : Organic solvents
FLUORINE NITRATE, FNO_3 : Organic materials
PERMANGANIC ACID, $HMnO_4$: Organic materials
NITRIC ACID, HNO_3 : Diethyl ether
PEROXODISULPHURIC ACID, $H_2O_8S_2$: Organic liquids
SODIUM PEROXIDE, Na_2O_2 : Organic liquids, etc.
OZONE, O_3 : Diethyl ether

Sulphur and compounds

See THIOTRITHIAZYL PERCHLORATE, $ClN_3O_4S_4$: Organic solvents
SULPHONYL DICHLORIDE, Cl_2O_2S: Diethyl ether
SULPHUR, S: Diethyl ether

tert-BUTYL HYDROPEROXIDE $C_4H_{10}O_2$

1. Milas, N. A. *et al., J. Amer. Chem. Soc.*, 1946, **68**, 205
2. Castrantas, 1965, 15

Though relatively stable, explosions have been caused by distillation to dryness [1], or attempted distillation at atmospheric pressure [2].
See other ALKYL HYDROPEROXIDES

DIETHYL PEROXIDE $C_4H_{10}O_2$

1. Baeyer, A. *et al., Ber.*, 1900, **33**, 3387
2. Castrantas, 1965, 15; Baker, G. *et al., Chem. & Ind.*, 1964, 1988
3. Gray, P. *et al., Proc. R. Soc.*, 1971, **A325**, 175

While it is acknowledged as rather explosive, the stated lack of shock-sensitivity at room temperature [1] is countered by its alternative description as shock-sensitive and detonable [2]. The vapour explodes above a certain critical pressure, at temperatures above 190°C [3].

See other DIALKYL PEROXIDES

1,1-DIMETHOXYETHANE $C_4H_{10}O_2$

Fl.P., 1°C (o)

1,2-DIMETHOXYETHANE $C_4H_{10}O_2$

Fl.P., below 21°C

Lithium tetrahydroaluminate
See LITHIUM TETRAHYDROALUMINATE, AlH_4Li: 1,2-Dimethoxyethane

1-HYDROXYBUTYL-3-HYDROPEROXIDE $C_4H_{10}O_3$

Rieche, A., *Ber.*, 1930, **63**, 2642
Explodes on heating.
See other 1-OXYPEROXY COMPOUNDS

TRIMETHYL ORTHOFORMATE $C_4H_{10}O_3$

Fl.P., 15°C
Preparative hazard.
See CHLOROFORM, $CHCl_3$: Sodium, Methanol
: Sodium methoxide

DIETHYL SULPHATE $C_4H_{10}O_4S$

2,7-Dinitro-9-phenylphenanthridine,
Water

Hodgson, J. F., *Chem. & Ind.*, 1968, 1399

Accidental ingress of water to the heated mixture liberated sulphuric acid or its half ester which caused a violent reaction with generation of a very large volume of solid black foam.

See 4-NITROANILINE-2-SULPHONIC ACID, $C_6H_6N_2O_5S$
SULPHURIC ACID, H_2O_4S: Nitroaryl bases, etc.

Potassium *tert*-butoxide

See POTASSIUM *tert*-BUTOXIDE, C_4H_9KO: Acids, etc.

1-BUTANETHIOL $C_4H_{10}S$

Fl.P., 2°C

Nitric acid

See NITRIC ACID, HNO_3: Butanethiol

2-BUTANETHIOL $C_4H_{10}S$

Fl.P., –23°C

DIETHYL SULPHIDE $C_4H_{10}S$

Fl.P., –10°C

2-METHYLPROPANETHIOL $C_4H_{10}S$

Fl.P., –10°C

2-METHYL-2-PROPANETHIOL $C_4H_{10}S$

Fl.P., below −29°C

DIETHYLZINC $C_4H_{10}Zn$

Aluminum Alkyls and other Organometallics, 5, Ethyl Corp., New York, 1967
Immediately pyrophoric in air, reacts violently with water.
See other ALKYLMETALS

DIETHYLALUMINIUM HYDRIDE $C_4H_{11}Al$

See ALKYLALUMINIUM ALKOXIDES AND HYDRIDES

DIETHYLARSINE $C_4H_{11}As$

Von Schwartz, 1918, 322,
Sidgwick, 1950, 762
Inflames in air, even at 0°C
See other ALKYLNON-METAL HYDRIDES

BUTYLAMINE $C_4H_{11}N$

Fl.P., 7°C; E.L., 1.7–9.8%

Perchloryl fluoride
See PERCHLORYL FLUORIDE, $ClFO_3$: Nitrogenous bases

***sec*-BUTYLAMINE** $C_4H_{11}N$

Fl.P., −9°C

tert-BUTYLAMINE $C_4H_{11}N$

Fl.P., –7°C; E.L., 1.7–9.8% (both at 100°C)

DIETHYLAMINE $C_4H_{11}N$

Fl.P., –18°C; E.L., 1.8–10.1%
MCA SD-97, 1971

Cellulose nitrate
See CELLULOSE NITRATE: Amines

ISOBUTYLAMINE $C_4H_{11}N$

Fl.P., –9°C

N-2-HYDROXYETHYLDIMETHYLAMINE $C_4H_{11}NO$

Cellulose nitrate
See CELLULOSE NITRATE: Amines

DIETHYLPHOSPHINE $C_4H_{11}P$

Houben-Weyl, 1963, Vol. 12(1), 69
von Schwartz, 1918, 323
Secondary lower-alkylphosphines readily ignite in air.
See other ALKYLNON-METAL HYDRIDES

ETHYLDIMETHYLPHOSPHINE $C_4H_{11}P$

Smith, J. F., private comm., 1970
May ignite in air.
See other ALKYLNON-METALS

TETRAMETHYLDIARSANE $C_4H_{12}As_2$

Sidgwick, 1950, 770
Inflames in air.
See other ALKYLNON-METALS

BIS-DIMETHYLARSINYL OXIDE $C_4H_{12}As_2O$

von Schwartz, 1918, 322
Inflames in air.
See related ALKYLNON-METALS

BIS-DIMETHYLARSINYL SULPHIDE $C_4H_{12}As_2S$

von Schwartz, 1918, 322
Inflames in air.
See related ALKYLNON-METALS

TETRAMETHYLDIGOLD DIAZIDE $C_4H_{12}Au_2N_6$

Beck, W. *et al., Inorg. Chim. Acta,* 1968, 2, 468
The dimeric azide is extremely sensitive and may explode under water if touched.
See other GOLD COMPOUNDS
METAL AZIDES

1,2-DIAMINO-2-METHYLPROPANEOXODIPEROXO-CHROMIUM (VI) $C_4H_{12}CrN_2O_5$

House, D. A. *et al.*, *Inorg. Chem.*, 1967, **6**, 1078, footnote 6
This blue precipitate is formed intermediately during preparation of 1,2-diamino-2-methylpropaneaquadiperoxochromium(IV) monohydrate, and its analogues are dangerously explosive and should not be isolated without precautions.
See other AMMINECHROMIUM PEROXOCOMPLEXES

2-DIMETHYLAMINOETHYLAMINE $C_4H_{12}N_2$

Fl.P., 11°C

N-HYDROXYETHYL-1,2-DIAMINOETHANE $C_4H_{12}N_2O$

Cellulose nitrate
See CELLULOSE NITRATE: Amines

BIS-DIMETHYLSTIBINYL OXIDE $C_4H_{12}OSb_2$

Sidgwick, 1950, 777
Ignites in air.
See related ALKYLMETALS

TETRAMETHYLLEAD $C_4H_{12}Pb$

Sidgwick, 1950, 463
Liable to explode violently above 90°C.
See other ALKYLMETALS

TETRAMETHYLPLATINUM $C_4H_{12}Pt$

Gilman, H. *et al., J. Amer. Chem. Soc.,* 1953, **75**, 2065
It explodes weakly on heating.
See other ALKYLMETALS
PLATINUM COMPOUNDS

TETRAMETHYLDISTIBANE $C_4H_{12}Sb_2$

Sidgwick, 1950, 779
Ignites in air.
See other ALKYLMETALS

TETRAMETHYLSILANE $C_4H_{12}Si$

Fl.P., below 0°C (gas above 26°C)
See other ALKYLNON-METALS

TETRAMETHYLTIN $C_4H_{12}Sn$

Fl.P., below 21°C; E.L., 1.9– %
See other ALKYLMETALS

TETRAMETHYLAMMONIUM CHLORITE $C_4H_{13}ClNO_2$

Levi, G. R., *Gazz. Chim. Ital.,* 1922[I], **52**, 207
The dry solid explodes on impact.
See other CHLORITE SALTS
OXOSALTS OF NITROGENOUS BASES

DIETHYLENETRIAMINEDIPEROXOCHROMIUM(IV)

$C_4H_{13}CrN_3O_4$

House, D. A. *et al., Inorg. Chem.*, 1966, **5**, 840
The monohydrate explodes at 109–110°C during slow heating.
See other AMMINECHROMIUM PEROXOCOMPLEXES

BIS(2-AMINOETHYL)AMINE (DIETHYLENETRIAMINE) $C_4H_{13}N_3$

Cellulose nitrate
See CELLULOSE NITRATE : Amines

TETRAMETHYLDIALUMINIUM DIHYDRIDE $C_4H_{14}Al_2$

Wiberg, W. E. *et al., Angew. Chem.*, 1939, **52**, 372
It ignites and burns explosively in air.
See other ALKYLALUMINIUM ALKOXIDES, etc.

1,2-DIAMINO-2-METHYLPROPANEAQUADIPEROXOCHROMIUM(IV) $C_4H_{14}CrN_2O_5$

House, D. A. *et al., Inorg. Chem.*, 1967, **6**, 1078
The monohydrate exploded at 83–84°C during slow heating, and is potentially explosive at 20–25°C. There is also a preparative hazard.
See 1,2-DIAMINO-2-METHYLPROPANEOXODIPEROXOCHROMIUM(V) $C_4H_{12}CrN_2O_5$
See other AMMINECHROMIUM PEROXOCOMPLEXES

TETRAMETHYLAMMONIUM AMIDE $C_4H_{14}N_2$

Ammonia
Musker, W. K., *J. Org. Chem.*, 1967, **32**, 3189

During the preparation, the liquid ammonia used as solvent must be completely removed at –45°C. The compound decomposes explosively at ambient temperature in presence of ammonia.
See related N-METAL DERIVATIVES

BIS-1,2-DIAMINOETHANEDICHLOROCOBALT(III) CHLORATE
$C_4H_{16}Cl_3CoN_4O_3$

It explodes at 320°C.
See other AMMINEMETAL OXOSALTS

BIS-1,2-DIAMINOETHANEDICHLOROCOBALT(III) PERCHLORATE
$C_4H_{16}Cl_3CoN_4O_4$

It explodes at 300°C; low impact sensitivity.
See other AMMINEMETAL OXOSALTS

cis-BIS-1,2-DIAMINOETHANEDINITROCOBALT(III) IODATE
$C_4H_{16}CoIN_6O_7$

Lobanov, N. I., *Zh. Neorg. Khim.*, 1959, **4**, 151
It dissociates explosively on heating.
See other AMMINEMETAL OXOSALTS

BIS(1,2-DIAMINOETHANE)HYDROXOOXORHENIUM(V) DIPERCHLORATE
$C_4H_{17}Cl_2N_4O_{10}Re$

Murmann, R. K. *et al., Inorg. Synth.*, 1966, 8, 174–175
It explodes violently when dried at above room temperature and is also shock-sensitive. Several explosions occurred during analytical combustion.
See other AMMINEMETAL OXOSALTS

PENTAAMMINEPYRAZINERUTHENIUM(III) DIPERCHLORATE $C_4H_{19}Cl_2N_7O_8Ru$

Creutz, C. A., private comm., 1969

After ether washing, 30 mg of the salt exploded violently when disturbed.

See other AMMINEMETAL OXOSALTS

BIS(DIMETHYLAMINOBORANE)ALUMINIUM TETRAHYDROBORATE $C_4H_{22}AlB_3N_2$

Burg., A. B. *et al., J. Amer. Chem. Soc.,* 1951, **73**, 957

The impure oily product ignites in air and reacts violently with water.

See other COMPLEX HYDRIDES

POTASSIUM TETRACYANOMERCURATE (2–) $C_4HgK_2N_4$

Ammonia

Pieters, 1957, 30

Contact may be explosive.

See also POTASSIUM HEXACYANOFERRATE (3–), $C_6FeK_3N_6$: Ammonia

See other METAL CYANIDES

DICYANOACETYLENE C_4N_2

Ciganek, E. *et al., J. Org. Chem.,* 1968, **33**, 542

Highly endothermic, it is potentially explosive in the pure state or in concentrated solutions, but fairly stable in dilute solution.

See related HALOACETYLENE DERIVATIVES

See other CYANO COMPOUNDS

TETRACARBONYLNICKEL C_4NiO_4

Fl.P., below −20°C; E.L., 2.0 –   %

Bromine
Blanchard, A. A. *et al., J. Amer. Chem. Soc.,* 1926, **48**, 872
The two interact explosively in the liquid state, but smoothly as vapour.

Mercury,
Oxygen
Mellor, 1946, Vol. 5, 955
A mixture of the dry carbonyl and oxygen will explode on vigorous shaking with mercury.

Oxygen,
Butane
1. Egerton, A. *et al., Proc. R. Soc.,* 1954, **A225**, 427
2. Badin, E. J. *et al., J. Amer. Chem. Soc.,* 1948, **70**, 2055

The carbonyl on exposure to air produces a deposit which becomes peroxidised and may ignite. Mixtures with air or oxygen at low partial and total pressures explode after a variable induction period [1]. Addition of the carbonyl to a butane–oxygen mixture at 20–40°C caused explosive reaction in some cases [2].
See other CARBONYLMETALS

HEXACHLOROCYCLOPENTADIENE C_5Cl_6

Sodium
See SODIUM Na: Halocarbons

SODIUM PENTACYANONITROSYLFERRATE (2–) $C_5FeN_6Na_2O$

Sodium nitrite
See SODIUM NITRITE, $NNaO_2$: Metal cyanides

PENTACARBONYLIRON C_5FeO_5

Fl.P., −15°C; may ignite in air

Acetic acid,
Water
Braye, E. H. *et al., Inorg. Synth.,* 1966, **8**, 179
A brown pyrophoric powder is produced if the carbonyl is dissolved in acetic acid containing above 5% of water.

Nitrogen oxide
See NITROGEN OXIDE, NO: Pentacarbonyliron

Transition metal halides,
Zinc
Lawrenson, M. J., private comm., 1970
The preparation of metal carbonyls by treating a transition metal halide either with carbon monoxide and zinc or with iron pentacarbonyl is well-known and smooth. However, a violent eruptive reaction occurs if a methanolic solution of cobalt, rhodium or ruthenium halide is treated with both zinc and iron pentacarbonyl.
See other CARBONYLMETALS

1,3-PENTADIYN-1-YLSILVER C_5H_3Ag

Schluhbach, H. H. *et al., Ann.,* 1950, **568**, 155
Very sensitive to impact or friction and explodes when moistened with sulphuric acid.

1,3-PENTADIYN-1-YLCOPPER C_5H_3Cu

Schluhbach, H. H. *et al., Ann.,* 1950, **568**, 155
Explodes on impact or friction.
See other METAL ACETYLIDES

1,3-PENTADIYNE C_5H_4

Brandsma, 1971, 7, 36
It explodes on distillation at atmospheric pressure.
See other ACETYLENIC COMPOUNDS

3-PYRIDINEDIAZONIUM TETRAFLUOROBORATE $C_5H_4BF_4N_3$

1. Johnson, E. P. *et al., Chem. Eng. News,* 1967, **45**(44), 44
2. Roe, A. *et al., J. Amer. Chem. Soc.,* 1947, **69**, 2443

A sample, air-dried on aluminium foil, exploded spontaneously, while another sample exploded on heating to 47°C [1]. The earlier reference describes the instability of the salt above 15°C if freed of solvent, and both the 2- and 4-isomeric salts were found to be very unstable and incapable of isolation [2].
See other DIAZONIUM TETRAHALOBORATES

2-FLUOROPYRIDINE C_5H_4FN

See BROMINE TRIFLUORIDE, BrF_3: Pyridine

DIAZOCYCLOPENTADIENE $C_5H_4N_2$

Wedd, A. G., *Chem. & Ind.,* 1970, 109
The material exploded violently during distillation at 48–53°C/66 mbar. Use of solutions of undistilled material is recommended.
See other DIAZO COMPOUNDS

1,2-DIHYDROPYRIDO[2,1,*e*]TETRAZOLE $C_5H_4N_4$

Fargher, R. G. *et al., J. Chem. Soc.,* 1915, **107**, 695

Explodes on touching with a hot rod.
See other TETRAZOLES

CYCLOPENTADIENYL SILVER PERCHLORATE $C_5H_5AgClO_4$

Ulbricht, T. L. V., *Chem. & Ind.*, 1961, 1570
The complex explodes on heating.
See other HEAVY METAL DERIVATIVES

CYCLOPENTADIENYLGOLD(I) C_5H_5Au

Huttel, R. *et al.*, *Angew. Chem. (Intern. Ed.)*, 1967, **6**, 862
It is sensitive to friction and heat, often deflagrating on gentle warming.
See other GOLD COMPOUNDS

2-BROMOMETHYLFURAN C_5H_5BrO

Dunlop, 1953, 231
It is very unstable, and the liberated hydrogen bromide accelcrates further decomposition to explosive violence.
See also 2-CHLOROMETHYLFURAN, C_5H_5ClO

2-CHLOROMETHYLFURAN C_5H_5ClO

ABCM Quart. Safety Summ., 1962, **33**, 2

A small sample of freshly prepared and distilled material, when stored over a weekend, exploded violently owing to polymerisation or decomposition. Material should be prepared and used immediately, or if brief storage is inevitable, refrigeration is necessary.
See also 2-BROMOMETHYLFURAN, C_5H_5BrO
2-CHLOROMETHYLTHIOPHENE, C_5H_5ClS

2-CHLOROMETHYLTHIOPHENE **C_5H_5ClS**

1. Bergeim, F. H., *Chem. Eng. News,* 1952, **30**, 2546; Meyer, F.C., ibid., 3352
2. Wiberg, K. B., *Org. Synth.,* 1955, Coll. Vol. 3, 197

The material is unstable and gradually decomposes, even when kept cold and dark, with liberation of hydrogen chloride, which accelerates the decomposition. If kept in closed containers, pressure increase may cause an explosion [1]. Amines stabilise the material, which can then be kept cold in a vented container for several months [2].
See also 2-CHLOROMETHYLFURAN, C_5H_5ClO

1-IODO-3-PENTEN-1-YNE **C_5H_5I**

Vaughn, J. A. *et al., J. Amer. Chem. Soc.,* 1934, **56**, 1208
After distillation at 72°C/62 mbar, the residue always exploded if heating was continued.
See other HALOACETYLENE DERIVATIVES

PYRIDINE **C_5H_5N**

Fl.P., 20°C; E.L., 1.8–12.4%

Formamide,
Iodine,
Sulphur trioxide
See FORMAMIDE, CH_3NO: Iodine, etc.

Maleic anhydride

See MALEIC ANHYDRIDE, $C_4H_2O_3$: Cations, etc.

Oxidants

See BROMINE TRIFLUORIDE, BrF_3: Pyridine
CHROMIUM TRIOXIDE, CrO_3: Pyridine
FLUORINE, F_2: Nitrogenous bases
DINITROGEN TETRAOXIDE, N_2O_4: Heterocyclic bases

CYCLOPENTADIENYLSODIUM C_5H_5Na

King, R. B. *et al.*, *Inorg. Synth.*, 1963, **7**, 101
The solid obtained by evaporation of the air-sensitive solution is pyrophoric in air.

See ALKALI-METAL DERIVATIVES OF HYDROCARBONS
See other ARYL METALS

CYCLOPENTADIENE C_5H_6

Fl.P., below 25°C

Kirk-Othmer, 1965, Vol. 6, 690
Dimerisation is highly exothermic, the rate increasing rapidly with temperature, and may cause rupture of a closed uncooled container. The monomer may be largely prevented from dimerising by storage at –80°C or below.

See PROPENE, C_3H_6

Nitric acid

See NITRIC ACID, HNO_3: Hydrocarbons

Oxides of nitrogen

See NITROGEN OXIDE, NO: Dienes, Oxygen
DINITROGEN TETRAOXIDE, N_2O_4: Hydrocarbons

Oxygen

Hock, H. *et al.*, *Chem. Ber.*, 1951, **84**, 349
Exposure of the diene to oxygen gives peroxidic products containing

some monomeric, but largely polymeric, peroxides, which explode strongly on contact with a flame.
See other POLYPEROXIDES

Sulphuric acid
See SULPHURIC ACID, H_2O_4S: Cyclopentadiene
See other DIENES

2-METHYL-1-BUTEN-3-YNE C_5H_6

Fl.P., below −7°C (o)
See other ACETYLENIC COMPOUNDS

3,5-DIBROMOCYCLOPENTENE $C_5H_6Br_2$

Lithium tetrahydroaluminate
See LITHIUM TETRAHYDROALUMINATE, AlH_4Li: 3,5-Dibromocyclopentene

PYRIDINIUM PERCHLORATE $C_5H_6ClNO_4$

1. Kuhn, R. *et al., Chem. Ztg.,* 1950, **74**, 139
2. Anon., *Chemiearbeit,* 1963, **15**(3), 19
3. Arndt, F. *et al., Chem. Ztg.,* 1950, **74**, 140
4. Schumacher, 1960, 213

It can be detonated on impact, but is normally considered a stable intermediate (m.p., 288°C), suitable for purification of pyridine [1]. Occasionally explosions have occurred when the salt was disturbed [2], which have been variously attributed to presence of ethyl perchlorate, ammonium perchlorate or chlorates. A safer preparative modification is described [3]. It explodes on heating to above 335°C, or at a lower temperature if ammonium perchlorate is present [4].
See other OXOSALTS OF NITROGENOUS BASES

1,3-DICHLORO-5,5-DIMETHYL-2,4-IMIDAZOLIDINDIONE

$C_5H_6Cl_2N_2O_2$

Xylene

Anon., *Chem. Trade J.*, 1951, **129**, 136

An attempt to chlorinate xylene with the 'dichlorohydantoin' caused a violent explosion.

See other *N*-HALOIMIDES

4,4-BIS(DIFLUOROAMINO)-3-FLUOROIMINO-1-PENTENE

$C_5H_6F_5N_3$

Parker, C. O. *et al., J. Org. Chem.*, 1972, **37**, 922

Samples of this and related poly-difluoroamino compounds exploded during analytical combustion.

See other *N*-HALOGEN COMPOUNDS

GLUTARYL DIAZIDE

$C_5H_6N_6O_2$

Curtiss, T. *et al., J. Prakt. Chem.*, 1900, **62**, 196

A very small sample exploded sharply on heating.

See other ACYL AZIDES

DIVINYL KETONE

C_5H_6O

Trinitromethane

See TRINITROMETHANE, CHN_3O_6 : Divinyl ketone

2-METHYLFURAN

C_5H_6O

Fl.P., –30°C

2-PENTEN-4-YN-3-OL C_5H_6O

Brandsma, 1971, 73
The residue from distillation at 20 mbar exploded vigorously at above 90°C.
See other ACETYLENIC COMPOUNDS

FURFURYL ALCOHOL $C_5H_6O_2$

Acids
1. Tobie, W. C., *Chem. Eng. News,* 1940, 18, 72
2. *MCA Case History No. 858*
3. Dunlop 1953, 214, 221, 783

A mixture of the alcohol with formic acid rapidly self-heated, then reacted violently [1]. A stirred mixture with cyanoacetic acid exploded violently after application of heat [2]. Contact with acids or acidic materials causes self-condensation of the alcohol, which may be explosively violent under unsuitable physical conditions. The general mechanism has been discussed [3].
See NITRIC ACID, HNO_3: Furfuryl alcohol

Hydrogen peroxide
See HYDROGEN PEROXIDE, H_2O_2: Alcohols

2-METHYLTHIOPHENE C_5H_6S

Fl.P., 8°C

2-PROPYNYL VINYL SULPHIDE C_5H_6S

Brandsma, 1971, 7, 182
It decomposes explosively above 85°C.
See other ACETYLENIC COMPOUNDS

4-BROMOCYCLOPENTENE C_5H_7Br

Preparative hazard.

See LITHIUM TETRAHYDROALUMINATE, AlH_4Li: 3,5-Dibromocyclopentene

ETHYL 2,2,3-TRIFLUOROPROPIONATE $C_5H_7F_3O_2$

Sodium hydride

See SODIUM HYDRIDE, HNa: Ethyl 2,2,3-trifluoropropionate

1-METHYLPYRROLE C_5H_7N

Fl.P., 16°C

CYCLOPENTENE C_5H_8

Fl.P., –29°C

2-METHYL-1,3-BUTADIENE (ISOPRENE) C_5H_8

Fl.P., –53°C; E.L., 2.0–8.9%; A.I.T., 220°C

See other DIENES

Air

Kirk-Othmer, 1967, Vol. 12, 79

In absence of inhibitors, isoprene absorbs air with formation of peroxides, which do not separate from solution. Although the solution is not detonable, the gummy peroxide polymer obtained by evaporation can be detonated on impact under standard conditions.

See also BUTADIENE, C_4H_6: Alone, or Air

Ozone

Loveland, J. W., *Chem. Eng. News*, 1956, **34**(3), 292

Isoprene (1 g) dissolved in heptane was ozonised at −78°C. Soon after cooling was stopped, a violent explosion, followed by a lighter one, occurred. This was attributed to high concentrations of peroxides and ozonides building up at the low temperature employed. Operation at a higher temperature would permit the ozonides and peroxides to decompose, avoiding high concentrations in the reaction mixture.

See other OZONIDES

1,3-PENTADIENE C_5H_8

Fl.P., −43°C; E.L., 2–8.3%

1,4-PENTADIENE C_5H_8

Fl.P., below 0°C

1-PENTYNE C_5H_8

Fl.P., −20°C

TETRAKIS-(*N,N*-DICHLOROAMINOMETHYL)METHANE $C_5H_8Cl_8N_4$

Lévy, R. S. *et al., Mem. Poudres,* 1958, **40**, 109

Shock- and heat-sensitive, it is a brisant more powerful than mercury fulminate.

See other N-HALOGEN COMPOUNDS

ALLYL VINYL ETHER C_5H_8O

Fl.P., below 21°C (o)

CYCLOPENTANONE C_5H_8O

Hydrogen peroxide,
Nitric acid
See HYDROGEN PEROXIDE, H_2O_2: Ketones, etc.

CYCLOPROPYL METHYL KETONE C_5H_8O

Fl.P., 13°C

2,3-DIHYDROPYRAN C_5H_8O

Fl.P., –16°C

1-ETHOXY-2-PROPYNE C_5H_8O

Brandsma, 1971, 172
Distillation of a 1 kg quantity at 80°C/1 bar led to a violent explosion. As the compound had not been stored under nitrogen during the 3 weeks since preparation, peroxides were suspected.
See other ACETYLENIC COMPOUNDS

2-METHYL-3-BUTYN-2-OL C_5H_8O

Fl.P., below 21°C

METHYL ISOPROPENYL KETONE C_5H_8O

Fl.P., 21°C; E.L., 1.8–9.0% (both at 50°C)

ALLYL ACETATE $C_5H_8O_2$

Fl.P., 22°C

ETHYL ACRYLATE $C_5H_8O_2$

Fl.P., 9°C; E.L., 1.8 – %

METHYL CROTONATE $C_5H_8O_2$

Fl.P., –1°C

METHYL METHACRYLATE $C_5H_8O_2$

Fl.P., 11°C; E.L., 2.1–12.5%

Air
Barnes, C. E. *et al., J. Amer. Chem. Soc.,* 1950, **72**, 210
Exposure of the purified (unstabilised) monomer to air at room temperature for two months generated an ester–oxygen interpolymer, which exploded on evaporation of the surplus monomer at 60°C (but not 40°C).
See other POLYPEROXIDES

Dibenzoyl peroxide
See DIBENZOYL PEROXIDE, $C_{14}H_{10}O_4$: Methyl methacrylate

ISOPROPENYL ACETATE $C_5H_8O_2$

Fl.P., 16°C; E.L., 1.9 – %

ISOPRENE DIOZONIDE $C_5H_8O_6$

See 2-METHYL-1,3-PENTADIENE, C_5H_8 : Ozone

CHLOROCYCLOPENTANE C_5H_9Cl

Fl.P., 16°C

PIVALOYL CHLORIDE C_5H_9ClO

Fl.P., 24°C

PIVALONITRILE C_5H_9N

Fl.P., 21°C

1,2,5,6-TETRAHYDROPYRIDINE C_5H_9N

Fl.P., 16°C

CYCLOPENTANONE OXIME C_5H_9NO

Sulphuric acid
See SULPHURIC ACID, H_2O_4S: Cyclopentanone oxime

PIVALOYL AZIDE $C_5H_9N_3O$

Bühler, A. *et al., Helv. Chim. Acta,* 1943, **26**, 2123
The azide exploded very violently on warming, and on one occasion, on standing.
See other ACYL AZIDES

tert-BUTYL AZIDOFORMATE $C_5H_9N_3O_2$

1. Carpino, L. A. *et al., Org. Synth.,* 1964, **44**, 15
2. Sakai, K. *et al., J. Org. Chem.,* 1971, **36**, 2387

Explosion during distillation at 74°C/92 mbar has been recorded on only one out of several hundred occasions [1]. An alternative preparation avoiding distillation has been described [2].
See other ACYL AZIDES

DIMETHYL-1-PROPYNLTHALLIUM C_5H_9Tl

Nast, R. *et al., J. Organomet. Chem.,* 1966, **6**, 461
Explodes on heating and is sensitive to stirring or impact.
See other ALKYLMETALS
METAL ACETYLIDES

CYCLOPENTANE C_5H_{10}

Fl.P., −7°C

1,1-DIMETHYLCYCLOPROPANE C_5H_{10}

Gas above 20°C

ETHYLCYCLOPROPANE C_5H_{10}

Fl.P., below 10°C

2-METHYL-1-BUTENE C_5H_{10}

Fl.P., –20°C

2-METHYL-2-BUTENE C_5H_{10}

Fl.P., –20°C

3-METHYL-1-BUTENE C_5H_{10}

Fl.P., below –7°C (gas above 20°C); E.L., 1.5–9.1%

METHYLCYCLOBUTANE C_5H_{10}

Fl.P., below 10°C

1-PENTENE C_5H_{10}

Fl.P., –18°C; E.L., 1.5–8.7%

2-PENTENE C_5H_{10}

Fl.P., –18°C

CYANODIETHYLGOLD $C_5H_{10}AuN$

See the tetramer, TETRACYANOOCTAETHYLTETRAGOLD, $C_{20}H_{40}Au_4N_4$

1-PERCHLORYLPIPERIDINE $C_5H_{10}ClNO_3$

Alone,
or Piperidine
Gardner, D. L. *et al., J. Org. Chem.*, 1964, **29**, 3738–3739
It is a dangerously sensitive oil which has exploded violently on storage, heating or contact with piperidine. Absorption in alumina was necessary to desensitise it to allow of non-explosive analytical combustion.
See other PERCHLORYL COMPOUNDS

DICHLOROMETHYLENEDIETHYLAMMONIUM CHLORIDE $C_5H_{10}Cl_3N$

Bader, A. R., private comm., 1971
Viehe, H. G. *et al., Angew. Chem. (Intern. Ed.)*, 1971, **10**, 573
The compound does not explode on heating; the published statement is in error.

3-DIMETHYLAMINOPROPIONITRILE $C_5H_{10}N_2$

Fl.P., below 22°C

3,7-DINITROSO-1,3,5,7-TETRAAZABICYCLO[3.3.1]NONANE $C_5H_{10}N_6O_2$

MCA Case History No. 841
A cardboard drum of this blowing agent (as a 40% dispersion in fine

silica) ignited when roughly handled in storage.
See other NITROSO COMPOUNDS

ALLYL ETHYL ETHER $C_5H_{10}O$

Fl.P., below 24°C
ABCM Quart. Safety Summ., 1963, **34**, 7
A commercial sample was distilled without previously being tested for peroxide. An explosion occurred towards the end of the distillation, and the residue was later shown to contain peroxide.
See other PEROXIDISABLE COMPOUNDS

ETHYL PROPENYL ETHER $C_5H_{10}O$

Fl.P., below –5°C

ISOPROPYL VINYL ETHER $C_5H_{10}O$

Fl.P., –32°C

ISOVALERALDEHYDE $C_5H_{10}O$

Fl.P., –5°C

3-METHYL-2-BUTANONE $C_5H_{10}O$

Fl.P., below 22°C

3-METHYL-1-BUTEN-3-OL $C_5H_{10}O$

Fl.P., 18°C

2-METHYLTETRAHYDROFURAN $C_5H_{10}O$

Fl.P., −12°C

2-PENTANONE $C_5H_{10}O$

Fl.P., 7°C; E.L., 1.6–8.2%

Bromine trifluoride
See BROMINE TRIFLUORIDE, BrF_3 : 2-Pentanone

3-PENTANONE $C_5H_{10}O$

Fl.P., 13°C; E.L., 1.6 – %

Hydrogen peroxide,
Nitric acid
See HYDROGEN PEROXIDE, H_2O_2 : Ketones, etc.

4-PENTEN-1-OL $C_5H_{10}O$

Fl.P., below 23°C
See other PEROXIDISABLE COMPOUNDS

TETRAHYDROPYRAN $C_5H_{10}O$

Fl.P., −20°C
See other PEROXIDISABLE COMPOUNDS

VALERALDEHYDE $C_5H_{10}O$

Fl.P., 12°C

BUTYL FORMATE $C_5H_{10}O_2$

Fl.P., 18°C; E.L., 1.7–8%

3,3-DIMETHOXYPROPENE $C_5H_{10}O_2$

Fl.P., 19°C
See other PEROXIDISABLE COMPOUNDS

2,2-DIMETHYL-1,3-DIOXOLAN $C_5H_{10}O_2$

Fl.P., –1°C
See other PEROXIDISABLE COMPOUNDS

ETHYL PROPIONATE $C_5H_{10}O_2$

Fl.P., 12°C; E.L., 1.9–11%

ISOBUTYL FORMATE $C_5H_{10}O_2$

Fl.P., below 21°C; E.L., 2.0–8.9%

ISOPROPYL ACETATE $C_5H_{10}O_2$

Fl.P., 4°C; E.L., 1.7–7.8%

2-METHOXYETHYL VINYL ETHER $C_5H_{10}O_2$

Fl.P., 18°C (o)
See other PEROXIDISABLE COMPOUNDS

METHYL BUTYRATE $C_5H_{10}O_2$

Fl.P., 14°C

4-METHYL-1,3-DIOXAN $C_5H_{10}O_2$

Fl.P., 16°C
See other PEROXIDISABLE COMPOUNDS

METHYL ISOBUTYRATE $C_5H_{10}O_2$

Fl.P., 13°C (o)

PROPYL ACETATE $C_5H_{10}O_2$

Fl.P., 14°C; E.L., 1.7–8%

TETRAHYDROFURFURYL ALCOHOL $C_5H_{10}O_2$

3-Nitro-*N*-bromophthalimide
See *N*-HALOIMIDES: Alcohols

DIETHYL CARBONATE $C_5H_{10}O_3$

Fl.P., 25°C

***trans*-2-PENTENE OZONIDE** $C_5H_{10}O_3$

See 2-HEXENE OZONIDE, $C_6H_{12}O_3$
See other OZONIDES

1-BROMO-3-METHYL BUTANE $C_5H_{11}Br$

Fl.P., 21°C

2-BROMOPENTANE $C_5H_{11}Br$

Fl.P., 20°C

1-CHLORO-3-METHYLBUTANE $C_5H_{11}Cl$

Fl.P., below 21°C; E.L., 1.5–7.4%

2-CHLORO-2-METHYLBUTANE $C_5H_{11}C$

Fl.P., 12°C; E.L., 1.5–7.4%

1-CHLOROPENTANE $C_5H_{11}C$

Fl.P., 13°C (o); E.L., 1.6–8.6%

2-IODOPENTANE $C_5H_{11}I$

Fl.P., below 23°C

CYCLOPENTYLAMINE $C_5H_{11}N$

Fl.P., 13°C

1-METHYLPYRROLIDINE $C_5H_{11}N$

Fl.P., 3°C

PIPERIDINE $C_5H_{11}N$

Fl.P., 3°C

1-Perchlorylpiperidine
See 1-PERCHLORYLPIPERIDINE, $C_5H_{10}ClNO_3$: Piperidine

ISOPENTYL NITRITE $C_5H_{11}NO_2$

Fl.P., 10°C; E.L., 1.0 – % (calc.); A.I.T., 209°C

PENTYL NITRITE $C_5H_{11}NO_2$

Fl.P., below 23°C

2,2-DIMETHYLPROPANE (NEOPENTANE) C_5H_{12}

Fl.P., below –7°C (gas above 9°C); E.L., 1.4–7.5%

Preparative hazard.
See ALUMINIUM, Al: Halocarbons

3-METHYLBUTANE (ISOPENTANE) C_5H_{12}

Fl.P., –51°C; E.L., 1.4–7.6%

PENTANE C_5H_{12}

Fl.P., –49°C; E.L., 1.4–8.0%

N-BROMOTETRAMETHYLGUANIDINE $C_5H_{12}BrN_3$

Papa, A. J., *J. Org. Chem.*, 1966, **31**, 1426
The material is unstable even at 0°C, and explodes if heated above 50°C at atmospheric pressure.
See other *N*-HALOGEN COMPOUNDS

N-CHLOROTETRAMETHYLGUANIDINE $C_5H_{12}ClN_3$

Papa, A. J., *J. Org. Chem.*, 1966, **31**, 1426

The material is unstable even at 0°C, and explodes if heated above 50°C at atmospheric pressure.
See other *N*-HALOGEN DERIVATIVES

BUTYL METHYL ETHER $C_5H_{12}O$

Fl.P., below 18°C

ETHYL ISOPROPYL ETHER $C_5H_{12}O$

Fl.P., below –15°C

See other PEROXIDISABLE COMPOUNDS

ETHYL PROPYL ETHER $C_5H_{12}O$

Fl.P., below –20°C; E.L., 1.7–9%

ISOPENTANOL $C_5H_{12}O$

Fl.P., below 23°C; E.L., 1.4–9.0% (both at 100°C)

Hydrogen trisulphide

See HYDROGEN TRISULPHIDE, H_2S_3 : Pentyl alcohol

sec-PENTANOL $C_5H_{12}O$

Fl.P., below 23°C

tert-PENTANOL $C_5H_{12}O$

Fl.P., 19°C; E.L., 1.2–9%

DIETHOXYMETHANE $C_5H_{12}O_2$

Fl.P., below 21°C

1,1-DIMETHOXYPROPANE $C_5H_{12}O_2$

Fl.P., below 10°C

2,2-DIMETHOXYPROPANE $C_5H_{12}O_2$

Fl.P., –7°C

Metal perchlorates

1. Dickinson, R. C. *et al., Chem. Eng. News,* 1970, **48**(28), 6;
   Cramer, R., ibid., **48**(45), 7
2. Mikulski, S. M., *Chem. Eng. News,* 1971, **49**(4), 8

During dehydration of the hydrated manganese and nickel salts with dimethoxypropane, heating above 65°C caused violent explosions, probably involving oxidation by the anion [1]. Triethyl orthoformate is recommended as a safer dehydrating agent [2].

2(2-METHOXYETHOXY)ETHANOL $C_5H_{12}O_3$

Calcium hypochlorite
See CALCIUM HYPOCHLORITE, $CaCl_2O_2$: Hydroxy compounds

PENTAERYTHRITOL $C_5H_{12}O_4$

Thiophosphoryl chloride
See THIOPHOSPHORYL CHLORIDE, Cl_3PS : Pentaerythritol

3-METHYLBUTANETHIOL $C_5H_{12}S$

Fl.P., below 23°C

PENTANETHIOL $C_5H_{12}S$

Fl.P., 18°C

ISOPENTYLAMINE $C_5H_{13}N$

Fl.P., –1°C; E.L., 2.3–22%

N-METHYLBUTYLAMINE $C_5H_{13}N$

Fl.P., 2°C

1-PENTYLAMINE $C_5H_{13}N$

Fl.P., –1°C

DIETHYLMETHYLPHOSPHINE $C_5H_{13}P$

Personal experience
May ignite in air with long exposure.
See other ALKYLNON-METALS

2-DIMETHYLAMINO-*N*-METHYLETHYLAMINE $C_5H_{14}N_2$

Fl.P., 14°C

3-DIMETHYLAMINOPROPYLAMINE $C_5H_{14}N_2$

Cellulose nitrate
See CELLULOSE NITRATE: Amines

DIMETHYLTRIMETHYLSILOXOGOLD $C_5H_{15}AuOSi$

See the dimer, TETRAMETHYLBIS (TRIMETHYLSILOXO)DIGOLD, $C_{10}H_{30}Au_2O_2Si_2$

PENTAAMMINEPYRIDINERUTHENIUM(III) DIPERCHLORATE $C_5H_{20}Cl_2N_6O_8Ru$

Creutz, C. A., private comm., 1969
The dry salt exploded on touching.
See other AMMINEMETAL OXOSALTS

PENTAFLUOROPHENYLALUMINIUM DIBROMIDE $C_6AlBr_2F_5$

Chambers, R. D. *et al., J. Chem. Soc. (C)* 1967, 2185; *Tetrahedron Lett.*, 1965, 2389

Ignites in air; explodes violently on rapid heating to 195°C or during uncontrolled hydrolysis.
See other HALO-ARYLMETALS

HEXACARBONYLCHROMIUM C_6CrO_6

Weast, 1972, B-83
It explodes at 210°C.
See other CARBONYLMETALS

PENTAFLUOROPHENYLLITHIUM C_6F_5Li

Deuterium oxide
Kinsella, E. *et al., Chem. & Ind.*, 1971, 1017
During addition of an ethereal solution of deuterium oxide (containing some peroxide) to a suspension of the organolithium reagent in pentane, a violent explosion occurred. This may have been initiated by the peroxide present.
See other HALO-ARYLMETALS

HEXAFLUOROBENZENE C_6F_6

Fl.P., 10°C

PERFLUOROHEXYL IODIDE $C_6F_{13}I$

Sodium
See SODIUM, Na : Halocarbons

POTASSIUM HEXACYANOFERRATE (3–) ('FERRICYANIDE')

$C_6FeK_3N_6$

Ammonia

Pieters, 1957, 30

Sidgwick, 1950, 1359

Contact may be explosive, possibly owing to rapid oxidation of ammonia by alkaline 'ferricyanide'.

See also POTASSIUM TETRACYANOMERCURATE (2–), $C_4HgK_2N_4$

Chromium trioxide

See CHROMIUM TRIOXIDE, CrO_3 : Potassium hexacyanoferrate(3–)

Sodium nitrite

See SODIUM NITRITE, $NNaO_2$: Metal cyanides

See other METAL CYANIDES

POTASSIUM HEXACYANOFERRATE (4–) ('FERROCYANIDE')

$C_6FeK_4N_6$

Sodium nitrite

See SODIUM NITRITE, $NNaO_2$: Metal cyanides

See other METAL CYANIDES

IRON(III) OXALATE

$C_6Fe_2O_{12}$

Weinland, R. *et al., Z. Anorg. Chem.,* 1929, 178, 219

The salt, probably complex, decomposes at 100°C.

See other METAL OXALATES

4-CHLORO-2,5-DINITROBENZENEDIAZONIUM-6-OLATE

$C_6HClN_4O_5$

See NITRIC ACID, HNO_3 : 4-Chloro-2-nitroaniline

See other ARENEDIAZONIUMOLATES

3,4-DIFLUORO-2-NITROBENZENEDIAZONIUM-6-OLATE
3,6-DIFLUORO-2-NITROBENZENEDIAZONIUM-4-OLATE

$C_6HF_2N_3O_3$

Finger, G. C. *et al., J. Amer. Chem. Soc.,* 1951, **73**, 148
Both isomers, produced under different conditions of diazotisation of 3,4,6-trifluoro-2-nitroaniline, exploded on heating, ignition or impact.
See other ARENEDIAZONIUMOLATES

LEAD 2,4,6-TRINITRORESORCINOLATE $C_6HN_3O_8Pb$

MCA Case History No. 957
Three beakers of lead styphnate were being heated in a laboratory oven to dry the salt. On moving one of the beakers, all three detonated. Several heavy metal salts of polynitro-phenols are known to be detonable, so these materials should not be handled while dry.
See other HEAVY METAL DERIVATIVES

2,6-DIBROMOBENZOQUINONE-4-CHLOROIMINE $C_6H_2Br_2ClNO$

1. Hartmann, W. W. *et al., Org. Synth.,* 1943, Coll. Vol. 2, 177
2. Horber, D. F. *et al., Chem. & Ind.,* 1967, 1551
3. Taranto, B. J., *Chem. Eng. News,* 1967, **45**(52), 54

The chloroimine decomposed violently on a drying tray at 60°C [1], and a bottle accidentally heated to 50°C exploded [2]. Decomposition of the heated solid occurs after a temperature-dependent induction period. Unheated material eventually exploded after storage at ambient temperature [3].
See also 2,6-DICHLOROBENZOQUINONE-4-CHLOROIMINE, $C_6H_2Cl_3NO$
See other *N*-HALOGEN COMPOUNDS

2,6-DICHLOROBENZOQUINONE-4-CHLOROIMINE $C_6H_2Cl_3NO$

Taranto, B. J., *Chem. Eng. News,* 1967, **45**(52), 54

The heated, solid material decomposed violently after an induction period dependent upon temperature. Unheated material exploded after storage at ambient temperature.
See also 2,6-DIBROMOBENZOQUINONE-4-CHLOROIMINE., $C_6H_2Br_2ClNO$
See other *N*-HALOGEN COMPOUNDS

1,2,4,5-TETRACHLOROBENZENE $C_6H_2Cl_4$

Sodium hydroxide
1. *MCA Case History No. 620*
2. *ABCM Quart. Safety Summ.*, 1952, **23**, 5
3. Calnan, A. B., private comm., 1972

Several serious incidents have been reported about the commercial preparation of 2,4,5-trichlorophenol by alkaline hydrolysis of tetrachlorobenzene. On two occasions the hydrolysis under pressure with methanolic alkali at 125°C went out of control, one reaching 400°C [1]. After hydrolysis in ethylene glycol solution, the residue from vacuum stripping exploded, probably owing to overheating. Electric heating was used without knowledge of the liquid temperature [2].

In another incident in 1968, a violent explosion occurred during hydrolysis under pressure. Serious after-effects were caused by the widespread distribution of the most powerful dermatitic compound known, di(4,5-dichlorobenzo)dioxin, which was produced by cyclisation during the explosion. Traces remaining on contaminated plant after 3 years' weathering still caused dermatitis at a concentration of 10 p.p.m. [3].
See other HALOARYL COMPOUNDS

1,2,4,5-TETRAFLUOROBENZENE $C_6H_2F_4$

Fl.P., 4°C (o)

3,4,5-TRIIODOBENZENEDIAZONIUM NITRATE $C_6H_2I_3N_3O_3$

Kalb, L. *et al.*, *Ber.*, 1926, **59**, 1867

Unstable on warming, explodes on heating in a flame.
See other DIAZONIUM SALTS

4,6-DINITROBENZENEDIAZONIUM-2-OLATE $C_6H_2N_4O_5$

Urbanski, 1967, Vol. 3, 204
This priming explosive, as sensitive as mercuric fulminate, is much more powerful than metal-containing initiators.
See other ARENEDIAZONIUMOLATES

PICRYL AZIDE $C_6H_2N_6O_6$

Schrader, E., *Ber.*, 1917, **50**, 778
Explodes weakly under impact, but not on heating.
See other ORGANIC AZIDES

TRIETHYNYLALUMINIUM C_6H_3Al

Dioxan,
or Trimethylamine
Chini, P. *et al., Chim. Ind. (Milan)*, 1962, **44**, 1220
The residue from sublimation of the complex with dioxan is explosive and the complex should not be dried by heating. The trimethylamine complex may also explode on sublimation.
See other METAL ACETYLIDES

TRIETHYNYLARSINE C_6H_3As

Voskuil, W. *et al., Rec. Trav. Chim.*, 1964, 83, 1301
Explodes on strong friction.
See also METAL ACETYLIDES

1-CHLORO-2,4-DINITROBENZENE $C_6H_3ClN_2O_4$

See 1-FLUORO-2,4-DINITROBENZENE, $C_6H_3FN_2O_4$

2,4-DINITROBENZENESULPHENYL CHLORIDE $C_6H_3ClN_2O_4S$

Kharasch, N. *et al., Org. Synth.*, 1964, **44**, 48
During vacuum removal of solvent, the residual chloride must not be overheated, as it may explode.
See other NITRO-ACYL HALIDES

2,6-DINITRO-4-PERCHLORYLPHENOL $C_6H_3ClN_2O_8$

Gardner, D. M. *et al., J. Org. Chem.*, 1963, **28**, 2652
This analogue of picric acid is dangerously explosive and very shock-sensitive.
See other PERCHLORYL COMPOUNDS

2,4,5-TRICHLOROPHENOL $C_6H_3Cl_3O$

Preparative hazard
See 1,2,4,5-TETRACHLOROBENZENE, $C_6H_2Cl_4$: Sodium hydroxide

1-FLUORO-2,4-DINITROBENZENE $C_6H_3FN_2O_4$

1. Halpern, B. D., *Chem. Eng. News*, 1951, **29**, 2666
2. Muir, G. D., private comm., 1969

The residue left from conversion of the chloro to the fluoro compound exploded during distillation at 1.3 mbar [1]. Reheating the residue from distillation caused violent decomposition [2].

Ether peroxides

Shafer, P. R., private comm., 1967

When air was admitted after vacuum evaporation of a (peroxidic) ether solution of the dinitro compound (5 g), a violent explosion occurred. Exploding ether peroxide may have initiated the dinitro compound.

See other HALOARYL COMPOUNDS

1,2,4-TRIFLUOROBENZENE **$C_6H_3F_3$**

Fl.P., –5°C (o)

POTASSIUM 4,6-DINITROBENZOFUROXAN HYDROXIDE COMPLEX **$C_6H_3KN_4O_7$**

Boulton, A. J. *et al., J. Chem. Soc.*, 1965, 5414

This explosive complex, similar to, but not formally an *aci*-nitro salt, has been evaluated as an explosive.

See other POLYNITROARYL COMPOUNDS

TRINITROSOPHLOROGLUCINOL **$C_6H_3N_3O_6$**

Heavy metal salts may be hazardous.

See TRILEAD TRINITROSOPHLOROGLUCINOLATE, $C_{12}N_6O_{12}Pb_3$

See other NITROSO COMPOUNDS

PICRIC ACID **$C_6H_3N_3O_7$**

Kirk-Othmer, 1965, Vol. 8, 617

Urbanski, 1964, Vol. 1, 518

Picric acid, in common with several other poly-nitrophenols, is an explosive material in its own right and is usually stored as a water-wet paste. It forms salts with many metals, some of which (lead, mercury,

copper or zinc) are rather sensitive to heat, friction or impact. The salts with ammonia and amines and molecular complexes with hydrocarbons, etc. are, in general, not so sensitive. Contact of picric acid with concrete floors may form the friction-sensitive calcium salt.
See other POLYNITROARYL COMPOUNDS

TRIETHYNYLPHOSPHINE C_6H_3P

Voskuil, W. *et al., Rec. Trav. Chim.*, 1964, **83**, 1301
Explodes on strong friction and, on standing, it decomposes and may explode spontaneously.
See other ACETYLENIC COMPOUNDS

TRIETHINYLANTIMONY C_6H_3Sb

Voskuil, W., *et al., Rec. Trav. Chim.*, 1964, **83**, 1301
Explodes on strong friction.
See other METAL ACETYLIDES

SILVER BENZO-1,2,3-TRIAZOLE-1-OLATE $C_6H_4AgN_3O$

Deorha, D. S. *et al., J. Indian Chem. Soc.*, 1964, **41**, 793
It is unsuitable as a gravimetric precipitate, as it explodes on heating rather than giving weighable silver.
See other HEAVY METAL DERIVATIVES

o-NITROBENZENEDIAZONIUM TETRACHLOROBORATE $C_6H_4BCl_4N_3O_2$

See DIAZONIUM TETRAHALOBORATES

m-BROMOPHENYLLITHIUM C_6H_4BrLi

See ORGANOLITHIUM REAGENTS

p-BROMOPHENYLLITHIUM C_6H_4BrLi

Anon., *Angew. Chem. (Nachr.)*, 1962, **10**, 65
It explodes if traces of oxygen are present in the inert atmosphere needed for the preparation.
See ORGANOLITHIUM REAGENTS
See other HALO-ARYLMETALS

p-BROMOBENZENEDIAZONIUM SALTS $C_6H_4BrN_2^+X^-$

Hydrogen sulphide
See DIAZONIUM SULPHIDES

m–, o– or *p*-CHLOROFLUOROBENZENE C_6H_4ClF

Fl.P.s, all 18°C

p-CHLOROBENZENEDIAZONIUM TRIIODIDE $C_6H_4ClI_3N_2$

See DIAZONIUM TRIIODIDES

m-CHLOROPHENYLLITHIUM C_6H_4ClLi

See ORGANOLITHIUM REAGENTS

p-CHLOROPHENYLLITHIUM C_6H_4ClLi

Anon., *Angew. Chem., (Nachr.)*, 1962, **10**, 65
It explodes if traces of oxygen are present in the inert atmosphere needed for the preparation.
See ORGANOLITHIUM REAGENTS
See other HALO-ARYLMETALS

p-CHLORONITROBENZENE $C_6H_4ClNO_2$

Sodium methoxide
ABCM Quart. Safety Summ., 1944, **15**, 15
Addition of the chloro compound to a solution of sodium methoxide in methanol caused an unusually exothermic reaction to occur. The lid of the 450 litre vessel was blown off, and a fire and explosion followed. No cause for the unusual vigour of the reaction was found.
See other HALOARYL COMPOUNDS

NITROPERCHLORYLBENZENE $C_6H_4ClNO_5$

McCoy, G., *Chem. Eng. News*, 1960, **38**(4), 62
The nitration product of perchlorylbenzene is explosive, comparable in shock sensitivity with lead azide, with a very high propagation rate.
See other PERCHLORYL COMPOUNDS

m-CHLOROBENZENEDIAZONIUM SALTS $C_6H_4ClN_2^+X^-$

Potassium thiophenolate,
or Sodium polysulphide
See DIAZONIUM SULPHIDES

o-CHLOROBENZENEDIAZONIUM SALTS $C_6H_4ClN_2^+X^-$

Potassium 2-chlorothiophenolate
See DIAZONIUM SULPHIDES

1-CHLOROBENZOTRIAZOLE $C_6H_4ClN_3$

Hopps, H. B., *Chem. Eng. News,* 1971, **49**(30), 3
Spontaneous ignition during packaging.
See other N-HALOGEN COMPOUNDS

4-CHLORO-2,6-DINITROANILINE $C_6H_4ClN_3O_4$

Preparative hazard.
See NITRIC ACID, HNO_3 : 4-Chloro-2-nitroaniline

Nitrosylsulphuric acid
Anon., *Angew. Chem. (Nachr.),* 1970, 18, 62
During large-scale diazotisation of the amine, severe local overheating is thought to have caused the explosion observed. The effect could not be reproduced in the laboratory.
See other POLYNITROARYL COMPOUNDS

m-NITROBENZENEDIAZONIUM PERCHLORATE $C_6H_4ClN_3O_6$

Schumacher, 1960, 205
Explosive, very sensitive to heat and shock.
See other DIAZONIUM PERCHLORATES

o-DICHLOROBENZENE $C_6H_4Cl_2$

Aluminium
See ALUMINIUM, Al: Halocarbons

1,6-DICHLORO-2,4-HEXADIYNE $C_6H_4Cl_2$

1. Driedger, P. E. *et al., Chem. Eng. News,* 1972, **50**(12), 51
2. Whiting, M. C., *Chem. Eng. News,* 1972, **50**(23), 86
3. Ford, M. *et al., Chem. Eng. News,* 1972, **50**(30), 67

It is extremely shock-sensitive, a 4.0 kg cm shock causing detonation in 50% of test runs (cf. 3.5 kg cm for propargyl bromide; 2.0 kg cm for glyceryl nitrate). The intermediate bis-chlorosulphite involved in the preparation needs low temperatures to prevent vigorous decomposition. The corresponding diiodo derivative was expected to be similarly hazardous [1], and this has been confirmed [2]. Improvements in preparative techniques (use of dichloromethane solvent at –30°C) to avoid violent reaction have also been described [3].

See other HALOACETYLENE DERIVATIVES

DICHLOROPHENOLS $C_6H_4Cl_2O$

Laboratory Chemical Disposal Co. Ltd., confidential information

During vacuum fractionation of the mixed dichlorophenols produced by hydrolysis of trichlorobenzene, admission of air caused the column contents to be forced down into the boiler at 210°C, when a violent explosion ensued.

See also 1,2,4,5-TETRACHLOROBENZENE, $C_6H_2Cl_4$: Sodium hydroxide

2,4-HEXADIYNYLENE BISCHLOROSULPHITE $C_6H_4Cl_2O_4S_2$

Driedger, P. E. *et al., Chem. Eng. News,* 1972, **50**(12), 51

This is probably an intermediate in the preparation of the 1,6-dichloro compound from 2,4-hexadiyne-1,6-diol and thionyl chloride in dimethylformamide. The reaction mixture must be kept at a low temperature to avoid vigorous decomposition and charring. This may be due to interaction of the intermediate with the solvent.

See also HALOACETYLENE DERIVATIVES
DIMETHYLFORMAMIDE, C_3H_7NO
2,4-HEXADIYNYLENE BISCHLOROFORMATE, $C_8H_4Cl_2O_4$

p-FLUOROPHENYLLITHIUM C_6H_4FLi

See ORGANOLITHIUM REAGENTS

m-DIFLUOROBENZENE $C_6H_4F_2$

Fl.P., below 0°C

p-DIFLUOROBENZENE $C_6H_4F_2$

Fl.P., –5°C (o)

OCTAFLUOROADIPAMIDE $C_6H_4F_8N_2O_2$

Lithium tetrahydroaluminate
See LITHIUM TETRAHYDROALUMINATE, AlH_4Li: Fluoroamides

1,6-DIIODO-2,4-HEXADIYNE $C_6H_4I_2$

See 1,6 DICHLORO-2,4-HEXADIYNE, $C_6H_4Cl_2$

POTASSIUM *p*-NITROPHENOLATE $C_6H_4KNO_3$

Personal experience
A sample of this potassium salt exploded after long storage in a non-evacuated desiccator. This may have been caused by the presence of the potassium salt of the tautomeric *aci-p*-quinone imine.
See *aci*-NITRO SALTS

m- or *p*-DILITHIOBENZENE $C_6H_4Li_2$

See ORGANOLITHIUM REAGENTS

SODIUM *o*-NITROTHIOPHENOLATE $C_6H_4NNaO_2S$

Davies, H. J., *Chem. & Ind.*, 1966, 257
The material exploded when the temperature of evaporation of the slurry was increased by adding a higher-boiling solvent. This may have been caused by the presence of some sodium salt of the tautomeric *aci-o*-thioquinone imine. Such salts are shock- or thermally unstable.
See *aci*-NITRO SALTS

DISODIUM *p*-NITROPHENYL ORTHOPHOSPHATE $C_6H_4NO_6P$

Preparative hazard.
See NITRIC ACID, HNO_3: Disodium phenyl orthophosphate

BENZO-1,2,3-THIADIAZOLE-1,1-DIOXIDE $C_6H_4N_2O_2S$

Wittig, G. *et al.*, *Chem. Ber.*, 1962, **95**, 2718; *Org. Synth.*, 1967, **47**, 6
The solid explodes at 60°C, on impact or friction, or sometimes spontaneously.

BENZENEDIAZONIUM-4-SULPHONATE $C_6H_4N_2O_3S$

1. Wichelhaus, H., *Ber.*, 1901, **34**, 11
2. *Chemiearbeit*, 1959, **11**(9), 59

This internal salt of diazotised sulphanilic acid (Pauly's reagent) exploded violently on touching when thoroughly dry [1]. Use of a metal spatula to remove a portion of a refrigerated sample of the solid diazonium compound caused a violent explosion. All solid diazo compounds must be stored in small quantities under refrigeration in loosely plugged containers. Handle gently with non-metallic spatulae using personal protection [2].

See other DIAZONIUM SALTS

*m-, o- or p-*DINITROBENZENE $C_6H_4N_2O_4$

Nitric acid

See NITRIC ACID, HNO_3: Nitroaromatics

See other POLYNITROARYL COMPOUNDS

*m-*DINITROBENZENE $C_6H_4N_2O_4$

Tetranitromethane

See TETRANITROMETHANE, CN_4O_8: Aromatic nitrocompounds

*o-*NITROBENZENEDIAZONIUM SALTS $C_6H_4N_3O_2^+X^-$

Disodium disulphide,
or Disodium polysulphide

See DIAZONIUM SULPHIDES

*p-*NITROBENZENEDIAZONIUM SALTS $C_6H_4N_3O_2^+X^-$

Hydrogen sulphide,
or Disodium sulphide

See DIAZONIUM SULPHIDES

4-NITROBENZENEDIAZONIUM NITRATE $C_6H_4N_4O_5$

Bamberger, E., *Ber.*, 1895, **28**, 239
The isolated dry salt explodes on heating, but not on friction.
See other DIAZONIUM SALTS

N,2,4,6-TETRANITROANILINE $C_6H_4N_5O_8$

Olsen, R. E. *et al., ACS 54*, 1966, 50
Impure tetryl must be recrystallised with care as it may deflagrate at only 50°C.
See other *N*-NITRO COMPOUNDS
POLYNITROARYL COMPOUNDS

4-NITROBENZENEDIAZONIUM AZIDE $C_6H_4N_6O_2$

Hantzsch, A., *Ber.*, 1903, **36**, 2058
The dry azide exploded violently with a brilliant flash.
See other DIAZONIUM SALTS

2-ETHYNYLFURAN C_6H_4O

Dunlop, 1953, 77
It explodes on heating or in contact with conc. nitric acid.
See other ACETYLENIC COMPOUNDS

PHENYLSILVER C_6H_5Ag

1. Huben-Weyl, 1970 **13**(1), 774
2. Krause, E. *et al., Ber.*, 1923, **56**, 2064

The dry solid explodes on warming to room temperature, or on light friction or stirring [1]. It is more stable wet with ether [2].
See other ARYLMETALS

SILVER PHENYLSELENONATE $C_6H_5AgO_3Se$

See PHENYLSELENONIC ACID, $C_6H_6O_3Sc$
See other HEAVY METAL DERIVATIVES

4-HYDROXY-3,5-DINITROBENZENEARSONIC ACID $C_6H_5AsN_2O_8$

Phillips, M. A., *Chem. & Ind.*, 1947, 61
A heated, unstirred, water-wet sludge of this analogue of picric acid exploded violently, apparently owing to local overheating in the containing flask.
See other POLYNITROARYL COMPOUNDS

BENZENEDIAZONIUM TETRAFLUOROBORATE $C_6H_5BF_4N_2$

Caesium fluoride,
Difluoroamine
See DIFLUOROAMINE, F_2HN: Benzenediazonium tetrafluoroborate, etc.

BENZENEDIAZONIUM CHLORIDE $C_6H_5ClN_2$

Hantzsch, A. *et al., Ber.*, 1901, **34**, 3338
The dry salt is more or less explosive, depending on method of preparation.
See also BIS-BENZENEDIAZONIUM ZINC TETRACHLORIDE, $C_{12}H_{10}Cl_4N_4Zn$

Potassium *O*-methyldithiocarbonate
See THIOPHENOL, C_6H_6S
DIAZONIUM SULPHIDES
See other DIAZONIUM SALTS

BENZENSULPHINYL CHLORIDE C_6H_5ClOS

Anon., *Chem. Eng. News.* 1957, **35**, 57
A glass bottle, undisturbed for several months, exploded, probably from gas pressure caused by photolytic or hydrolytic decomposition.

BENZENESULPHONYL CHLORIDE $C_6H_5ClO_2S$

Dimethyl sulphoxide
See DIMETHYL SULPHOXIDE, C_2H_6OS: Acyl halides

PERCHLORYLBENZENE $C_6H_5ClO_3$

Aluminium trichloride
Bruce, W. F., *Chem. Eng. News*, 1960, **38**(4), 63
A mixture is quiescent for some time, then suddenly explodes. Many perchloryl aromatics appear to be inherently shock-sensitive.
See other PERCHLORYL COMPOUNDS

N,N-DICHLOROANILINE $C_6H_5Cl_2N$

Sidgwick, 1950, 708
It is explosive.
See other *N*-HALOGEN COMPOUNDS

[(CHROMYLDIOXY)IODO] BENZENE $C_6H_5CrIO_4$

Sidgwick, 1950, 1250
It explodes at 66°C.
See other IODINE COMPOUNDS

FLUOROBENZENE C_6H_5F

Fl.P., –15°C

BENZENEDIAZONIUM IODIDE $C_6H_5IN_2$

See DIAZONIUM TRIIODIDES

IODOSYLBENZENE C_6H_5IO

1. Saltzman, A. *et al., Org. Synth.*, 1963, **43**, 60
2. Banks, D. F., *Chem. Rev.*, 1966, **66**, 255

It explodes at 210°C [1] and homologues generally explode on melting [2].

See other IODINE COMPOUNDS

IODYLBENZENE $C_6H_5IO_2$

1. Sharefkin, J. G. *et al., Org. Synth.*, 1963, **43**, 65
2. Banks, D. F., *Chem. Rev.*, 1966, **66**, 259

It explodes at 230°C [1]. Extreme care should be used in heating, compressing or grinding iodyl compounds, as heat or impact may cause detonation of homologues [2].

See other IODINE COMPOUNDS

BENZENEDIAZONIUM TRIIODIDE $C_6H_5I_3N_2$

See DIAZONIUM TRIIODIDES

POTASSIUM 3,5-DINITRO-2(1-TETRAZENYL)PHENOLATE $C_6H_5KN_6O_5$

Kenney, J. F., US Pat. 2 728 760, 1952
It is explosive.
See other HIGH-NITROGEN COMPOUNDS

POTASSIUM BENZENESULPHONYLPEROXOSULPHATE $C_6H_5KO_7S_2$

Davies, 1961, 65
It explodes on friction or warming.
See other DIACYL PEROXIDES

PHENYLLITHIUM C_6H_5Li

Titanium tetraethoxide
Gilman, H. *et al., J. Org. Chem.*, 1945, **10**, 507
The product of interaction at 0°C (of unknown composition) ignited in air and reacted violently with water.
See other ARYLMETALS

***p*-BENZOQUINONE MONOIMINE** C_6H_5NO

Willstätter, R. *et al., Ber.*, 1906, **37**, 4607
The solid decomposes with near-explosive violence.

NITROBENZENE $C_6H_5NO_2$

MCA SD-21, 1967

Aluminium trichloride,
Phenol
See ALUMINIUM TRICHLORIDE, $AlCl_3$: Nitrobenzene, etc.

Aniline,
Gycerol,
Sulphuric acid
See QUINOLINE, C_9H_7N

Oxidants
See TETRANITROMETHANE, CN_4O_8 : Aromatic nitrocompounds
SODIUM CHLORATE, $ClNaO_3$: Nitrobenzene
NITRIC ACID, HNO_3 : Nitroaromatics
: Nitrobenzene
PEROXODISULPHURIC ACID, $H_2O_8S_2$: Organic liquids
DINITROGEN TETRAOXIDE, N_2O_4 : Nitrobenzene

Phosphorus pentachloride
See PHOSPHORUS PENTACHLORIDE, Cl_5P: Nitrobenzene

Potassium
See POTASSIUM, K: Nitrogen-containing explosives

Potassium hydroxide,
Benzanthrone
ABCM Quart. Safety Summ., 1953, **24**, 42
Accidental substitution of nitrobenzene for aniline as diluent during large-scale fusion of benzanthrone with potassium hydroxide caused a violent explosion.

Sulphuric acid
MCA Case History No. 678
Nitrobenzene was washed with dilute (5%) sulphuric acid to remove amines, and became contaminated with some acid emulsion which had formed. After distillation, the hot, acid, tarry residue attacked the iron vessel, evolving hydrogen, and eventually exploded. It was later found that addition of the nitrobenzene to the diluted acid did not give emulsions, while the reversed addition did. A final wash with sodium carbonate solution was added to the process.
See other C-NITRO COMPOUNDS

o-NITROSOPHENOL **$C_6H_5NO_2$**

Alone,
or Acids

Baeyer, A. *et al., Ber.*, 1902, **35**, 3037
It explodes on heating or in contact with concentrated acids.
See other NITROSO COMPOUNDS

p-NITROSOPHENOL $C_6H_5NO_2$

Milne, W. D., *Ind. Eng. Chem.*, 1919, **11**, 489
Stored barrels heated spontaneously and caused a fire. Contamination of the bulk material by acid or alkali may cause ignition.
See other NITROSO COMPOUNDS

m-NITROBENZENESULPHONIC ACID $C_6H_5NO_5S$

Sulphuric acid
MCA Case Histories Nos. 1482, 944
A 270 litre batch of a solution in sulphuric acid exploded violently after storage at ~150°C for several hours. An exotherm develops at 145°C, and the acid is known to decompose at ~200° C. The earlier Case History describes a similar incident when water, leaking from a cooling coil into the fuming sulphuric acid medium, caused an exotherm to over 150°C, and subsequent violent decomposition.

BENZENEDIAZONIUM SALTS $C_6H_5N_2^+X^-$

Alone
See THE CHLORIDE, $C_6H_5ClN_2$
THE NITRATE, $C_6H_5N_3O_3$

Ammonium sulphide,
or Hydrogen sulphide,
or Disodium sulphide
See DIAZONIUM SULPHIDES

BENZOTRIAZOLE $C_6H_5N_3$

Anon., *Chem. Eng. News*, 1956, **34**, 2450
A 1 tonne quantity, during distillation at 160°C/2.5 mbar, exothermally decomposed, then detonated. No cause was found, and similar batches had previously distilled satisfactorily. The multiple N–N bonding may tend to cause instability in the molecule, particularly if heavy metals are present. They were absent in this case.
See other HIGH-NITROGEN COMPOUNDS
N-METAL DERIVATIVES

PHENYL AZIDE $C_6H_5N_3$

Lindsay, R. O. *et al., Orgn. Synth.*, 1955, Coll. Vol. 3, 710
Though distillable at considerably reduced pressure, phenyl azide explodes when heated at atmospheric pressure, and occasionally at lower pressures.

Lewis acids
Boyer, J. H. *et al., Chem. Rev.*, 1954, **54**, 29
The ready decomposition of most aryl azides with sulphuric acid and Lewis acids may be vigorous or violent, depending on structure and conditions. In absence of a diluent (carbon disulphide), phenyl azide and aluminium trichloride exploded violently.
See other ORGANIC AZIDES

BENZENESULPHONYL AZIDE $C_6H_5N_3O_2S$

Dermer, O. C. *et al., J. Amer. Chem. Soc.*, 1955, **77**, 71
Whereas the pure azide decomposes rapidly but smoothly at 105°C, the crude material explodes violently on heating.
See other ACYL AZIDES

BENZENEDIAZONIUM NITRATE $C_6H_5N_3O_3$

Urbanski, 1967, Vol. 3, 201
Although not a practical explosive, the isolated salt is highly sensitive to friction and impact, and explodes at 90°C.
See other DIAZONIUM SALTS

PHENYLPHOSPHONIC DIAZIDE $C_6H_5N_6OP$

Baldwin, R. A., *J. Org. Chem.*, 1965, **30**, 3866
Although a small sample had been distilled at 72–74°C/0.13 mbar without incident, it decomposed vigorously on exposure to flame, and the liquid was impact-sensitive, exploding vigorously. The material can be prepared and handled safely in pyridine solution.
See other ACYL AZIDES

PHENYLTHIOPHOSPHONIC DIAZIDE $C_6H_5N_6PS$

Baldwin, R. A., *J. Org. Chem.*, 1965, **30**, 3866
During an attempt to distil crude material at 80°C/0.13 mbar, a violent explosion occurred. The material can be prepared and handled safely in pyridine solution.
See other ACYL AZIDES

BENZENE C_6H_6

Fl.P., –11°C; E.L., 1.4–8.0%
MCA SD-2, 1960

Diborane
See DIBORANE, B_2H_6: Air, etc.

Interhalogens

See BROMINE TRIFLUORIDE, BrF_3 : Halogens, etc.
BROMINE PENTAFLUORIDE, BrF_5 : Hydrogen-containing materials
IODINE PENTAFLUORIDE, F_5I : Benzene
IODINE HEPTAFLUORIDE, F_7I : Organic solvents

See Uranium hexafluoride, below

Oxidants

See SILVER PERCHLORATE, $AgClO_4$: Aromatic compounds
DIOXYGENYL TETRAFLUOROBORATE, BF_4O_2 : Organic materials
NITRYL PERCHLORATE, $ClNO_6$: Organic solvents
DIOXYGEN DIFLUORIDE, F_2O_2
PERMANGANIC ACID, $HMnO_4$: Organic materials
NITRIC ACID, HNO_3 : Hydrocarbons
PEROXOMONOSULPHURIC ACID, H_2O_5S: Aromatics
PEROXODISULPHURIC ACID, $H_2O_8S_2$: Organic liquids
SODIUM PEROXIDE, Na_2O_2 : Organic liquids, etc.
OXYGEN (Liquid), O_2 : Hydrocarbons
OZONE, O_3 : Aromatic compounds
: Benzene, etc.

Uranium hexafluoride

See URANIUM HEXAFLUORIDE, F_6U: Aromatic hydrocarbons

BENZVALENE C_6H_6

Katz, T. J. *et al., J. Amer. Chem. Soc.*, 1971, **93**, 3782
When isolated, this highly strained valence tautomer of benzene exploded violently when scratched. It may be handled safely in solution in ether.

1,3-HEXADIEN-5-YNE C_6H_6

Brandsma, 1971, 132

Distillation of this rather unstable material at normal pressure involves risk of an explosion.
See other ACETYLENIC COMPOUNDS

1,5-HEXADIEN-3-YNE (DIVINYLACETYLENE) C_6H_6

Fl.P., below –20°C; E.L., 1.5– %

Air

Handy, C. T. *et al., J. Org. Chem.*, 1962, **27**, 41

The dienyne reacts readily with atmospheric oxygen, forming an explosively unstable polymeric peroxide. Equipment used with it should be rinsed with a dilute solution of a polymerisation inhibitor to prevent formation of unstable residual films. Adequate shielding of operations is essential.

See other ACETYLENIC COMPOUNDS
POLYPEROXIDES

p-BROMOANILINE C_6H_6BrN

See *p*-BROMOBENZENEDIAZONIUM SALTS, $C_6H_4BrN_2^+X^-$

IODYLBENZENE PERCHLORATE $C_6H_6ClIO_6$

Masson, I., *J. Chem. Soc.,* 1935, 1674

A small sample exploded very violently while still damp.

See other IODINE COMPOUNDS

m-CHLOROANILINE C_6H_6ClN

See *m*-CHLOROBENZENEDIAZONIUM SALTS, $C_6H_4ClN_2^+X^-$

o-CHLOROANILINE C_6H_6ClN

See 2-CHLOROBENZENEDIAZONIUM SALTS $C_6H_4ClN_2^+X^-$

p-AMINOBENZENEDIAZONIUM PERCHLORATE $C_6H_6ClN_3O_4$

Hofmann, K. A. *et al.*, *Ber.*, 1910, **43**, 2624
Extremely explosive.
See DIAZONIUM PERCHLORATES

1,2,3,4,5,6-HEXACHLOROCYCLOHEXANE $C_6H_6Cl_6$

N,N-Dimethylformamide
'DMF Brochure', Billingham, ICI, 1965
There is a potentially dangerous reaction of hexachlorocyclohexane with DMF in presence of iron. The same occurs with carbon tetrachloride, but not with dichloromethane or 1,2-dichloroethane under the same conditions.

N-ETHYLHEPTAFLUOROBUTYRAMIDE $C_6H_6F_7NO$

See LITHIUM TETRAHYDROALUMINATE, AlH_4Li: Fluoroamides

p-BENZOQUINONE DIIMINE $C_6H_6N_2$

Acids
Willstätter, R. *et al.*, *Ber.*, 1906, **37**, 4607
The unstable solid decomposes explosively in contact with concentrated hydrochloric or sulphuric acids.

1,4-DICYANO-2-BUTENE $C_6H_6N_2$

MCA Case History No. 1747
Overheating in a vacuum evaporator initiated accelerating polymerisation-decomposition of dicyanobutene, and the rapid gas evolution eventually caused pressure-failure of the process equipment.
See other CYANO COMPOUNDS

BIS(ACRYLONITRILE)NICKEL(0) $C_6H_6N_2Ni$

Schrauzer, G. N., *J. Amer. Chem. Soc.*, 1959, **81**, 5310
Pyrophoric in air.
See other CYANO COMPOUNDS

m-NITROANILINE $C_6H_6N_2O_2$

Ethylene oxide
See ETHYLENE OXIDE, C_2H_4O: *m*-Nitroaniline

o-NITROANILINE $C_6H_6N_2O_2$

See *o*-NITROBENZENEDIAZONIUM SALTS, $C_6H_4N_3O_2^+X^-$

Sulphuric acid
See SULPHURIC ACID, H_2O_4S: Nitroaryl bases

p-NITROANILINE $C_6H_6N_2O_2$

See *p*-NITROBENZENEDIAZONIUM SALTS, $C_6H_4N_3O_2^+X^-$

Sulphuric acid
See SULPHURIC ACID, H_2O_4S: Nitroaryl bases

4-NITROANILINE-2-SULPHONIC ACID $C_6H_6N_2O_5S$

See SULPHURIC ACID, H_2O_4S: Nitroaryl bases

AMMONIUM PICRATE $C_6H_6N_4O_7$

Unpublished observations
A small sample, isolated incidentally during decomposition of a picrate with ammonia, exploded during analytical combustion. Presence of traces of metallic picrates (arising from metal contact) increases temperature sensitivity. The salt may also explode on impact.
See PICRIC ACID, $C_6H_3N_3O_7$

4,5-HEXADIEN-2-YN-1-OL C_6H_6O

Brandsma, 1971, 8, 54
Dilution with white mineral oil before distillation is recommended to prevent explosion of the concentrated distillation residue.
See other ACETYLENIC COMPOUNDS
PEROXIDISABLE COMPOUNDS

PHENOL C_6H_6O

MCA SD-4, 1964

Aluminium trichloride,
Nitrobenzene
See ALUMINIUM TRICHLORIDE, $AlCl_3$: Nitrobenzene, etc.

Peroxodisulphuric acid
See PEROXODISULPHURIC ACID, $H_2O_8S_2$: Organic liquids

Peroxomonosulphuric acid
See PEROXOMONOSULPHURIC ACID, H_2O_5S: Aromatics

1,2-DIHYDROXYBENZENE (PYROCATECHOL) $C_6H_6O_2$

Nitric acid
See NITRIC ACID, HNO_3: Pyrocatechol

PHENYLSELENONIC ACID $C_6H_6O_3Se$

Stoecker, M. *et al., Ber.*, 1906, **39**, 2197
It explodes feebly at 180°C but the silver salt is more explosive.

BENZENE TRIOZONIDE $C_6H_6O_9$

Harries, C. *et al., Ber.*, 1904, **37**, 3431
Extremely explosive at the slightest touch.
See OZONE, O_3: Aromatic compounds
See other OZONIDES

BENZENETHIOL (THIOPHENOL) C_6H_6S

Fl.P., below 21°C
Preparative hazard.
1. Leuckart, R., *J. Prakt. Chem.*, 1890, **41**, 179
2. Graesser, R., private comm., 1968

During preparation of thiophenol by addition of a cold solution of potassium *O*-methyldithiocarbonate to a cold solution of benzenediazonium chloride, a violent explosion, accompanied by an orange flash, occurred. This was attributed to the formation and decomposition of bis-benzenediazodisulphide. A preparation in which the diazonium salt was added to the 'xanthate' solution proceeded smoothly.
See also DIAZONIUM SULPHIDES

2,4-DINITROPHENYLHYDRAZINIUM PERCHLORATE $C_6H_7ClN_4O_8$

Anon., *Angew. Chem. (Nachr.)*, 1967, **15**, 78
Although solutions of the perchlorate are stable as prepared, explosive decomposition may occur during concentration by evaporation.
See also HYDRAZINIUM SALTS
See other OXOSALTS OF NITROGENOUS BASES

ANILINE C_6H_7N

MCA SD-17, 1963
See BENZENEDIAZONIUM SALTS, $C_6H_5N_2^+X^-$

Benzenediazonium-2-carboxylate
See BENZENEDIAZONIUM-2-CARBOXYLATE, $C_7H_4N_2O_2$

Boron trichloride
See BORON TRICHLORIDE, BCl_3: Aniline

Nitromethane
See NITROMETHANE, CH_3NO_2: Acids, etc.

Oxidants
See *N*-HALOIMIDES: Alcohols, Amines
PEROXYFORMIC ACID, CH_2O_3: Organic materials
DIISOPROPYL PEROXYDICARBONATE, $C_8H_{14}O_6$
PERCHLORYL FLUORIDE, $ClFO_3$: Nitrogenous bases
PERCHLORIC ACID, $ClHO_4$: Aniline, etc.
FLUORINE NITRATE, FNO_3: Organic materials
NITROSYL PERCHLORATE, $ClNO_5$: Organic materials
FLUORINE, F_2: Nitrogenous bases
NITRIC ACID, HNO_3: Aromatic amines
PEROXOMONOSULPHURIC ACID, H_2O_5S: Aromatics
PEROXODISULPHURIC ACID, $H_2O_8S_2$: Organic liquids
SODIUM PEROXIDE, Na_2O_2: Organic liquids, etc.
OZONE, O_3: Aromatic compounds

Trichloronitromethane
See TRICHLORONITROMETHANE, CCl_3NO_2: Aniline

2-METHYLPYRIDINE C_6H_7N

Hydrogen peroxide,
Iron(II) sulphate,
Sulphuric acid
See HYDROGEN PEROXIDE, H_2O_2: Iron(II) sulphate, etc.

N-PHENYLHYDROXYLAMINE C_6H_7NO

Preparative hazard.
See ZINC, Zn: Nitrobenzene

1,3-CYCLOHEXADIENE C_6H_8

Fl.P., below 23°C
See also PROPENE, C_3H_6

Air
1. Bodendorf, K., *Arch. Pharm.*, 1933, **271**, 11
2. Hock, H. *et al., Chem. Ber.*, 1951, **84**, 349

Cyclohexadiene autoxidises slowly in air, but the residual (largely polymeric) peroxide explodes very violently on ignition [1]. The monomeric peroxide has also been isolated [2].
See other DIENES
POLYPEROXIDES

1,4-CYCLOHEXADIENE C_6H_8

Fl.P., −11°C
See also PROPENE, C_3H_6
See other DIENES

PHENYLHYDRAZINE $C_6H_8N_2$

Lead(IV) oxide
See LEAD(IV) OXIDE, O_2Pb: Nitrogen compounds

Perchloryl fluoride
See PERCHLORYL FLUORIDE, $ClFO_3$: Nitrogenous bases

2-DIAZOCYCLOHEXANONE $C_6H_8N_2O$

Regitz, M. *et al., Org. Synth.*, 1971, **51**, 86
It may explode on being heated, and a relatively large residue should be left during vacuum distillation to prevent overheating.
See other DIAZO COMPOUNDS

2-BUTEN-1-YL DIAZOACETATE $C_6H_8N_2O_2$

Blankley, C. J. *et al., Org. Synth.*, 1969, **49**, 25
Precautions are necessary during vacuum distillation of this potentially explosive material.
See other DIAZO COMPOUNDS

ANILINIUM NITRATE $C_6H_8N_2O_3$

Nitric acid
See NITRIC ACID, HNO_3: Anilinium nitrate

2,5-DIMETHYLFURAN C_6H_8O

Fl.P., 16°C

3-METHYL-2-PENTEN-4-YN-1-OL C_6H_8O

Lorentz, F., *Loss Prevention,* 1967, **1**, 1–5
Temperature control during pressure hydrogenation of *cis*- or *trans*-isomers is essential, since at 155°C violent decomposition to carbon, hydrogen and carbon monoxide with development of over 1000 bar pressure will occur. The material should not be heated above 100°C, particularly if acid or base is present, to avoid exothermic polymerisation.

Sodium hydroxide
1. *MCA Case History No. 363*
2. Silver, L., *Chem. Eng. Progr.*, 1967, **63**(8), 43
Presence of traces of sodium hydroxide probably caused formation of the acetylenic sodium salt, which exploded during high-vacuum distillation in a metal still [1]. A laboratory investigation which duplicated the explosion, without revealing the precise cause, was also reported [2].
See other ACETYLENIC COMPOUNDS

1,2-CYCLOHEXANEDIONE $C_6H_8O_2$

Preparative hazard.
See NITRIC ACID, HNO_3: Cyclohexanol

CITRIC ACID $C_6H_8O_7$

Metal nitrates
See METAL NITRATES: Citric acid

TRIVINYLBISMUTH C_6H_9Bi

Coates, 1967, Vol. 1, 538
Ignites in air.
See other ALKYL METALS

IODINE TRIACETATE $C_6H_9IO_6$

1. Schutzenberger, P., *Compt. Rend.*, 1861, **52**, 135
2. Oldham, F. W. H. *et al.*, *J. Chem. Soc.*, 1941, 368
The acetate explodes at 140°C [1], and higher homologues decompose at about 120°C, the violence decreasing with ascent of the homologous series [2].
See other IODINE COMPOUNDS

tert-BUTYLNITROACETYLENE $C_6H_9NO_2$

Amines

Jäger, V. *et al.*, *Angew. Chem.* (*Intern. Ed.*), 1969, **8**, 273

Reaction with primary, secondary or tertiary amines proceeds explosively with ignition in absence of a solvent.

See other ACETYLENIC COMPOUNDS

BIS(2-CYANOETHYL)AMINE $C_6H_9N_3$

BCISC Quart. Safety Summ., 1967, **38**, 42

Some 18-month old bottles of iminodipropionitrile exploded, probably owing to slow hydrolysis and release of ammonia.

See other CYANO COMPOUNDS

N-NITROSOPHENYLHYDROXYLAMINE *O*-AMMONIUM SALT ('CUPFERRON') $C_6H_9N_3O_2$

Thorium salts

Pitwell, L. R., *J.R. Inst. Chem.*, 1956, **80**, 173

Solutions of this reagent are destabilised by the presence of thorium ions. If a working temperature of 10–15°C is much exceeded, the risk of decomposition, not slowed by cooling, and accelerating to explosion, exists. Titanium and zirconium salts also cause slight destabilisation, but decomposition temperatures are 35 and 40°C, respectively.

SODIUM ETHYL ACETOACETATE $C_6H_9NaO_3$

2-Iodo-3,5-dinitrobiphenyl

See 2-IODO-3,5-DINITROBIPHENYL, $C_{12}H_7IN_2O_4$: Sodium, etc.

CYCLOHEXENE C_6H_{10}

Fl.P., below –6°C; E.L., 1.2–   % (at 100°C)

2,3-DIMETHYL-1,3-BUTADIENE C_6H_{10}

Air
Bodendorf, K., *Arch. Pharm.*, 1933, **271**, 33
A few mg of the autoxidation residue, a largely polymeric peroxide, exploded violently on ignition.
See POLYPEROXIDES
See other DIENES

1,4-HEXADIENE C_6H_{10}

Fl.P., –21°C; E.L., 2.0–6.1%

1,5-HEXADIENE C_6H_{10}

Fl.P., –46°C

1-HEXYNE C_6H_{10}

Fl.P., below 10°C

2- or 3-HEXYNE C_6H_{10}

Fl.P.s, both below 23°C

2-METHYL-1,3-PENTADIENE C_6H_{10}

Fl.P., below –20°C

4-METHYL-1,3-PENTADIENE C_6H_{10}

Fl.P., –34°C

CADMIUM PROPIONATE $C_6H_{10}CdO_4$

Anon., *Chem. Age,* 1957, **77**, 794
The salt exploded during drying at 60–100°C in an electric oven, presumably because of overheating, and production and ignition of 3-pentanone vapour.

3,3-PENTAMETHYLENEDIAZIRINE $C_6H_{10}N_2$

Schmitz, E. *et al., Org. Synth*., 1965, **45**, 85; *Ber*., 1961, **94**, 2168
Small-scale preparation is recommended, in view of a previous explosion during superheating of the liquid during distillation at 109°C.
See DIAZIRINES

tert-BUTYL DIAZOACETATE $C_6H_{10}N_2O_2$

Regitz, M. *et al., Org. Synth*., 1968, **48**, 36
Vacuum distillation of the product is potentially hazardous.
See other DIAZO COMPOUNDS

DIETHYL AZODIFORMATE $C_6H_{10}N_2O_4$

U.S. Pat. 3 347 845, 1967
Kauer, J. C., *Org. Synth.*, 1963, Coll. Vol. 4, 412
Shock-sensitive, it burns explosively, like its lower homologue.
See DIMETHYL AZODIFORMATE, $C_4H_6N_2O_4$
See other AZO COMPOUNDS

N,N'-DIACETYL-*N,N'*-DINITRO-1,2-DIAMINOETHANE $C_6H_{10}N_4O_6$

Violent decomposition occurred at 142°C.
See DIFFERENTIAL THERMAL ANALYSIS
See other *N*-NITRO COMPOUNDS

BUTOXYACETYLENE $C_6H_{10}O$

Jacobs, T. L. *et al.*, *J. Amer. Chem. Soc.*, 1942, **64**, 223
Small samples rapidly heated in sealed tubes to around 100°C exploded.
See other ACETYLENIC COMPOUNDS

CYCLOHEXANONE $C_6H_{10}O$

Hydrogen peroxide,
Nitric acid
See HYDROGEN PEROXIDE, H_2O_2: Ketones, etc.

Nitric acid
See NITRIC ACID, HNO_3: 4-Methylcyclohexanone

DIALLYL ETHER $C_6H_{10}O$

Fl.P., –6°C

MCA Case History No. 412

The ether was left exposed to air and sunlight for two weeks before distillation, and became peroxidised. During distillation to small bulk, a violent explosion occurred.

See other ALLYL COMPOUNDS

ETHYL CROTONATE $C_6H_{10}O_2$

Fl.P., 2°C

ETHYL METHACRYLATE $C_6H_{10}O_2$

Fl.P., 21°C; E.L. 1.8– %

VINYL BUTYRATE $C_6H_{10}O_2$

Fl.P., 21°C (o); E.L., 1.4–8.8%

2-HYDROXY-2-METHYLGLUTARIC ACID $C_6H_{10}O_3$

Preparative hazard.

See 4-HYDROXY-4-METHYL-1,6-HEPTADIENE, $C_8H_{14}O$: Ozone

DIPROPIONYL PEROXIDE $C_6H_{10}O_4$

Swern, 1971, Vol. 2, 815

Pure material explodes on standing at room temperature.

See other DIACYL PEROXIDES

DIALLYL SULPHATE $C_6H_{10}O_4S$

von Braun, J. *et al., Ber.*, 1917, **50**, 293
Decomposed explosively during distillation (no details).
See other ALLYL COMPOUNDS

DIETHYL PEROXYDICARBONATE $C_6H_{10}O_6$

An oil of extreme instability when impure, and sensitive to heat or impact, exploding more powerfully than dibenzoyl peroxide.
See PEROXYCARBONATE ESTERS

DIALLYL SULPHIDE $C_6H_{10}S$

N-Bromosuccinimide
See *N*-HALOIMIDES: Alcohols, etc.
See other ALLYL COMPOUNDS

2-ETHOXY-1-IODO-3-BUTENE $C_6H_{11}IO$

Trent, J. *et al., Chem. Eng. News,* 1966, **44**(43), 7
During a 1g mol-scale preparation by a published method using butadiene, ethanol, iodine and mercury oxide, a violent explosion occurred while ethanol was being distilled off at 35°C under slight vacuum. The cause of the explosion could not be established, and several smaller-scale preparations had been uneventful.

DIALLYLAMINE $C_6H_{11}N$

Fl.P., 21°C

4-METHYLMORPHOLINE $C_6H_{11}NO$

Fl.P., 24°C

ETHYL 2-FORMYLPROPIONATE OXIME $C_6H_{11}NO_3$

Hydrogen chloride (?)

Loftus, F., private comm., 1972

The formyl ester and hydroxylamine were allowed to react in ethanol, which was then removed by vacuum evaporation at 40°C. The dry residue decomposed violently 5 min later. Analysis of the residue suggested that the oxime had undergone an exothermic Beckmann rearrangement, possibly catalysed by traces of hydrogen chloride in the reaction residue.

See also BUTYRALDOXIME, C_4H_9NO

DIALLYL PHOSPHITE $C_6H_{11}O_3P$

Houben-Weyl, 1964, Vol. 12(2), 22

The ester is liable to explode during distillation unless thoroughly dry allyl alcohol is used, and if more than two-thirds of the product is distilled over.

See other ALLYL COMPOUNDS

CYCLOHEXANE C_6H_{12}

Fl.P., −17°C; E.L., 1.3–8.35%

MCA SD-68, 1957

Dinitrogen tetraoxide

See DINITROGEN TETRAOXIDE, N_2O_4 : Hydrocarbons

ETHYLCYCLOBUTANE C_6H_{12}

Fl.P., below −15°C; E.L., 1.2–7.7%; A.I.T., 210°C

1-HEXENE C_6H_{12}

Fl.P., −26°C; E.L., 1.2–6.9%

2-HEXENE C_6H_{12}

Fl.P., −21°C

METHYLCYCLOHEXANE C_6H_{12}

Fl.P., −29°C

2-METHYL-1-PENTENE C_6H_{12}

Fl.P., −28°C

4-METHYL-1-PENTENE C_6H_{12}

Fl.P., −7°C

***cis*-4-METHYL-2-PENTENE** C_6H_{12}

Fl.P., −32°C

***trans*-4-METHYL-2-PENTENE** C_6H_{12}

Fl.P., −29°C

1,4-DIAZABICYCLO[2.2.2]OCTANE $C_6H_{12}N_2$

Cellulose nitrate
See CELLULOSE NITRATE: Amines

BUTYL VINYL ETHER $C_6H_{12}O$

Fl.P., –1°C

CYCLOHEXANOL $C_6H_{12}O$

Oxidants
See CHROMIUM TRIOXIDE, CrO_3: Alcohols
NITRIC ACID, HNO_3: Alcohols

2,2-DIMETHYL-3-BUTANONE $C_6H_{12}O$

Fl.P., 12°C

2-ETHYLBUTYRALDEHYDE $C_6H_{12}O$

Fl.P., 21°C (o)

3-HEXANONE $C_6H_{12}O$

Fl.P., 14°C

ISOBUTYL VINYL ETHER $C_6H_{12}O$

Fl.P., –9°C
See other PEROXIDISABLE COMPOUNDS

2-METHYLPENTANAL $C_6H_{12}O$

Fl.P., below 23°C

3-METHYLPENTANAL $C_6H_{12}O$

Fl.P., below 23°C

2-METHYL-3-PENTANONE $C_6H_{12}O$

Fl.P., 11°C

3-METHYL-2-PENTANONE $C_6H_{12}O$

Fl.P., 15°C

4-METHYL-2-PENTANONE $C_6H_{12}O$

Fl.P., 17°C; E.L., 1.2–8%

Air
1. Smith, G. H., *Chem. & Ind.*, 1972, 291
2. Bretherick, L., ibid., 363

4-Methyl-2-pentanone had not been considered prone to autoxidation, but an explosion during aerobic evaporation of the solvent [1] was attributed to formation and explosion of a peroxide [2].

Potassium *tert*-butoxide

See POTASSIUM *tert*-BUTOXIDE, C_4H_9KO: Acids, etc.

BUTYL ACETATE $C_6H_{12}O_2$

Fl.P., 23°C; E.L., 1.4–7.5%

Potassium *tert*-butoxide

See POTASSIUM *tert*-BUTOXIDE, C_4H_9KO: Acids, etc.

***sec*-BUTYL ACETATE** $C_6H_{12}O_2$

Fl.P., 18°C; E.L., 1.3–7.5%

2,6-DIMETHYL-1,4-DIOXAN $C_6H_{12}O_2$

Fl.P., 24°C

See other PEROXIDISABLE COMPOUNDS

ETHYL ISOBUTYRATE $C_6H_{12}O_2$

Fl.P., below 18°C

2-ETHYL-2-METHYL-1,3-DIOXOLANE $C_6H_{12}O_2$

Fl.P., 23°C (o)

4-HYDROXY-4-METHYL-2-PENTANONE $C_6H_{12}O_2$

Fl.P., 9°C

ISOBUTYL ACETATE $C_6H_{12}O_2$

Fl.P., 17°C; E.L., 2.4–10.5%

ISOPROPYL PROPIONATE $C_6H_{12}O_2$

Fl.P., below 20°C

METHYL ISOVALERATE $C_6H_{12}O_2$

Fl.P., below 22°C

METHYL PIVALATE $C_6H_{12}O_2$

Fl.P., 7°C

METHYL VALERATE $C_6H_{12}O_2$

Fl.P., below 25°C

2,5-DIMETHOXYTETRAHYDROFURAN $C_6H_{12}O_3$

Fl.P., below 10°C

tert-BUTYL PERACETATE $C_6H_{12}O_3$

1. Martin, J. H., *Ind. Eng. Chem.*, 1960, **52**(4), 49A
2. Castrantas, 1965, 16

The perester explodes with great violence when rapidly heated to a critical temperature. Previous standard explosivity tests had not shown this behaviour. The presence of benzene as solvent prevents the explosive decomposition, but if the solvent evaporates, the residue is dangerous [1]. The pure ester is also shock-sensitive and detonable, but commercial 75% solutions are not [2].

See other PEROXYESTERS

ISOBUTYL PERACETATE $C_6H_{12}O_3$

Fl.P., below 25°C

trans-2-HEXENE OZONIDE $C_6H_{12}O_3$

Loan, L. D. *et al., J. Amer. Chem. Soc.*, 1965, **87**, 741

Attempts to get carbon and hydrogen analyses by combustion of the ozonides of 2-butene, 2-pentene and 2-hexene caused violent explosions, though oxygen analyses were uneventful.

See other OZONIDES

PEROXYHEXANOIC ACID $C_6H_{12}O_3$

Swern, D., *Chem. Rev.*, 1949, **45**, 10

Fairly stable at ambient temperature, it explodes and ignites on rapid heating.

See other PEROXYACIDS

2,4,6-TRIMETHYLTRIOXANE (PARALDEHYDE) $C_6H_{12}O_3$

Fl.P., 17°C; E.L., 1.3– %

Nitric acid
See NITRIC ACID, HNO_3: 2,4,6-Trimethyltrioxane

TETRAMETHOXYETHYLENE $C_6H_{12}O_4$

Preparative hazard.
See 1,2,3,4-TETRACHLORO-7,7-DIMETHOXY-5-PHENYLBICYCLO [2.2.1]-2,5-HEPTADIENE, $C_{15}H_{12}Cl_4O_2$

3,3,6,6-TETRAMETHYL-1,2,4,5-TETRAOXANE $C_6H_{12}O_4$

Baeyer, A. *et al., Ber.*, 1900, **33**, 858
Action of hydrogen peroxide or permonosulphuric acid on acetone produces this dimeric acetone peroxide, which explodes violently on impact, friction or rapid heating.
See other CYCLIC PEROXIDES

BIS(DIMETHYLTHALLIUM) ACETYLIDE $C_6H_{12}Tl_2$

Nast, R. *et al., J. Organomet. Chem.*, 1966, **6**, 461
Extremely explosive, heat- and friction-sensitive.
See other ALKYLMETALS
METAL ACETYLIDES

CYCLOHEXYLAMINE $C_6H_{13}N$

Fl.P., 21°C

Nitric acid
See NITRIC ACID, HNO_3: Cyclohexylamine

1-METHYLPIPERIDINE $C_6H_{13}N$

Fl.P., below 23°C

2-METHYLPIPERIDINE $C_6H_{13}N$

Fl.P., 10°C

3-METHYLPIPERIDINE $C_6H_{13}N$

Fl.P., 8°C

4-METHYLPIPERIDINE $C_6H_{13}N$

Fl.P., 9°C

PERHYDROAZEPINE (HEXAMETHYLENEIMINE) $C_6H_{13}N$

Fl.P., 22°C

2,2-DIMETHYLBUTANE C_6H_{14}

Fl.P., –48°C; E.L., 1.2–7.0%

2,3-DIMETHYLBUTANE C_6H_{14}

Fl.P., −29°C; E.L., 1.2–7.0%

HEXANE C_6H_{14}

Fl.P., −23°C; E.L., 1.1–7.5%; A.I.T., 225°C

Dinitrogen tetraoxide
See DINITROGEN TETRAOXIDE, N_2O_4: Hydrocarbons

ISOHEXANE C_6H_{14}

Fl.P., −7°C; E.L., 1.2–7.0%

3-METHYLPENTANE C_6H_{14}

Fl.P., −7°C; E.L., 1.2–7.0%

DIISOPROPYLBERYLLIUM $C_6H_{14}Be$

Coates, G. E. *et al.*, *J. Chem. Soc.*, 1954, 22
Uncontrolled reaction with water is explosive.
See other ALKYLMETALS

DIPROPYL HYPONITRITE $C_6H_{14}N_2O_2$

See DIALKYL HYPONITRITES

BUTYL ETHYL ETHER $C_6H_{14}O$

Fl.P., 4°C

DIISOPROPYL ETHER $C_6H_{14}O$

Fl.P., –28°C; E.L., 1.4–21.0%

1. Anon., *Chem. Eng. News,* 1942, **20**, 1458
2. *MCA Case Histories Nos. 603, 1607*
3. Douglas, I.B., *J. Chem. Educ.*, 1963, **40**, 469
4. Hamstead, A. C. *et al., J. Chem. Eng. Data*, 1960, **5**, 383; *Ind. Eng. Chem.*, 1961, **53**(2), 63A; Rieche, A. *et al., Ber.*, 1942, **75**, 1016

There is a long history of violent explosions involving peroxidised di-isopropyl ether, with initiation by disturbing a drum [1], unscrewing a bottle cap [2], or accidental impact [3]. This ether, with two tertiary hydrogen atoms adjacent to the oxygen link, is extremely readily peroxidised after a few hours' exposure to air, initially to a di-hydroperoxide, which disproportionates to dimeric and trimeric acetone peroxides which are highly explosive [4]. It has been found that the ether may be inhibited completely against peroxide formation for a few years by addition of *N*-benzyl-*p*-aminophenol at 16 p.p.m., or by diethylenetriamine, triethylenetetramine, or tetraethylenepentamine at 50 p.p.m. [4].

See 3,3,6,6-TETRAMETHYL-1,2,4,5-TETRAOXANE, $C_6H_{12}O_4$
3,3,6,6,9,9-HEXAMETHYL-1,2,4,5,7,8-HEXAOXAONANE, $C_9H_{18}O_6$
See other PEROXIDISABLE COMPOUNDS

DIPROPYL ETHER $C_6H_{14}O$

Fl.P., below 21°C; A.I.T., 215°C

2-METHYL-2-PENTANOL $C_6H_{14}O$

Fl.P., 21°C

3-METHYL-3-PENTANOL $C_6H_{14}O$

Fl.P., 24°C

1,1-DIETHOXYETHANE (DIETHYLACETAL) $C_6H_{14}O_2$

Fl.P., below 4°C; E.L., 1.7–10.4%
Mee, A. J., *School Sci Rev.*, 1940, 22(86), 95
A peroxidised sample exploded violently during distillation.
See other PEROXIDISABLE COMPOUNDS

1,2-DIETHOXYETHANE $C_6H_{14}O_2$

A.I.T., 205°C
See other PEROXIDISABLE COMPOUNDS

DIPROPYL PEROXIDE $C_6H_{14}O_2$

Swern, 1972, Vol. 3, 21
Unexpected explosions have occurred with dipropyl peroxides.
See other DIALKYL PEROXIDES

BIS(2-METHOXYETHYL) ETHER $C_6H_{14}O_3$

'Diglyme Properties and Uses', Brochure, London, ICI, 1967
Details of properties, applications and safe handling. Light, heat and oxygen (air) promote formation of potentially explosive peroxides. These may be removed by stirring with a suspension of iron oxide in aqueous alkali.

Metal hydrides
See ALUMINIUM HYDRIDE, AlH_3 : Carbon dioxide, etc.
LITHIUM TETRAHYDROALUMINATE, AlH_4Li : Bis(2-methoxyethyl) ether

DIPROPYLALUMINIUM HYDRIDE $C_6H_{15}Al$

See other ALKYLALUMINIUM ALKOXIDES, etc.

TRIETHYLALUMINIUM $C_6H_{15}Al$

Fl.P., below −53°C; A.I.T., below −53°C

Alcohols
Mirviss, S. B. *et al., Ind. Eng. Chem.*, 1961, **53**(1), 54A
Interaction with methanol, ethanol and 2-propanol with the undiluted trialkyl is explosive, while *tert*-butanol reacts vigorously.
See ALKYLALUMINIUM DERIVATIVES

N,N-Dimethylformamide
'DMF Brochure', Billingham, ICI, 1965
A mixture of the amide and triethylaluminium explodes when heated.
See other ALKYLMETALS

ETHOXYDIETHYLALUMINIUM $C_6H_{15}AlO$

Dangerous Loads, 1972
The pure material, or solutions concentrated above 20%, ignite in air.
See other TRIALKYLALUMINIUM ALKOXIDES, etc.

TRIETHYLDIALUMINIUM TRICHLORIDE $C_6H_{15}Al_2Cl_3$

Carbon tetrachloride
Reineckel, H., *Angew. Chem. (Intern. Ed.)*, 1964, **3**, 65
A mixture exploded when warmed to room temperature.
See other ALKYLALUMINIUM HALIDES

TRIETHYLARSINE $C_6H_{15}As$

Sidgwick, 1950, 762
Inflames in air.
See other ALKYLNON-METALS

TRIETHYLBORANE $C_6H_{15}B$

Hurd, D. T., *J. Amer. Chem. Soc.*, 1948, **70**, 2053
Meerwein, H., *J. Prakt. Chem.*, 1937, **147**, 240
Ignites on exposure to air. Mixtures with triethylaluminium have been used as hypergolic igniters for rocket propulsion systems.
See other ALKYLNON-METALS

TRIETHYLBORATE $C_6H_{15}BO_3$

Fl.P., 11°C; Fl.P., 18°C for 30% v/v solution in ethanol

TRIETHYLBISMUTH $C_6H_{15}Bi$

Coates, 1967, Vol. 1, 537
Ignites in air, and explodes at about 150°C before distillation begins.
See other TRIALKYLBISMUTHS

TRIETHYLGALLIUM $C_6H_{15}Ga$

Dennis, L. M. *et al., J. Amer. Chem. Soc.*, 1932, **54**, 182
The compound spontaneously inflames in air. Breaking a bulb containing 0.2 g under cold water shattered the container. The monoetherate behaved similarly under 6 M nitric acid.
See other ALKYLMETALS

BUTYLETHYLAMINE $C_6H_{15}N$

Fl.P., 18°C (o)

DIISOPROPYLAMINE $C_6H_{15}N$

Fl.P., −7°C

1,3-DIMETHYLBUTYLAMINE $C_6H_{15}N$

Fl.P., 13°C

DIPROPYLAMINE $C_6H_{15}N$

Fl.P., 7°C

TRIETHYLAMINE $C_6H_{15}N$

Fl.P., −7°C; E.L., 1.2–8.0%

Dinitrogen tetraoxide
See DINITROGEN TETRAOXIDE, N_2O_4 : Triethylamine

Maleic anhydride
See MALEIC ANHYDRIDE, $C_4H_2O_3$: Cations, etc.

TRIETHYLAMINE HYDROGEN PEROXIDATE $C_6H_{15}N \,.\, 4H_2O_2$

See CRYSTALLINE HYDROGEN PEROXIDATES

2-DIETHYLAMMONIOETHYL NITRATE NITRATE $C_6H_{15}N_3O_6$

1. Barbière, J., *Bull. Soc. Chim. Fr.*, 1944, **11**, 470–480
2. Fakstorp, J. *et al., Acta Chem. Scand.*, 1951, **5**, 968–969

Repetition of the original preparation [1] involving interaction of diethylaminoethanol with fuming nitric acid, followed by vacuum distillation of excess acid, invariably caused explosions during this operation. A modified procedure without evaporation is described [2].

See related ALKYL NITRATES
OXOSALTS OF NITROGENOUS BASES

TRIETHYLPHOSPHINE $C_6H_{15}P$

Oxygen
Engler, C. *et al., Ber.*, 1901, **34**, 2935
Action of oxygen at low temperature on the phosphine produces an explosive product.

See other ALKYLNON-METALS

TRIETHYLANTIMONY $C_6H_{15}Sb$

von Schwartz, 1918, 322
Inflames in air.

See other ALKYLMETALS

TRIETHYLAMMONIUM NITRATE $C_6H_{16}NO_3$

Dinitrogen tetraoxide
See DINITROGEN TETRAOXIDE, N_2O_4: Triethylammonium nitrate

1,2-BIS(DIMETHYLAMINO)ETHANE $C_6H_{16}N_2$

Fl.P., 18°C

DIETHOXYDIMETHYLSILANE $C_6H_{16}O_2Si$

Fl.P., below 23°C
See related ALKYLNON-METALS

TRIETHYLTIN HYDROPEROXIDE $C_6H_{16}O_2Sn$

Aleksandrov, Y. A. *et al., Zh. Obshch. Khim.*, 1965, **35**, 115
While the hydroperoxide is stable at room temperature, the addition compound which it readily forms with excess hydrogen peroxide is not, decomposing violently.
See ORGANOMINERAL PEROXIDES

BIS-TRIMETHYLSILYL CHROMATE $C_6H_{18}CrO_4Si_2$

1. *ABCM Quart. Safety Summ.*, 1960, **31**, 16
2. Schmidt, M. *et al., Angew. Chem.*, 1958, **70**, 704

Small quantities can be distilled at *ca.* 75°C/1.3 mbar, but larger quantities are liable to explode violently owing to local overheating [1]. An attempt to prepare an analogous poly(dimethylsilyl) chromate by heating a dimethylpolysiloxane with chromium trioxide at 140°C exploded violently after 20 min at this temperature [2].

LITHIUM BIS(TRIMETHYLSILYL)AMIDE $C_6H_{18}LiNSi_2$

Amonoo-Neizer, E. H. *et al., Inorg. Synth.*, 1966, **8**, 21
It is unstable in air and ignites when compressed.

***N,N'*-BIS(2-AMINOETHYL)1,2-DIAMINOETHANE (TRIETHYLENETETRAMINE)** $C_6H_{18}N_4$

Cellulose nitrate
See CELLULOSE NITRATE: Amines

BIS-TRIMETHYLSILYL OXIDE $C_6H_{18}OSi_2$

Fl.P., $-1°C$
See related ALKYLNON-METALS

BIS-TRIMETHYLSILYL PEROXOMONOSULPHATE $C_6H_{18}O_5SSi_2$

Blaschette, A. *et al.*, *Angew. Chem. (Intern. Ed.)*, 1969, **8**, 450
It is stable at $-30°C$, but on warming to ambient temperature decomposes violently evolving sulphur trioxide.
See other PEROXYESTERS

HEXAMETHYLDIPLATINUM $C_6H_{18}Pt_2$

Gilman, H. *et al.*, *J. Amer. Chem. Soc.*, 1953, **75**, 2065
It explodes sharply in a shower of sparks on heating.
See other PLATINUM COMPOUNDS

HEXAMETHYLDISILAZANE $C_6H_{19}NSi_2$

Fl.P., 14°C

BIS-1,2-DIAMINOPROPANE-*cis*-DICHLOROCHROMIUM(III) PERCHLORATE $C_6H_{20}Cl_3CrN_4O_4$

Perchloric acid
Anon., *Chem. Eng. News*, 1963, **41**(27), 47
A mixture of the complex and 20 volumes of *ca.* 70% acid was being stirred at 22°C when it exploded violently. Traces of ether or ethanol may have been present. Extreme caution must be exercised when concentrated perchloric acid is contacted with organic materials with agitation or without cooling.
See other AMMINEMETAL OXOSALTS

HEXAUREAGALLIUM(III) PERCHLORATE $C_6H_{24}Cl_3GaN_{12}O_{18}$

Lloyd, D. J. *et al., J. Chem. Soc.*, 1943, 76
Decomposes violently when heated strongly above its m.p., 179°C.
See other AMMINEMETAL OXOSALTS

TRIS-1,2-DIAMINOETHANECOBALT(III) NITRATE $C_6H_{24}CoN_9O_9$

No explosion on heating, but medium impact-sensitivity.
See AMMINEMETAL OXOSALTS

HEXAUREACHROMIUM(III) NITRATE $C_6H_{24}CrN_{15}O_{15}$

Explodes at 265°C, medium impact-sensitivity.
See AMMINEMETAL OXOSALTS

HEXACARBONYLMOLYBDENUM C_6MoO_6

Diethyl ether
Owen, B. B. *et al., Inorg. Synth.,* 1950, **3**, 158
Solutions of hexacarbonylmolybdenum have exploded after extended storage.
See other CARBONYLMETALS

TETRAAZIDO-*p*-BENZOQUINONE $C_6N_{12}O_2$

Friess, K. *et al., Ber.*, 1923, **56**, 1304
Extremely explosive, sensitive to heat, impact or friction. A single crystal heated on a spatula exploded and bent it badly.
See other 2-AZIDOCARBONYL COMPOUNDS
HIGH-NITROGEN COMPOUNDS

HEXACARBONYLTUNGSTEN C_6O_6W

Hurd, D. T. *et al., Inorg. Chem.*, 1957, **5**, 136
It is dangerous to attempt the preparation from tungsten hexachloride, aluminium powder and carbon monoxide in an autoclave of greater than 0.3 litre capacity.
See other CARBONYLMETALS

4-IODOBENZENEDIAZONIUM-2-CARBOXYLATE $C_7H_3IN_2O_2$

Gommper, R. *et al., Chem. Ber.*, 1968, **101**, 2348
It is a highly explosive solid.
See other DIAZONIUM CARBOXYLATES

2,4,6-TRINITROTOLUENE $C_7H_3N_3O_6$

Added impurities
Haüptli, H., private comm., 1972
During investigation of the effect of presence of 1% of added impurities on the thermal explosion temperature of TNT (297°) it was found that fresh red lead, sodium carbonate and potassium hydroxide reduced the explosion temperatures to 192, 218 and 192°C, respectively.
See also POTASSIUM, K: Nitrogen-containing explosives

2,4,6-TRINITROBENZOIC ACID $C_7H_3N_3O_8$

Preparative hazard.
See SODIUM DICHROMATE, $CrNa_2O_7$: Sulphuric acid, TNT
See other POLYNITROARYL COMPOUNDS

3-PHENYL-1-TETRAZOLYL-1-TETRAZENE $C_7H_3N_8$

Thiele, J., *Ann.*, 1892, **270**, 60

It explodes on warming, as do *C*-substituted homologues.
See other HIGH-NITROGEN COMPOUNDS

o-, *m*- or *p*-TRIFLUOROMETHYLPHENYLMAGNESIUM BROMIDE $C_7H_4BrF_3Mg$

1. Appleby, J. C., *Chem. & Ind.*, 1971, 120
2. 'Benzotrifluorides Catalogue 6/15', West Chester, Pa., Marshallton Res. Labs., 1971
3. Taylor, R. *et al.*, private comm., 1973

During a 2 kg preparation under controlled conditions, the *m*-isomer exploded violently [1]. The same had occurred previously in a 1.25 kg preparation 30 min after all magnesium had dissolved. A small preparation of the *o*- isomer had also exploded. It was found that *m*- preparations in ether decomposed violently at 75°C and in benzene-tetrahydrofuran at 90°C. Molar preparations or less have been done several times uneventfully when temperatures below 40°C were maintained, but caution is urged. The *m*- and *p*- reagents prepared in ether exploded when the temperature was raised by adding benzene and distilling the ether out [3].
See other HALO-ARYLMETALS

p-BROMOBENZOYL AZIDE $C_7H_4BrN_3O$

Curtiss, T. *et al.*, *J. Prakt. Chem.*, 1898, **58**, 201
Explodes violently above its m.p., 46°C.
See other ACYL AZIDES

p-CHLOROTRIFLUOROMETHYLBENZENE $C_7H_4ClF_3$

Sodium dimethylsulphinate
Meschino, J. A. *et al.*, *J. Org. Chem.*, 1971, **36**, 3637, footnote 9
Interaction with dimethylsulphinate carbanion (from the sulphoxide and sodium hydride) at −5°C is very exothermic, and addition of the chloro compound must be slow to avoid violent eruption.

p-CHLOROPHENYLISOCYANATE C_7H_4ClNO

Preparative hazard.
See *p*-CHLOROBENZOYL AZIDE, $C_7H_4ClN_3O$

o-NITROBENZOYL CHLORIDE $C_7H_4ClNO_3$

1. Cook, N. C. *et al., Chem. Eng. News*, 1945, **23**, 2394
2. Bonner, W. D. *et al., J. Amer. Chem. Soc.*, 1946, **68**, 344; *Chem. & Ind.*, 1948, 89

The hot material remaining after vacuum stripping of solvent up to 130°C decomposed with evolution of gas and then exploded violently 50 min after heating had ceased. Further attempts to distil the acid chloride even in small quantities at below 1.3 mbar caused exothermic decomposition at 110°C. It was, however, possible to flash-distil the chloride in special equipment [1]. Two later similar publications recommend use in solution of the unisolated material [2].
See other NITRO-ACYL HALIDES

p-CHLOROBENZOYL AZIDE $C_7H_4ClN_3O$

1. Cobern, D. *et al., Chem. & Ind.*, 1965, 1625; ibid., 1966, 375
2. Tyabji, M. M., *Chem. & Ind.*, 1965, 2070

A violent explosion occurred during vacuum distillation of *p*-chlorophenylisocyanate, prepared by Curtiuys reaction on *p*-chlorobenzoyl azide. It was found by IR spectroscopy that this isocyanate (as well as others prepared analogously) contained some unchanged azide, to which the explosion was attributed. The use of IR spectroscopy to check for absence of azides in isocyanates is recommended before distillation [1]. Subsequently, the explosion was attributed to free hydrogen azide, produced by hydrolysis of unchanged azide [2].

TRIFLUOROMETHYLBENZENE $C_7H_4F_3$

Fl.P., below 21°C

o-, *m*- or *p*-TRIFLUOROMETHYLPHENYLLITHIUM $C_7H_4F_3Li$

See ORGANOLITHIUM REAGENTS
See other HALO-ARYLMETALS

BENZENEDIAZONIUM-2-CARBOXYLATE $C_7H_4N_2O_2$

Alone,
or Aniline,
or Isocyanides

1. Yaroslavsky, S., *Chem. & Ind.*, 1965, 765
2. Huisgen, R. *et al.*, *Ber.*, 1965, **98**, 4104
3. Sullivan, J. M., *Chem. Eng. News*, 1971, **49**(16), 5
4. Embree, H. D., *Chem. Eng. News*, **49**(30), 3
5. Mich, T. K. *et al.*, *J. Chem. Educ.*, 1968, **45**, 272
6. Matuszak, C. A., *Chem. Eng. News*, 1971, **49**(24), 39
7. Logullo, F. M. *et al.*, *Org. Synth.*, 1968, **48**, 17
8. Stiles, R. M. *et al.*, *J. Amer. Chem. Soc.*, 1963, **85**, 1795, footnote 30a

The isolated internal salt is explosive, and should only be handled in small amounts [1]. It reacts explosively with aniline, and violently with aryl isocyanides [2]. The explosive nature of the salt has been amply confirmed [3,4], and it may be precipitated during the generation of benzyne by diazotisation of anthranilic acid at ambient temperature; some heating is essential for complete decomposition [5]. During decomposition of the isolated salt, too-rapid addition to hot solvent caused a violent explosion [6]. The published procedure [7] must be closely followed for safe working. Contrary to earlier beliefs, diazonium carboxylate hydrohalide salts are also shock-sensitive explosives [8].

1-Pyrrolidinylcyclohexene
Kuehne, M. E., *J. Amer. Chem. Soc.*, 1962, **84**, 841
Decomposition of benzenediazonium-2-carboxylate in the enamine at 40°C caused a violent explosion.
See other DIAZONIUM CARBOXYLATES

o-NITROBENZONITRILE $C_7H_4N_2O_2$

1. Miller, C. S., *Org. Synth.*, Coll. Vol. 3, 646
2. Partridge, M. W., private comm., 1968

When the published method [1] for preparing the *p*- isomer is used to prepare *o*-nitrobenzonitrile, a moderate explosion often occurs towards the end of the reaction period. (This may be due to formation of nitrogen trichloride as a by-product.) Using an alternative procedure, involving heating *o*-chloronitrobenzene with copper(I) cyanide in pyridine for 7 h at 160°C, explosions occurred towards the end of the heating period in about one in five preparations [2].

See other CYANO COMPOUNDS

4-HYDROXYBENZENEDIAZONIUM-3-CARBOXYLATE $C_7H_4N_2O_3$

Anon., private comm., 1968

Explodes at 155°C.

See other DIAZONIUM CARBOXYLATES

m-NITROBENZOYL NITRATE $C_7H_4N_2O_6$

Francis, F. E., *J. Chem. Soc.*, 1906, **89**, 1

The nitrate explodes if heated rapidly, like benzoyl nitrate.

See other ACYL NITRATES

2,6-DINITROBENZYL BROMIDE $C_7H_5BrN_2O_4$

Potassium phthalimide

Reich, S. *et al., Bull. Soc. Chim. Fr.*[4], 1917, **21**, 119

Reaction temperature in absence of solvent must be below 130–135°C to avoid explosive decomposition.

See other POLYNITROARYL COMPOUNDS

PHENYLCHLORODIAZIRINE $C_7H_5ClN_2$

Wheeler, J. J., *Chem. Eng. News*, 1970, **48**(30), 10
The neat material is about three times as shock-sensitive as glyceryl nitrate, and should not be handled undiluted.
See other DIAZIRINES

BENZOYL CHLORIDE C_7H_5ClO

Dimethyl sulphoxide
See DIMETHYL SULPHOXIDE, C_2H_6OS: Acyl halides

POTASSIUM BENZOYLPEROXOSULPHATE $C_7H_5KO_6S$

Davies, 1961, 65
Explodes on friction or warming.
See other DIACYL PEROXIDES

m-NITROBENZALDEHYDE $C_7H_5NO_3$

Large, J. *et al.*, *Chem. & Ind.*, 1967, 1424
Delays in working up the crude product caused violent explosions during attempted vacuum distillation. An alternative method of crystallisation is described.
See other *C*-NITRO COMPOUNDS

BENZOYL NITRATE $C_7H_5NO_4$

1. Francis, F. E., *J. Chem. Soc.*, 1906, **89**, 1
2. Ferrario, E., *Gazz. Chim. Ital.*, 1901, [1], **40**, 99

An unstable liquid which is capable of distillation under reduced pressure, but which explodes violently on rapid heating at normal pressure [1]. It may also explode on exposure to light [2].

Water

Francis, F. E., *J. Chem. Soc.*, 1906, **89**, 1

Reactivity of benzoyl nitrate towards moisture is so great that attempted filtration through an undried filter paper causes explosive decomposition (possibly involving cellulose nitrate?).

See other ACYL NITRATES

2,5-PYRIDINEDICARBOXYLIC ACID $C_7H_5NO_4$

Preparative hazard.

See NITRIC ACID, HNO_3 : 5-Ethyl-2-methylpyridine

BENZOYL AZIDE $C_7H_5N_3O$

Bergel, F., *Angew. Chem.*, 1927, **40**, 974

The sensitivity towards heat of this explosive compound is increased by previous compression, confinement, and presence of impurities. Crude material exploded violently between 120 and 165°C.

See other ACYL AZIDES

2,4,6-TRINITROTOLUENE $C_7H_5N_3O_6$

Sodium dichromate,
Sulphuric acid

See SODIUM DICHROMATE, $CrNa_2O_7$: Sulphuric acid, etc.

N-2,4,6-TETRANITRO-*N*-METHYLANILINE (TETRYL) $C_7H_5N_5O_8$

Hydrazine

See HYDRAZINE, H_4N_2: Oxidants

Trioxygen difluoride
See TRIOXYGEN DIFLUORIDE, F_2O_3

1-NITRO-3(2,4-DINITROPHENYL)UREA $C_7H_5N_5O_7$

McVeigh, J. L. *et al., J. Chem. Soc.*, 1945, 621
It is an explosive of lower power, but of greater impact- or frictional sensitivity than picric acid, which should not be stored in bottles with ground-in stoppers.
See other *N*-NITRO COMPOUNDS

2-CHLORO-4-NITROTOLUENE $C_7H_6ClNO_2$

Sodium hydroxide
1. *ABCM Quart. Safety Summ.*, 1962, **33**, 20
2. *MCA Case History No. 907*

The residue from vacuum distillation of crude material (contaminated with sodium hydroxide) exploded after showing signs of decomposition. Experiment showed that 2-chloro-4-nitrotoluene decomposes violently when heated at 170–200°C in presence of alkali. Thorough water washing is therefore essential before distillation is attempted [1]. A similar incident was reported with mixed chloronitrotoluenes [2].
See other HALOARYL COMPOUNDS

4-CHLORO-2-METHYLBENZENEDIAZONIUM SALTS $C_7H_6ClN_2^+X^-$

Sodium hydrogensulphide,
or Disodium sulphide,
or Disodium polysulphide
See DIAZONIUM SULPHIDES

2,4-DINITROTOLUENE $C_7H_6N_2O_4$

MCA SD-93, 1966
The commercial material, containing some 20% of the 2,6- isomer, decomposes at 250°, but at 280°C decomposition becomes self-sustaining. Prolonged heating below these temperatures may also cause some decomposition, and the presence of impurities may decrease the decomposition temperatures.

Nitric acid,
See NITRIC ACID, HNO_3: Nitroaromatics
See other POLYNITROARYL COMPOUNDS

5-PHENYLTETRAZOLE $C_7H_6N_4$

Wedekind, E., *Ber.*, 1898, **31**, 948
Explodes on attempted distillation.
See other TETRAZOLES

N-AZIDOCARBONYLAZEPINE $C_7H_6N_4O$

Chapman, L. E. *et al.*, *Chem. & Ind.*, 1966, 1266
Explodes on attempted distillation.
See other ACYL AZIDES

BENZALDEHYDE C_7H_6O

Peroxyformic acid
See PEROXYFORMIC ACID, CH_2O_3: Organic materials

PEROXYBENZOIC ACID $C_7H_6O_3$

1. Baeyer, A. *et al., Ber.*, 1900, **33**, 1577
2. Silbert, L. S. *et al., Org. Synth.*, 1963, **43**, 95

Explodes weakly on heating [1]. A chloroform solution of peroxybenzoic acid exploded during evaporation of solvent. Evaporation should be avoided if possible, or conducted with full precautions [2].

See other PEROXYACIDS

METHYL 4-BROMOBENZENEDIAZOATE $C_7H_7BrN_2O$

Bamberger, E., *Ber.*, 1895, **28**, 233

Explodes on heating.

See other ARENEDIAZOATES

o-TOLUENEDIAZONIUM PERCHLORATE $C_7H_7ClN_2O_4$

Hofmann, K. A. *et al., Ber.*, 1906, **39**, 3146

The solid salt is explosive even when wet.

See other DIAZONIUM PERCHLORATES

4-CHLORO-2-METHYLPHENOL C_7H_7ClO

Sodium hydroxide

ABCM Quart. Safety Summ., 1957, **28**, 39

A large quantity (700 kg) of the chlorophenol, left in contact with concentrated sodium hydroxide solution for 3 days, decomposed, reaching red heat and evolving fumes which ignited explosively. Although this could not be reproduced under laboratory conditions, it is believed that exothermic hydrolysis to the hydroquinone occurred, the high viscosity of the liquid preventing dissipation of heat.

See other HALOARYL COMPOUNDS

TROPYLIUM PERCHLORATE $C_7H_7ClO_4$

Ferrini, P. G. *et al., Angew. Chem.*, 1962, **74**, 488
Pressing the salt through a funnel with a glass rod caused a violent explosion.
See other NON-METAL PERCHLORATES

TOLYLCOPPER C_7H_7Cu

Camus, A. *et al., J. Organomet. Chem.*, 1968, 442–443
The solid *o*-, *m*- and *p*- isomers usually exploded strongly on exposure to oxygen at 0°C, or weakly on heating above 100°C *in vacuo.*
See other ARYLMETALS

m-FLUOROTOLUENE C_7H_7F

Fl.P., 12°C

o-FLUOROTOLUENE C_7H_7F

Fl.P., 8°C

p-FLUOROTOLUENE C_7H_7F

Fl.P., 10°C

m-TOLUENEDIAZONIUM IODIDE $C_7H_7IN_2$

See *m*-TOLUENEDIAZONIUM SALTS, $C_7H_7N_2^+ X^-$: Potassium iodide
See other DIAZONIUM TRIIODIDES

o-TOLUENEDIAZONIUM IODIDE $C_7H_7IN_2$

See DIAZONIUM TRIIODIDES

p-TOLUENEDIAZONIUM TRIIODIDE $C_7H_7I_3N_2$

See DIAZONIUM TRIIODIDES

o-METHOXYBENZENEDIAZONIUM TRIIODIDE $C_7H_7I_3N_2O$

See DIAZONIUM TRIIODIDES

p-METHOXYBENZENEDIAZONIUM TRIIODIDE $C_7H_7I_3N_2O$

See DIAZONIUM TRIIODIDES

o-NITROTOLUENE $C_7H_7NO_2$

Sodium hydroxide
ABCM Quart. Safety Summ., 1945, **16**, 2
Crude *o*-nitrotoluene, containing some hydrochloric and acetic acids, was charged into a vacuum still with flake sodium hydroxide to effect neutralisation prior to distillation. An explosion occurred later. Similar treatments had been uneventfully used previously, using sodium carbonate or lime for neutralisation. It seems likely that the explosion was caused by formation of the solid *aci*-sodium salt of *o*-nitrotoluene and its heating to explosive decomposition by exothermic reaction of the acids present, then the liberated water, with solid sodium hydroxide.
See other aci-NITRO SALTS

p-NITROTOLUENE $C_7H_7NO_2$

ABCM Quart. Safety Summ., 1938, **9**, 65; 1939, **10**, 2
The residue from large-scale vacuum distillation of *p*-nitrotoluene was left to cool after turning off heat and vacuum, and 8 h later a violent explosion occurred. This was variously attributed to: presence of an excessive proportion of dinitrotoluenes in the residue, which would decompose on prolonged heating; presence of traces of alkali in the crude material, which would form unstable *aci*-nitro salts during distillation; ingress of air to the hot residue on shutting off the vacuum supply, and subsequent accelerating oxidation of the residue. All these factors may have contributed, and several such incidents have occurred.

Sulphuric acid
Hunt, J. K., *Chem. Eng. News*, 1949, **27**, 2504
Solutions of *p*-nitrotoluene in 93% sulphuric acid decompose very violently if heated to 160°C. This happened on plant scale when automatic temperature control failed.

Tetranitromethane
See TETRANITROMETHANE, CN_4O_8: Aromatic nitrocompounds
See other *C*-NITRO COMPOUNDS

BENZYL NITRATE $C_7H_7NO_3$

Nef, J. U., *Ann.*, 1899, **309**, 172
Decomposed explosively at 180–200°C.

Lewis acids
See ALKYL NITRATES

4-METHYL-2-NITROPHENOL $C_7H_7NO_3$

Sodium hydroxide,
Sodium carbonate,
Methanol

MCA Case History No. 701
Failure to agitate a large-scale mixture of the reagents caused an eruption due to action of the exotherm when mixing did occur.

o-NITROANISOLE $C_7H_7NO_3$

Hydrogen
1. Carswell, T. S., *J. Amer. Chem. Soc.*, 1931, **53**, 2417
2. Adkins, H., ibid, 2808

Catalytic hydrogenation on 400 g scale at 34 bar under excessively vigorous conditions (250°C, 12% catalyst, no solvent) caused the thin autoclave (without bursting disc) to rupture [1]. Under more appropriate conditions the hydrogenation is safe [2].

Sodium hydroxide,
Zinc
Anon., *Angew. Chem. (Nachr.)*, 1955, **3**, 186
In preparation of 2,2′-dimethoxyazoxybenzene, solvent ethanol was distilled out of the mixture of *o*-nitroanisole, zinc and sodium hydroxide, before reaction was complete. The exothermic reaction continued unmoderated, and finally exploded.
See other *C*-NITRO COMPOUNDS

m-TOLUENEDIAZONIUM SALTS $C_7H_7N_2^+X^-$

Ammonium sulphide,
or Hydrogen sulphide
See DIAZONIUM SULPHIDES

Potassium *O*-ethyl dithiocarbonate
1. Tarbell, D. S. *et al., Org. Synth.*, 1955, Coll. Vol. 3, 810
2. Parham, W. E. *et al., Org. Synth.*, 1967, **47**, 107

During interaction of the diazonium chloride and the *O*-ethyl dithiocarbonate ('xanthate') solutions, care must be taken to ensure that the intermediate diazonium dithiocarbonate decomposes to

m-thiocresol as fast as it is formed [1]. This can be assured by presence of a trace of nickel in the solution to effect immediate catalytic decomposition [2]. When the two solutions were mixed cold and then heated to effect decomposition, a violent explosion occurred [2].
See other DIAZONIUM SULPHIDES, etc.

m-TOLUENEDIAZONIUM SALTS $C_7H_7N_2^+X^-$

Potassium iodide
Trumbull, E. R., *J. Chem. Educ.*, 1971, **48**, 640
Addition of potassium iodide solution to diazotised *m*-toluidine was accompanied on three occasions by an eruption of the beaker contents. This was not observed with the *o*- and *p*- isomers or other aniline derivatives.
See DIAZONIUM TRIIODIDES

o- or *p*- TOLUENEDIAZONIUM SALTS $C_7H_7N_2^+X^-$

Ammonium sulphide,
or Hydrogen sulphide
See DIAZONIUM SULPHIDES

p-TOLUENESULPHONYL AZIDE $C_7H_7N_3O_2S$

Rewicki, D. *et al.*, *Angew. Chem. (Intern. Ed.)*, 1972, **11**, 44
The impure compound, present as a majority in the distillation residue from preparation of 1-diazoindene, will explode if the bath temperature exceeds 120°C.
See other ACYL AZIDES

METHYL 2-NITROBENZENEDIAZOATE $C_7H_7N_3O_3$

Bamberger, E., *Ber.*, 1895, **28**, 237

Explodes violently on heating, or on disturbing after 24 h confinement in a sealed tube at ambient temperature. The 4-nitro analogue is more stable.
See other ARENEDIAZOATES

SODIUM *p*-CRESOLATE C_7H_7ONa

May and Baker Ltd., private comm., 1968
Sodium *p*-cresolate solution was dehydrated azeotropically with boiling chlorobenzene and the filtered solid was dried in an oven, where it soon ignited and glowed locally. This continued for half an hour after it was removed from the oven. A substituted potassium phenolate, prepared differently, also ignited on heating. Finely divided and moist alkali phenolates may be prone to vigorous oxidation when heated in air.

BICYCLO-[2.2.1]-2,5-HEPTADIENE C_7H_8

Fl.P., −21°C

1,3,5-CYCLOHEPTATRIENE C_7H_8

Fl.P., 4°C

TOLUENE C_7H_8

Fl.P., 4°C; E.L., 1.3–7.0%
MCA SD-63, 1956

Bromine trifluoride
See BROMINE TRIFLUORIDE, BrF_3: Solvents

1,3-Dichloro-5,5-dimethyl-2,4-imidazolidinidione
See 1,3-DICHLORO-5,5-DIMETHYL-2,4-IMIDAZOLIDINDIONE, $C_5H_6Cl_2N_2O_2$: Toluene

Dinitrogen tetraoxide
See DINITROGEN TETRAOXIDE, N_2O_4: Hydrocarbons

Nitric acid
See NITRIC ACID, HNO_3: Hydrocarbons

Uranium hexafluoride
See URANIUM HEXAFLUORIDE, F_6U: Aromatic hydrocarbons

4-CHLORO-2-METHYLANILINE C_7H_8ClN

See 4-CHLORO-2-METHYLBENZENEDIAZONIUM SALTS, $C_7H_6ClN_2^+X^-$

METHYL BENZENEDIAZOATE $C_7H_8N_2O$

Bamberger, E., *Ber.*, 1895, **28**, 228
Explodes on heating or after 1 h storage in a sealed tube at ambient temperature.
See other ARENEDIAZOATES

m-THIOCRESOL C_7H_8S

Preparative hazard.
See *m*-TOLUENEDIAZONIUM SALTS, $C_7H_7N_2^+X^-$: Potassium *O*-ethyl dithiocarbonate

BENZYLAMINE C_7H_9N

N-Chlorosuccinimide
See *N*-HALOIMIDES: Alcohols, etc.

m-TOLUIDINE C_7H_9N

See *m*-TOLUENEDIAZONIUM SALTS, $C_7H_7N_2^+X^-$

o-TOLUIDINE C_7H_9N

See *o*-TOLUENEDIAZONIUM SALTS, $C_7H_7N_2^+X^-$

Nitric acid
See NITRIC ACID, HNO_3: Aromatic amines

p-TOLUIDINE C_7H_9N

See *p*-TOLUENEDIAZONIUM SALTS, $C_7H_7N_2^+X^-$

2,6-DIMETHYLPYRIDINE-1-OXIDE C_7H_9NO

Phosphoryl chloride
See PHOSPHORYL CHLORIDE, Cl_3OP: 2,6-Dimethylpyridine-1-oxide

CYCLOHEPTENE C_7H_{12}

Fl.P., below 23°C

1-HEPTYNE C_7H_{12}

Fl.P., below 23°C

3-HEPTYNE C_7H_{12}

Fl.P., below 23°C

4-METHYLCYCLOHEXENE C_7H_{12}

Fl.P., –1°C (o)

2-HEPTYN-1-OL $C_7H_{12}O$

Brandsma, 1971, 69
Dilution of the alcohol with white mineral oil before vacuum distillation is recommended to avoid the possibility of explosion of the undiluted distillation residue.
See other ACETYLENIC COMPOUNDS

3-METHYLCYCLOHEXANONE $C_7H_{12}O$

Hydrogen peroxide,
Nitric acid
See HYDROGEN PEROXIDE, H_2O_2: Ketones, etc.

4-METHYLCYCLOHEXANONE $C_7H_{12}O$

Nitric acid
See NITRIC ACID, HNO_3 : 4-Methylcyclohexanone

1-ETHOXY-3-METHYL-1-BUTYN-3-OL $C_7H_{12}O_2$

Brandsma, 1971, 78

Traces of acids adhering to glassware are sufficient to induce explosive decomposition of the alcohol during distillation, and must be neutralised by pre-treatment with ammonia gas. Low pressures and temperatures are essential during distillation. Explosions during distillation using a water pump and bath temperatures above 115°C were frequent.

See ETHOXYETHYNYL ALCOHOLS

See other ACETYLENIC COMPOUNDS

1,2-DIMETHYLCYCLOPENTENE OZONIDE $C_7H_{12}O_3$

Criegee, R. *et al.*, *Ber.*, 1953, **86**, 3

Though the ozonide appeared stable to small-scale high-vacuum distillation, the distillation residue exploded violently at 130°C.

See other OZONIDES

N-CYANO-2-BROMOETHYLBUTYLAMINE $C_7H_{13}BrN$

BCISC Quart. Safety Summ., 1964, **35**, 23

A small sample heated to 160°C decomposed exothermically, reaching 250°C. It had been previously distilled at 95°C/0.65 mbar without decomposition.

See also *N*-CYANO-2-BROMOETHYLCYCLOHEXYLAMINE, $C_9H_{15}BrN$

See other CYANO COMPOUNDS

CYCLOHEPTANE C_7H_{14}

Fl.P., 15°C

ETHYLCYCLOPENTANE C_7H_{14}

Fl.P., below 21°C; E.L., 1.1–6.7%

1-HEPTENE C_7H_{14}

Fl.P., –1°C

2-HEPTENE C_7H_{14}

Fl.P., below –1°C

3-HEPTENE C_7H_{14}

Fl.P., below –7°C

METHYLCYCLOHEXANE C_7H_{14}

Fl.P., –4°C; E.L., 1.2–6.7%

2,3,3-TRIMETHYLBUTENE C_7H_{14}

Fl.P., below 0°C

2,4-DIMETHYL-3-PENTANONE $C_7H_{14}O$

Fl.P., 15°C

ISOPENTYL ACETATE $C_7H_{14}O_2$

Fl.P., below 23°C; E.L., 1.1–7.0% (both at 100°C)

PENTYL ACETATE $C_7H_{14}O_2$

Fl.P., 25°C; E.L., 1.1–7.5%

2-PENTYL ACETATE $C_7H_{14}O_2$

Fl.P., 23°C; E.L., 1.1–7.5%

3,3-DIETHOXYPROPENE $C_7H_{14}O_2$

Fl.P., below 23°C
See other PEROXIDISABLE COMPOUNDS

ETHYL ISOVALERATE $C_7H_{14}O_2$

Fl.P., 25°C

ISOBUTYL PROPIONATE $C_7H_{14}O_2$

Fl.P., below 22°C

ISOPROPYL BUTYRATE $C_7H_{14}O_2$

Fl.P., 24°C

ISOPROPYL ISOBUTYRATE $C_7H_{14}O_2$

Fl.P., below 21°C

1,4-BUTANEDIOL MONO-2,3-EPOXYPROPYL ETHER $C_7H_{14}O_3$

Trichloroethylene
See TRICHLOROETHYLENE, C_2HCl_3: Epoxides

2,6-DIMETHYLPIPERIDINE $C_7H_{15}N$

Fl.P., 16°C

1-ETHYLPIPERIDINE $C_7H_{15}N$

Fl.P., 19°C

2-ETHYLPIPERIDINE $C_7H_{15}N$

Fl.P., − 12°C

2,3-DIMETHYLPENTANE C_7H_{16}

Fl.P., below −6°C; E.L., 1.1–6.7%

2,4-DIMETHYLPENTANE C_7H_{16}

Fl.P., − 12°C

HEPTANE C_7H_{16}

Fl.P., –4°C; E.L., 1.2–6.7%; **A.I.T.**, 223°C

2-METHYLHEXANE C_7H_{16}

Fl.P., – 1°C; E.L., 1.0–6.0%

3-METHYLHEXANE C_7H_{16}

Fl.P., – 1°C; E.L., 1.0–6.0%

2,2,3-TRIMETHYLBUTANE C_7H_{16}

Fl.P., below 0°C; E.L., 1.0– %

3-DIETHYLAMINOPROPYLAMINE $C_7H_{18}N_2$

Cellulose nitrate
See CELLULOSE NITRATE: Amines

TETRA(CHLOROETHYNYL)SILANE C_8Cl_4Si

Viehe, H. G., *Chem. Ber.*, 1959, **92**, 3075
Though thermally stable, it exploded violently on grinding (with potassium bromide) or on impact.
See other HALOACETYLENE DERIVATIVES

OCTACARBONYLDICOBALT $C_8Co_2O_8$

Wender, I. *et al., Inorg. Synth.*, 1957, **5**, 191
When the carbonyl is prepared (by rapid cooling) as a fine powder, it decomposes in air to give pyrophoric dodecacarbonyltetracobalt.
See other CARBONYLMETALS

3-BROMO-2,7-DINITRO-5-BENZO[*b*]-THIOPHENEDIAZONIUM-4-OLATE $C_8HBrN_4O_5S$

Brown, I. *et al., Chem. & Ind.* 1962, 982
It is an explosive compound.
See NITRIC ACID, HNO_3: Benzo[*b*]thiophene derivatives

N-BROMO-3-NITROPHTHALIMIDE $C_8H_3BrN_2O_4$

Tetrahydrofurfuryl alcohol
See *N*-HALOMIDES: Alcohols

ISOPHTHALOYL DICHLORIDE $C_8H_4Cl_2O_2$

Methanol
Morrell, S. H., private comm., 1968
Violent reaction occurred between isophthaloyl dichloride and methanol when they were accidentally added in succession to the same waste solvent bottle.

2,4-HEXADIYNYLENE BISCHLOROFORMATE $C_8H_4Cl_2O_4$

Driedger, P. E. *et al., Chem. Eng. News*, 1972, **50**(12), 51
One sample exploded with extreme violence at 15°C/0.2 mbar,

while another smaller sample had been distilled uneventfully at 114–115°C/0.2 mbar.
See other HALOACETYLENE DERIVATIVES

TETRAETHYNYLGERMANIUM C_8H_4Ge

1. Chokiewicz, W. *et al., Compt. Rend.*, 1960, **250**, 866
2. Shikkiev, I. A. *et al., Dokl. Akad. Nauk SSSR,* 1961, **139**, 1138 (Eng. transl. 830)

It explodes on rapid heating [1] or friction [2].
See other METAL ACETYLIDES

PHTHALOYL DIAZIDE $C_8H_4N_6O_2$

Lindemann, H. *et al., Ann.*, 1928, **464**, 237
The *symm* diazide is extremely explosive, while the *asymm* isomer is less so, melting at 56°C, but exploding on rapid heating.
See other ACYL AZIDES

PHTHALIC ANHYDRIDE $C_8H_4O_3$

Copper oxide
See COPPER OXIDE, CuO: Phthalic acid, etc

Nitric acid
See NITRIC ACID, HNO_3: Phthalic anhydride

Sodium nitrite
See SODIUM NITRITE, $NNaO_2$: Phthalic acid, etc.

PHTHALOYL PEROXIDE $C_8H_4O_4$

Jones, M. *et al., J. Org. Chem.*, 1971, **36**, 1536
Detonable by impact or by melting at 123°C. It is probably polymeric.
See other DIACYL PEROXIDES

TETRAETHYNYLTIN C_8H_4Sn

Jenkner, H., Ger. Pat. 115 736, 1961
It explodes on rapid heating.
See other METAL ACETYLIDES

2,4-DINITROPHENYLACETYL CHLORIDE $C_8H_5ClN_2O_5$

Bonner, T. G., *J. R. Inst. Chem.*, 1957, **81**, 596
Explosive decomposition occurred during attempted vacuum distillation, attributed to either presence of some trinitro compound in the unpurified dinitrophenylacetic acid used, or to the known instability of *o*-nitro acid chlorides. A previous publication (ibid., 407) erroneously described the decomposition of 2,4-dinitrobenzoyl chloride.
See other NITRO-ACYL HALIDES

3-NITROPHTHALIC ACID $C_8H_5NO_6$

Preparative hazard.
See NITRIC ACID, HNO_3: Phthalic anhydride, etc.

NITROTEREPHTHALIC ACID $C_8H_5NO_6$

Preparative hazard.
See NITRIC ACID, HNO_3: Sulphuric acid, Terephthalic acid

SODIUM PHENYLACETYLIDE C_8H_5Na

Nef, J. U., *Ann.*, 1899, **308**, 264
The ether-moist powder ignites in air.
See ALKALI-METAL DERIVATIVES OF HYDROCARBONS
See other METAL ACETYLIDES

3-METHYL-2-NITROBENZOYL CHLORIDE $C_8H_6ClNO_3$

Dahlbom, R., *Acta Chem. Scand.*, 1960, **14**, 2049
Attempted distillation caused an explosion; other *o*-nitro acid chlorides behave similarly.
See other NITRO-ACYL HALIDES

***o*-NITROPHENYLACETYL CHLORIDE** $C_8H_6ClNO_3$

Hayao, S., *Chem. Eng. News*, 1964, **42**(13), 39
Distillation of solvent chloroform from the preparation caused the residue to explode violently on two occasions. Previous publications had recommended use of solutions of the chloride, in preference to isolated material. Other *o*-nitro acid chlorides behave similarly.
See other NITRO-ACYL HALIDES

PHENOXYACETYLENE C_8H_6O

Jacobs, T. L. *et al.*, *J. Amer. Chem. Soc.*, 1942, **64**, 223
Small samples rapidly heated in sealed tubes to around 100°C exploded.
See other ACETYLENIC COMPOUNDS

PHENYLGLYOXAL $C_8H_6O_2$

Fl.P., below 23°C

PHTHALIC ACID $C_8H_6O_4$

Sodium nitrite

See SODIUM NITRITE, $NNaO_2$: Phthalic acid

TEREPHTHALIC ACID $C_8H_6O_4$

Preparative hazard.

See NITRIC ACID, HNO_3 : Hydrocarbons (reference 6)

BENZYL CHLOROFORMATE $C_8H_7ClO_2$

Water

See ARYL CHLOROFORMATES: Water

PHENYLACETONITRILE C_8H_7N

Sodium hypochlorite

See SODIUM HYPOCHLORITE, ClNaO : Phenylacetonitrile

See other CYANO COMPOUNDS

p-NITROACETANILIDE $C_8H_7N_2O_3$

Sulphuric acid

See SULPHURIC ACID, H_2O_4S : Nitroaryl bases, etc.

1,3,5,7-CYCLOOCTATETRAENE C_8H_8

Fl.P., below 22°C

STYRENE C_8H_8

1. Unpublished information
2. *MCA SD-37*, 1971

The autocatalytic exothermic polymerisation reaction exhibited by styrene was involved in a plant-scale incident where accidental heating caused violent ejection of liquid and vapour from a storage tank [1]. Polymerisation becomes self-sustaining above 65°C [2].

Alkali metal–graphite compounds

Williams, N. E., private comm., 1968

Interlaminar compounds of sodium or potassium in graphite will ionically polymerise styrene (and other monomers) smoothly. The occasional explosions experienced were probably due to rapid collapse of the layer structure and liberation of very finely divided metal.

Oxygen

Barnes, C. E. *et al., J. Amer. Chem. Soc.*, 1950, 72, 210

Exposure of unstabilised styrene to oxygen at 40–60°C generated a styrene–oxygen interpolymeric peroxide which, when isolated, exploded violently on gentle heating

See other POLYPEROXIDES

5-AMINO-3-PHENYL-1,2,4-TRIAZOLE $C_8H_8N_4$

Polya, J. B., *Chem. & Ind.*, 1965, 812

The solution, obtained by conventional diazotisation of the above, contained some solid which was removed by filtration through sintered glass. The solid, thought to be precipitated diazonium salt (possibly an internal salt), exploded violently on being disturbed with a metal spatula.

See 5-AMINOTETRAZOLE, CH_3N_5

See other DIAZONIUM SALTS

p-METHOXYBENZYL CHLORIDE C_8H_9ClO

1. Carroll, D. W., *Chem. Eng. News*, 1960, 38(34), 40

2. Ager, J. H. *et al., Chem. Eng. News*, **38**(43), 5

Two incidents involving explosion of bottles of chloride stored at room temperature are described [1,2]. Safe preparation with storage at 5°C is detailed [2].

2,4-DIMETHYLBENZENEDIAZONIUM TRIIODIDE $C_8H_9I_3N_2$

See DIAZONIUM TRIIODIDES

5(1,1-DINITROETHYL)-2-METHYLPYRIDINE $C_8H_9N_3O_4$

See NITRIC ACID, HNO_3: 5-Ethyl-2-methylpyridine
See other POLYNITROALKYL COMPOUNDS

1,3,5-CYCLOOCTATRIENE C_8H_{10}

See DIETHYL ACETYLENEDICARBOXYLATE, $C_8H_{10}O_4$: Cyclooctatriene
See also DIENES

6,6-DIMETHYLFULVENE C_8H_{10}

Air
Engler, C. *et al., Ber.*, 1901, **34**, 2935
Dimethylfulvene readily peroxidised, giving an insoluble (polymeric?) peroxide which exploded violently on heating to 130°C, and also caused ether used to triturate it to ignite.
See other POLYPEROXIDES

ETHYLBENZENE C_8H_{10}

Fl.P., 15°C; E.L., 1.2–6.8%

m-XYLENE C_8H_{10}

Fl.P., 25°C; E.L., 1.1–7.0%

o-XYLENE C_8H_{10}

Fl.P., 17°C; E.L., 1.0–6.0%

p-XYLENE C_8H_{10}

Fl.P., 25°C; E.L., 1.1–7.0%

Acetic acid,
Air
Shraer, B. I., *Khim. Prom.*, 1970, **46**(10), 747–750
In liquid phase aerobic oxidation of *p*-xylene in acetic acid to terephthalic acid, it is important to eliminate the inherent hazards of this fuel–air mixture. Effects of temperature, pressure and presence of steam on explosive limits of the mixture have been investigated.

1,3-Dichloro-5,5-dimethyl-2,4-imidazolidindione
See 1,3-DICHLORO-5,5-DIMETHYL-2,4-IMIDAZOLIDINDIONE, $C_5H_6Cl_2N_2O_2$: Xylene

Nitric acid
See NITRIC ACID, HNO_3 : Hydrocarbons

p-BROMO-*N*,*N*-DIMETHYLANILINE $C_8H_{10}BrN$

Anon., *Chem. Eng. News*, 1961, **39**(13), 37
During vacuum distillation, self-heating could not be controlled and proceeded to explosion. Internal dehydrohalogenation to a benzyne seems a possibility.

N,N-DIMETHYL-*p*-NITROSOANILINE $C_8H_{10}N_2O$

Acetic anhydride,
Acetic acid
Bain, P. J. S. *et al., Chem. Brit.*, 1971, 7, 81
An exothermic reaction, sufficiently violent to expel the flask contents, occurs after an induction period of 15–30 s when acetic anhydride is added to a solution of the nitroso compound in acetic acid. 4,4′-Azobis-(*N,N*-dimethylaniline), isolated from the reaction tar, may have been formed in a redox reaction, possibly involving an oxime derived from the nitroso compound.

DICROTONOYL PEROXIDE $C_8H_{10}O_4$

Guillet, J. E. *et al., Ger. Pat.* 1 131 407, 1962
This diacyl peroxide and its higher homologues, though of relatively high thermal stability, are very shock-sensitive. Replacement of the hydrogen atoms adjacent to the carbonyl groups with alkyl groups renders the group non-shock-sensitive.
See other DIACYL PEROXIDES

DIETHYL ACETYLENEDICARBOXYLATE $C_8H_{10}O_4$

1,3,5-Cyclooctatriene
Foote, C. S., private comm., 1965
A mixture of the reactants being heated at 60°C to effect the Diels–Alder addition exploded. Onset of this vigorously exothermic reaction was probably delayed by an induction period, and presence of a solvent and/or cooling would have moderated it.
See other ACETYLENIC COMPOUNDS

DIALLYL PEROXYDICARBONATE $C_8H_{10}O_6$

The distilled oil exploded on storage at ambient temperature.
See PEROXYCARBONATE ESTERS
See other ALLYL COMPOUNDS

BIS(3-CARBOXYPROPIONYL)PEROXIDE $C_8H_{10}O_8$

Lombard, R. *et al. Bull. Soc. Chim. Fr.*, 1963, **12**, 2800
Explodes on contact with flame. The commercial dry 95% material ('succinic acid peroxide') is highly hazard-rated.
See COMMERCIAL ORGANIC PEROXIDES
DIACYL PEROXIDES

2,4,6-TRIMETHYLPYRILIUM PERCHLORATE $C_8H_{11}ClO_5$

Hafner, K. *et al., Org. Synth.*, 1964, **44**, 102
The crystalline solid is an impact- and friction-sensitive explosive and must be handled with precautions. These include use of solvent-moist material and storage in a corked rather than glass-stoppered vessel.
See other NON-METAL PERCHLORATES

N,*N*-DIMETHYLANILINE $C_8H_{11}N$

Dibenzoyl peroxide
See DIBENZOYL PEROXIDE, $C_{14}H_{10}O_4$: *N*,*N*-Dimethylaniline

Diisopropyl peroxydicarbonate
See DIISOPROPYL PEROXYDICARBONATE, $C_8H_{14}O_6$

N-ETHYLANILINE $C_8H_{11}N$

Nitric acid
See NITRIC ACID, HNO_3 : Aromatic amines

5-ETHYL-2-METHYLPYRIDINE $C_8H_{11}N$

Nitric acid

See NITRIC ACID, HNO_3 : 5-Ethyl-2-methylpyridine

3,3-DIMETHYL-1-PHENYLTRIAZENE $C_8H_{11}N_3$

Baeyer, O. *et al., Ber.*, 1875, **8**, 149
Heusler, F., *Ann.*, 1890, **260**, 249
It decomposes explosively on attempted distillation at atmospheric pressure, but may be distilled uneventfully at reduced pressure.
See other TRIAZENES

4-VINYLCYCLOHEXENE C_8H_{12}

Fl.P., 16°C
See other PEROXIDISABLE COMPOUNDS

1-*p*-CHLOROPHENYLBIGUANIDIUM HYDROGENDICHROMATE $C_8H_{12}ClCr_2N_5O_7$

See DICHROMATE SALTS OF NITROGENOUS BASES
See other OXOSALTS OF NITROGENOUS BASES

tert-BUTYL-2-DIAZOACETOACETATE $C_8H_{12}N_2O_3$

Regitz, M. *et al., Org. Synth.*, 1968, **48**, 38
During low-temperature crystallisation, scratching to induce seeding of the solution must be discontinued as soon as the sensitive solid separates.
See other DIAZO COMPOUNDS

AZOISOBUTYRONITRILE $C_8H_{12}N_4$

Acetone

Carlisle, P. J., *Chem. Eng. News,* 1949, **27**, 150

During recrystallisation of technical material from acetone, explosive decomposition occurred. Non-explosive decomposition occurred when the nitrile was heated alone, or in presence of methanol.

See other AZO COMPOUNDS
CYANO COMPOUNDS

*O,O-tert-*BUTYL HYDROGEN MONOPEROXYMALEATE $C_8H_{12}O_5$

Castrantas, 1965, 17

Slightly shock-sensitive, the commercial dry 95% material is highly hazard-rated.

See other COMMERCIAL ORGANIC PEROXIDES
PEROXYESTERS

TETRAVINYLLEAD $C_8H_{12}Pb$

Holliday, A. K. *et al., Chem. & Ind*., 1968, 1699

In the preparation from lead (II) chloride and vinylmagnesium bromide in tetrahydrofuran–hexane, violent explosions occurred during isolation of the product by distillation of solvent. This could be avoided by a procedure involving steam distillation of the tetravinyllead, no significant loss of yield (80%) by hydrolysis being noted.

See other ALKYLMETALS

3-BUTEN-1-YNYLDIETHYLALUMINIUM $C_8H_{13}Al$

Petrov, A. A. *et al., Zh. Obsch. Khim*., 1962, **32**, 1349

Ignites in air.

See other METAL ACETYLIDES
TRIALKYLALUMINIUMS

1-PHENYLBIGUANIDIUM HYDROGENDICHROMATE

$C_8H_{13}Cr_2N_5O_7$

See DICHROMATE SALTS OF NITROGENOUS BASES
See other OXOSALTS OF NITROGENOUS BASES

1-DIETHYLAMINO-1-BUTEN-3-YNE $C_8H_{13}N$

Brandsma, 1971, 163
Dilution of the amine with white mineral oil is recommended before vacuum distillation, to avoid explosive decomposition of the residue.
See other ACETYLENIC COMPOUNDS

1,7-OCTADIENE C_8H_{14}

Fl.P., 25°C

1-OCTYNE C_8H_{14}

Fl.P., 16°C

2-,3- or 4-OCTYNE C_8H_{14}

Fl.Ps., all below 23°C

VINYL CYCLOHEXANE C_8H_{14}

Fl.P., 16°C

4-HYDROXY-4-METHYL-1,6-HEPTADIENE $C_8H_{14}O$

Ozone

1. Rabinowitz, J. L., *Biochem Prep.*, 1958, **6**, 25
2. Miller, F. W., *Chem. Eng. News*, 1973, **51**(6), 29

Following a published procedure [1] but using 75 g of diene instead of the 12.6 g specified, the ozonisation product from the diene exploded during vacuum desiccation. Eight previous preparations on the specified scale had been uneventful.

See other OZONIDES

DIISOBUTYRYL PEROXIDE $C_8H_{14}O_4$

Swern, 1971, Vol. 2, 815

Pure material explodes on standing at room temperature.

See other DIACYL PEROXIDES

ACETYL CYCLOHEXANESULPHONYL PEROXIDE $C_8H_{14}O_5S$

'Code of Practice for Storage of Organic Peroxides', GC14, Widnes, Interox Chemicals Ltd., 1970

While the commercial material damped with 30% water is not shock- or friction-sensitive, if it dries out it may become a high-hazard material. As a typical low-melting diacyl peroxide, it may be expected to decompose vigorously or explosively on slight heating or on mechanical initiation.

See other DIACYL PEROXIDES

DIISOPROPYL PEROXYDICARBONATE $C_8H_{14}O_6$

Alone,
or Amines,
or Potassium iodide

1. Strong, W. A., *Ind. Eng. Chem.*, 1964, **56**(12), 33
2. Strain, F., *J. Amer. Chem. Soc.*, 1950, **72**, 1254

3. 'Bulletin T. S. 350', Pittsburgh, Pittsburgh Plate Glass Co., 1963

When warmed slightly above its m.p. (10°C) the ester undergoes slow self-accelerating decomposition which may become dangerously violent under confinement [1]; at 25–30°C decomposition occurred in 10–30 min [2]. Addition of 1% of aniline, 1,2-diaminoethane or potassium iodide caused instant decomposition, and of *N,N*-dimethylaniline, instant explosion [2]. Bulk solutions of the ester (45%) in benzene-cyclohexane stored at 5°C developed sufficient heat to decompose explosively after a day, and 50–90% solutions were found to be impact-sensitive [1]. The solid is normally stored and transported at below –18°C in loose-topped trays [3].

See other DIACYL PEROXIDES
PEROXYCARBONATE ESTERS

4-HYDROXY-4-METHYL-1,6-HEPTADIENE DIOZONIDE $C_8H_{14}O_7$

See 4-HYDROXY-4-METHYL-1,6-HEPTADIENE, $C_8H_{14}O$: Ozone
See other OZONIDES

DI(2-METHOXYETHYL) PEROXYDICARBONATE $C_8H_{14}O_8$

Exploded at 34°C.
See other PEROXYCARBONATE ESTERS

***cis*-1,2-DIMETHYLCYCLOHEXANE** C_8H_{16}

Fl.P., 11°C

***trans*-1,2-DIMETHYLCYCLOHEXANE** C_8H_{16}

Fl.P., 7°C

1,3-DIMETHYLCYCLOHEXANE C_8H_{16}

Fl.P., 6°C

cis-**1,4-DIMETHYLCYCLOHEXANE** C_8H_{16}

Fl.P., 10°C

trans-**1,4-DIMETHYLCYCLOHEXANE** C_8H_{16}

Fl.P., 16°C

1-OCTENE C_8H_{16}

Fl.P., 8°C

2-OCTENE C_8H_{16}

Fl.P., 14°C

2,3,4-TRIMETHYL-1-PENTENE C_8H_{16}

Fl.P., below 21°C

2,4,4-TRIMETHYL-1-PENTENE C_8H_{16}

Fl.P., − 29°C

2,2,4-TRIMETHYL-2-PENTENE C_8H_{16}

Fl.P., 1°C

2,4,4-TRIMETHYL-2-PENTENE C_8H_{16}

Fl.P., 2°C

3,4,4-TRIMETHYL-2-PENTENE C_8H_{16}

Fl.P., below 21°C

3-AZONIABICYCLO[3.2.2]NONANE NITRATE $C_8H_{16}N_2O_3$

Violent decomposition occurred at 258°C.
See DIFFERENTIAL THERMAL ANALYSIS
See other OXOSALTS OF NITROGENOUS BASES

1,3,6,8-TETRAAZONIATRICYCLO [6.2.1.1^{3,6}] DODECANE $C_8H_{16}N_8O_{12}$

Violent decomposition occurred at 260°C.
See DIFFERENTIAL THERMAL ANALYSIS
See other OXOSALTS OF NITROGENOUS BASES

2-ETHYLHEXANAL $C_8H_{16}O$

A.I.T., 197°C

3,6-DIETHYL-3,6-DIMETHYL-1,2,4,5-TETRAOXANE $C_8H_{16}O_4$

Castrantas, 1965, 18
Swern, 1970, Vol. 1, 37
By analogy, this dimeric 2-butanone peroxide and the corresponding trimer probably contribute largely to the high shock-sensitivity of the commercial 'MEK peroxide' mixture, which contains these and other peroxides.
See other CYCLIC PEROXIDES
KETONE PEROXIDES

2,3-DIMETHYLHEXANE C_8H_{18}

Fl.P., 7°C (o)

2,4-DIMETHYLHEXANE C_8H_{18}

Fl.P., 10°C (o)

3-ETHYL-2-METHYLPENTANE C_8H_{18}

Fl.P., below 21°C

2-METHYLHEPTANE C_8H_{18}

Fl.P., below 10°C

3-METHYLHEPTANE C_8H_{18}

Fl.P., −23°C

OCTANE C_8H_{18}

Fl.P., 13°C; E.L., 1.0–4.7%; A.I.T., 220°C

2,2,3-TRIMETHYLPENTANE C_8H_{18}

Fl.P., below 21°C

2,2,4-TRIMETHYLPENTANE C_8H_{18}

Fl.P., −12°C; E.L., 1.1–6.0%

2,3,3-TRIMETHYLPENTANE C_8H_{18}

Fl.P., below 21°C

2,3,4-TRIMETHYLPENTANE C_8H_{18}

Fl.P., 4°C

DIISOBUTYLALUMINIUM CHLORIDE $C_8H_{18}AlCl$

See ALKYLALUMINIUM HALIDES

DI-*tert*-BUTYL CHROMATE $C_8H_{18}CrO_4$

ABCM Quart. Safety Summ., 1953, **24**, 2
A preparation by addition of *tert*-butanol to chromium trioxide on

large scale in full unstirred flask with poor cooling detonated owing to local overheating. Effective cooling and stirring essential.

DIBUTYL HYPONITRITE $C_8H_{18}N_2O_2$

See DIALKYL HYPONITRITES

DIBUTYL ETHER $C_8H_{18}O$

Fl.P., 25°C; E.L., 1.5–7.6%; A.I.T., 194°C

DI-*tert*-BUTYL PEROXIDE $C_8H_{18}O_2$

Fl.P., 18°C
Preparative hazard.
See HYDROGEN PEROXIDE, H_2O_2: *tert*-Butanol, etc.
See other DIALKYL PEROXIDES

DIISOBUTYRYL PEROXIDE $C_8H_{18}O_4$

1. Kharasch, S. *et al., J. Amer. Chem. Soc.*, 1941, **63**, 526
2. *MCA Case History No. 579*

A sample of the peroxide in ether (prepared according to a published procedure [1]) was being evaporated to dryness with a stream of air when it exploded violently. Handling the peroxide as a dilute solution at low temperature is recommended.
See other DIACYL PEROXIDES

DIISOBUTYLALUMINIUM HYDRIDE $C_8H_{19}Al$

1. Mirviss, S. B. *et al., Ind. Eng. Chem.*, 1961, **53**(1), 54A

2. 'Specialty Reducing Agents, Brochure TA-2002/1, New York, Texas Alkyls, 1971

The higher thermal stability of dialkylaluminium hydrides over the corresponding trialkylaluminiums is particularly marked in this case with 2-branched alkyl groups [1]. Used industrially as a powerful reducant, it is supplied as a solution in hydrocarbon solvents. The undiluted material ignites in air unless diluted to below 25% concentration.

See other ALKYLALUMINIUM ALKOXIDES AND HYDRIDES

DIBUTYLAMINE $C_8H_{19}N$

Cellulose nitrate

See CELLULOSE NITRATE: Amines

DI-*sec*-BUTYLAMINE $C_8H_{19}N$

Fl.P., 24°C

DIISOBUTYLAMINE $C_8H_{19}N$

Fl.P., 21°C

DIBUTYL HYDROGENPHOSPHITE $C_8H_{19}O_3P$

Air

Morrell, S. H., private comm., 1968

The phosphite was being distilled under reduced pressure. At the end of distillation the air bleed was opened more fully, when spontaneous combustion took place inside the flask. It was suggested that phosphine had been formed.

TETRAETHYLDIARSANE $C_8H_{20}As_2$

Sidgwick, 1950, 770
Inflames in air.
See other ALKYLNON-METALS

3,6,9-TRIAZA-11-AMINOUNDECANOL ('HYDROXYETHYLTRIETHYLENETETRAMINE') $C_8H_{22}N_4O$

Cellulose nitrate
See CELLULOSE NITRATE: Amines

1,11-DIAMINO-3,6,9-TRIAZAUNDECANE ('TETRAETHYLENEPENTAMINE') $C_8H_{23}N_5$

Carbon tetrachloride
See CARBON TETRACHLORIDE, CCl_4: 1,11-Diamino-3,6,9-triazaundecane

Cellulose nitrate
See CELLULOSE NITRATE: Amines

TETRAMETHYLAMMONIUM PENTAPEROXODICHROMATE(2–) $C_8H_{24}Cr_2N_2O_{12}$

Mellor, 1943, Vol. 11, 358
It explodes on moderate heating or in contact with sulphuric acid.
See other OXOSALTS OF NITROGENOUS BASES

1,2-DIAMINOETHANEBIS-TRIMETHYLGOLD $C_8H_{26}Au_2N_2$

Gilman, H. *et al., J. Amer. Chem. Soc.*, 1948, **70**, 550
The solid complex is very sensitive to light, and explodes violently

on heating in an open crucible. A drop of concentrated nitric acid added to the dry material causes detonation.
See other GOLD COMPOUNDS

BIS-DIETHYLENETRIAMINECOBALT(III) PERCHLORATE $C_8H_{26}Cl_3CoN_6O_{12}$

Explodes at 325°C; high impact-sensitivity.
See other AMMINEMETAL OXOSALTS

RUBIDIUM OCTACARBIDE C_8Rb

See RUBIDIUM, Rb : Non-metals

NONACARBONYLDIIRON $C_9Fe_2O_9$

ABCM Quart. Safety Summ., 1943, **14**, 18
Commercial iron carbonyl (fl.p., 35°C) has an auto-ignition temperature in contact with brass of 93°C, lower than that of carbon disulphide.
See other CARBONYLMETALS

2-NONEN-4,6,8-TRIYN-1-AL C_9H_4O

Bohlman, F. *et al., Chem. Ber.*, 1963, **96**, 2586
Extremely unstable, explodes after a few minutes at room temperature and ignites at 110°C.
See other ACETYLENIC COMPOUNDS

1-DIAZOINDENE $C_9H_6N_2$

Preparative hazard.
See *p*-TOLUENESULPHONYL AZIDE, $C_7H_7N_3O_2S$

2,4-DIISOCYANATOTOLUENE $C_9H_6N_2O_2$

Acyl chlorides,
or Bases
MCA SD-73, 1971
The diisocyanate may undergo exothermic polymerisation in contact with bases or more than traces of acyl chlorides, sometimes used as stabilisers.

1-IODO-3-PHENYL-2-PROPYNE C_9H_7I

Whiting, M. C., *Chem. Eng. News,* 1972, **50**(23), 86
It detonated on distillation at *ca.* 180°C.
See other HALOACETYLENE DERIVATIVES

QUINOLINE C_9H_7N

1. Blumann, A., *Proc. R. Aust. Chem. Inst.*, 1964, **31**, 286
2. *MCA Case History No. 1008*

The traditional unpredictably violent nature of the Skraup reaction (the preparation of quinoline and its derivatives by treating aniline, etc., with glycerol, sulphuric acid and an oxidant, usually nitrobenzene) is attributed to lack of stirring and adequate temperature control in many published descriptions [1].

A large-scale (450 litre) reaction, in which sulphuric acid was added to a stirred mixture of aniline, glycerol, nitrobenzene, ferrous sulphate and water, went out of control soon after the addition. A 150 mm rupture disc blew out first, followed by the manhole cover of the vessel. The violent reaction is attributed to doubling the scale of the reaction, an unusually high ambient temperature (reaction contents at 32°C) and the accidental addition of excess sulphuric acid. Experiment showed that a critical temperature of 120°C was reached immediately on addition of excess acid under these conditions [2].

Dinitrogen tetraoxide
See DINITROGEN TETRAOXIDE, N_2O_4: Heterocyclic bases

Linseed oil,
Thionyl chloride
See SULPHINYL CHLORIDE, Cl_2OS: Quinoline, etc.

Maleic anhydride
See MALEIC ANHYDRIDE, $C_4H_2O_3$: Cations, etc.

3,6-DIMETHYLBENZENEDIAZONIUM-2-CARBOXYLATE

$C_9H_8N_2O_2$

Hart, H. *et al., J. Org. Chem.*, 1972, **37**, 4272
The hydrochloride appears to be stable for considerable periods at ambient temperature, but explodes on melting at 88°C.
See BENZENEDIAZONIUM-2-CARBOXYLATE, $C_7H_4N_2O_2$
See other DIAZONIUM CARBOXYLATES

4,6-DIMETHYLBENZENEDIAZONIUM-2-CARBOXYLATE

$C_9H_8N_2O_2$

Gommper, R. *et al., Chem. Ber.*, 1968, **101**, 2348
It is a highly explosive solid.

BENZYLOXYACETYLENE C_9H_8O

Olsman, H. *et al., Rec. Trav. Chim.*, 1964, **83**, 305
If heated above 60°C during vacuum distillation, explosive rearrangement occurs.
See other ACETYLENIC COMPOUNDS

CINNAMMALDEHYDE C_9H_8O

Sodium hydroxide
Morrell, S. H., private comm., 1968
Rags soaked in sodium hydroxide and in the aldehyde overheated and ignited when they came into contact in a waste bin.
See other PEROXIDISABLE COMPOUNDS

1-PHENYL-1,2-PROPANEDIONE $C_9H_8O_2$

Fl.P., below 21°C

NITROINDANE $C_9H_9NO_2$

1. Lindner, J. *et al.*, *Ber.*, 1927, **60**, 435
2. Gribble, G. W., *Chem. Eng. News,* 1973, **51**(6), 30, 39

The crude mixture of 4- and 5-nitroindanes produced by mixed acid nitration of indane following a literature method [1] is hazardous to purify by distillation. The warm residue from distillation of 0.015 g mol a 80°C/1.3 mbar exploded on admission of air, and a 1.3 g mol batch exploded as distillation under the same conditions began. Removal of higher-boiling poly-nitrated material before distillation is recommended [2].
See other C-NITRO COMPOUNDS

3-PHENYLPROPIONYL AZIDE $C_9H_9N_3O$

Curtiss, T. *et al., J. Prakt. Chem.*, 1901, **64**, 297
A sample exploded on a hot water bath.
See other ACYL AZIDES

1,3,5-TRIS(NITROMETHYL)BENZENE $C_9H_9N_3O_6$

See NITRIC ACID, HNO_3 : Mesitylene

2-CHLORO-1-NITROSO-2-PHENYLPROPANE $C_9H_{10}ClNO$

MCA Case History No. 747
A sample of the air-dried material decomposed vigorously on keeping in a closed bottle at ambient temperature overnight.
See other NITROSO COMPOUNDS

3,5-DIMETHYLBENZOIC ACID $C_9H_{10}O_2$

Preparative hazard.
See NITRIC ACID, HNO_3: Hydrocarbons

ALLYL BENZENESULPHONATE $C_9H_{10}O_3S$

Dye, W. T. *et al., Chem. Eng. News*, 1950, **28**, 3452
The residue, from vacuum distillation at 92–135°C/2.6 mbar, darkened, thickened, then exploded after removal of the heat source.
For precautions:
See TRIALLYL PHOSPHATE, $C_9H_{15}O_4P_2$
See other ALLYL COMPOUNDS

NITROMESITYLENE $C_9H_{11}NO_2$

Preparative hazard.
See NITRIC ACID, HNO_3 : Hydrocarbons (references 7,8)

5(4-DIMETHYLAMINOBENZENEAZO) TETRAZOLE $C_9H_{11}N_7$

Thiele, J., *Ann.*, 1892, **270**, 54
Explodes at 155°C.
See other TETRAZOLES

1,2,3,4-TETRAHYDROISOQUINOLINIUM NITRATE $C_9H_{12}N_2O_3$
1,2,3,4-TETRAHYDROQUINOLINIUM NITRATE

Violent decomposition at 268 and 236°C, respectively.
See DIFFERENTIAL THERMAL ANALYSIS
See other OXOSALTS OF NITROGENOUS BASES

2(2-PHENYL)PROPYL HYDROPEROXIDE $C_9H_{12}O_2$

Leroux, A., *Mém. Poudres*, 1955, **37**, 49
Simon, A. H. *et al., Chem. Ber.*, 1957, **90**, 1024
The explosibility of this unusually stable compound 'cumene hydroperoxide', has been investigated. It is very difficult, but not impossible, to induce explosive decomposition.
See other ALKYL HYDROPEROXIDES

BENZYLDIMETHYLAMINE $C_9H_{13}N$

Cellulose nitrate
See CELLULOSE NITRATE: Amines

1,3,5-TRIMETHYLANILINE $C_9H_{13}N$

Nitrosyl perchlorate
See NITROSYL PERCHLORATE, $ClNO_5$: Organic materials

3,3,6,6-TETRAKIS(BROMOMETHYL)-9,9-DIMETHYL-1,2,4,5,7,8-HEXAOXAONANE $C_9H_{14}Br_4O_6$

Schulz, M. *et al., Chem. Ber.*, 1967, **100**, 2245
Explodes on impact or friction, as do the tetrachloro and 9-ethyl-9-methyl analogues.
See other CYCLIC PEROXIDES

2,6-DIMETHYL-2,5-HEPTADIEN-4-ONE DIOZONIDE $C_9H_{14}O_7$

Harries, G. *et al., Ann.*, 1910, **374**, 338
'Phorone' diozonide ignites at room temperature.
See other OZONIDES

N-CYANO-2-BROMOETHYLCYCLOHEXYLAMINE $C_9H_{15}BrN$

BCISC Quart. Safety Summ., 1964, **35**, 23
The reaction product from *N*-cyclohexylaziridine and cyanogen bromide (believed to be the title compound) exploded violently on attempted distillation at 160°C/0.5 mbar.
See also *N*-CYANO-2-BROMOETHYLBUTYLAMINE, $C_7H_{13}BrN$
See other CYANO COMPOUNDS

p-TOLYLBIGUANIDIUM HYDROGENDICHROMATE $C_9H_{15}Cr_2N_5O_7$

See DICHROMATE SALTS OF NITROGENOUS BASES
See other OXOSALTS OF NITROGENOUS BASES

TRIALLYL PHOSPHATE $C_9H_{15}O_4P$

1. Dye, W. T. *et al., Chem. Eng. News*, 1950, **28**, 3452
2. Steinberg, G. M., ibid., 3755

Alkali-washed material, stabilised with 0.25% of pyrogallol, was distilled at 103°C/4 mbar until slight decomposition began. The heating mantle was removed and the still-pot temperature had fallen below its maximum value of 135°C when the residue exploded violently [1]. The presence of solid alkali [2] or 5% of phenolic inhibitor is recommended, together with low-temperature high-vacuum distillation, to avoid formation of acidic decomposition products, which catalyse rapid polymerisation.
See other ALLYL COMPOUNDS

2,6-DIMETHYL-3-HEPTENE C_9H_{18}

Fl.P., 21°C (o)

1,3,5-TRIMETHYLCYCLOHEXANE C_9H_{18}

Fl.P., 19°C

DI-*tert*-BUTYL DIPEROXYCARBONATE $C_9H_{18}O_5$

A partially decomposed sample exploded violently at 135° C.
See PEROXYCARBONATE ESTERS

3,3,6,6,9,9-HEXAMETHYL-1,2,4,5,7,8-HEXAOXAONANE $C_9H_{18}O_6$

Dilthey, W. *et al., J. Prakt. Chem.*, 1940, **154**, 219
This trimeric acetone peroxide is powerfully explosive and will perforate a steel plate when heated on it.
See other CYCLIC PEROXIDES

3,6,9-TRIETHYL-1,2,4,5,7,8-HEXAOXAONANE $C_9H_{18}O_6$

Rieche, A. *et al., Ber.*, 1939, **72**, 1938
This trimeric 'propylidene peroxide' is an extremely explosive and friction-sensitive oil.
See other CYCLIC PEROXIDES

2,2,4,4,6,6-HEXAMETHYLTRITHIANE $C_9H_{18}S_3$

Nitric acid

See NITRIC ACID, HNO_3 : Hexamethyltrithiane

3,3-DIETHYLPENTANE C_9H_{20}

Fl.P., below 21°C; E.L., 0.7–7.7%

2,5-DIMETHYLHEPTANE C_9H_{20}

Fl.P., 24°C

3,5-DIMETHYLHEPTANE C_9H_{20}

Fl.P., 23°C

4,4-DIMETHYLHEPTANE C_9H_{20}

Fl.P., 21°C

3-ETHYL-2,3-DIMETHYLPENTANE C_9H_{20}

Fl.P., 8°C

3-ETHYL-4-METHYLHEXANE C_9H_{20}

Fl.P., 24°C

4-ETHYL-2-METHYLHEXANE C_9H_{20}

Fl.P., below 21°C

2-METHYLOCTANE C_9H_{20}

A.I.T., 220°C

3-METHYLOCTANE C_9H_{20}

A.I.T., 220°C

4-METHYLOCTANE C_9H_{20}

A.I.T., 225°C

NONANE C_9H_{20}

A.I.T., 190°C

2,2,5-TRIMETHYLHEXANE C_9H_{20}

Fl.P., 13°C

2,2,3,3-TETRAMETHYLPENTANE C_9H_{20}

Fl.P., below 21°C; E.L., 0.8–4.9%

2,2,3,4-TETRAMETHYLPENTANE C_9H_{20}

Fl.P., below 21°C

TRIISOPROPYLALUMINIUM $C_9H_{21}Al$

See TRIALKYLALUMINIUMS

TRIPROPYLALUMINIUM $C_9H_{21}Al$

See TRIALKYLALUMINIUMS

TRIISOPROPYLPHOSPHINE $C_9H_{21}P$

Chloroform,
or Oxidants
Catalogue entry, Frankfurt, Deutsche Advance Produktion, 1968
Particularly this phosphine reacts, when undiluted, rather vigorously with most peroxides, ozonides, *N*-oxides, and also chloroform. It may be safely destroyed by pouring into a solution of bromine in carbon tetrachloride.
See other ALKYLNON-METALS

6-QUINOLINECARBONYL AZIDE $C_{10}H_6N_4O$

Houben-Weyl, 1952, Vol. 8, 682

A sample heated above its m.p. (88°C) exploded violently.
See other ACYL AZIDES

DI-2-FUROYL PEROXIDE $C_{10}H_6O_6$

Castrantas, 1965, 17
Explodes violently on friction and heating.
See other DIACYL PEROXIDES

MERCURY 2-NAPHTHALENEDIAZONIUM TRICHLORIDE $C_{10}H_7Cl_3HgN_2$

Nesmeyanow, A. M., *Org. Synth.*, 1943, Coll. Vol. 2, 433
The isolated double salt explodes violently if heated during drying.
See other DIAZONIUM SALTS

1-NITRONAPHTHALENE $C_{10}H_7NO_2$

Tetranitromethane
See TETRANITROMETHANE, CN_4O_8: Aromatic nitrocompounds

1-NAPHTHALENEDIAZONIUM SALTS $C_{10}H_7N_2^+X^-$

Ammonium sulphide,
or Hydrogen sulphide
See DIAZONIUM SULPHIDES

2-NAPHTHALENEDIAZONIUM SALTS $C_{10}H_7N_2^+X^-$

Ammonium sulphide,
or Hydrogen sulphide,
or Sodium sulphides
See DIAZONIUM SULPHIDES

NAPHTHYLSODIUM $C_{10}H_7Na$

Chlorinated diphenyl
MCA Case History No. 565
To help extinguish a burning batch of naphthylsodium, a chlorinated diphenyl heat-transfer liquid was added. An exothermic reaction, followed by an explosion, occurred.

Sodium is known to react violently with many halogenated materials.

See SODIUM, Na: Halocarbons
See other ARYLMETALS

NAPHTHALENE $C_{10}H_8$

MCA SD-58, 1956
Dinitrogen pentaoxide
See DINITROGEN PENTAOXIDE, N_2O_5: Naphthalene

1-NAPHTHYLAMINE $C_{10}H_9N$

See 1-NAPHTHALENEDIAZONIUM SALTS, $C_{10}H_7N_2^+X^-$

2-NAPHTHYLAMINE $C_{10}H_9N$

See 2-NAPHTHALENEDIAZONIUM SALTS, $C_{10}H_7N_2^+X^-$

DIPYRIDINESILVER(I) PERCHLORATE $C_{10}H_{10}AgClN_2O_4$

Acids
Kauffman, G. B. *et al., Inorg. Synth.*, 1960, **6**, 7, 8
Contact with acids, especially hot, must be avoided to prevent the possibility of violent explosion.
See other AMMINEMETAL OXOSALTS

FERROCENIUM PERCHLORATE $C_{10}H_{10}ClFeO_4$

See 1,3-BIS(DI-η-CYCLOPENTADIENYLIRON)-2-PROPEN-1-ONE, $C_{23}H_{22}Fe_2O$: Perchloric acid, etc.

OXODIPEROXODIPYRIDINECHROMIUM(VI) $C_{10}H_{10}CrN_2O_5$

1. Caldwell, S. H. *et al., Inorg. Chem.*, 1969, **8**, 151
2. Adams, D. M. *et al., J. Chem. Educ.*, 1966, **43**, 94
3. Collins, J. C. *et al., Org. Synth.*, 1972, **52**, 5–8

This complex, formerly called 'pyridine perchromate' and now finding wide application as a powerful and selective oxidant, is violently explosive when dry [1]. Use while moist on the day of preparation and destroy any surplus with dilute alkali [2]. Preparation and use of the reagent have been detailed further [3]. The corresponding complexes of aniline, piperidine and quinoline may be similarly hazardous [2]. Dipyridinium dichromate is a much safer similarly powerful oxidant.

See also CHROMIUM TRIOXIDE, CrO_3: Pyridine
DIPYRIDINIUM DICHROMATE, $C_{10}H_{12}Cr_2NO_7$

See other AMMINECHROMIUM PEROXOCOMPLEXES

BIS(η-CYCLOPENTADIENYLDINITROSYLCHROMIUM) $C_{10}H_{10}Cr_2N_4O_4$

Flitcroft, N. *et al., Chem. & Ind.*, 1969, 201

A small sample exploded violently upon laser irradiation for Raman spectroscopy.

See other ORGANOMETALLICS

BIS(η-CYCLOPENTADIENYL)MAGNESIUM $C_{10}H_{10}Mg$

Barber, W. A., *Inorg. Synth.*, 1960, **6**, 15

It may ignite on exposure to air.

See other ORGANOMETALLICS

DIPYRIDINESODIUM $C_{10}H_{10}N_2Na$

Sidgwick, 1950, 89
The addition product of sodium and pyridine ignites in air.
See also DI(2-METHYLPYRIDINE)SODIUM, $C_{12}H_{14}N_2Na$

1-KETO-1,2,3,4-TETRAHYDRONAPHTHALENE $C_{10}H_{10}O$

Preparative hazards.
See CHROMIUM TRIOXIDE, CrO_3 : Acetic anhydride, etc.
HYDROGEN PEROXIDE, H_2O_2 : Acetone, etc.

DIMETHYL-PHENYLETHYNYLTHALLIUM $C_{10}H_{11}Tl$

Nast, R. *et al., J. Organomet. Chem.*, 1966, **6**, 461
May explode on heating, stirring or impact.
See other ALKYLMETALS
METAL ACETYLIDES

DICYCLOPENTADIENE $C_{10}H_{12}$

Fl.P., –7°C(o)

ETHYLPHENYLTHALLIC ACETATE PERCHLORATE $C_{10}H_{12}ClO_6Tl$

See PERCHLORIC ACID, $ClHO_4$: Ethylbenzene

DIPYRIDINIUM DICHROMATE $C_{10}H_{12}Cr_2N_2O_7$

Coates, W. M. *et al., Chem. & Ind.*, 1969, 1594

Though an oxidant of comparable power to 'pyridine perchromate', the dichromate is free of the explosive properties of the former.
See OXODIPEROXODIPYRIDINECHROMIUM(VI), $C_{10}H_{12}CrNO_5$
See other DICHROMATE SALTS OF NITROGENOUS BASES

3,4-DIMETHYL-4-(3,4-DIMETHYL-5-ISOXAZOLYAZO)-ISOXAZOLIN-5-ONE $C_{10}H_{12}N_4O_3$

Boulton, A. J. *et al., J. Chem. Soc.*, 1965, 5415
It invariably decomposed explosively if heated rapidly to 100°C but was stable to impact or friction.
See other AZO COMPOUNDS

1,2,3,4-TETRAHYDRO-1-NAPHTHYL HYDROPEROXIDE $C_{10}H_{12}O_2$

Hock, H. *et al., Ber.*, 1933, **66**, 61
Explodes on superheating the liquid.
See other ALKYL HYDROPEROXIDES

BIS(2,4-PENTANEDIONATO)CHROMIUM $C_{10}H_{14}CrO_4$

Ocone, L. R. *et al., Inorg. Synth.*, 1966, **8**, 131
It ignites in air.
See also CHROMIUM DIACETATE, $C_4H_6CrO_4$
See other ORGANOMETALLICS

2-METHYL-1-PHENYL-2-PROPYL HYDROPEROXIDE $C_{10}H_{14}O_2$

Preparative hazard.

See HYDROGEN PEROXIDE, H_2O_2 : 2-Methyl-1-phenyl-2-propanol, etc.

1,5-*p*-MENTHADIENE $C_{10}H_{16}$

Air

Bodendorf, K., *Arch. Pharm.*, 1933, **271**, 28

The terpene readily peroxidises with air, and the (polymeric) peroxidic residue exploded violently on attempted distillation at 100°C/0.4 mbar

See other POLYPEROXIDES

2-PINENE $C_{10}H_{16}$

Nitrosyl perchlorate

See NITROSYL PERCHLORATE, $ClNO_5$: Organic materials

SEBACOYL DICHLORIDE $C_{10}H_{16}Cl_2O_2$

Hüning, S. *et al.*, *Org. Synth.*, 1963, **43**, 37

During vacuum distillation of the chloride at 173°C/20 mbar, the residue frequently decomposes spontaneously, producing a voluminous black foam.

DISODIUM 1,3-DIHYDROXY-1,3-BIS-(*aci*-NITROMETHYL)-2,2,4,4-TETRAMETHYLCYCLOBUTANE $C_{10}H_{16}N_2Na_2O_6$

Dauben, H. J., Jr., *Org. Synth.*, 1963, Coll. Vol. 4, 223

The dry powdered condensation product of sodium *aci*-nitromethane (2 mol) with dimethylketene dimer exploded violently when added to crushed ice.

See SODIUM *aci*-NITROMETHANE, CH_2NNaO_2 : Water

See other *aci*-NITRO SALTS

1,4-EPIDIOXY-2-*p*-MENTHENE (ASCARIDOLE) $C_{10}H_{16}O_2$

Castrantas, 1965, 15
Explosive decomposition on heating from 130 to 150°C.
See other CYCLIC PEROXIDES

1-PYRROLIDINYLCYCLOHEXENE $C_{10}H_{17}N$

Benzenediazonium-2-carboxylate
See BENZENEDIAZONIUM-2-CARBOXYLATE, $C_7H_4N_2O_2$: 1-Pyrrolidinylcyclohexene

1,4-BUTANEDIOL DI-2,3-EPOXYPROPYL ETHER $C_{10}H_{18}O_4$

Trichloroethylene
See TRICHLOROETHYLENE, C_2HCl_3 : Epoxides

DI-2-METHYLBUTYRYL PEROXIDE $C_{10}H_{18}O_4$

Swern, 1971, Vol. 2, 815
Pure material explodes on standing at room temperature.
See other DIACYL PEROXIDES

DI-*tert*-BUTYL DIPEROXYOXALATE $C_{10}H_{18}O_6$

Castrantas, 1965, 17
Exploded on removing from a freezing mixture.

3-BUTEN-1-YNYLTRIETHYLLEAD $C_{10}H_{18}Pb$

Zavagorodnii, S. V. *et al., Dokl. Akad. Nauk SSSR*, 1962, **143**, 855 (Eng. transl. 268)
It explodes on rapid heating.
See other METAL ACETYLIDES

2-ETHYLHEXYL VINYL ETHER $C_{10}H_{20}O$

A.I.T., 202°C

ISOPENTYL ISOVALERATE $C_{10}H_{20}O_2$

Fl.P., 24°C

DECANE $C_{10}H_{22}$

A.I.T., 210°C

2-METHYLNONANE $C_{10}H_{22}$

A.I.T., 210°C

OXODIPEROXODIPIPERIDINECHROMIUM(VI) $C_{10}H_{22}CrN_2O_5$

See OXODIPEROXODIPYRIDINECHROMIUM(VI), $C_{10}H_{10}CrN_2O_5$

ETHOXYDIISOBUTYLALUMINIUM $C_{10}H_{23}AlO$

May ignite in air.
See other ALKYLALUMINIUM ALKOXIDES

1,4,8,11-TETRAAZACYCLOTETRADECANENICKEL(II) PERCHLORATE $C_{10}H_{24}Cl_2NiO_8$

Barefield, E. K., *Inorg. Chem.*, 1972, **11**, 2274
It exploded violently during analytical combustion.
See other AMMINEMETAL OXOSALTS

TETRAMETHYLBIS(TRIMETHYLSILOXO)DIGOLD $C_{10}H_{30}Au_2O_2Si_2$

Schmidbaur, H. *et al., Inorg. Chem.*, 1966, **5**, 2069
Sublimed crystals decomposed explosively at 120°C.
See other GOLD COMPOUNDS

1(2-NAPHTHYL)-3(5-TETRAZOLYL)TRIAZENE $C_{11}H_8N_7$

Thiele, J., *Ann.*, 1892, **270**, 54; 1893, **273**, 144
Explodes at 184°C.
See other TETRAZOLES

η-BENZENE-η-CYCLOPENTADIENYLIRON(II) PERCHLORATE $C_{11}H_{11}ClFeO_4$

Denning, R. G. *et al., J. Organomet. Chem.*, 1966, **5**, 292
The dry material is shock-sensitive and detonated on touching with a spatula.
See other ORGANOMETALLICS

3,3-DIMETHYL-1(3-QUINOLYL)TRIAZENE $C_{11}H_{12}N_4$

Rondestvedt, C. S. *et al., J. Org. Chem.*, 1957, **22**, 201
The crude material decomposes violently if allowed to dry, and purified material explodes at 131.5°C or during analytical combustion.
See other TRIAZENES

3-BUTYN-1-YL *p*-TOLUENESULPHONATE $C_{11}H_{12}O_3S$

Eglington, G. *et al., J. Chem. Soc.*, 1950, 3653
The material could be distilled in small amounts at below 0.01 mbar, but exploded on attempted distillation at 0.65 mbar.
See other ACETYLENIC COMPOUNDS

tert-BUTYL *p*-NITROPEROXYBENZOATE $C_{11}H_{13}NO_5$

Criegee, R. *et al., Ann.*, 1948, **560**, 135
This and *p*-nitrobenzoates of homologous *tert*-alkyl hydroperoxides explode in a flame.

tert-BUTYL PERBENZOATE $C_{11}H_{14}O_3$

Fl.P., 19°C

Criegee, R. *Angew. Chem.*, 1953, **65**, 398–399
Shortly after interruption of vacuum distillation from an oil bath at 115°C, to change a thermometer, the ester exploded violently. This was attributed to overheating.
See other PEROXYESTERS

2-*tert*-BUTYL-3-PHENYLOXAZIRANE $C_{11}H_{15}NO$

Emmons, W. D. *et al., Org. Synth.*, 1969, **49**, 13
Vacuum distillation of this active oxygen compound is potentially hazardous and precautions are necessary.
See other 1,2-EPOXIDES

N-PHENYLAZOPIPERIDINE $C_{11}H_{15}N_3$

Hydrofluoric acid
Wallach, O., *Ann.*, 1886, **235**, 258; 1888, **243**, 219
Interaction to give fluorobenzene is violent and is not suitable for above 10 g quantities
See other HIGH-NITROGEN COMPOUNDS

3-METHYL-3-BUTEN-1-YNYLTRIETHYLLEAD $C_{11}H_{20}Pb$

Zavgorodnii, S. V. *et al., Dokl. Akad. Nauk SSSR*, 1962, **143**, 855 (Eng. transl. 268)
It explodes on rapid heating.
See other METAL ACETYLIDES

3-DIBUTYLAMINOPROPYLAMINE $C_{11}H_{26}N_2$

Cellulose nitrate
See CELLULOSE NITRATE: Amines

BIS(PENTAFLUOROPHENYL)ALUMINIUM BROMIDE $C_{12}AlBrF_{10}$

1. Chambers, R. D. *et al., J. Chem. Soc.* (*C*), 1967, 2185; *Tetrahedron Lett.*, 1965, 2389
2. Cohen, S. C. *et al., Advan. Fluorine Chem.*, 1970, **6**, 156

Ignites in air, explodes during uncontrolled hydrolysis and chars during controlled hydrolysis [1]. When isolated as the etherate, attempts to remove solvent ether caused violent decompositions [2].
See other HALO-ARYLMETALS

DODECACARBONYLTETRACOBALT $C_{12}Co_4O_{12}$

See OCTACARBONYLDICOBALT, $C_8Co_2O_8$
See other CARBONYLMETALS

DODECACARBONYLTRIIRON $C_{12}Fe_3O_{12}$

King, R. B. *et al., Inorg. Synth.*, 1963, **7**, 195
On prolonged storage, pyrophoric decomposition products are formed.
See other CARBONYLMETALS

CALCIUM PICRATE $C_{12}H_4CaN_6O_{14}$
See PICRIC ACID, $C_6H_3N_3O_7$

BIS-2,4,5-TRICHLOROBENZENEDIAZO OXIDE $C_{12}H_4Cl_6N_4O$
Alone,
or Benzene
Kaufmann, T. *et al., Ann.*, 1960, **634**, 77
The dry solid explodes under a hammer blow, or on moistening with benzene.
See other BIS-ARENEDIAZO OXIDES

COPPER DIPICRATE $C_{12}H_4CuN_6O_{14}$

See PICRIC ACID, $C_6H_3N_3O_7$
See other HEAVY METAL DERIVATIVES

MERCURY DIPICRATE $C_{12}H_4HgN_6O_{14}$

See PICRIC ACID, $C_6H_3N_3O_7$
See other HEAVY METAL DERIVATIVES

LEAD DIPICRATE $C_{12}H_4N_6O_{14}Pb$

Belcher, R., *J. R. Inst. Chem.*, 1960, **84**, 377
During the usual qualitative inorganic analytical procedure, samples containing the lead and salicylate radicals can lead to the formation and possible detonation of lead dipicrate. This arises during evaporation of the filtrate with nitric acid, after precipitation of the copper–tin group metals with hydrogen sulphide. Salicylic acid is converted under these

conditions to picric acid, which, in presence of lead, gives explosive lead dipicrate.

An alternative (MAQA) scheme is described which avoids this possibility.

See PICRIC ACID, $C_6H_3N_3O_7$
See other HEAVY METAL DERIVATIVES

ZINC DIPICRATE **$C_{12}H_4N_6O_{14}Zn$**

See PICRIC ACID, $C_6H_3N_3O_7$
See other HEAVY METAL DERIVATIVES

1,3,5-TRIETHYNYLBENZENE **$C_{12}H_6$**

Shvartsberg, M. S. *et al., Izv. Akad. Nauk SSSR, Ser. Khim.*, 1963, **110**, 1836
Polymerised explosively on rapid heating and compression.
See other ACETYLENIC COMPOUNDS

2,6-DIPERCHLORYL-4,4′-DIPHENOQUINONE **$C_{12}H_6Cl_2O_8$**

Gardner, D. M. *et al., J. Org. Chem.*, 1963, **28**, 2650
A shock-sensitive explosive.
See other PERCHLORYL COMPOUNDS

POTASSIUM HEXAETHYNYLCOBALTATE(4–) **$C_{12}H_6CoK_4$**

Nast, R. *et al., Z. Anorg. Chem.*, 1955, **282**, 210
It is moderately stable at below –30°C, very shock- and friction-sensitive, and explodes violently on contact with water. At ambient temperature, it rapidly forms explosive decomposition products. Its addition compound with ammonia behaves similarly, exploding on contact with air.
See other ORGANOMETALLICS

2-IODO-3,5-DINITROBIPHENYL $C_{12}H_7IN_2O_4$

Sodium salt of ethyl acetoacetate
Zaheer, S. H. *et al., J. Indian Chem. Soc.*, 1955, **32**, 491
Interaction of 2-halo-3,5-dinitrobiphenyls with the sodium salt should be limited to 5–6 g of the title compound to avoid explosions observed with larger quantities.
See other HALOARYL COMPOUNDS

THIANTHRENIUM PERCHLORATE $C_{12}H_8ClO_4S_2$

Shine, H. J. *et al., Chem. & Ind.*, 1969, 782; *J. Org. Chem.*, 1971, **36**, 2925
A small portion (1–2 g) of the freshly prepared suction-dried material exploded violently during transfer from a sintered filter. Initiation may have been caused by friction of transfer or rubbing with a glass rod. Preparation of only 50–100 mg quantities is recommended.
See other NON-METAL PERCHLORATES

BIS-*p*-NITROBENZENEDIAZO SULPHIDE $C_{12}H_8Cl_2N_4O$

Alone,
or Benzene
Bamberger, E., *Ber.*, 1896, **29**, 464
More stable than unsubstituted analogues, it may be desiccated at 0°C, but is then extremely sensitive and violently explosive. Contact with benzene (even at 0°C) is violent and the reaction may become explosive.
See other BIS-ARENEDIAZO OXIDES

2-*trans*-1-AZIDO-1,2-DIHYDROACENAPHTHYL NITRATE $C_{12}H_8N_4O_3$

Trahanovsky, W. S. *et al., J. Amer. Chem. Soc.*, 1971, **93**, 5257
Although several other 1-azido-2-nitrato-alkanes appeared thermally stable, the acenaphthane derivative exploded violently on heating (probably during analytical combustion).
See other ALKYL NITRATES
ORGANIC AZIDES

BIS-*p*-NITROBENZENEDIAZO SULPHIDE $C_{12}H_8N_6O_4S$

Tomlinson, W. R., *Chem. Eng. News*, 1951, **29**, 5473
The dry material is extremely sensitive and can be exploded by very light friction. The material is too sensitive to handle other than as a solution, or dilute slurry in excess solvent, and then only on 1 g scale.
See DIAZONIUM SULPHIDES

DI(BENZENEDIAZONIUM) ZINC TETRACHLORIDE $C_{12}H_{10}Cl_4N_4Zn$

Muir, G. D., *Chem. & Ind.*, 1956, 58
A batch of the double salt exploded, either spontaneously or from slight vibration, after thorough drying under vacuum at ambient temperature overnight. Although dry diazonium salts are known to be light-, heat- and shock-sensitive when dry, the double salts with zinc chloride were considered to be more stable. Presence of traces of solvent reduces the risk of frictional heating and deterioration.
See other DIAZONIUM SALTS

FERROCENE-1,1-DICARBOXYLIC ACID $C_{12}H_{10}FeO_4$

Phosphoryl chloride
See PHOSPHORYL CHLORIDE, Cl_3OP: Ferrocene-1,1-dicarboxylic acid

DIPHENYLMERCURY $C_{12}H_{10}Hg$

Non-metal oxides
1. Dreher, E. *et al.*, *Ann.*, 1870, **154**, 127
2. Otto, R., *J. Prakt. Chem.* [2], 1870, **1**, 183

Chlorine monoxide reacts violently [1] and sulphur trioxide very violently [2] with diphenylmercury.
See other ARYLMETALS

DIPHENYLMAGNESIUM $C_{12}H_{10}Mg$

Air,
Water
Sidgwick, 1950, 234
It ignites in moist (but not dry) air, and reacts violently with water, reaching incandescence.
See other ARYLMETALS

BIS-BENZENEDIAZO OXIDE $C_{12}H_{10}N_4O$

Bamberger, E., *Ber.*, 1896, **29**, 460
Extremely unstable, it explodes on attempted isolation from the liquor, or on allowing the latter to warm to $-18°C$.
See other BIS-ARENEDIAZO OXIDES

DI(BENZENEDIAZO) SULPHIDE $C_{12}H_{10}N_4S$

Tomlinson, W. R., *Chem. Eng. News*, 1951, **29**, 5473
The wet solid can be exploded by impact or heating, and explodes while drying in air at ambient temperature. The material is too sensitive to handle other than as a solution or dilute slurry in excess solvent, and then only on 1 g scale.
See DIAZONIUM SULPHIDES

DIPHENYLSELENONE $C_{12}H_{10}O_2Se$

Krafft, F. *et al.*, *Ber.*, 1896, **29**, 424
Explodes feebly on heating in a test-tube.

DIBENZENESULPHONYL PEROXIDE $C_{12}H_{10}O_6S_2$

Davies, 1961, 65
Explodes at 53–54°C.
See other DIACYL PEROXIDES

DIPHENYLDISTIBENE (STIBOBENZENE) $C_{12}H_{10}Sb_2$

Air,
or Nitric acid
Schmidt, H., *Ann.*, 1920, **421**, 235
This antimony analogue of azobenzene ignites in air and is oxidised explosively by nitric acid.

DIPHENYLTIN $C_{12}H_{10}Sn$

Nitric acid
See NITRIC ACID, HNO_3: Diphenyltin

1,3-DIPHENYLTRIAZENE $C_{12}H_{11}N_3$

Müller, E. *et al.*, *Chem. Ber.*, 1962, **95**, 1257
It decomposes explosively at the m.p., 98°C.

Acetic anhydride
Heusler, F., *Ber.*, 1891, **24**, 4160
A mixture exploded with extraordinary violence on warming.
See other TRIAZENES

1,5-DIPHENYL-1,4-PENTAZDIENE $C_{12}H_{11}N_5$

Griess, P., *Ann.*, 1866, **137**, 81
The dry solid explodes violently on warming, impact or friction.
C-homologues behave similarly.
See other HIGH-NITROGEN COMPOUNDS

TETRAACRYLONITRILECOPPER(I) PERCHLORATE $C_{12}H_{12}ClCuN_4O_4$

Ondrejovic, G., *Chem. Zvesti*, 1964, **18**, 281
Decomposes explosively on heating.
See AMMINEMETAL OXOSALTS
See other CYANO COMPOUNDS

TETRAACRYLONITRILECOPPER(II) PERCHLORATE $C_{12}H_{12}Cl_2CuN_4O_8$

Ondrejovic, G., *Chem. Zvesti*, 1964, **18**, 281
Decomposes explosively on heating.
See AMMINEMETAL OXOSALTS
See other CYANO COMPOUNDS

BIS(η-BENZENE)CHROMIUM(0) $C_{12}H_{12}Cr$

Oxygen
Anon., *Chem. Eng. News*, 1964, **42**(38), 55
The orange-red complex formed with oxygen in benzene decomposes vigorously on friction or heating in air.
See other ORGANOMETALLICS

BIS(η-BENZENE)IRON(0) $C_{12}H_{12}Fe$

Timms, P. L., *Chem. Eng. News*, 1969, **47**(18), 43
Prepared in the vapour phase at low temperature, the solid explodes at −40°C.
See other ORGANOMETALLICS

1,2-DIPHENYLHYDRAZINE $C_{12}H_{12}N_2$

Perchloryl fluoride
See PERCHLORYL FLUORIDE, $ClFO_3$: Nitrogenous bases

DIANILINEOXODIPEROXOCHROMIUM(VI) $C_{12}H_{14}CrN_2O_5$

See OXODIPEROXODIPYRIDINECHROMIUM(VI), $C_{10}H_{10}CrN_2O_5$

BIS(2-METHYLPYRIDINE)SODIUM $C_{12}H_{14}N_2Na$

Sidgwick, 1950, 89
The addition product of sodium and 2-methylpyridine ignites in air.
See also DIPYRIDINESODIUM, $C_{10}H_{10}N_2Na$

DIANILINIUM DICHROMATE $C_{12}H_{16}Cr_2N_2O_7$

Gibson, G. M., *Chem. & Ind.*, 1966, 553
It is unstable on storage.
See other DICHROMATE SALTS OF NITROGENOUS BASES

3,6-DI(SPIROCYCLOHEXANE)TETRAOXANE $C_{12}H_{20}O_4$

Dilthey, W. *et al.*, *J. Prakt. Chem.*, 1940, **154**, 219
This dimeric cyclohexanone peroxide explodes on impact.
See other CYCLIC PEROXIDES

3-BUTEN-1-YNYLDIISOBUTYLALUMINIUM $C_{12}H_{21}Al$

Petrov, A. A. *et al.*, *Zh. Obsch. Khim.*, 1962, **32**, 1349
Ignites in air.
See other METAL ACETYLIDES
TRIALKYLALUMINIUMS

BIS(1-HYDROXYCYCLOHEXYL) PEROXIDE $C_{12}H_{22}O_4$

Stoll, M. *et al.*, *Helv. Chim. Acta*, 1930, **13**, 142
Normally stable, it explodes on attempted vacuum distillation.
See other 1-OXYPEROXY COMPOUNDS

DIHEXANOYL PEROXIDE $C_{12}H_{22}O_4$

Castrantas, 1965, 17
Explodes at 85°C.
See other DIACYL PEROXIDES

1(1′-HYDROPEROXY-1′-CYCLOHEXYLPEROXY)-CYCLOHEXANOL $C_{12}H_{22}O_5$

Davies, 1961, 74
This appears to be a main constituent of commercial 'cyclohexanone peroxide' together with the symmetrical bis-hydroxy peroxide (below), known to be hazardous.
See COMMERCIAL ORGANIC PEROXIDES
BIS(1-HYDROXYCYCLOHEXYL) PEROXIDE, $C_{12}H_{22}O_6$
See other 1-OXYPEROXY COMPOUNDS

BIS(1-HYDROPEROXYCYCLOHEXYL) PEROXIDE $C_{12}H_{22}O_6$

Criegee, R. *et al., Ann.*, 1949, **565**, 7
Explodes strongly in a flame.
See COMMERCIAL ORGANIC PEROXIDES
See other 1-OXYPEROXY COMPOUNDS

DI[TRIS-1,2-DIAMINOETHANECOBALT(III)] TRIPEROXODISULPHATE $C_{12}H_{24}Co_2N_{12}O_{24}S_6$

Beacom, S. E., *Nature*, 1959, **183**, 38
It explodes upon ignition, or after application of UV irradiation and heating to 120°C.
See other AMMINEMETAL OXOSALTS

DI[TRIS-1,2-DIAMINOETHANECHROMIUM(III)] TRIPEROXODISULPHATE $C_{12}H_{24}Cr_2N_{12}O_{24}S_6$

Beacom, S. E., *Nature*, 1959, **183**, 38
It explodes upon ignition, or after application of UV irradiation and then heating to 115°C.
See other AMMINEMETAL OXOSALTS

3,6,9-TRIETHYL-3,6,9-TRIMETHYL-1,2,4,5,7,8-HEXAOXAONANE $C_{12}H_{24}O_6$

See 3,6-DIETHYL-3,6-DIMETHYL-1,2,4,5-TETRAOXANE, $C_8H_{16}O_4$
See other CYCLIC PEROXIDES

DODECANE $C_{12}H_{26}$

A.I.T., 205°C

DIHEXYL ETHER $C_{12}H_{26}O$

A.I.T., 185°C

2,2-DI(*tert*-BUTYLPEROXY)BUTANE $C_{12}H_{26}O_4$

Dickey, F. H. *et al., Ind. Eng. Chem.*, 1949, **41**, 1673
The pure material explodes on heating to about 130°C, on sparking or on impact.
See other DIALKYLPEROXIDES

TRIISOBUTYLALUMINIUM $C_{12}H_{27}Al$

Fl.P., below 0°C; A.I.T., below 4°C
'Specialty Reducing Agents', Brochure TA-2002/1, New York, Texas Alkyls, 1971
Used industrially as a powerful reducant, it is supplied as a solution in

hydrocarbon solvents. The undiluted material is of relatively low thermal stability (decomposing above 50°C) and ignites in air unless diluted to below 25% concentration.
See other TRIALKYLALUMINIUMS

mixo-TRIBUTYLBORANE $C_{12}H_{27}B$

Hurd, D. T., *J. Amer. Chem. Soc.*, 1948, **70**, 2053
A mixture of the *n*- and iso- isomers ignited on exposure to air.
See ALKYLBORANES
See other ALKYLNON-METALS

TRIBUTYLBISMUTH $C_{12}H_{27}Bi$

Gilman, H. *et al.*, *J. Amer. Chem. Soc.*, 1939, **61**, 1170
It explodes violently in oxygen and ignites in air.
See other TRIALKYLBISMUTHS

TRIBUTYLPHOSPHINE $C_{12}H_{27}P$

A.I.T., 200°C
See other ALKYLNON-METALS

TITANIUM TETRAPROPOXIDE $C_{12}H_{28}O_4Ti$

Fl.P., below 22°C

LEAD(II) TRINITROSOPHLOROGLUCINOLATE $C_{12}N_6O_{12}Pb_3$

Freund, H. E., *Angew. Chem.*, 1961, **73**, 433
An air-dried sample exploded when disturbed, possibly owing to aerobic oxidation to the trinitro compound.
See other NITROSO COMPOUNDS

DODECACARBONYLDIVANADIUM $C_{12}O_{12}V_2$

Pruett, R. L. *et al., Chem. & Ind.*, 1960, 119
Ignites in air.
See other CARBONYLMETALS

1-BROMO-1,2-CYCLOTRIDECADIEN-4,8,10-TRIYNE $C_{13}H_9Br$

Leznoff, C. C. *et al., J. Amer. Chem. Soc.*, 1968, **90**, 731
It explodes at 65°C and slowly decomposes in the dark at 0°C.
See other HALOACETYLENE DERIVATIVES

NITROPHENYL-2-CARBOXYBENZENEDIAZOATE (4–) $C_{13}H_9N_3O_5$

Griess, P., *Ber.*, 1884, **17**, 338
It explodes on heating.
See other ARENEDIAZO ARYL OXIDES

DIPHENYLTHALLIUM FULMINATE $C_{13}H_{10}NOTl$

See DIMETHYLTHALLIUM FULMINATE, C_3H_6NOTl

α-BENZENEDIAZOBENZYL HYDROPEROXIDE $C_{13}H_{12}N_2O_2$

Busch, M. *et al., Ber.*, 1914, **47**, 3277
Swern, 1971, Vol. 2, 19
The phenylhydrazones of benzaldehyde and its homologues, or of acetone, are readily autoxidised in solution and rearrange to give the diazo-hydroperoxides, isolable as solids which may explode after a short time on standing, though not on friction or impact. Contact with a flame or with concentrated sulphuric or nitric acids also initiates explosion.
See other ALKYL HYDROPEROXIDES

DIBUTYL-3-METHYL-3-BUTEN-1-YNYLBORANE $C_{13}H_{23}B$

Davidsohn, W. E., *Chem. Rev.*, 1967, **67**, 75
It ignites in air.
See other ACETYLENIC COMPOUNDS
ALKYLNON-METALS

BIS-2,4-DICHLOROBENZOYL PEROXIDE $C_{14}H_6Cl_4O_4$

'Lucidol Data Sheet', Buffalo, Wallace and Tiernan, 1963
Whereas the pure compound is extremely shock-sensitive and decomposes rapidly at 80°C, the commercial 50% dispersion in plasticiser is not shock-sensitive.
See COMMERCIAL ORGANIC PEROXIDES
See other DIACYL PEROXIDES

CALCIUM BIS-*p*-IODYLBENZOATE $C_{14}H_8CaI_2O_8$

Unpublished information, 1948
Formulated granules accidentally dried to below normal moisture content exploded violently.
See other IODINE COMPOUNDS

BIS-*o*-AZIDOBENZOYL PEROXIDE $C_{14}H_8N_6O_4$

1. Leffler, J. E., *Chem. Eng. News*, 1963, **41**(48), 45
2. Hoffman, J., *Chem. Eng. News*, **41**(52), 5

A small sample of crystalline material on a sintered glass funnel detonated with extreme violence when touched with a metal spatula [1]. Static electrical initiation may have been involved [2].
See other DIACYL PEROXIDES

9,10-EPIDIOXYANTHRACENE $C_{14}H_8O_2$

Dufraisse, C. *et al., Compt. Rend.*, 1935, **201**, 428
Decomposes explosively at 120°C.
See other CYCLIC PEROXIDES

2,2-BIPHENYLDICARBONYL PEROXIDE $C_{14}H_8O_4$

Ramirez, F., *J. Amer. Chem. Soc.*, 1964, **86**, 4394
It explodes violently on heating to 70°C, or on impact, but can be preserved at low temperature.
See other DIACYL PEROXIDES

ANTHRACENE $C_{14}H_{10}$

Fluorine
See FLUORINE, F_2: Hydrocarbons

MERCURIC PEROXYBENZOATE $C_{14}H_{10}HgO_6$

Castrantas, 1965, 19
Explodes if heated above its normal decomposition temperature of 100–110°C.
See other METAL PEROXOACID SALTS

1,1-BENZOYLPHENYLDIAZOMETHANE $C_{14}H_{10}N_2O$

Nenitzescu, C. D. *et al., Org. Synth.*, 1943, Coll. Vol. 2, 497
The material may explode if heated to above 40°C.
See other DIAZO COMPOUNDS

DIBENZOYL PEROXIDE $C_{14}H_{10}O_4$

1. *MCA SD-81,* 1960
2. Lappin, G. R., *Chem. Eng. News,* 1948, **26**, 3518; Taub, D., *Chem. Eng. News,* 1949, **27**, 46
3. Nozaki, K. *et al., J. Amer. Chem. Soc.*, 1946, **68**, 1692

The dry material is readily ignited, burns very rapidly and is moderately sensitive to heat, shock, friction or contact with combustible materials. When heated above m.p. (103–105°C), instantaneous and explosive decomposition occurs without flame, but the decomposition products are flammable. If under confinement (or in large bulk), decomposition may be violently explosive [1]. An explosion which occurred when a screw-capped bottle of the peroxide was opened was attributed to friction initiating a mixture of peroxide and organic dust in the cap-threads. Waxed paper tubs are recommended to store this and other sensitive solids [2]. Crystallisation of dibenzoyl peroxide from hot chloroform solution involves a high risk of explosion. Precipitation from cold chloroform solution by methanol is safer [3]. Water- or plasticiser-containing pastes of dibenzoyl peroxide are much safer for industrial use.

Carbon tetrachloride,
Ethylene
Bolt, R. O. *et al., Chem. Eng. News*, 1947, **25**, 1866
Interaction of ethylene and carbon tetrachloride at elevated temperatures and pressures, initiated with benzoyl peroxide as radical source, caused violent explosions on several occasions. Precautions recommended include use of minimum pressure and quantity of initiator, maximum agitation, and presence of water as an inert moderator of high specific heat.
See also WAX FIRE

N,*N*-Dimethylaniline
Anon., *Angew. Chem.(Nachr.)*, 1954, **2**, 83
The solid peroxide exploded on contact with a drop of dimethylaniline.

Dimethyl sulphide
Pyror, W. A. *et al., J. Org. Chem.*, 1972, **37**, 2885
The rapid decomposition of benzoyl peroxide by dimethyl sulphide is explosive in absence of solvent.

Lithium tetrahydroaluminate
Sutton, D. A., *Chem. & Ind.*, 1951, 272
One of two attempts to reduce the diacyl peroxide in ether led to a moderately violent explosion.

Methyl methacrylate
MCA Case History No. 996
Local overheating and ignition occurred when solid benzoyl peroxide was put into a beaker which had been rinsed out with methyl methacrylate. Contact between the peroxide, a powerful oxidising agent and potential source of free radicals, and oxidisable or polymerisable materials should be under controlled conditions.
See other DIACYL PEROXIDES

N-*m*-TOLYL-*o*-NITROBENZIMIDYL CHLORIDE $C_{14}H_{11}ClN_2O_2$

Preparative hazard.
See PHOSPHORUS PENTACHLORIDE, Cl_5P: *o*-Nitrobenzoyl-*m*-toluidide

1,1-DIPHENYLETHYLENE $C_{14}H_{12}$

Oxygen
Staudinger, H., *Ber.*, 1925, **58**, 1075
Exposure of the alkene to oxygen at ambient temperature and pressure produces an alkene–oxygen interpolymeric peroxide, which explodes lightly on heating. An attempt to react the alkene with oxygen at 100 bar and 40–50°C caused a violent explosion in the autoclave.
See other POLYPEROXIDES

BIS-5-CHLOROTOLUENEDIAZONIUM ZINC TETRACHLORIDE

$C_{14}H_{12}Cl_6N_4Zn$

ABCM Quart. Safety Summ., 1953, **24**, 42
A batch containing only half the normal water content (60%) exploded violently during ball-milling. Tests later showed the dry material to be shock-sensitive.
See other DIAZONIUM SALTS

3-METHYL-2-NITROBENZANILIDE

$C_{14}H_{12}N_2O_3$

Phosphorus pentachloride
See PHOSPHORUS PENTACHLORIDE, Cl_5P: 3′-Methyl-2-nitrobenzanilide

2-AZOXYANISOLE

$C_{14}H_{14}N_2O_3$

Preparative hazard.
See 2-NITROANISOLE, $C_7H_6NO_3$: Sodium hydroxide, etc.

BIS-TOLUENEDIAZO OXIDE

$C_{14}H_{14}N_4O$

Alone,
or Toluene
Bamberger, E., *Ber.*, 1896, **29**, 452, 458
Extremely unstable, it explodes under its reaction liquor at above −4°C. Very shock- and friction-sensitive, a small sample exploded when dried on a porous tile and set off the moist material some distance away. Contact with toluene, even at −5°C, causes an explosive reaction with flame.
See other BIS(ARENEDIAZO) OXIDES

DIBENZYL ETHER $C_{14}H_{14}O$

Aluminium dichloride hydride

See ALUMINIUM DICHLORIDE HYDRIDE ETHERATE, $AlCl_2H \cdot C_4H_{10}O$

See other PEROXIDISABLE COMPOUNDS

DIBENZYL PHOSPHITE $C_{14}H_{15}O_3P$

Atherton, F. R. *et al., J. Chem. Soc.*, 1945, 382; 1948, 1106

It decomposes at 160°C, but prolonged heating at 120°C may have the same effect. Not more than 50 g should be distilled at once, using high-vacuum conditions (b.p., 100–120°C/0.001 mbar) unless a preliminary treatment to remove acidic impurities has been used.

α-PENTYLCINNAMALDEHYDE $C_{14}H_{18}O$

Anon., *Chem. Trade J.*, 1937, **100**, 362

This is very prone to spontaneous oxidative heating. A mixture with absorbent cotton reached 230°C 4 min after exposure to air.

See other PEROXIDISABLE COMPOUNDS

2,6-DI-*tert*-BUTYL-4-NITROPHENOL $C_{14}H_{21}NO_3$

1. *ASESB Expl. Incid. Report 1961,* **24**
2. Barnes, T. J. *et al., J. Chem. Soc.*, 1961, 953

A sample of the compound exploded violently after short heating to 100°C. Although this was attributed to presence of polynitro derivatives [1], the thermal decomposition of this type of nitro compound is known [2].

See other *C*-NITRO COMPOUNDS

DICYCLOHEXYLCARBONYL PEROXIDE $C_{14}H_{22}O_4$

Castrantas, 1965, 17
Larger quantities may explode without apparent reason.
See other DIACYL PEROXIDES

1-ACETOXY-1-HYDROPEROXY-6-CYCLODODECANONE $C_{14}H_{24}O_5$

Criegee, R. *et al., Ann.*, 1949, **564**, 9
Explodes on removal from a freezing mixture.
See other 1-OXYPEROXY COMPOUNDS

TETRADECANE $C_{14}H_{30}$

A.I.T., 200°C

BIS(TRIETHYLTIN)ACETYLENE $C_{14}H_{30}Sn_2$

Stannic chloride
Beerman, C. *et al., Z. Anorg. Chem.*, 1954, **276**, 20
The product of interaction is highly explosive.
See other METAL ACETYLIDES

HEPTAKIS(DIMETHYLAMINO)TRIALUMINIUM TRIBORON PENTAHYDRIDE $C_{14}H_{47}Al_3B_3N_7$

Hall, R. E. *et al., Inorg. Chem.*, 1969, **8**, 270
The crystalline solid is spontaneously flammable in air.
See other COMPLEX HYDRIDES

1,2,3,4-TETRACHLORO-7,7-DIMETHOXY-5-PHENYLBICYCLO-[2.2.1]-2,5-HEPTADIENE $C_{15}H_{12}Cl_4O_2$

Hoffmann, R. W. *et al., Tetrahedron*, 1965, **21**, 900
Pyrolysis of the material at 130°C under nitrogen at low pressure to give tetramethoxyethylene may be explosive if more than 25g is used.

TRIS(2,4-PENTANEDIONATO)MOLYBDENUM(III) $C_{15}H_{21}MoO_6$

Larson, M. L. *et al., Inorg. Synth.*, 1966, **8**, 153
Rapidly oxidises in air, sometimes igniting.
See other ORGANOMETALLICS

tert-BUTYL 1-ADAMANTANEPEROXYCARBOXYLATE $C_{15}H_{24}O_3$

Razuvajev, G. A. *et al., Tetrahedron*, 1969, **25**, 4925
Explodes on heating to 90–100°C.
See other PEROXYESTERS

TRIS(SPIROCYCLOPENTANE)-1,1,4,4,7,7-HEXAOXAONANE $C_{15}H_{24}O_6$

Bjorklund, G. H. *et al., Trans. R. Soc. Can.(Sect.III)*, 1950, **44**, 25
A violent explosive, very sensitive to shock, friction and rapid heating.
See HYDROGEN PEROXIDE, H_2O_2 : Ketones, etc.
See other CYCLIC PEROXIDES

TRIS-2,4,6(DIMETHYLAMINOMETHYL)PHENOL $C_{15}H_{27}N_3O$

Cellulose nitrate
See CELLULOSE NITRATE: Amines

HEXAETHYLTRIALUMINIUM TRITHIOCYANATE $C_{15}H_{30}Al_3N_3S_3$

Dehnicke, K., *Angew. Chem. (Intern. Ed.)*, 1967, **6**, 947
On heating at 210°C *in vacuo* it disproportionates explosively, but smoothly at 180°C.
See related ALKYLALUMINIUM HALIDES

mixo-DIMETHOXYDINITROANTHRAQUINONE $C_{16}H_{10}N_2O_8$

Sulphuric acid
Hildreth, J. D., *Chem. & Ind.*, 1970, 1592
During hydrolysis of crude dimethoxydinitroanthraquinone by heating in sulphuric acid, a runaway exothermic decomposition occurred causing vessel failure. Experiment showed a threshold decomposition temperature of 150–155°C, and oxidising effect of nitro groups, yielding CO and CO_2 above 162°C.
See other POLYNITROARYL COMPOUNDS

BIS(η-CYCLOOCTATETRANENE)URANIUM(0) $C_{16}H_{16}U$

Streitweiser, A. *et al.*, *J. Amer. Chem. Soc.*, 1968, **90**, 7364
Inflames in air.
See other ORGANOMETALLICS

DIBUTYL PHTHALATE $C_{16}H_{22}O_4$

Chlorine
See CHLORINE, Cl_2 : Dibutyl phthalate

DI-*tert*-BUTYL DIPEROXYPHTHALATE $C_{16}H_{22}O_6$

Castrantas, 1965, 17
Shock-sensitive.
See other PEROXYESTERS

BROMO-5,7,7,12,14,14-HEXAMETHYL-1,4,8,11-TETRAAZA-4,11- CYCLOTETRADECADIENEIRON(II) PERCHLORATE $C_{16}H_{32}BrClFeN_4O_4$

See [14] DIENE-N_4 IRON COMPLEXES

IODO-5,7,7,12,14,14-HEXAMETHYL-1,4,8,11-TETRAAZA-4,11-CYCLOTETRADECADIENEIRON(II) PERCHLORATE $C_{16}H_{32}ClFeIN_4O_4$

See [14] DIENE-N_4 IRON COMPLEXES

CHLORO-5,7,7,12,14,14-HEXAMETHYL-1,4,8,11-TETRAAZA-4,11-CYCLOTETRADECADIENEIRON(II) PERCHLORATE $C_{16}H_{32}Cl_2FeN_4O_4$

See [14] DIENE-N_4 IRON COMPLEXES

DICHLORO-5,7,7,12,14,14,HEXAMETHYL-1,4,8,11-TETRAAZA-4,11-CYCLOTETRADECADIENEIRON(III) PERCHLORATE $C_{16}H_{32}Cl_3FeN_4O_4$

See [14] DIENE-N_4 IRON COMPLEXES

HEXADECANE $C_{16}H_{34}$

A.I.T., 205°C

DIOCTYL ETHER $C_{16}H_{34}O$

A.I.T., 205°C

BENZANTHRONE $C_{17}H_{10}O$

Nitrobenzene,
Potassium hydroxide
See NITROBENZENE, $C_6H_5NO_2$: Potassium hydroxide, etc.

IRON(3+) HEXACYANOFERRATE(4−) $C_{18}Fe_7N_{18}$

Lead chromate
See LEAD CHROMATE, CrO_4Pb: Iron hexacyanoferrate
See other METAL CYANIDES

9-PHENYL-9-IODAFLUORENE $C_{18}H_{13}I$

Banks, D. F., *Chem. Rev.*, 1966, **66**, 248
It explodes at 105°C.
See other IODINE COMPOUNDS

TRIPHENYLALUMINIUM $C_{18}H_{15}Al$

Neely, T. A. *et al.*, *Org. Synth.*, 1965, **45**, 107
Triphenylaluminium and its etherate evolved heat and sparks on contact with water.
See other ARYLMETALS

1,3-BIS(PHENYLTRIAZENO)BENZENE $C_{18}H_{16}N_6$

Kleinfeller, H., *J. Prakt. Chem.* [2], 1928, **119**, 61
It explodes if rapidly heated.
See other TRIAZENES

TRIPHENYLTIN HYDROPEROXIDE $C_{18}H_{16}O_2Sn$

Dannley, R. L. *et al., J. Org. Chem.*, 1965, **30**, 3845
It explodes reproducibly at 75°C.
See ORGANOMINERAL PEROXIDES

2,2,4-TRIMETHYLDECAHYDROQUINOLINE PICRATE $C_{18}H_{26}N_4O_7$

2-(2-Butoxyethoxy)ethanol
Franklin, N. C., private comm., 1967
Evaporation of a solution of the picrate in the diether caused a violent explosion. The solvent had probably peroxidised during open storage and residual mixture of peroxide and picrate had exploded during evaporation. Use of peroxide-free solvent and lower evaporation temperature appeared to be safe.
See other POLYNITROARYL COMPOUNDS

1,4-OCTADECANOLACTONE $C_{18}H_{34}O_2$

Preparative hazard.
See PERCHLORIC ACID, $ClHO_4$: Oleic acid

BIS(DIBUTYLBORINO)ACETYLENE $C_{18}H_{36}B_2$

Hartmann, H. *et al., Z. Anorg. Chem.*, 1959, **299**, 174
Ignites in air.
See other ACETYLENIC COMPOUNDS
ALKYLNON-METALS

OLEIC ACID $C_{18}H_{36}O_2$

Aluminium

See ALUMINIUM, Al: Oleic acid

Perchloric acid

See PERCHLORIC ACID, $ClHO_4$: Oleic acid

HEXAMETHYLENETETRAMMONIUM TETRAPEROXO-CHROMATE(V) (?) $C_{18}H_{48}Cr_4N_{12}O_{32}$

House, D. A. *et al., Inorg. Chem.*, 1966, **6**, 1078, footnote 8

Material recrystallised from water, and washed with methanol to dry by suction, ignited and then exploded on the filter funnel.

2,7-DINITRO-9-PHENYL PHENANTHRIDINE $C_{19}H_{11}N_3O_4$

Diethyl sulphate

See DIETHYL SULPHATE, $C_4H_{10}O_4S$: 2,7-Dinitro-9-phenylphentanthridine

TRIPHENYLMETHYL NITRATE $C_{19}H_{15}NO_3$

Lewis acids

See ALKYL NITRATES

p-CHLORO(BIS-*p*NITROBENZOYLDIOXYIODO)BENZENE $C_{20}H_{12}ClIN_2O_{10}$

See (DIBENZOYLDIOXYIODO)BENZENES

p-CHLORO(BIS-*m*-CHLOROBENZOYLDIOXYIODO)BENZENE $C_{20}H_{12}Cl_3IO_6$

See (DIBENZOYLDIOXYIODO)BENZENES

(BIS-*m*-CHLOROBENZOYLDIOXYIODO)BENZENE $C_{20}H_{13}Cl_2IO_6$

See (DIBENZOYLDIOXYIODO)BENZENES

(BIS-*p*-NITROBENZOYLDIOXYIODO)BENZENE $C_{20}H_{13}IN_2O_{10}$

See (DIBENZOYLDIOXYIODO)BENZENES

OXODIPEROXODIQUINOLINECHROMIUM(VI) $C_{20}H_{14}CrN_2O_5$

See OXODIPEROXODIPYRIDINECHROMIUM(VI), $C_{10}H_{10}CrN_2O_5$

1,3-DIPHENYL-1,3-EPIDIOXY-1,3-DIHYDROISOBENZOFURAN $C_{20}H_{14}O_3$

Dufraisse, C. *et al., Compt. Rend.*, 1946, **223**, 735
Formally an ozonide, this photo-peroxide explodes at 18°C.
See other OZONIDES

1,1-BIS(*p*-NITROBENZOYLPEROXY)CYCLOHEXANE $C_{20}H_{18}N_2O_{10}$

Criegee, R. *et al., Ann.*, 1948, **560**, 135
Explodes at 120°C.
See other PEROXYESTERS

1,1-BIS(BENZOYLPEROXY)CYCLOHEXANE $C_{20}H_{20}O_6$

Criegee, R. *et al., Ann.*, 1948, **560**, 135
Explodes sharply in a flame.
See other PEROXYESTERS

SODIUM ABIETATE $C_{20}H_{29}NaO_2$

See METAL ABIETATES

1,1,6,6-TETRAKIS(ACETYLPEROXY)CYCLODODECANE $C_{20}H_{32}O_{12}$

Criegee, R. *et al., Ann.*, 1948, **560**, 135
Weak friction causes strong explosion.
See other PEROXYESTERS

DIACETONITRILE-5,7,7,12,14,14-HEXAMETHYL-1,4,8,11-TETRA-AZA-4,11-CYCLOTETRADECADIENEIRON(II) PERCHLORATE $C_{20}H_{36}Cl_2FeN_6O_8$

See [14] DIENE-N_4 IRON COMPLEXES

TETRACYANOOCTAETHYLTETRAGOLD $C_{20}H_{40}Au_4N_4$

Burawoy, A. *et al., J. Chem. Soc.*, 1935, 1026
This tetramer of cyanodiethylgold is friction-sensitive and also decomposes explosively above 80°C. The propyl homologue is not friction-sensitive, but also decomposes explosively on heating in bulk.
See other CYANO COMPOUNDS
GOLD COMPOUNDS

o-METHOXY(BIS-*p*-NITROBENZOYLDIOXYIODO)BENZENE $C_{21}H_{15}IN_2O_{11}$

See (DIBENZOYLDIOXYIODO)BENZENES

TRIBENZYLARSINE $C_{21}H_{21}As$

Dondorov, J. *et al.*, *Ber.*, 1935, **68**, 1255
The pure material oxidises slowly at first in air at ambient temperature, but reaction becomes violent through autocatalysis.
See other ALKYLNON-METALS

TRI-*p*-TOLYLAMMONIUM PERCHLORATE $C_{21}H_{22}ClNO_4$

Weitz, E. *et al.*, *Ber.*, 1926, **59**, 2307
Explodes violently when heated above its m.p., 123°C.
See other OXOSALTS OF NITROGENOUS BASES

2,2-BIS[4(2′,3′-EPOXYPROPOXY)PHENYL]PROPANE $C_{21}H_{24}O_4$

Trichloroethylene
See TRICHLOROETHYLENE, C_2HCl_3: Epoxides

ACETONITRILEIMIDAZOLE-5,7,7,12,14,14-HEXAMETHYL-1,4,8,11-TETRAAZA-4,11-CYCLOTETRADECADIENEIRON(II) PERCHLORATE $C_{21}H_{39}Cl_2FeN_7O_8$

See [14]DIENE-N_4IRON COMPLEXES

BIS(CYCLOPENTADIENYL)BIS(PENTAFLUOROPHENYL)ZIRCONIUM $C_{22}H_{10}F_{10}Zr$

Chaudhari, M. A. *et al.*, *J. Chem. Soc. (A)*, 1966, 838
Explodes in air (but not nitrogen) above its m.p., 219°C.
See other HALO-ARYLMETALS

DI-1-NAPHTHOYL PEROXIDE $C_{22}H_{14}O_4$

Castrantas, 1965, 17
Explodes on friction.
See other DIACYL PEROXIDES

4-[2-(4-HYDRAZINO-1-PHTHALAZINYL)HYDRAZINO]-4-METHYL-2-PENTANONE (4-HYDRAZINO-1-PHTHALAZINYL) HYDRAZONE DINICKEL(II) TETRAPERCHLORATE

$C_{22}H_{28}N_{12}Ni_2Cl_4O_{16}$

Rosen, W., *Inorg. Chem.*, 1971, **10**, 1833
The green complex isolated directly from the reaction mixture explodes fairly violently at elevated temperatures. The tetrahydrate produced by recrystallisation from aqueous methanol is also moderately explosive when heated rapidly.
See other AMMINEMETAL OXOSALTS

1,3-BIS(DI-η-CYCLOPENTADIENYLIRON)-2-PROPEN-1-ONE

$C_{23}H_{20}Fe_2O$

Perchloric acid,
Acetic anhydride,
Ether,
Methanol
Anon., *Chem. Eng. News,* 1966, **44**(49), 50
Condensation of the iron complex with cyclopentanone in perchloric acid–acetic anhydride–ether medium had been attempted. The non-crystalline residue, after methanol-washing and drying in air for several weeks, exploded on being disturbed. This was attributed to possible presence of a derivative of ferrocenium perchlorate, a powerful explosive and detonator. However, methyl or ethyl perchlorates may have been involved.
See ALKYL PERCHLORATES
PERCHLORIC ACID, $ClHO_4$: Diethyl ether
See other ORGANOMETALLICS

1,3,6,8-TETRAPHENYLOCTAZATRIENE $C_{24}H_{20}N_8$

Wohl, A. *et al., Ber.*, 1900, **33**, 2741
This compound (and several *C*-homologues) is unstable and explodes sharply on heating, impact or friction. In a sealed tube, the explosion is violent.
See other HIGH-NITROGEN COMPOUNDS

BIS(DI-η-BENZENECHROMIUM(IV) DICHROMATE $C_{24}H_{24}Cr_4O_7$

Anon., *Chem. Eng. News*, 1964, **42**(38), 55
This catalyst exists as explosive orange-red crystals.
See other ORGANOMETALLICS

5,7,7,12,14,14-HEXAMETHYL-1,4,8,11-TETRAAZA-4,11-CYCLO-TETRADECADIENE-1,10-PHENANTHROLINEIRON(II) PERCHLORATE $C_{28}H_{40}Cl_2FeN_6O_8$

See [14] DIENE-N_4 IRON COMPLEXES

TRIS-2,2′-BIPYRIDINESILVER(II) PERCHLORATE $C_{30}H_{24}AgCl_2N_6O_8$

Morgan, G. T. *et al., J. Chem. Soc.*, 1930, 2594
Explodes on heating.
See other AMMINEMETAL OXOSALTS

TRIS-2,2′-BIPYRIDINECHROMIUM(II) PERCHLORATE $C_{30}H_{24}Cl_2CrN_6O_8$

Holah, D. G. *et al., Inorg. Synth.*, 1967, **10**, 34
Explodes violently on slow heating to 250°C and can be initiated by static sparks, but not apparently by impact.
See other AMMINEMETAL OXOSALTS

TRIS-2,2′-BIPYRIDINECHROMIUM(0) $C_{30}H_{24}CrN_6$

Herzog, S. *et al., Z. Naturforsch.*, 1957, **12**, 809
Ignites in air.
See related AMMINEMETAL OXOSALTS

(5,7,7,12,14,14-HEXAMETHYL-1,4,8,11-TETRAAZA-4,11-CYCLO-BISTETRADECADIENE) HYDROXODIIRON(II) TRIPER-CHLORATE $C_{32}H_{65}Cl_3Fe_2N_8O_{13}$

See [14] DIENE-N_4 IRON COMPLEXES

BIS[AQUA-5,7,7,12,14,14-HEXAMETHYL-1,4,8,11-TETRAAZA-4,11-CYCLOTETRADECADIENEIRON(II)] OXIDE TETRAPER-CHLORATE $C_{32}H_{68}Cl_4Fe_2N_8O_{19}$

See [14] DIENE-N_4 IRON COMPLEXES

CALCIUM ABIETATE $C_{40}H_{58}CaO_4$

See METAL ABIETATES

MANGANESE ABIETATE $C_{40}H_{58}MnO_4$

See METAL ABIETATES

LEAD ABIETATE $C_{40}H_{58}O_4P_4$

See METAL ABIETATES

ZINC ABIETATE $C_{40}H_{58}O_4Zn$

See METAL ABIETATES

HEXAPYRIDINEIRON(II) TRIDECACARBONYLTETRAFERRATE(2–) $C_{43}H_{30}Fe_5N_6O_{13}$

Brauer, 1965, Vol. 2, 1758
Extremely pyrophoric.
See other AMMINEMETAL OXOSALTS

OXYBIS(*N,N*-DIMETHYLACETAMIDETRIPHENYLSTIBONIUM) DIPERCHLORATE $C_{44}H_{50}Cl_2N_2O_{11}Sb$

Goel, R. G. *et al., Inorg. Chem.*, 1972, **11**, 2143
It exploded on several occasions during handling and attempted analysis.
See other AMMINEMETAL OXOSALTS

ALUMINIUM ABIETATE $C_{60}H_{87}AlO_6$

See METAL ABIETATES

CALCIUM Ca

Air
'Product Information Sheet No. 212', Sandwich, Pfizer Chemicals, 1969
Calcium is pyrophoric when finely divided.

Asbestos cement
Scott, P. J., *School Sci. Rev.,* 1967, **49**(167), 252
Drops of molten calcium falling on to hard asbestos cement sheeting caused a violent explosion which perforated the sheet. Interaction with sorbed water in the cement seems likely to have occurred.
See Water, below

Halogens
Mellor, 1941, Vol. 3, 638
Massive calcium ignites in fluorine at ambient temperature, and finely divided (but not massive) calcium ignites in chlorine similarly.

Lead dichloride
Mellor, 1941, Vol. 3, 639
Interaction is explosive on warming.

Phosphorus(V) oxide
See PHOSPHORUS(V) OXIDE, O_5P_2 : Metals

Silicon
Mellor, 1940, Vol. 6, 176–177
Interaction is violently incandescent above 1050°C after a short delay.

Sodium,
Mixed oxides
BCISC Quart. Safety Summ., 1966, **37**(145), 6
An operator working above the charging hole of a sludge reactor was severely burned when a quantity of a burning sludge containing calcium and sodium metals and their oxides was ejected. This very reactive mixture is believed to have been ignited by drops of perspiration falling from the operator.
See Water, below

Sulphur
Mellor, 1941, Vol. 3, 639
A mixture reacts explosively when ignited.

Water
'Product Information Sheet No. 212', Sandwich, Pfizer Chemicals, 1969
Calcium or its alloys react violently with water (or dilute acids) and the heat of reaction may ignite evolved hydrogen under appropriate contact conditions.
See Asbestos cement, above
See other METALS

CALCIUM CHLORIDE $CaCl_2$

Methyl vinyl ether
See METHYL VINYL ETHER, C_3H_6O: Acids

Water
MCA Case History No. 69
The exotherm produced by adding solid calcium chloride to hot water caused violent boiling.

Zinc
ABCM Quart. Safety Summ., 1932, **3**, 35
Prolonged action of calcium chloride solution upon the zinc coating of galvanised iron caused slow evolution of hydrogen, which became ignited and exploded.
See other METAL HALIDES

CALCIUM HYPOCHLORITE $CaCl_2O_2$

Sidgwick, 1950, 1217
This powerful oxidant is technically of great importance for bleaching and sterilisation applications, and contact with reducants or combustible materials must be under controlled conditions. It is present in diluted state in bleaching powder, which is a less powerful oxidant with lower available chlorine content.
See BLEACHING POWDER

Ammonium chloride
1. Morris, D. L., *The Science Teacher,* 1968, **35**(6), 4
2. Anderson, M. B., *The Science Teacher,* 1968, **35**(9), 4

Report of an explosion through unintentional use of calcium hypochlorite instead of calcium hydroxide in the preparation of ammonia gas. Nitrogen trichloride was produced [1,2].
See Nitrogenous bases, below

Carbon
Mellor, 1941, Vol. 2, 262

A confined intimate mixture of hypochlorite and finely divided charcoal exploded on heating.

Contaminants

MCA Case History No. 666

The contents of a drum erupted and ignited during intermittent use. This was attributed to contamination of the soldered metal scoop (normally kept in the drum) by oil, grease or water, or all three, and a subsequent exothermic reaction with the hypochlorite.

See Organic matter, below

N,N-Dichloromethylamine

See *N,N*-DICHLOROMETHYLAMINE, CH_3Cl_2N

Hydroxy compounds

Fawcett, H. H., *Ind. Eng. Chem.*, 1959, **51**(4), 90A

Contact of the solid hypochlorite with glycerol, digol monomethyl ether or phenol causes ignition within a few minutes, accompanied by irritant smoke, particularly with phenol. Ethanol may cause an explosion, as may methanol.

Nitrogenous bases

Kirk-Othmer, 1963, Vol. 2, 105

Primary aliphatic or aromatic amines react with calcium (or sodium) hypochlorite to form *N*-mono- or di-chloro- amines which are explosive, but less so than nitrogen trichloride.

See Ammonium chloride, above

Nitromethane

Fawcett, H. H., *Trans. Nat. Safety Cong. Chem. Fertilizer Ind.*, Vol. 5, 32, Chicago, NSC, 1963

They interact, after a delay, with extreme vigour.

Organic matter

'Halane' Information Sheet, Wyandotte Chem. Co., Michigan, 1958

Mixtures of the solid hypochlorite with 1% of admixed organic contaminants are sensitive to heat in varying degree. Wood caused ignition at 176°C, while oil caused violent explosion at 135°C.

See Contaminants, above

Sodium hydrogensulphate,
Starch,
Sodium carbonate
Anon., *Ind. Eng. Chem. (News Ed.)*, 1937, **15**, 282
Shortly after a mixture of the four ingredients had been compressed into tablets, incandescence and an explosion occurred. This may have been due to interaction of the hypochlorite and starch, accelerated by the acid sulphate, and may also have involved dichlorine monoxide.

Sulphur
Katz, S. A. *et al.*, *Chem. Eng. News*, 1965, **46**(29), 6
Admixture of damp sulphur and solid 'swimming pool chlorine' caused a violently exothermic reaction, and ejection of molten sulphur.
See other METAL OXOHALOGENATES

CALCIUM DIHYDRIDE CaH_2

Metal halogenates
Mellor, 1946, Vol. 3, 651
Mixtures of the hydride with various bromates, chlorates or perchlorates explode on grinding.
See METAL HALOGENATES

Silver fluoride
See SILVER FLUORIDE, AgF: Calcium dihydride

Tetrahydrofuran
See LITHIUM TETRAHYDROALUMINATE, AlH_4Li: Tetrahydrofuran

CALCIUM BIS-HYDROXYLAMIDE $CaH_4N_2O_2$

See HYDROXYLAMINE, H_3NO: Metals

CALCIUM PERMANGANATE $CaMn_2O_8$

Acetic acid,
or Acetic anhydride
See POTASSIUM PERMANGANATE, $KMnO_4$: Acetic acid

Hydrogen peroxide
See HYDROGEN PEROXIDE, H_2O_2 : Metals, etc.

CALCIUM DIAZIDE CaN_6

Mellor, 1940, Vol. 8, 349
Calcium, strontium and barium diazides are not shock-sensitive, but explode on heating at about 150, 170 and 225 (or 152)°C, respectively. In sealed tubes the explosion temperatures are greater.
See other METAL AZIDES

CALCIUM OXIDE CaO

Hydrogen fluoride
See HYDROGEN FLUORIDE, FH: Oxides

Interhalogens
See BROMINE PENTAFLUORIDE, BrF_5 : Acids, etc.
CHLORINE TRIFLUORIDE, ClF_3 : Metals, etc.

Phosphorus(V) oxide
See PHOSPHORUS(V) OXIDE, O_5P_2 : Inorganic bases

Water
1. Anon., *Fire*, 1935, **28**, 30
2. Amos, T., *Zentr. Zuckerind.*, 1923, **32**, 103
3. *BCISC Quart. Safety Summ.*, 1967, **38**, 15
4. *BCISC Quart. Safety Summ.*, 1971, **42**(168), 4

Quickline, when mixed with $\frac{1}{3}$ of its weight of water, will reach 150–

300°C (depending on quantity) and may ignite combustible material. Occasionally 800–900°C has been attained [1]. Moisture present in wooden storage bins caused ignition of the latter [2]. A water jet was used unsuccessfully to try to clear a pump hose blocked with quicklime. On standing, the exothermic reaction proceeded far enough to generate enough steam to explosively clear the blocked pipe [3].

Two glass bottles of calcium oxide burst while in a laboratory owing to the considerable increase in bulk which occurs on hydration. Storage in plastics bottles under dessication is recommended. Granular oxide should cause fewer problems in this respect than the powder.

See other METAL OXIDES

CALCIUM PEROXIDE — CaO_2

Oxidisable materials

Castrantas, 1965, 4

Grinding the peroxide in contact with oxidisable materials may cause fire.

See other METAL PEROXIDES

CALCIUM SULPHATE — CaO_4S

Aluminium

See ALUMINIUM, Al: Metal oxosalts

Diazomethane

See DIAZOMETHANE, CH_2N_2: Calcium sulphate

Phosphorus

See PHOSPHORUS, P: Metal sulphates

CALCIUM PEROXODISULPHATE — CaO_8S_2

Castrantas, 1965, 6

Shock-sensitive; explodes violently.

See other PEROXOACID SALTS

CALCIUM SULPHIDE CaS

Oxidants
Mellor, 1941, Vol. 3, 745
Alkaline-earth sulphides react vigorously with chromyl chloride, lead dioxide, potassium chlorate (explodes lightly) and potassium nitrate (explodes violently).
See other METAL SULPHIDES

CALCIUM SILICIDE $CaSi$

Acids
Mellor, 1940, Vol. 6, 177
Interaction is vigorous, and the silanes evolved ignite in air.

CALCIUM DISILICIDE $CaSi_2$

Carbon tetrachloride
Zirconium Fire and Explosion Incidents, TID-5365, Washington, USAEC, 1956
Calcium disilicide exploded when milled in the solvent.

Diiron trioxide
See DIIRON TRIOXIDE, Fe_2O_3 : Calcium disilicide

Metal fluorides
Berger, E., *Compt. Rend.,* 1920, **170**, 29
Calcium silicide ignites in close contact with alkali metal fluorides (forming silicon tetrafluoride).

Potassium nitrate
See POTASSIUM NITRATE, KNO_3 : Calcium disilicide
See other METAL NON-METALLIDES

CALCIUM PEROXOCHROMATE(3-) $Ca_3Cr_2O_{12}$

Raynolds, J. H. *et al., J. Amer. Chem. Soc.*, 1930, **52**, 1851
It explodes at 100°C.
See other PEROXOACID SALTS

TRICALCIUM DINITRIDE Ca_3N_2

von Schwartz, 1918, 322
Spontaneously flammable in air (probably when finely divided in moist air).

Halogens
Mellor, 1940, Vol. 8, 103
Reaction with incandescence in chlorine gas or bromine vapour.
See other N-METAL DERIVATIVES

TRICALCIUM DIPHOSPHIDE Ca_3P_2

Dichlorine oxide
See DICHLORINE OXIDE, Cl_2O: Oxidisable materials

Oxygen
Van Wazer, 1958, Vol. 1, 145
Calcium and other alkaline earth phosphides incandesce in oxygen at about 300°C.

Water
Mellor, 1940, Vol. 8, 841
Calcium and other phosphides on contact with water liberate phosphine, which is usually spontaneously flammable in air, owing to the diphosphane content.
See other METAL NON-METALLIDES

CADMIUM Cd

Oxidants

See NITRYL FLUORIDE, FNO_2 : Metals
AMMONIUM NITRATE, $H_4N_2O_3$: Metals

Selenium,
or Tellurium
Mellor, 1940, Vol. 4, 480
Reaction on warming powdered cadmium with selenium or tellurium is exothermic, but less vigorous than that of zinc.
See other METALS

CADMIUM CHLORATE $CdCl_2O_6$

Copper(II) sulphide
See CHLORIC ACID, $ClHO_3$: Copper(II) sulphide

CADMIUM DIAMIDE CdH_4N_2

Alone,
or Water
Mellor, 1940, Vol. 8, 261
When heated rapidly, the amide may explode. Reaction with water is violent.
See other *N*-METAL DERIVATIVES

TETRAAMMINECADMIUM(II) PERMANGANATE $CdH_{12}Mn_2N_4O_8$

Mellor, 1942, Vol. 12, 335
Explodes on impact.
See other AMMINEMETAL OXOSALTS

CADMIUM DIAZIDE CdN_6

1. Turney, T. A., *Chem. & Ind.,* 1965, 1295
2. Mellor, 1967, Vol. 8, Suppl. 2.2, 25, 50

A solution, prepared by mixing saturated solutions of cadmium sulphate and sodium azide in a 10 ml glass tube, exploded violently several hours after preparation [1]. The dry solid is extremely hazardous, exploding on heating or light friction. A violent explosion occurred with cadmium rods in contact with aqueous hydrogen azide [2].

See other METAL AZIDES

CADMIUM OXIDE CdO

Magnesium

See MAGNESIUM, Mg: Metal oxides

CADMIUM SELENIDE $CdSe$

Preparative hazard.

See SELENIUM, Se: Metals (reference 6)

TRICADMIUM DINITRIDE Cd_3N_2

Fischer, F. *et al., Ber.,* 1910, **43**, 1469

The shock of the violent explosion caused by heating a sample of the nitride caused an unheated adjacent sample to explode.

Acids,
or Bases

Mellor, 1964, Vol. 8, Suppl. 2.1, 161

It reacts explosively with dilute acids or bases.

Water
Mellor, 1940, Vol. 8, 261
Explodes on contact.
See other N-METAL DERIVATIVES

TRICADMIUM DIPHOSPHIDE **Cd_3P_2**

Nitric acid
See NITRIC ACID, HNO_3: Tricadmium diphosphide

CERIUM **Ce**

Alone,
or Metals
Mellor, 1945, Vol. 5, 602–603
Cerium, or its alloys, readily give incendive sparks (pyrophoric particles) on frictional contact, this effect of iron alloys being widely used in various forms of 'flint' lighters. The massive metal ignites and burns brightly at 160°C, and cerium wire burns in a Bunsen flame more brilliantly than magnesium.

Of its alloys with aluminium, antimony, arsenic, bismuth, cadmium, calcium, copper, magnesium, mercury, sodium and zinc, those containing major proportions of cerium are often extremely pyrophoric. The mercury amalgams ignite spontaneously in air without the necessity for frictional generation of small particles. The interaction of cerium with zinc is explosively violent, and with antimony or bismuth very exothermic.

Halogens
Mellor, 1946, Vol. 5, 603
Cerium filings ignite in chlorine or bromine vapour at about 215°C.

Phosphorus
See PHOSPHORUS, P: Metals

Silicon
Mellor, 1946, Vol. 5, 605
Interaction at 1400°C to form cerium silicide is violently exothermic, often destroying the containing vessel.
See other METALS

CERIUM TRIHYDRIDE CeH_3

Muthmann, W. *et al., Ann.*, 1902, **325**, 261
The hydride is stable in dry air, but may ignite in moist air.
See other METAL HYDRIDES

CERIUM NITRIDE CeN

Water
Mellor, 1940, Vol. 8, 121
Contact with water vapour slowly causes incandescence, while a limited amount of water or dilute acid causes rapid incandescence with ignition of evolved ammonia and hydrogen.
See other N-METAL DERIVATIVES

DICERIUM TRISULPHIDE Ce_2S_3

Mellor, 1946, Vol. 5, 649
Pyrophoric in ambient air when finely divided.
See other METAL SULPHIDES

CHROMYL AZIDE CHLORIDE $ClCrN_3O_2$

Explosive solid.
See other METAL AZIDE HALIDES

CAESIUM TETRAFLUOROCHLORATE(1–) $ClCsF_4$

See METAL POLYHALOHALOGENATES

CAESIUM CHLOROXENATE $ClCsO_3Xe$

Jaselskis, B. *et al., J. Amer. Chem. Soc.*, 1967, **89**, 2770
It explodes at 205°C *in vacuo*.
See other XENON COMPOUNDS

CHLORINE FLUORIDE ClF

Sidgwick, 1950, 1149
'Product Information Sheet ClF', Tulsa, Ozark-Mahoning Co., 1970
This powerful oxidant reacts with other materials similarly to chlorine trifluoride or fluorine, but more readily than the latter.
See CHLORINE TRIFLUORIDE, ClF_3
FLUORINE, F_2

Aluminium

Mellor, 1956, Vol. 2, Suppl. 1, 63
Aluminium burns more readily in chlorine fluoride than in fluorine.

Fluorocarbon polymers

'Product Information Sheet ClF', Tulsa, Ozark-Mahoning Co., 1970
Chlorine fluoride can probably ignite Teflon and Kel-F at high temperatures or under friction or flow conditions.

Tellurium

Mellor, 1943, Vol. 11, 26
Interaction is incandescent.
See other INTERHALOGENS

FLUORONIUM PERCHLORATE $ClFH_2O_4$

Water

Hantzsch, A., *Ber.,* 1930, **63**, 97

This hydrogen fluoride–perchloric acid complex reacts explosively with water.

CHLORYL HYPOFLUORITE $ClFO_3$

Hoffman, C. J., *Chem., Rev.,* 1964, **64**, 97

Not then completely purified or characterised, its explosive nature was in contrast to the stability of the isomeric perchloryl fluoride.

See other HALOGEN OXIDES
HYPOHALITES

PERCHLORYL FLUORIDE $ClFO_3$

'Booklet DC-1819', Philadelphia, Pennsalt Chem. Corp., 1957

Anon., *Chem. Eng. News*, 1960, **38**, 62

Procedures relevant to safe handling and use are discussed. Perchloryl fluoride is stable to heat, shock and moisture, but is a powerful oxidiser comparable with liquid oxygen. It forms flammable and/or explosive mixtures with combustible gases or vapours.

Calcium acetylide,
or Potassium cyanide,
or Potassium thiocyanate,
or Sodium iodide

Kirk-Othmer, 1966, Vol. 9, 602

Unreactive at 25°C, these solids react explosively in the gas at 100–300°C.

Finely divided solids

McCoy, G., *Chem. Eng. News,* 1960, **38**(4), 62

Oxidisable organic materials of high surface to volume ratio (carbon powder, foamed elastomers, lampblack, sawdust) may react very violently, even at −78°C, with perchloryl fluoride, which should be handled with the same precautions as liquid oxygen.

Hydrocarbons,
or Hydrogen sulphide,
or Nitrogen oxide,
or Sulphur dichloride
or Vinylidene chloride
Braker, 1971, 459
At ambient temperature, perchloryl fluoride is unreactive with the above compounds, but reaction is explosive at 100–300°C, or if the mixtures are ignited.

3α-Hydroxy-5β-androstane-11, 17-dione 17-hydrazone
Nomine, G. *et al., J. Chem. Educ.*, 1969, **46**, 329
A reaction mixture in aqueous methanol exploded violently at −65°C. (Hydrazine–perchloryl fluoride redox reaction?) A previous reaction at 20°C had been uneventful, and the low-temperature explosion could not be reproduced.

Nitrogenous bases
Scott, F. L. *et al., Chem. & Ind.*, 1960, 528
Interaction, in presence of diluent below 0°C, with isopropylamine or isobutylamine caused separation of explosive liquids, and, with aniline, phenylhydrazine and 1,2-diphenylhydrazine, explosive solids.

Sodium methoxide,
Methanol
Papesch, V., *Chem. Eng. News*, 1959, **36**, 60
Addition of solid methoxide to a reaction vessel containing methanol vapour and gaseous perchloryl fluoride caused ignition and explosion. This could be avoided by adding all the methoxide first, or by nitrogen purging before addition of methoxide.

See other HALOGEN OXIDES
PERCHLORYL COMPOUNDS

FLUORINE PERCHLORATE $ClFO_4$

Alone,
or Hydrogen,
or Potassium iodide

1. Rohrback, G. H. *et al., J. Amer. Chem. Soc.*, 1949, **69**, 677
2. Hoffman, C. J., *Chem. Rev.*, 1964, **64**, 97

The pure liquid explodes on freezing at –167°C, and the gas is readily initiated by sparks, flame or contact with grease, dust or rubber tube. Contact of the gas with aqueous potassium iodide also caused an explosion [1] and ignition occurs in excess hydrogen gas [2].

See HYPOHALITES
See other HALOGEN OXIDES

NITROGEN CHLORIDE DIFLUORIDE ClF_2N

Petry, R. C., *J. Amer. Chem. Soc.*, 1960, **82**, 2401

Caution in handling is recommended for this *N*-halogen compound.

See other *N*-HALOGEN COMPOUNDS

PHOSPHORUS CHLORIDE DIFLUORIDE ClF_2P

Hexafluoroisopropylideneaminolithium

See HEXAFLUOROISOPROPYLIDENEAMINOLITHIUM, C_3F_6LiN: Non-metal halides

CHLORINE TRIFLUORIDE ClF_3

1. Anon., *J. Chem. Educ.*, 1967, **44**, A1057–1062
2. O'Connor, D. J. *et al., Chem. & Ind.*, 1957, 1155

Handling procedures for this highly reactive oxidant gas have been detailed [1]. Surplus gas is best burnt with town or natural gas, followed by absorption in alkali [2].

Acids
Mellor, 1956, Vol. 2, Suppl. 1, 157
Strong nitric and sulphuric acids reacted violently.

Ammonium fluoride
Gardner, D. M. *et al., Inorg. Chem.*, 1963, 2, 413
The reaction gases (containing chlorodifluoramine) must be handled below −5°C to avoid explosion. Ammonium hydrogenfluoride reacts similarly.

Carbon tetrachloride
Mellor, 1956, Vol. 2, Suppl. 1, 156
Chlorine trifluoride will dissolve in carbon tetrachloride at low temperatures without interaction. Such solutions are dangerous, being capable of detonation. If it is used as a solvent for fluorination with the trifluoride, it is therefore important to prevent build-up of high concentrations.

Fluorinated polymers
'Chlorine Trifluoride', Tech. Bull. TA 8522-3, Morristown, N.J., Baker & Adamson Div. of Allied Chemicals Corp., 1968
Bulk surfaces of polytetrafluoroethylene or polychlorotrifluoroethylene are resistant to the liquid or vapour under static conditions, but breakdown and ignition may occur under flow conditions.
See Polychlorotrifluoroethylene, below

Fuels
See ROCKET PROPELLANTS

Hydrogen-containing materials
Mellor, 1956, Vol. 2, Suppl. 1, 157
Explosive reactions occur with ammonia, coal-gas, hydrogen or hydrogen sulphide.

Iodine
Mellor, 1956, Vol. 2, Suppl.1, 157
Ignites on contact

Metals,
or Metal oxides,
or Metal salts
or Non-metals,
or Non-metal oxides

Mellor, 1956, Vol. 2, Suppl.1, 155–157
Sidgwick, 1950, 1156
'Chlorine Trifluoride', Tech. Bull. TA 8532-3, Morristown, N.J., Baker & Adamson Div. of Allied Chem Corp., 1968
Chlorine trifluoride is a hypergolic oxidiser with recognised fuels, and contact with the following at ambient or slightly elevated temperatures is violent, ignition often occurring. The state of subdivision may affect the results.

Antimony, arsenic, selenium, silicon, tellurium; iridium, iron, molybdenum, osmium, potassium, rhodium, tungsten; (and when primed with charcoal, aluminium, copper, lead, magnesium, silver, tin, zinc).

Aluminium oxide, arsenic trixide, bismuth trioxide, calcium oxide, chromic oxide, lanthanum oxide, lead dioxide, magnesium oxide, manganese dioxide, molybdenum trioxide, phosphorus pentoxide, stannic oxide, sulphur dioxide (explodes), tantalum pentoxide, tungsten trioxide, vanadium pentoxide.

Red phosphorus, sulphur; but with carbon, the observed ignition has been attributed to presence of impurities.

Iodides of mercury, potassium; silver nitrate, potassium carbonate.

Nitro compounds
Mellor, 1956, Vol. 2, Suppl.1, 156
Several nitro compounds are soluble in chlorine trifluoride, but the solutions are extremely shock-sensitive. These include trinitrotoluene, hexanitrodiphenyl, hexanitrodiphenyl-amine, -sulphide or -ether. Highly chlorinated compounds behave similarly.

Organic materials
Mellor, 1956, Vol. 2, Suppl.1, 155
Violence of the reaction, sometimes explosive, with e.g. acetic acid, benzene, ether, is associated with both their carbon and hydrogen contents. If nitrogen is also present, explosive fluoroamino compounds may also be involved. Fibrous materials–cotton, paper, wood – invariably ignite.

Polychlorotrifluoroethylene
Anon., *Chem. Eng. News*, 1965, **43**(20), 41
An explosion occurred while chlorine trifluoride was being bubbled

through the fluorocarbon oil at −4°C. Moisture (snow) may have fallen into the mixture, reacted exothermically with the trifluoride and initiated interaction of the mixture.

Refractory materials
Cloyd 1965, 58
Mellor 1956, Vol. 2, Suppl.1, 157
Fibrous or finely divided refractory materials, asbestos, glass wool, sand or tungsten carbide, may ignite with the liquid and continue to burn in the gas. The presence of adsorbed or lattice-water seems necessary for attack on the siliceous materials to occur.

Water
1. Sidgwick, 1950, 1156
2. Mellor, 1956, Vol. 2, Suppl.1, 156, 158
3. Ruff, O. *et al., Z. Anorg. Chem.*, 1930, **190**, 270

Interaction is violent and may be explosive, even with ice, oxygen being evolved [1]. Part of the water dropped into a flask of the gas was expelled by the violent reaction ensuing [2]. An analytical procedure, involving absorption of chlorine trifluoride into 10% sodium hydroxide solution from the open capillary neck of a quartz ampoule to avoid explosion, was described [3].
See other INTERHALOGENS

CHLORINE DIOXYGEN TRIFLUORIDE — ClF_3O_2

Streng, A. G., *Chem. Rev.*, 1963, **63**, 607
It is a very powerful oxidant, though of low stability.

See other HALOGEN OXIDES

POTASSIUM TETRAFLUOROCHLORATE(1−) — ClF_4K

See METAL POLYHALOHALOGENATES

RUBIDIUM TETRAFLUOROCHLORATE(1−) — ClF_4Rb

See METAL POLYHALOHALOGENATES

CHLORINE PENTAFLUORIDE ClF_5

Nitric acid
Christe, K. O., *Inorg. Chem.*, 1972, **11**, 1220
Interaction of anhydrous nitric acid with chlorine pentafluoride vapour at –40°C, or with the liquid at above –100°C, is very vigorous.

Water
1. Pilipovich, D. *et al., Inorg. Chem.*, 1967, **6**, 1918
2. Christe, K. O., *Inorg. Chem.*, 1972, **11**, 1220
Interaction of liquid chlorine pentafluoride with ice at –100°C [1], or of the vapour with water vapour above 0°C [2], is extremely vigorous.
See other INTERHALOGENS

CHLOROGERMANE $ClGeH_3$

Ammonia
Johnson, O. H., *Chem. Rev.*, 1951, **48**, 274
Both mono- and di-chlorogermanes react with ammonia to give involatile products which explode on heating.
See related N-METAL DERIVATIVES

HYDROGEN CHLORIDE ClH

Preparative hazard.
See Sulphuric Acid, below

Aluminium
See ALUMINIUM, Al: Hydrogen chloride
See also ALUMINIUM–TITANIUM ALLOYS, Al–Ti: Oxidants

Fluorine
See FLUORINE, F_2 : Hydrogen halides

Hexalithium disilicide

See HEXALITHIUM DISILICIDE, Li_6Si_2: Acids

Metal acetylides or carbides

See DICAESIUM ACETYLIDE, C_2Cs_2 : Acids
DIRUBIDIUM ACETYLIDE, C_2Rb_2 : Acidic materials
URANIUM DICARBIDE, C_2U: Hydrogen chloride

Potassium permanganate

See POTASSIUM PERMANGANATE, $KMnO_4$: Hydrochloric acid

Sodium

See SODIUM, Na: Acids

Sulphuric acid

1. *MCA Case History No. 1785*
2. Libman, D. D., *Chem. & Ind.*, 1948, 728
3. Smith, G. B. L. *et al., Inorg. Synth.*, 1950, **3**, 132

Accidental addition of 6500 litres of concentrated hydrochloric acid to a bulk sulphuric acid storage tank released sufficient hydrogen chloride by dehydration to cause the tank to explode violently [1]. Complete dehydration of hydrochloric acid solution releases some 250 volumes of gas. A laboratory apparatus for effecting this safely has been described [2], which avoids the possibility of layer formation in unstirred flask generators [3].

Tetraselenium tetranitride

See TETRASELENIUM TETRANITRIDE, N_4Se_4

HYPOCHLOROUS ACID **ClHO**

Alcohols

Mellor, 1956, Vol. 2, Suppl.1, 560

Contact of these, or of chlorine and alcohols, readily forms unstable alkyl hypochlorites.

See HYPOHALITES

Ammonia
Mellor, 1940, Vol. 8, 217
The violent explosion occurring on contact with ammonia gas is due to formation of nitrogen trichloride and its probable initiation by the heat of solution of ammonia.

Arsenic
Mellor, 1941, Vol. 2, 254
Ignition on contact.
See other HYPOHALITES
OXOHALOGEN ACIDS

CHLORIC ACID **$ClHO_3$**

Muir, G. D., private comm., 1968
Aqueous chloric acid solutions decompose explosively if evaporative concentration is carried too far.

Cellulose
Mellor, 1946, Vol. 2, 310
Filter paper ignites after soaking in chloric acid.
See Oxidisable materials, below

Copper sulphide
Mellor, 1956, Vol. 2, Suppl. 1, 584
Copper sulphide explodes with concentrated chloric acid solution, or cadmium, magnesium or zinc chlorates.
See also METAL HALOGENATES

Oxidisable materials
In contact with oxidisable substances, reactions are similar to those of the metal chlorates.
See Cellulose, above
METAL HALOGENATES
See other OXOHALOGEN ACIDS

CHLOROSULPHURIC ACID $ClHO_3S$

Phosphorus,
or Water

1. Heumann, K. *et al., Ber.*, 1882, **15**, 417
2. *MCA SD-33,* 1968

The acid is a strong oxidising agent, and above 25–30°C, interaction with yellow phosphorus is vigorous and accelerates to explosion. A higher temperature is needed to start reaction with red phosphorus [1]. Reaction with water is highly exothermic and violent, owing to combined heat of hydrolysis to sulphuric and hydrochloric acids and their heats of dilution. Handling precautions are detailed [2].

Silver nitrate
See SILVER NITRATE, $AgNO_3$: Chlorosulphuric acid

PERCHLORIC ACID $ClHO_4$

1. *MCA SD-11,* 1965
2. Lazerte, D., *Chem. Eng. News,* 1971, **49**(3), 33
3. Muse, L. A., *J. Chem. Educ.,* 1972, **49**, A463
4. Graf, F. A., *Chem. Eng. Progr.,* 1966, **62**(10), 109

Most of the numerous and frequent hazards experienced with perchloric acid have been associated with either its exceptional oxidising power or the inherent instability of its covalent compounds, some of which form readily. Although the 70–72% acid of commerce behaves as a very strong but non-oxidising acid, it becomes an extreme oxidant and powerful dehydrator at elevated temperatures (160°C) or when anhydrous [1].
See Dehydrating agents, below

Where an equally strong but non-oxidising acid can be used, trifluoromethanesulphonic acid is recommended [2]. Safe laboratory handling procedures have been detailed [3] and an account of safe handling of perchloric acid in the large-scale preparation of hydrazinium diperchlorate, with recommendations for materials of construction, has been published [4].

Acetic anhydride,
Acetic acid,
Organic materials

1. Burton, H. *et al., Analyst*, 1955, **80**, 4
2. Schumacher, 1960, 187, 193
3. Kuney, J. H., *Chem. Eng. News,* 1947, **25**, 1659
4. Tech. Survey No. 2, *Fire Hazards and Safeguards for Metalworking Industries,* US Board of Fire Underwriters, 1954

Mixtures of hydrated perchloric acid with enough acetic anhydride produce a solution of anhydrous perchloric acid in acetic acid/ anhydride, which is of high catastrophic potential [1]. Sensitivity to shock and heat depends on composition of the mixture, and vapour evolved on heating is flammable [2]. Such solutions have been used for electropolishing operations, and during modifications to an electro-polishing process, a cellulose acetate rack was introduced into a large volume of an uncooled mixture of perchloric acid and acetic anhydride. Dissolution of the rack introduced organic material into the virtually anhydrous acid and caused it to explode disastrously. This cause was confirmed experimentally [3,4].

See Solvents, below
See also ACETIC ANHYDRIDE, $C_4H_6O_3$: Perchloric acid

Acetic anhydride,
Carbon tetrachloride,
2-Methylcyclohexanone
Gall, M. *et al., Org. Synth.*, 1972, **52**, 40
During acetylation of the enolised ketone, the 70% perchloric acid must be added last to the reaction mixture to provide maximum dilution and cooling effect.

Alcohols
See ALKYL PERCHLORATES
Cellulose derivatives, below
Glycols, below

Aniline,
Formaldehyde
Aniline, 86, New York, Allied Chem. Corp., 1964
Aniline reacts with perchloric acid and then formaldehyde to give an explosively combustible condensed resin.

Antimony(III) compounds
Burton, H. *et al., Analyst,* 1955, **80**, 4
Treatment of tervalent compounds of antimony with perchloric acid can be very hazardous.
See also Bismuth, below

Azo-pigment,
Orthoperiodic acid
1. *ABCM Quart. Safety Summ.*, 1961, **32**, 125
2. Smith, F. G. *et al., Talanta,* 1960, **4**, 185
During the later stages of the wet oxidation of an azo- pigment with mixed perchloric and orthoperiodic acids, a violent reaction, accompanied by flashes of light, set in, and terminated in an explosion [1]. The general method upon which this oxidation was based is described as hazard-free [2].

1,3-Bis(di-η-cyclopentadienyliron)-2-propen-1-one
See 1,3-BIS(DI-η-CYCLOPENTADIENYLIRON)-2-PROPEN-1-ONE, $C_{23}H_{20}Fe_2O$: Perchloric acid

Bis-1,2-diaminopropane-*cis*-dichlorochromium(III) perchlorate
See BIS-1,2-DIAMINOPROPANE-*cis*-DICHLOROCHROMIUM(III) PERCHLORATE, $C_6H_{20}Cl_3CrN_4O_4$: Perchloric acid

Bismuth
Nicholson, D. G. *et al., J. Amer. Chem. Soc.,* 1935, **57**, 817
Attempts to dissolve bismuth and its alloys in hot perchloric acid carry a very high risk of explosion. At 110°C a dark brown coating is formed, and if left in contact with the acid (hot or cold), explosion occurs sooner or later. The same is true of antimony and its tervalent compounds.

Carbon
Mellor, 1946, Vol. 2, 380
Contact of a drop of the anhydrous acid with wood charcoal causes a very violent explosion.

Cellulose and derivatives
1. Schumacher, 1960, 187, 195
2. Harris, E. M., *Chem. Eng.*, 1949, **56**, 116

3. Sutcliffe, G. R., *J. Textile Ind.*, 1950, **41**, 196T

Contact of the hot concentrated acid or of the cold anhydrous acid with cellulose (as paper, wood fibre or sawdust, etc.) is very dangerous and may cause a violent explosion. Many fires have been caused by long-term contact of diluted acid with wood with subsequent evaporation and ignition [1,2]. Contact of cellulose acetate with 1200 litres of uncooled anhydrous perchloric acid in acetic anhydride caused an extremely violent explosion [1] and interaction of benzyl cellulose with boiling 72% acid was also explosive [3]. Perchlorate esters of cellulose may have been involved in all these incidents.

Dehydrating agents

1. Mellor, 1941, Vol. 2, 373, 380; 1956, Vol. 2, Suppl. 2.1, 598, 603
2. Kuney, J. H., *Chem. Eng. News,* 1947, **25**, 1659
3. Burton, H. *et al., Analyst,* 1955, **80**, 4
4. Schumacher, 1960, 71, 187, 193
5. Wirth, C. M. P., *Lab. Practice*, 1966, **15**, 675
6. Plesch, P. H. *et al., Chem.& Ind.,* 1971, 1043
7. Musso, H. *et al., Angew Chem.,* 1970, **82**, 46

Although commercial 70–72% perchloric acid (approximating to the dihydrate) itself is stable, incapable of detonation and readily stored, it may be fairly readily dehydrated by contact with dehydrating agents to anhydrous perchloric acid. This is not safe when stored at room temperature, since it slowly decomposes, even in the dark, with accumulation of chlorine dioxide in the solution, which darkens and finally explodes after about 30 days [1,2,5]. The 72% acid (or perchlorate salts) may be converted to the anhydrous acid by heating with sulphuric acid, phosphorus pentoxide or phosphoric acid, or by distillation under reduced pressure [1,4,5]. In contact with cold acetic anhydride, mixtures of the anhydrous acid with excess anhydride and acetic acid are formed, which are particularly dangerous, being sensitive to mechanical shock, heating or the introduction of organic contaminants [2–4].

See Acetic anhydride, above

A solution of the monohydrate in chloroform exploded in contact with phosphorus pentoxide [1]. A safer method of preparing anhydrous solutions of perchloric acid in methylene chloride, which largely avoids the risk of explosion, has been described [6]. Further precautions are detailed in an account of an explosion during a similar preparation [7].

Solutions of the anhydrous acid of less than 55% concentration in acetic acid or anhydride are relatively stable [8].

Diethyl ether
Michael, A. T. *et al.*, *Amer. Chem. J.*, 1900, **23**, 444
The explosions sometimes observed on contact of the anhydrous acid with ether are probably due to formation of ethyl perchlorate by scission of the ether, or possibly to formation of diethyloxonium perchlorate.

Ethylbenzene,
Thallium triacetate
Uemura, S. *et al.*, *Bull. Chem. Soc. Japan*, 1971, **44**, 2571
Application of a published method of thallation to ethylbenzene caused a violent explosion. A reaction mixture of thallium triacetate, acetic acid, perchloric acid and ethylbenzene was stirred at 65°C for 5 h, then filtered from precipitated thallous salts. Vacuum evaporation of the filtrate at 60°C gave a pasty residue, which exploded. This preparation of ethylphenylthallic acetate perchlorate monohydrate had been done twice previously uneventfully, as had been analogous preparations involving thallation of benzene, toluene, 3 isomeric xylenes and anisole in a total of 150 runs, where excessive evaporation had been avoided.

Glycerol,
Lead oxide
MCA Case History No. 799
During maintenance work on casings of fans used to extract perchloric acid fumes, 7 violent explosions occurred when flanges sealed with lead oxide–glycercol cement were disturbed. Explosions, attributed to formation of explosive compounds by interaction of the cement with perchloric acid, may have involved perchlorate esters and/or lead salts. Use of an alternative silicate–hexafluorosilicate cement is recommended.

Glycols and their ethers
Schumacher, 1960, 195, 214
Glycols and their ethers undergo violent decomposition in contact with ~70% perchloric acid. This seems likely to involve formation of the

glycol perchlorate esters (after scission of ethers) which are explosive, those of ethylene glycol and 1-chloro-2,3-propanediol being more powerful than glyceryl nitrate, and the former so sensitive that is explod on addition of water.

Hydrogen
Schumacher, 1960, 189
Dietz, W., *Angew Chem.*, 1939, **52**, 616
Occasional explosions experienced during use of hot perchloric acid to dissolve steel samples for analysis is attributed to formation of hydrogen –perchloric acid mixtures and their ignition by steel particles at temperatures as low as 215°C.

Iodides
Michael, A. *et al.*, *Amer.Chem.J.*, 1900, **23**, 444
The anhydrous acid ignites in contact with sodium iodide or hydriodic acid.

Iron(II) sulphate
Tod, H., private comm., 1968
During preparation of iron diperchlorate, a mixture of iron sulphate and perchloric acid was being strongly heated when a most violent explosion occurred. Heating should be gentle to avoid initiating this redox system.

Ketones
Schumacher, 1960, 195
Ketones may undergo violent decomposition in contact with ~70% acid.

Nitric acid,
Organic matter
1. Anon., *Ind. Eng. Chem. (News Ed.)*, 1937, **15**, 214
2. Lambie, D. A., *Chem. & Ind.*, 1962, 1421
3. Mercer, E. R., private comm., 1967
4. Muse, L. A., *Chem. Eng. News*, 1973, **51**(6), 29–30
5. Cooke, G. W., private comm., 1967

The mixed acids have been used to digest organic material prior to analysis, but several explosions have been reported, including those with vegetable oil [1], milk [2] and calcium oxalate precipitates from plants [3]. To avoid trace metal contamination by homogenising rat carcases in a blender, the carcases were dissolved in nitric acid. After separation of fat and addition of perchloric acid (125 ml), evaporation of samples to near-dryness caused a violent explosion [4].

Finely ground plant material in contact with perchloric/nitric acid mixture on a heated sand bath became hot before all the plant material was saturated with the acid and exploded. Subsequent digests left overnight in contact with cold acids proceeded smoothly [5]. Cellulose nitrate and/or perchlorate may have been involved.

Nitrogenous epoxides
Harrison, G. E., private comm., 1966
Traces of perchloric acid used as hydration catalyst for ring opening of nitrogenous epoxides caused precipitation of organic perchlorate which was highly explosive. Concentration of acid was less than 1% by volume.
See other OXOSALTS OF NITROGENOUS BASES

Oleic acid
1. Swern, D. *et al.*, US Pat. 3 054 804; *Chem. Eng. News*, 1963, **41**(12), 39
2. Anon., *Chem. Eng. News*, 1963, **41**(27), 47

The improved preparation of 1,4-octadecanolactone [1] involves heating oleic acid (or other C_{18} acids) with *ca.* 70% perchloric acid to 115°C. This is considered to be a potentially dangerous method [2].

Phosphine
See PHOSPHONIUM PERCHLORATE, ClH_4O_4P

Pyridine
See PYRIDINIUM PERCHLORATE, $C_5H_6ClNO_4$

Sodium phosphinate ('hypophosphite')
Smith, F. G., *Analyst*, 1955, **80**, 16
Though no interaction occurs in the cold, these powerful reducing and oxidising agents violently explode on heating.

Sulphoxides
1. Therésa, J. de B., *Anales Soc. Españ. Fis. y Quim.*, 1949, **45B**, 235
2. Graf, F. A., *Chem. Eng. Progr.*, 1966, **62**(10), 109
3. Uemura, S. *et al.*, *Bull. Chem. Soc. Japan*, 1971, **44**, 2571

Lower members of the series of salts formed between organic sulphoxides and perchloric acid are unstable and explosive when dry. The salt formed from dibenzyl sulphoxide explodes at 125°C [1]. Dimethyl sulphoxide explodes on contact with 70% perchloric acid solution [2]; one drop of acid added to 10 ml of sulphoxide at 20°C caused a violent explosion [3].
See also DIMETHYL SULPHOXIDE, C_2H_6OS: Oxosalts

Trichloroethylene
Prieto, M. A. *et al., Research on the Stabilization and Characterization of Highly Concentrated Perchloric Acid*, 36, Whittier, Calif., American Potash and Chem. Corp., 1962
The solvent reacts violently with the anhydrous acid.

Trizinc diphosphide
Muir, G. D., private comm., 1968
Use of perchloric acid to assist solution of a sample for analysis caused a violent reaction.
See other OXIDANTS
OXOHALOGEN ACIDS

CHLOROAMINE ClH_2N

1. Marckwald, W. *et al., Ber.*, 1923, **56**, 1323
2. Coleman, G. H., *et al., Inorg. Synth.*, 1939, **1**, 59

The solvent-free material, isolated at –70°C, violently disproportionates (sometimes explosively) at –50°C to ammonium chloride and nitrogen trichloride [1]. Ethereal solutions of chloroamine are readily handled [2].
See other N-HALOGEN COMPOUNDS

AMMONIUM CHLORIDE ClH_4N

Interhalogens
See BROMINE TRIFLUORIDE, BrF_3: Ammonium halides
BROMINE PENTAFLUORIDE, BrF_5: Acids, etc.

HYDROXYLAMMONIUM CHLORIDE ClH_4NO

See HYDROXYLAMMONIUM SALTS

AMMONIUM CHLORATE ClH_4NO_3

1. Brauer, 1963, Vol. 1, 314
2. Urbanski, 1965, Vol. 2, 476

It occasionally explodes spontaneously, and invariably above 100°C [1]. It will explode after 11 h at 40°C, and after 45 min at 70°C. Ammonium and chlorate salts should not be mixed together [2].

Water
Mellor, 1941, Vol. 2, 339, 1956, Vol. 2, Suppl.1, 591
A cold saturated solution may decompose explosively after a few days if much excess salt is present. Hot aqueous solutions have exploded during evaporation in steam-heated vessels.
See other OXOSALTS OF NITROGENOUS BASES

AMMONIUM PERCHLORATE ClH_4NO_4

MCA Case History No. 1002
Materials for a batch of ammonium perchlorate castable propellant were charged into a mechanical mixer. A metal spatula was left in accidentally, and the contents ignited when the mixer was started, owing to local friction caused by the spatula. A tool-listing safety procedure has been instituted.
See Impurities, below

Carbon
Galwey, A. K. *et al., Trans. Faraday. Soc.*, 1960, **56**, 581
Below 240°C intimate mixtures with sugar charcoal undergo exothermic decomposition, while mild explosions occur above 240°C.

Copper
Anon., *Chem. Eng.*, 1955, **62**(12), 335
Crystalline ammonium perchlorate ignited in contact with hot copper pipes.

Impurities
Jacobs, P. W. M. *et al., Chem. Rev.*, 1969, **69**, 590
The medium impact-sensitivity of this solid propellant component is greatly increased by co-crystallisation of certain impurities, notably nitryl perchlorate, potassium periodate and potassium permanganate.

Metals,
or Organic materials,
or Sulphur
Haz. Chem. Data, 1969, 41
This powerful oxidant functions as an explosive when mixed with finely divided metals, organic materials or sulphur, which increase the shock-sensitivity up to that of picric acid.
See other OXOSALTS OF NITROGENOUS BASES

PHOSPHONIUM PERCHLORATE **ClH_4O_4P**

Fichter, F. *et al., Helv. Chim. Acta,* 1934, **17**, 222
The crystalline salt obtained by action of phosphine on 68% perchloric acid at –20°C is dangerously explosive, and sensitive to contact with moist air, increase in temperature, or friction.
See related OXOSALTS OF NITROGENOUS BASES

HYDRAZINIUM CHLORITE **$ClH_5N_2O_2$**

Levi, G. R., *Gazz. Chim. Ital.,* 1923, [2], **53**, 105–108
It is spontaneously flammable when dry.
See other CHLORITE SALTS

AMMONIUM PERCHLORYLAMIDE **$ClH_5N_2O_3$**

See PERCHLORYLAMIDE SALTS

HYDRAZINIUM CHLORATE $ClH_5N_2O_3$

Salvadori, J., *Gazz. Chim. Ital.*, 1907, [2], **37**, 32–40
It explodes violently at its m.p., 80°C.
See other OXOSALTS OF NITROGENOUS BASES

HYDRAZINIUM PERCHLORATE $ClH_5N_2O_4$

Alone,
or Copper dichloride
1. Levy, J. B. *et al., ACS 54*, 1966, 55; Grelecki, C. J. *et al., ibid.*, 73
2. Shidlovskii, A. A. *et al., Zh. Priklad. Chim.*, 1962, **35**, 756

The deflagration and thermal decomposition of the salt, a component of solid rocket propellants, have been studied [1]. Presence of 5% of copper dichloride caused explosion to occur at 170°C [2].
See other OXOSALTS OF NITROGENOUS BASES

MERCURY(I) CHLORITE $ClHgO_2$

Levi, G. R., *Gazz. Chim. Ital.*, 1915, [2], **45**, 161
It is extremely unstable when dry, exploding spontaneously.
See other CHLORITE SALTS

IODINE CHLORIDE ClI

Metals
Mellor, 1940, Vol. 2, 119; 1956, Vol. 2, Suppl. 1, 452; 1963, Vol. 2, Suppl. 2.2, 1563
Mixtures containing sodium explode only on impact, while potassium explodes on contact with the chloride. Aluminium foil ignites after prolonged contact.
See other INTERHALOGENS

Ammonia,
or Ammonium sulphate
Mellor, 1941, Vol. 2, 702; 1940, Vol. 8, 217
High concentrations of ammonia in air react so vigorously with potassium chlorate as to be dangerous. Mixtures with ammonium sulphate when heated decompose with incandescence.

Aqua regia,
Ruthenium
Sidgwick, 1950, 1459
Ruthenium is insoluble in aqua regia, but addition of potassium chlorate causes explosive oxidation.

Carbon
Read, C. W. W., *School Sci. Rev.,* 1941, **22**(87), 341
Accidental substitution of powdered carbon for manganese dioxide in 'oxygen mixture' caused a violent explosion when the mixture was heated.

Cyanides
See METAL CYANIDES: Oxidants

Dinickel trioxide
Mellor 1942, Vol. 15, 395
Interaction at 300°C is violently exothermic, red-heat being attained.

Fabric
Anon., *Accidents,* 1968, **74**, 24
Fabric gloves (wrongly used in place of impervious plastics gloves), became impregnated during handling operations and subsequently ignited from cigarette ash.

Hydrogen iodide
Mellor, 1941, Vol. 2, 310
Molten potassium chlorate ignites hydrogen iodide gas.

Manganese dioxide
Mellor MIC, 1961, 333
When oxygen is generated in the laboratory by heating potassium chlorate with manganese dioxide as catalyst, the latter must be free of organic matter or an explosion will occur.

Manganese dioxide,
Potassium hydroxide
Molinari, E. *et al.*, *Inorg. Chem.*, 1964, **3**, 898
The oxidation of manganese dioxide to manganate by solid alkali–chlorate mixtures becomes explosive above 80–90°C at pressures above 19 kbar.

Metal phosphides
Mellor, 1940, Vol. 8, 839, 844
Tricopper diphosphide and trimercury tetraphosphide form impact-sensitive mixtures with potassium chlorate. By analogy, the phosphides of aluminium, magnesium, silver and zinc, etc., would be expected to form similar mixtures with metal halogenates.

Metal phosphinates ('hypophosphites')
Mellor, 1940, Vol. 8, 881, 883
Dry mixtures of barium phosphinate and potassium chlorate burn rapidly with a feeble report if unconfined, but even under the slight confinement of enclosing in paper, a sharp explosion occurs. The mixture is readily initiated by sparks, impact or friction. A mixture of calcium phosphinate, potassium chlorate and quartz exploded during mixing. Mixtures of various phosphinates and chlorates have been proposed as explosives, but they are very sensitive to initiation by sparks, friction or shock. Admixture of powdered magnesium causes a brilliant flash on initiation of the mixture.
See also SODIUM PHOSPHINATE, H_2NaO_2P: Oxidants

Metals
1. Mellor, 1941, Vol. 2, 310; Anon., *Chem. Eng. News*, 1936, **14**, 451
2. Mellor, 1940, Vol. 4, 480

3. Mellor, 1943, Vol. 11, 163
4. Mellor, 1941, Vol. 7, 20, 116, 260

Mixtures of finely divided aluminium, copper, magnesium [1] and zinc [2] with potassium chlorate (or other metal halogenates) are explosives and may be initiated by heat, impact or light friction. Chromium incandesces in the molten salt [3] and germanium explodes on heating with potassium chlorate. Titanium explodes on heating, while zirconium gives mild explosions on heating, and ignites when the mixture is impacted [4].

See also METAL HALOGENATES

Metal sulphides

1. Mellor, 1941, Vol. 2, 310
2. Mellor, 1939, Vol. 9, 523
3. Mellor, 1941, Vol. 3, 447

Many metal sulphides when mixed intimately with metal halogenates form heat-, impact- or friction-sensitive explosive mixtures [1]. That with diantimony trisulphide can be initiated by a spark [2] and with disilver sulphide a violent reaction occurs on heating [3].

See also METAL HALOGENATES

Metal thiocyanates

1. von Schwartz, 1918, 299–300, 328
2. Anon., *Chem. Age,* 1936, **35**, 42

Mixtures of thiocyanates with chlorates (or nitrates) are friction- and heat-sensitive, and explode on rubbing, heating to 400°C, or initiation by spark or flame [1]. A violent explosion occurred when a little chlorate was ground in a mortar contaminated with ammonium thiocyanate. A similar larger-scale explosion involving traces of barium thiocyanate is also described [2].

Non-metals

1. Mellor, 1941, Vol. 2, 310
2. Mellor, 1940, Vol. 8, 785–786
3. Mellor, 1946, Vol. 5, 15

Potassium chlorate (or other metal halogenates) intimately mixed with arsenic, carbon, phosphorus, sulphur or other readily oxidised materials give friction-, impact- and heat-sensitive mixtures which may explode violently [1]. When potassium chlorate is moistened with a solution of phosphorus in carbon disulphide, it eventually explodes as the solvent evaporates and oxidation proceeds [2]. Boron burns in molten potassium chlorate with dazzling brilliance [3].

See Sulphur, below
METAL CHLORATES
See also METAL HALOGENATES

Sulphur
Tanner, H. G., *J. Chem. Educ.*, 1959, **36**, 59
A review of the chemistry involved in this explosively unstable system.

Sulphur dioxide
Mellor, 1947, Vol. 10, 217; 1941, Vol. 2, 311
Contact at temperatures above 60°C causes flashing of the evolved chlorine dioxide. Solutions of sulphur dioxide in ethanol or ether cause an explosion on contact at ambient temperature.

Sulphuric acid
Mellor, 1947, Vol. 10, 435
Addition of potassium chlorate in portions to sulphuric acid maintained at below 60°C or above 200°C causes brisk effervescence. At intermediate temperatures, explosions are caused by the chlorine dioxide produced, and these reach maximum intensity at 120–130°C. Uncontrolled contact of any chlorate with sulphuric acid may be explosive.

See METAL CHLORATES: Acids

Sodium amide
Mellor, 1940, Vol. 8, 258
A mixture explodes.

Thorium dicarbide
See THORIUM DICARBIDE, C_2Th: Non-metals, etc.
See other METAL OXOHALOGENATES

POTASSIUM PERCHLORATE $ClKO_4$

Burton, H. *et al.*, *Analyst*, 1955, **80**, 16

Many explosions have been experienced during the gravimetric determination of either perchlorates or potassium as potassium perchlorate by a standard method involving an ethanol extraction. During subsequent heating, formation and explosion of ethyl perchlorate is very probable.

Aluminium powder,
Titanium dioxide

Fire Prot. Assoc. J., 1957, (36), 9

A mixture of the 3 compounds exploded violently during mixing. Previously the mixture had been accidentally ignited by a spark. Aluminium powder is incompatible with oxidants.

See ALUMINIUM, Al: Metal oxides

Metal powders

Schumacher, 1960, 210

The mixture of aluminium and/or magnesium powders with potassium perchlorate (a photo-flash composition) is very readily ignited, and three industrial explosions have occurred. Mixtures of nickel and titanium powders with the perchlorate and infusorial earth are very friction-sensitive, causing severe explosions, and easily ignited by very small (static) sparks.

Reducants

See PERCHLORATES: Reducants

Sulphur

Schumacher, 1960, 211–212

Mixtures of sulphur and potassium perchlorate, used in pyrotechnic devices, can be exploded by moderate impact. All other inorganic perchlorates form such impact-sensitive mixtures.

See other METAL OXOHALOGENATES

LITHIUM PERCHLORATE $ClLiO_4$

Nitromethane

See NITROMETHANE, CH_3NO_2: Lithium perchlorate

NITROSYL CHLORIDE $ClNO$

Acetone,
Platinum
Kaufmann, G. B., *Chem. Eng. News,* 1957, **35**(43), 60
A cold sealed tube containing nitrosyl chloride, platinum wire and traces of acetone exploded violently on being allowed to warm up.

Hydrogen,
Oxygen
See NITROGEN OXIDE, NO: Hydrogen, etc.
See other N-HALOGEN COMPOUNDS

N-CHLOROSULPHINYLIMIDE $ClNOS$

Anon., *Angew. Chem. (Nachr.),* 1970, **18**, 318
The ampouled solid exploded violently on melting. Distillation at normal pressure and impact tests had not previously indicated instability.
See other N-HALOGEN COMPOUNDS

NITRYL CHLORIDE $ClNO_2$

Inorganic materials,
or Organic matter
1. Batey, H. H. *et al., J. Amer. Chem. Soc.,* 1952, **74**, 3408
2. Kaplan, R. *et al., Inorg. Synth.,* 1954, **4**, 54

Interaction of the chloride with ammonia or sulphur trioxide is very violent, even at $-75°C$, and is vigorous with stannic bromide or iodide [1]. It attacks organic matter rapidly, sometimes explosively [2].
See other N-HALOGEN COMPOUNDS

CHLORINE NITRATE $ClNO_3$

Organic materials
Schmeisser, M., *Inorg. Synth.,* 1967, **9**, 129
Not inherently explosive, but it reacts explosively with alcohols, ethers and most organic materials.
See also FLUORINE NITRATE, FNO_3
See other OXIDANTS

NITROSYL PERCHLORATE $ClNO_5$

Pentaammineazidocobalt(III) perchlorate,
Phenylisocyamate
Burmeister, J. L. *et al., Chem. Eng. News,* 1968, **46**(8), 39
During an attempt to introduce phenylisocyanate into the Co co-ordination sphere, a mixture of the 3 components exploded when stirring was stopped.

Organic materials
Hoffman, K. A. *et al., Ber.,* 1909, **42**, 2031
As the anhydride of nitrous and perchloric acids, it is a very powerful oxidant. Pinene explodes sharply; acetone and ethanol ignite, then explode. Ether evolves gas, then explodes after a few seconds' delay. Small amounts of primary aromatic amines–aniline, toluidines, xylidines, mesidine–ignite on contact, while larger quantities explode dangerously, probably owing to rapid formation of diazonium perchlorates. Urea ignites on stirring with the perchlorate.
See other OXIDANTS

NITRYL PERCHLORATE $ClNO_6$

Ammonium perchlorate
See AMMONIUM PERCHLORATE, ClH_4NO_4: Impurities

Organic solvents
1. Spinks, J. W. T., *Chem. Eng. News,* 1960, **38**(15), 5
2. Gordon, W. E. *et al., Can. J. Res.,* 1940, **18B**, 358
Interaction with benzene gave a slight explosion and flash [1], while sharp explosions with ignition were observed with acetone and ether [2].
See other OXIDANTS

CHLORINE AZIDE ClN_3

Alone,
or Ammonia,

or Phosphorus,
or Silver azide,
or Sodium

1. Frierson, W. J. *et al., J. Amer. Chem. Soc.*, 1943, **65**, 1696, 1698
2. Rice, W. J. *et al., J. Chem. Educ.*, 1971, **48**, 659

The undiluted material is extremely unstable, usually exploding violently without cause at any temperature, even as solid at – 100°C. It gives an explosive yellow liquid with liquid ammonia; when condensed on to yellow phosphorus at –78°C an extremely violent explosion soon occurs. Addition of phosphorus to a solution of the azide in carbon tetrachloride at 0°C causes a series of mild explosions if the mixture is stirred or a violent explosion without stirring. Contact of the liquid or gaseous azide with silver azide at –78°C gave a blue colour, soon followed by explosion, and sodium reacted similarly under the same conditions [1].

When chlorine azide (25 mol %) is used as a thermally activated explosive initiator in a chemical gas laser tube, the partial pressure of the azide should never exceed 16 mbar [2].

See HALOGEN AZIDES

SULPHURYL AZIDE CHLORIDE ClN_3O_2S

Shozda, R. J. *et al., J. Org. Chem.*, 1967, **32**, 2876

During the preparation of this explosive liquid by interaction of sulphuryl chloride fluoride and sodium azide, traces of chlorine must be eliminated from the former to avoid detonation. The product is nearly as shock-sensitive as glyceryl trinitrate and may explode on rapid heating. Solutions (25% wt.) in solvents may be handled safely. The corresponding fluoride is believed to behave similarly.

See other NON-METAL AZIDES

THIOTRITHIAZYL PERCHLORATE $ClN_3O_4S_4$

Organic solvents

Goehring, 1957, 74

The precipitated perchlorate salt exploded on washing with acetone or ether.

THIOTRITHIAZYL CHLORIDE ClN_3S_4

Ammonia

Mellor, 1940, Vol. 8, 631–2.

The dry chloride (of uncertain structure; explodes on heating in air) rapidly absorbs ammonia gas and then explodes. Other thiotrithiazyl salts are explosive.

See also THIOTRITHIAZYL NITRATE, $N_4O_3S_4$
THIOTRITHIAZYL PERCHLORATE, $ClN_3O_4S_4$

TRIAZIDOCHLOROSILANE ClN_9Si

See TETRAAZIDOSILANE, $N_{12}Si$

SODIUM CHLORIDE $ClNa$

Lithium

See LITHIUM, Li: Sodium carbonate, etc.

SODIUM HYPOCHLORITE $ClNaO$

Alone

Brauer, 1963, Vol. 1, 311

The anhydrous solid, obtained by desiccation of the pentahydrate, tends to decompose explosively.

Amines

Kirk-Othmer, 1963, Vol. 2, 105

Primary aliphatic or aromatic amines react with sodium (or calcium) hypochlorite to form *N*-mono- or di-chloroamines which are explosive but less so than nitrogen trichloride.

Aziridine
Graefe, A. F. *et al., J. Amer. Chem. Soc.*, 1958, **80**, 3939
Interaction with sodium (or other) hypochlorite gives the explosive *N*-chloro compound.
See 1-CHLOROAZIRIDINE, C_2H_4ClN

Methanol
ICI Mond Div., private comm., 1968
Several explosions involving methanol and sodium hypochlorite were attributed to formation of methyl hypochlorite, especially in presence of acid or other esterification catalyst.
See HYPOHALITES

Phenylacetonitrile
Libman, D. D., private comm., 1968
Use of sodium hypochlorite solution to destroy acidified benzyl cyanide residues caused a violent explosion, thought to have been due to formation of nitrogen trichloride.
See other METAL OXOHALOGENATES

SODIUM CHLORITE **$ClNaO_2$**

Alone
Brauer, 1963, Vol. 1, 312
The anhydrous salt explodes on impact.

Organic matter
'The Diox Process', Newark, N. J., Wallace and Tiernan, 1949
Intimate mixtures of the solid chlorite with finely divided or fibrous organic matter may be very sensitive to heat, impact or friction.

Oxalic acid
MCA Case History No. 839
A bleach solution was being prepared by mixing solid sodium chlorite, oxalic acid and water, in that order. As soon as water was added, chlorine dioxide was evolved, and later exploded. The lower explosive limit of the latter is 10%, and the mixture is photo- and heat-sensitive.

Sodium dithionite
Anon., *Chem. Trade J.*, 1953, **132**, 564
Use of a scoop contaminated with sodium dithionite for sodium chlorite caused ignition of the latter. Materials containing sulphur (dithionite, natural rubber gloves) cause decomposition of sodium chlorite and contact should be avoided.
See other METAL OXOHALOGENATES

SODIUM CHLORATE **$ClNaO_3$**

Ammonium salts,
or Metals,
or Non-metals,
or Sulphides
MCA SD-42, 1952
Mixtures of the chlorate with ammonium salts, powdered metals, phosphorus, silicon, sulphur or sulphides are readily ignited and potentially explosive.
See METAL HALOGENATES

Nitrobenzene
Hodgson, J. F., private comm., 1973
The combination is powerfully explosive and has been widely used in recent guerrilla activities.

Organic matter
MCA SD-42, 1952
Cook, W. H., *Can. J. Res.,* 1933, **8**, 509
Mixtures of sodium (and other) chlorates with fibrous or absorbent organic materials, (charcoal, flour, shellac, sawdust, sugar), are hazardous. If the chlorate concentration is high, the mixtures may be ignited or caused to explode by static, friction or shock. Even at 10–15% concentration, low relative humidity may allow easy ignition and rapid combustion to occur.
See Paper, Static electricity; Wood, below
See also OXIDANTS AS HERBICIDES

Paper,
Static electricity
Ewing, O. R., *Chem. Eng. News,* 1952, **30**, 3210

Paper impregnated with sodium chlorate and dried, can be ignited by static sparks, but not by friction or impact. Paper bags or card cartons are unsuitable packing materials.

Phosphorus
Anon., *Angew. Chem. (Nachr.)*, 1957, **5**, 78
A mixture of red phosphorus and sodium chlorate exploded violently.

Sodium phosphinate
See SODIUM PHOSPHINATE, H_2NaO_2P: Oxidants

Sulphuric acid
1. *ABCM Quart. Safety Summ.*, 1944, **15**, 3
2. *MCA Case History No. 282*
Erroneous addition of concentrated sulphuric acid to solid sodium chlorate instead of sodium chloride caused an explosion due to formation of chlorine dioxide [1]. Accidental contact of 93% acid on clothing previously splashed with sodium chlorate caused immediate ignition [2].

Wood
1. *ABCM Quart. Safety Summ.*, 1947, **18**, 25
2. *BCISC Quart. Safety Summ.*, 1967, **38**, 42
Various fires and explosions caused by use of wooden containers with chlorates, and precautions necessary during handling and storage, are discussed [1,2].
See Organic matter, above

SODIUM PERCHLORATE $ClNaO_4$

See METAL PERCHLORATES

ANTIMONY(III) CHLORIDE OXIDE ClOSb

Bromine trifluoride
See BROMINE TRIFLUORIDE, BrF_3: Antimony(III) chloride oxide

1. Sidgwick, 1950, 1203
2. Mellor, 1941, Vol. 2, 288
3. Stedman, R. F., *Chem. Eng. News,* 1951, **29**, 5030
4. Cameron, A. E., ibid., 3196
5. Fawcett, H. H., ibid., 4459
6. McHale, E. T. *et al., J. Phys. Chem.,* 1968, **72**, 1849
7. *Hazards of Chlorine Dioxide,* New York, National Board of Fire Underwriters, 1950; *Chem. Eng. News,* 1950, **28**, 611
8. Derby, R. I. *et al., Inorg. Synth.,* 1954, **4**, 152

This is a powerful oxidant and explodes violently on the slightest provocation as gas or liquid [1]. It is initiated by contact with several materials (below), on heating rapidly to 100°C or on sparking [2], or by impact as solid at –100°C [3]. A small sample exploded during vacuum distillation at below –50°C [4], and it was stated that decomposition by sparking begins to become hazardous at concentrations of 7–8% in air [3], and that at 10% concentrations in air or at 0.1 bar partial pressure explosion may occur from any source of initiation such as sunlight, heat or electrostatic discharge [5].

A kinetic study of the decomposition shows that it is explosive above 45°C even in absence of light, and subject to long induction periods due to formation of intermediate dichlorine trioxide. UV irradiation greatly sensitises the dioxide to explosion [6].

A guide on fire and explosion hazards in industrial use of chlorine dioxide is available [7], and preparative precautions have been detailed [8].

Carbon monoxide
Mellor, 1941, Vol. 2, 288
Explosion on mixing.

Hydrogen
Mellor, 1941, Vol. 2, 288
Near-stoicheiometric mixtures detonate on sparking, or on contact with platinum sponge.

Mercury
Mellor, 1941, Vol. 2, 288
Chlorine dioxide explodes on shaking with mercury.

Non-metals
Mellor, 1941, Vol. 2, 289; 1956, Vol. 2, Suppl. 1, 532
Phosphorus, sulphur, sugar or combustible materials ignite on contact and may cause explosion.

Phosphorus pentachloride
See PHOSPHORUS PENTACHLORIDE, Cl_5P: Chlorine dioxide

Potassium hydroxide
Mellor, 1941, Vol. 2, 289
The liquid or gaseous oxide will explode in contact with solid potassium hydroxide or its concentrated solution.
See other HALOGEN OXIDES

CHLORINE TRIOXIDE ClO_3

Schmeisser, M., *Angew. Chem.*, 1955, **67**, 498
The dimeric form is formulated as $ClO_2^+ ClO_4^-$.
See CHLORYL PERCHLORATE, Cl_2O_6

THALLIUM(I) CHLORIDE ClTl

Fluorine
See FLUORINE, F_2 : Metal salts

CHLORINE Cl_2

MCA SD-80, 1970

Alcohols
See *tert*-Butanol, below
HYPOHALITES

Aluminium

Ann. Rep. Chief. Insp. Factories, 1953 (Cmd. 9330), 171

Corrosive failure of a vaporiser used in manufacture of aluminium chloride caused liquid chlorine to contact molten aluminium. A series of explosions occurred.

See ALUMINIUM, Al: Halogens
: Oxidants

See other Metals, below

Bromine pentafluoride

See BROMINE PENTAFLUORIDE, BrF_5: Acids, etc.

tert-Butanol

1. Bradshaw, C. P. C. *et al., Proc. Chem. Soc.*, 1963, 213
2. Mintz, M. J. *et al., Org. Synth.*, 1969, **49**, 9

Rate of admission of chlorine into the alcohol during preparation of *tert*-butyl hypochlorite must be regulated to keep temperature below 20°C to prevent explosion [1]. A safer and simpler preparation uses hypochlorite solution in place of chlorine [2].

See also tert-BUTYL HYPOCHLORITE, C_4H_9ClO

Carbon disulphide

MCA Case History No. 971

When liquid chlorine was added to carbon disulphide in an iron cylinder, an explosion occurred, due to the iron-catalysed chlorination of carbon disulphide to carbon tetrachloride. The operation had been done previously in glassware without incident.

Dibutyl phthalate

Statesir, W. A., *Chem. Eng. Progr.*, 1973, **69**(4), 54

A mixture of the ester and liquid chlorine confined in a stainless steel bomb reacted explosively at 118°C.

Dicaesium oxide

See DICAESIUM OXIDE, Cs_2O: Halogens

Diethyl ether

1. Harrison, G. E., private comm., 1966
2. Unpublished observation, 1949

Chlorine caused ignition of ether on contact [1]. Exposure of an ethereal solution of chlorine to daylight caused a mild, photocatalysed, explosion [2].

See also BROMINE, Br_2 : Diethyl ether

Dioxygen difluoride

See DIOXYGEN DIFLUORIDE, F_2O_2

Fluorine

See FLUORINE, F_2 : Halogens

Glycerol

Statesir, W. A., *Chem. Eng. Progr.*, 1973, **69**(4), 54

A mixture of glycerol and liquid chlorine confined in a stainless steel bomb exploded at 70–80°C

Hexachlorodisilane

Martin, G., *J. Chem. Soc.*, 1914, **105**, 2859

Hexachlorodisilane vapour ignited in chlorine above 300°C; violent explosions sometimes occurred.

Hydrocarbons

1. Mellor, 1956, Vol. 2, Suppl. 1, 380
2. Eisenlohr, D. H., US Pat. 2 989 571, 1961
3. von Schwartz, 1918, 142, 321
4. von Schwartz, 1918, 324
5. Mamadaliev, Y. G. *et al.*, *Chem. Abs.*, 1937, **31**, 8502[5]
6. Brooks, B. T., *Ind. Eng. Chem.*, 1924, **17**, 752
7. Johnson, J. H. *et al.*, *Hydrocarbon Proc. Petr. Ref.*, 1963, **42**(2), 174
8. Statesir, W. A., *Chem. Eng. Progr.*, 1973, **69**(4), 52–54
9. de Oliveria, D. B., *Hydrocarbon Proc.*, 1973, **53**(3), 112–126

Interaction of chlorine with methane is explosive at room temperature over yellow mercury oxide [1], and mixtures containing above 20 vol. % of chlorine are explosive [2]. Mixtures of acetylene and chlorine may explode on initiation by sunlight, other UV source, or high temperatures, sometimes very violently [3]. Mixtures with ethylene explode on initiation by sunlight, etc., or over mercury, dimercury or disilver oxides at ambient temperature, or over lead oxide at 100°C [1,4]. Interaction with ethane over activated carbon at 350°C has caused explosions, but added carbon dioxide reduces the risk [5].

Accidental introduction of gasoline into a cylinder of liquid chlorine caused a slow exothermic reaction which accelerated to detonation. This effect was verified [6]. Injection of liquid chlorine into a naphtha–sodium hydroxide mixture (to generate hypochlorite *in situ*) caused a violent explosion. Several other incidents involving violent reactions of saturated hydrocarbons with chlorine were noted [7].

In a review of incidents involving explosive reactivity of liquid chlorine with various organic auxiliary materials, two involved hydrocarbons. A polypropylene filter element fabricated with zinc oxide filler reacted explosively, rupturing the steel case previously tested to over 300 bar. Zinc chloride derived from the oxide may have initiated the runaway reaction. Hydrocarbon-based diaphragm pump oils or metal-drawing waxes were violently or explosively reactive [8]. A violent explosion in a wax chlorination plant may have involved unplanned contact of liquid chlorine with wax or chlorinated wax residues in a steel trap. Corrosion products in the trap may have catalysed the runaway reaction, but hydrogen (also liberated by corrosion in the trap) may also have been involved [9].

See TURPENTINE: Halogens

Hydrogen

1. Mellor, 1956, Vol. 2, Suppl. 1, 373–375
2. Weissweiler, A., *Z. Elektrochem.*, 1936, **42**, 499
3. Eichelberger, W. C. *et al., Chem. Eng. Progr.*, 1961, **57**(8), 94
4. Wood, J. L., *Loss Prevention,* 1969, **3**, 45–47
5. Stephens, T. J. R. *et al.*, Paper 62B, *65th A.I. Chem. Eng. Meeting, New York, 1972*
6. de Oliveria, D. B., *Hydrocarbon Proc.*, 1973, **52**(3), 112–126

Combination of the elements may be explosive over a wide range of physical conditions, with initiation by sparks, radiant energy or catalysis, e.g. by yellow mercuric oxide at ambient temperature [1]. There is a narrow range of concentrations in which the mixture is super-sensitive to initiation [2]. Explosion–detonation phenomena in chlorine production cells have been investigated [3,4]. The explosive limits of the mixture vary with the container shape and method of initiation, but are usually within the range 5–89% of hydrogen by volume [1]. After an explosion in a chlorine distillate receiver where hydrogen had been produced by corrosion, no initiation source could be identified or reasonably postulated following a thorough investigation

[5]. Several other hydrogen–chlorine explosions without identifiable ignition sources are also mentioned. Hydrogen may have been involved in a severe wax chlorinator explosion [6].

Metal acetylides and carbides
The mono- and di-alkali metal acetylides, copper acetylides, iron, uranium and zirconium carbides all ignite in chlorine, the former often at ambient temperature.

See TRIIRON CARBIDE, CFe_3: Halogens
DICAESIUM ACETYLIDE, C_2Cs_2: Halogens
COPPER ACETYLIDE, C_2Cu: Halogens
DICOPPER(I) ACETYLIDE, C_2Cu_2: Halogens
CAESIUM ACETYLIDE, C_2HCs
RUBIDIUM ACETYLIDE, C_2HRb
DILITHIUM ACETYLIDE, C_2Li_2: Halogens
STRONTIUM ACETYLIDE, C_2Sr: Halogens
URANIUM DICARBIDE, C_2U: Halogens
ZIRCONIUM DICARBIDE, C_2Zr: Halogens
See also LITHIUM ACETYLIDE-AMMONIA, $C_2HLi{\cdot}H_3N$: Gases

Metal hydrides
Mellor, 1941, Vol. 2, 483; Vol. 3, 73
Potassium, sodium and copper hydrides all ignite in chlorine at ambient temperatures.

Metal phosphides
See TRICOPPER DIPHOSPHIDE, Cu_3P_2: Oxidants

Metals
Mellor 1941, Vol. 2, 92, 95; 1956, Vol. 2, Suppl. 1, 380, 469; 1941, Vol. 3, 638; 1940, Vol. 4, 267, 480; 1941, Vol. 7, 208, 260, 436; 1941, Vol. 9, 379, 626; 1942, Vol. 12, 312; Vol. 15, 146
Tin ignites in liquid chlorine at $-34°C$, aluminium powder in the gas at $-20°C$, while vanadium powder explodes on contact at $0°C$ with the pressurised liquid. A solution in heptane ignites in contact with powdered copper well below $0°C$. Aluminium, brass foil, calcium powder, copper foil, iron wire, manganese powder and potassium all ignite in the dry gas at ambient temperature, as do powdered antimony, bismuth and germanium sprinkled into the gas, while magnesium, sodium and zinc ignite in the moist gas. Thorium, tin and uranium ignite and incandesce on warming (uranium to $150°C$), and powdered

nickel burns at 600°C. Aluminium–titanium alloys also ignite on heating in chlorine.

See Aluminium, above
Steel, below

Nitrogen compounds

1. Mellor, 1941, Vol. 2, 95; 1940, Vol. 8, 99, 288, 313, 607
2. Bowman, W. R. *et al., Chem. & Ind.,* 1963, 979
3. Bainbridge, E. G., ibid., 1350
4. Folkers, K. H. *et al., J. Amer. Chem. Soc.,* 1941, **63**, 3530

Ammonia–chlorine mixtures are explosive if warmed or if chlorine is in excess, owing to formation of nitrogen trichloride. Hydrazine, hydroxylamine and calcium nitride ignite in chlorine, and nitrogen triiodide may explode on contact with chlorine [1]. During chlorination of impure biuret in water at 20°C, a violent explosion occurred [2]. This was attributed to conversion of the cyanuric acid impurity (3%) to nitrogen trichloride and spontaneous explosion of the latter [3]. During interaction of chlorine and alkylthiouronium salts to give alkanesulphonyl chlorides, the dangerously explosive nitrogen trichloride may be produced if excess chlorine or slow chlorination is used. General precautions are discussed [4]. Aziridine readily forms the explosive *N*-chloro derivative.

See 1-CHLOROAZIRIDINE, C_2H_4ClN
N-HALOGEN COMPOUNDS

Non-metal hydrides

Mellor, 1939, Vol. 9, 55, 396; 1939, Vol. 8, 65; 1940, Vol. 6, 219; 1941, Vol. 5, 37

Arsine, phosphine and silane all ignite in contact with chlorine at ambient temperature, while diborane and stibine react explosively, the latter also with chlorine water.

See also ETHYLPHOSPHINE, C_2H_7P: Halogens

Non-metals

Mellor, 1941, Vol. 2, 92; 1956, Vol. 2, Suppl. 1, 380; 1943, Vol. 11, 26

Liquid chlorine at –34°C explodes with white phosphorus, and a solution in heptane at 0°C ignites red phosphorus. Boron, active carbon, silicon and phosphorus all ignite in contact with gaseous chlorine at ambient temperature. Arsenic incandesces on contact with liquid chlorine at –34°C, and the powder ignites when sprinkled into the gas at ambient temperature. Tellurium must be warmed slightly before incandescence occurs.

Oxygen difluoride
See OXYGEN DIFLUORIDE, F_2O: Halogens

Phosphorus compounds
1. Mellor, 1940, Vol. 8, 812, 842, 844–845, 897
2. von Schwartz, 1918, 324

Boron diiodophosphide, phosphine, phosphorus trioxide and trimercury tetraphosphide all ignite on contact with chlorine at ambient temperature. Trimagnesium diphosphide and trimanganese diphosphide ignite in warm chlorine [1], while ethylphosphine explodes with chlorine [2].

Polychlorobiphenyl
Statesir, W. A., *Chem. Eng. Progr.*, 1973, **69**(4), 53–54
A mixture of a polychlorobiphenyl process oil and liquid chlorine confined in a stainless steel bomb reacted exothermically between 25 and 81°C.

Silicones
Statesir, W. A., *Chem. Eng. Progr.*, 1973, **69**(4), 53–54
Silicone process oils mixed with liquid chlorine confined in a stainless steel bomb reacted explosively on heating: polydimethylsiloxane at 88–118°C, and polymethyltrifluoropropylsiloxane at 68–114°C. Previously, leakage of a silicone pump oil into a liquid chlorine feed system had caused rupture of a stainless steel ball valve under a pressure surge of about 2 kbar.

Steel
1. *MCA Case History No. 608*
2. Stephens, T. J. R. *et al.*, Paper 62B, 5, *65th A. I. Chem. Eng. Meeting, New York, 1972*

Chlorine leaking into a steam-heated mild steel pipe caused ignition of the latter at *ca.* 250°C [1]. Sheet steel in contact with chlorine usually ignites at 200–250°C, but the presence of soot, rust, carbon or other catalysts may reduce the ignition temperature to 100°C. Dry steel wool ignites with chlorine at only 50°C [2].

Sulphides
Mellor, 1940, Vol. 4, 952; 1946, Vol. 5, 144; 1939, Vol. 9, 270
Diarsenic disulphide, diboron trisulphide and mercuric sulphide all ignite in chlorine at ambient temperature, the first only in a rapid stream.

Synthetic rubber
Murray, R. L., *Chem. Eng. News,* 1948, **26**, 3369
During interaction of synthetic rubber and liquid chlorine, a violent explosion occurred. It is known that natural and synthetic rubbers will burn in liquid chlorine.

Tetraselenium tetranitride
See TETRASELENIUM TETRANITRIDE, N_4Se_4

Trialkylboranes
Coates, 1967, Vol. 1, 199
The lower homologues tend to ignite in chlorine or bromine.

Tungsten dioxide
Mellor, 1943, Vol. 11, 851
Incandescence on warming.
See other HALOGENS

COBALT(II) CHLORIDE Cl_2Co

Metals
See POTASSIUM, K: Metal halides
SODIUM, Na: Metal halides

DIHYDRAZINECOBALT(II) CHLORATE $Cl_2CoH_8N_4O_6$

Salvadori, R., *Gazz. Chim. Ital.*, 1910, [2], **40**, 9
Explodes powerfully on slightest impact or friction, or on heating to 90°C.
See other AMMINEMETAL OXOSALTS

HEXAAQUACOBALT(II) PERCHLORATE $Cl_2CoH_{12}O_{14}$

Salvadori, R., *Gazz. Chim. Ital.*, 1910, [2], **40**, 9
Explodes on impact; deflagrates on rapid heating.
See other AMMINEMETAL OXOSALTS

PENTAAMMINEPHOSPHINECOBALT(III) PERCHLORATE $Cl_2CoH_{17}N_5O_{10}P$

BCISC Quart. Safety Summ., 1965, **36**, 58
When a platinum wire (which may have been hot) was dipped for a flame test into a sintered funnel containing the air-dried complex, detonation occurred. This may have been due to heat and/or friction on a compound containing both strongly oxidising and strongly reducing radicals. Avoid dipping platinum wire into bulk samples of materials of unknown potential.
See other AMMINEMETAL OXOSALTS

CHROMYL CHLORIDE Cl_2CrO_2

1. Sidgwick, 1950, 1004
2. Pitwell, L. R., private comm., 1964

Though a powerful and often violent oxidant of inorganic and organic materials in absence of a diluent, it has found use as a solution in preparative organic chemistry for the controlled oxidation of alkyl aromatics [1]. In such a reaction, failure of a stirrer during addition of the chloride caused a build-up of unreacted material, followed by a violent explosion [2].

Ammonia
Mellor, 1943, Vol. 11, 394
Contact with ammonia causes incandescence.

Disulphur dichloride
Mellor, 1943, Vol. 11, 394
A jet of chromyl chloride vapour ignites in the vapour of disulphur dichloride.

Organic solvents
Mellor, 1943, Vol. 11, 396
Acetone, ethanol and ether ignite on contact with the chloride. Turpentine behaves similarly.

Phosphorus,
or Phosphorus trichloride
Mellor, 1943, Vol. 11, 395
Moist phosphorus explodes in contact with the liquid chloride. Addition of drops of chromyl dichloride to cooled phosphorus trichloride causes incandescence, and sometimes explosion.

Sodium azide
Mellor, 1967, Vol. 8, Suppl. 2.2, 36
Interaction of chromyl chloride and sodium azide to form chromyl azide is explosive in absence of a diluent.

Sulphur
Mellor, 1943, Vol. 11, 394
Contact of the liquid chloride with flowers of sulphur causes ignition.
See other OXIDANTS

CHROMYL PERCHLORATE Cl_2CrO_{10}

Alone,
or Organic solvents
Schmeisser, M., *Angew. Chem.*, 1955, **67**, 499
A powerful oxidant, which causes organic solvents to ignite on contact and which explodes violently above 80°C.
See other OXIDANTS

IRON(II) CHLORIDE Cl_2Fe

Ozonides
See OZONIDES: Metals, etc.

IRON(II) PERCHLORATE Cl_2FeO_8

Preparative hazard.
See PERCHLORIC ACID, $ClHO_4$: Iron sulphate

DICHLOROGERMANE Cl_2GeH_2

Ammonia
See CHLOROGERMANE, $ClGeH_3$: Ammonia

N,N-BIS(CHLOROMERCURI)HYDRAZINE $Cl_2H_2Hg_2N_2$

Hofmann, K. A. *et al., Ann.*, 1899, **305**, 217
An explosive compound.
See other N-METAL DERIVATIVES

CHLORONIUM PERCHLORATE $Cl_2H_2O_4$

Hantzsch, A., *Ber.*, 1930, **63**, 1789
This hydrogen chloride–perchloric acid complex spontaneously dissociates with explosive violence.

BISHYDROXYLAMINEZINC(II) CHLORIDE $Cl_2H_6N_2O_2Zn$

Walker, J. E. *et al., Inorg. Synth.*, 1967, **9**, 2
It explodes at 170°C.
See other AMMINEMETAL OXOSALTS

HYDRAZINIUM DIPERCHLORATE $Cl_2H_6N_2O_8$

Grelecki, C. J. *et al., ACS 54,* 1966, 73
The thermal decomposition of this solid rocket propellant component has been studied.
See other OXOSALTS OF NITROGENOUS BASES

BISHYDRAZINENICKEL(II) PERCHLORATE $Cl_2H_8N_4NiO_8$

Maissen, B. *et al., Helv. Chim. Acta*, 1951, **34**, 2084–2085
The salt known to be explosive when heated dry (but not under a hammer blow), exploded violently when stirred as a dilute aqueous suspension.
See other AMMINEMETAL OXOSALTS

BISHYDRAZINETIN(II) CHLORIDE $Cl_2H_8N_4Sn$

Mellor, 1941, Vol. 7, 430
Explodes on heating.
See other AMMINEMETAL OXOSALTS

TETRAAMINEBIS-DINITROGENOSMIUM DIPERCHLORATE $Cl_2H_{12}N_8O_8Os$

Creutz, C. A., private comm., 1969
The dry salt may explode on touching.
See other AMMINEMETAL OXOSALTS

MERCURY(II) CHLORIDE Cl_2Hg

Sodium *aci*-nitromethane
See SODIUM *aci*-NITROMETHANE, CH_2NNaO_2 : Mercury(II) chloride

MERCURY DICHLORITE Cl_2HgO_4

Levi, G. R., *Gazz. Chim. Ital.*, 1915, [2], **45**, 161
It is extremely unstable when dry, exploding spontaneously.
See other CHLORITE SALTS

MAGNESIUM CHLORATE Cl_2MgO_6

Copper(II) sulphide
See CHLORIC ACID, $ClHO_3$: Copper(II) sulphide

MAGNESIUM PERCHLORATE Cl_2MgO_8

Arylhydrazine
Belcher, R., private comm., 1968
Anhydrous magnesium perchlorate was used to thoroughly dry an ethereal solution of an arylhydrazine. During evaporation of the filtered solution it exploded completely and violently. Magnesium perchlorate is rather soluble in ether and may contain traces of free perchloric acid (probably in the anhydrous form, as the magnesium salt is a powerful dehydrator). It is entirely unsuitable for drying organic solvents.

Cellulose,
Dinitrogen tetraoxide,
Oxygen
ABCM Quart. Safety Summ., 1961, **32**, 6
Magnesium perchlorate contained in a glass tube between wads of cotton wool was used to dry a mixture of oxygen and dinitrogen tetraoxide. After several days, the drying tube exploded violently. It seems probable that the acidic fumes and cotton produced cellulose nitrate, aided by the dehydrating action of the perchlorate.

Dimethyl sulphoxide
1. Tobe, M. L. *et al.*, *J. Chem. Soc.*, 1964, 2991
2. Anon., *Chem. Eng. News*, 1965, **43**(47), 62

3. Dessy, R. E. *et al., J. Amer. Chem. Soc.*, 1964, **86**, 28
In the preparation of anhydrous dimethyl sulphoxide by a literature method [1] an explosion occurred during distillation from anhydrous magnesium perchlorate [2]. This may have been due to the presence of some free methanesulphonic acid as an impurity in the solvent, which could liberate perchloric acid. It is known that sulphoxides react explosively with 70% perchloric acid. The alternative procedure for drying dimethylsulphoxide with calcium hydride [3] seems preferable, as this would also remove any acidic impurities.

Hydrogen fluoride
Anon., *Ind. Eng. Chem. (News Ed.)*, 1939, **17**, 70
A violent explosion followed the use of magnesium perchlorate to dry wet fluorobutane. The latter had hydrolysed to give hydrogen fluoride which liberated anhydrous, explosive perchloric acid.* Magnesium perchlorate is unsuitable for drying acid or inflammable materials; calcium sulphate would be suitable.

Organic materials
1. Hodson, R. J., *Chem. & Ind.*, 1965, 1873
2. *MCA Case History No. 243*
The use of the perchlorate as desiccant in a drybag where contamination with organic compounds is possible is considered dangerous [1]. Magnesium perchlorate ('Anhydrone') was inadvertently used instead of calcium sulphate (anhydrite) to dry a reaction product (unstated) before vacuum distillation. The error was realised and all solid was filtered off. Towards the end of the distillation, decomposition and a violent explosion occurred, possibly due to the presence of dissolved magnesium perchlorate, or more probably to perchloric acid present as impurity in the magnesium perchlorate [2].

Phosphorus
See PHOSPHORUS, P: Magnesium perchlorate

Trimethyl phosphite
Jercinovic, L. M. *et al., J. Chem. Educ.*, 1968, **45**, 751

* This seems an unlikely explanation because of the large difference in dissociation constants of the acids.

Contact between the components caused violent explosions on several occasions.
See other METAL OXOHALOGENATES

MANGANESE(II) CHLORIDE Cl_2Mn

Zinc
See ZINC, Zn: Manganese dichloride

MANGANESE DIPERCHLORATE Cl_2MnO_8

Sidgwick, 1950, 1285
Explodes at 195°C

2,2-Dimethoxypropane
See 2,2-DIMETHOXYPROPANE, $C_5H_{12}O_2$: Metal perchlorates
See other METAL OXOHALOGENATES

VANADYL AZIDE DICHLORIDE Cl_2N_3OV

Explosive solid.
See METAL AZIDE HALIDES

DIAZIDODICHLOROSILANE Cl_2N_6Si

See TETRAAZIDOSILANE, $N_{12}Si$

NICKEL DIPERCHLORATE Cl_2NiO_8

2,2-Dimethoxypropane
See 2,2-DIMETHOXYPROPANE, $C_5H_{12}O_2$: Metal perchlorates

DICHLORINE OXIDE Cl_2O

Alone
Sidgwick, 1950, 1202
The liquid may explode on pouring or on boiling at 2°C, and the gas readily explodes on heating or sparking.

Carbon
Mellor, 1956, Vol. 5, 824
Addition of charcoal to the gas causes an immediate explosion, probably due to initiation by the heat of adsorption on the solid.

Dicyanogen
Brotherton, T. K. *et al.*, *Chem. Rev.*, 1959, **59**, 843
Contact causes ignition or explosion.

Diphenylmercury
See DIPHENYLMERCURY, $C_{12}H_{10}Hg$: Non-metal oxides

Nitrogen oxide
Mellor, 1940, Vol. 8, 433
Interaction is explosive.

Oxidisable materials
1. Mellor, 1941, Vol. 2, 241–242; 1946, Vol. 5, 824
2. Jacobs, P. W. M. *et al.*, *Chem. Rev.*, 1969, **69**, 559
3. Pilipovich, D. *et al.*, *Inorg. Chem.*, 1972, **11**, 2190
4. Cady, G. H., *Inorg. Synth.*, 1957, **5**, 156

The heat-sensitivity (above) may explain the explosions which occur on contact of many readily oxidisable materials with this powerful oxidant. Such materials include ammonia, potassium; arsenic, antimony; sulphur, charcoal (adsorptive heating may also contribute); calcium phosphide, phosphine, phosphorus; hydrogen sulphide and antimony, barium, mercury and tin sulphides. Various organic materials (paper, cork, rubber, turpentine, etc.) behave similarly [4]. Mixtures with hydrogen detonate on ignition [1]. The oxide explodes if heated rapidly or overheated locally (sparking, or adiabatic compression in a U-tube [2,3]), or often towards the end of slow thermal decompositior

Kinetic data are summarised [2]. Preparative precautions have been detailed [4].

Potassium
See POTASSIUM, K: Non-metal oxides
See other HALOGEN OXIDES

SULPHINYL CHLORIDE (THIONYL CHLORIDE) Cl_2OS

Ammonia
Foote, C. S., private comm., 1965
Addition of a solution of *p*-nitrobenzoyl chloride in excess thionyl chloride to ice-cold concentrated ammonia solution caused a violent explosion. This may have been due to formation of nitrogen trichloride.

Chloryl perchlorate
See CHLORYL PERCHLORATE, Cl_2O_6: Organic matter, etc.

Dimethyl sulphoxide
See DIMETHYL SULPHOXIDE, C_2H_6OS: Acyl halides

Hexafluoroisopropylideneaminolithium
See HEXAFLUOROISOPROPYLIDENEAMINOLITHIUM, C_3F_6LiN: Non-metal halides

Linseed oil
Quinoline,
Laws, G. F., private comm., 1966
It is important to add the quinoline and linseed oil, used in purifying thionyl chloride, to the chloride. If reversed addition is used, a vigorous decomposition may occur.

Sodium
See SODIUM, Na: Non-metal halides
See other NON-METAL HALIDES

SELENINYL CHLORIDE Cl_2OSe

Antimony
Mellor, 1947, Vol. 10, 906
Powdered antimony ignites on contact.

Metal oxides
Mellor, 1947, Vol. 10, 909
In contact with disilver oxide, light is evolved and sufficient heat to decompose some silver oxide. Similar effects were observed with the three lead oxides.

Potassium
Mellor, 1947, Vol. 10, 908
Potassium explodes violently in contact with the liquid.

Phosphorus
Mellor, 1947, Vol. 10, 906
Red phosphorus evolves light and heat in contact with the chloride, while white phosphorus explodes.
See other NON-METAL HALIDES

SULPHONYL DICHLORIDE (SULPHURYL CHLORIDE) Cl_2O_2S

Alkalies
Brauer, 1963, Vol. 1, 385
Reaction with alkalies may be explosively violent.

Diethyl ether
Dunstan, I. *et al., Chem. & Ind.,* 1966, 73
A solution of sulphuryl chloride in ether vigorously decomposed, evolvin hydrogen chloride. This was shown to be accelerated by the presence of peroxides. Peroxide-free ether should be used, and with care.

Dimethyl sulphoxide
See DIMETHYL SULPHOXIDE, C_2H_6OS: Acyl halides

Dinitrogen pentaoxide
See DINITROGEN PENTAOXIDE, N_2O_5: Sulphur dichloride, etc.

Lead dioxide
See LEAD DIOXIDE, O_2Pb: Non-metal halides

Phosphorus
See PHOSPHORUS, P: Non-metal halides
See other NON-METAL HALIDES

DICHLORINE TRIOXIDE Cl_2O_3

McHale, E. T. *et al., J. Amer. Chem. Soc.*, 1967, **89**, 2796
Very unstable, its vapour explodes at about 2 mbar and well below 0°C.
See other HALOGEN OXIDES

CHLORINE PERCHLORATE Cl_2O_4

Schack, C. J. *et al., Inorg. Chem.*, 1970, **9**, 1387; 1971, **10**, 1078
A shock-sensitive compound; bromine perchlorate is also unstable.
See other HALOGEN OXIDES

LEAD DICHLORITE Cl_2O_4Pb

Alone,
or Antimony sulphide,
or Sulphur
Mellor, 1941, Vol. 2, 283
It explodes on heating above 100°C or on rubbing with antimony sulphide or fine sulphur.
See other CHLORITE SALTS
METAL OXOHALOGENATES

DISULPHURYL DICHLORIDE $Cl_2O_5S_2$

Phosphorus
See PHOSPHORUS, P: Non-metal halides

Water
Sveda, M., *Inorg. Synth.*, 1950, **3**, 127
The pure chloride reacts slowly with water but, if more than a few % of chlorosulphuric acid (a usual impurity) are present, the reaction is rapid and could become violent with large quantities.
See other NON-METAL HALIDES

CHLORYL PERCHLORATE Cl_2O_6

Organic matter,
or Thionyl chloride,
or Water
1. Mellor, 1956, Vol. 2, Suppl. 1, 539
2. Schmeisser, M., *Angew. Chem.*, 1955, **67**, 499
3. Wechsberg, M. *et al., Inorg. Chem.*, 1972, **11**, 3066

Though the least explosive of the chlorine oxides, being insensitive to heat or shock, it is a very powerful oxidant and needs careful handling. It violently or explosively oxidises ethanol, stopcock grease, wood and organic matter generally [1]; it is liable to explode on contact with thionyl chloride in absence of solvent [2], or with water [1]. Recently its explosion on heating has been noted [3].
See other HALOGEN OXIDES

ZINC DICHLORATE Cl_2O_6Zn

Copper(II) sulphide
See CHLORIC ACID, $ClHO_3$: Copper(II) sulphide

PERCHLORYL PERCHLORATE (DICHLORINE HEPTAOXIDE) Cl_2O_7

1. Mellor, 1956, Vol. 2, Suppl. 1, 542

2. Schmeisser, M., *Angew. Chem.*, 1955, **67**, 498
It explodes violently on impact or rapid heating, but is a less powerful oxidant than other chlorine oxides [1]. It is now formulated as perchloryl perchlorate [2].
See other HALOGEN OXIDES

LEAD DIPERCHLORATE **Cl_2O_8Pb**

Methanol
Willard, H. H. *et al.*, *J. Amer. Chem. Soc.*, 1930, **52**, 2396
A saturated solution of anhydrous lead diperchlorate in dry methanol exploded violently when disturbed. Methyl perchlorate may have been involved.
See ALKYL PERCHLORATES
See other METAL OXOHALOGENATES

XENON(II) PERCHLORATE **Cl_2O_8Xe**

Wechsberg, M. *et al.*, *Inorg. Chem.*, 1972, **11**, 3066
During preparation from perchloric acid and xenon difluoride at $-50°C$, violent explosions occurred if the reaction mixture was allowed to warm up rapidly.
See other XENON COMPOUNDS

URANYL DIPERCHLORATE **$Cl_2O_{10}U$**

Ethanol
Erametsa, O., *Suomen Kemist*, 1942, **15B**, 1
Attempted recrystallisation of the salt from ethanol caused an explosion.
See ALKYL PERCHLORATES
See other METAL OXOHALOGENATES

LEAD DICHLORIDE **Cl_2Pb**

Calcium
See CALCIUM, Ca: Lead dichloride

SULPHUR DICHLORIDE Cl_2S

Acetone

Fawcett, F. S. *et al., Inorg. Synth.*, 1963, 7, 121

Acetone is an effective solvent for cleaning traces of sulphur chlorides from reaction vessels, but care is necessary, as the reaction is vigorous if more than traces are present.

Dimethyl sulphoxide

See DIMETHYL SULPHOXIDE, C_2H_6OS: Acyl halides, etc.

Hexafluoroisopropylideneaminolithium

See HEXAFLUOROISOPROPYLIDENEAMINOLITHIUM, C_3F_6LiN: Non-metal halides

Metals

See POTASSIUM, K: Non-metal halides
SODIUM, Na: Non-metal halides

Oxidants

See PERCHLORYL FLUORIDE, $ClFO_3$: Hydrocarbons, etc.
NITRIC ACID, HNO_3: Sulphur halides
DINITROGEN PENTAOXIDE, N_2O_5: Sulphur dichloride

Water

MCA SD-77, 1960

Exothermic reaction with water or steam.

See other NON-METAL HALIDES

DISULPHUR DICHLORIDE (THIOSULPHINYL CHLORIDE) Cl_2S_2

Antimony,
or Antimony or arsenic sulphides

Mellor, 1947, Vol. 10, 641

Interaction at ambient temperature is surprisingly energetic.

Chromyl chloride

See CHROMYL CHLORIDE, Cl_2CrO_2: Disulphur dichloride

Dimethyl sulphoxide
See DIMETHYL SULPHOXIDE, C_2H_6OS: Acyl halides, etc.

Diphosphorus trioxide
See PHOSPHORUS(III) OXIDE, O_3P_2: Disulphur dichloride

Mercury oxide
Mellor, 1947, Vol. 10, 643
Interaction is rapid and very exothermic.

Potassium
Mellor, 1947, Vol. 10, 642
A mixture of potassium and the liquid chloride is shock-sensitive and explodes violently on heating.

Sodium peroxide
See SODIUM PEROXIDE, Na_2O_2: Non-metal halides

Unsaturated materials
1. Mellor, 1947, Vol. 10, 641–642
2. Huestis, B. L., *Safety Eng.*, 1927, **54**, 95

Alkenes, terpenes and unsaturated glycerides react exothermically, some vigorously [1]. Ignition may occur with some organic materials [2].

Water
MCA SD-77, 1960
As with sulphinyl chloride, the exothermic reaction with limited amounts of water may be dangerously violent under confinement because of rapid gas evolution.
See other NON-METAL HALIDES

DISELENIUM DICHLORIDE Cl_2Se_2

Alkali metals or oxides
See POTASSIUM, K: Non-metal halides
POTASSIUM DIOXIDE, KO_2: Diselenium dichloride
SODIUM, Na: Non-metal halides
SODIUM PEROXIDE, Na_2O_2: Non-metal halides

TIN DICHLORIDE Cl_2Sn

Bromine trifluoride
See BROMINE TRIFLUORIDE, BrF_3: Tin dichloride

Calcium acetylide
See CALCIUM ACETYLIDE, C_2Ca: Tin dichloride

Hydrogen peroxide
See HYDROGEN PEROXIDE, H_2O_2: Tin(II) chloride

Metal nitrates
See METAL NITRATES: Esters, etc.

TITANIUM DICHLORIDE Cl_2Ti

1. Mellor, 1941, Vol. 7, 75; Brauer, 1965, Vol. 2, 1187
2. *NSC Data Sheet 485*, 1966

Readily ignites in air, particularly if moist [1]. The dichloride on heating under inert atmospheres disproportionates into the tetrachloride and pyrophoric titanium [2].
See other METAL HALIDES

ZIRCONIUM DICHLORIDE Cl_2Zr

Sidgwick, 1950, 652
When warm it ignites in air.
See other METAL HALIDES

COBALT(III) CHLORIDE Cl_3Co

Pentacarbonyliron,
Zinc
See PENTACARBONYLIRON, C_5FeO_5: Transition metal halides, etc.

PENTAAMMINECHLOROCOBALT(III) PERCHLORATE $Cl_3CoH_{15}N_5O_8$

Explodes at 320°C; high impact-sensitivity.
See AMMINEMETAL OXOSALTS

PENTAAMMINEAQUACOBALT(III) CHLORATE $Cl_3CoH_{17}N_5O_{10}$

Salvadori, R., *Gazz. Chim. Ital.*, 1910[II], **40**, 9
Explodes on impact, or heating to 130°C.
See other AMMINEMETAL OXOSALTS

HEXAAMMINECOBALT(III) CHLORITE $Cl_3CoH_{18}N_6O_6$

Levi, G. R., *Atti Acad. Lincei*, 1923[V], **32**(1), 623
It is impact-sensitive and explosive.
See other AMMINEMETAL OXOSALTS

HEXAAMINECOBALT(III) CHLORATE $Cl_3CoH_{18}N_6O_9$

Friederich, W. *et al., Z. ges. Schiess-Sprengstoffw.*, 1926, **21**, 49
It is explosive.
See other AMMINEMETAL OXOSALTS

HEXAAMINECOBALT(III) PERCHLORATE $Cl_3CoH_{18}N_6O_{12}$

Explodes at 360°C, highly impact-sensitive.
See AMMINEMETAL OXOSALTS

CHROMIUM(III) CHLORIDE Cl_3Cr

Lithium,
Nitrogen
See LITHIUM, Li: Metal chlorides, etc.

IRON(III) CHLORIDE Cl_3Fe

Ethylene oxide
See ETHYLENE OXIDE, C_2H_6O: Contaminants

Metals
See POTASSIUM, K: Metal halides
SODIUM, Na: Metal halides

GALLIUM TRIPERCHLORATE Cl_3GaO_{12}

Foster, L. S., *J. Amer. Chem. Soc.*, 1933, **61**, 3123; *Inorg. Synth.*, 1946, **2**, 28
During preparation by dissolving the metal in 72% perchloric acid, the hexahydrate separates as a crystalline solid. After filtration, the damp crystals must not contact any organic material (filter paper, horn spatula) since the adherent perchloric acid liquor is above 72% concentration owing to the hexahydrate formation.
See PERCHLORIC ACID, $ClHO_4$: Dehydrating agents
See other METAL OXOHALOGENATES

TRICHLOROSILANE Cl_3HSi

Fl. P., –28°C(o); A.I.T., 104°C
See other HALOSILANES

HEXAAMINETITANIUM(III) CHLORIDE $Cl_3H_{18}N_6Ti$

Schumb, W. C. *et al., J. Amer. Chem. Soc.*, 1938, **55**, 599
Reacts violently with water.
See related METAL HALIDES

IODINE(III) PERCHLORATE Cl_3IO_{12}

Christe, K. O. *et al., Inorg. Chem.*, 1972, **11**, 1683
The solid was stable at –45°C but exploded on laser irradiation at low temperature.
See INORGANIC PERCHLORATES
See other IODINE COMPOUNDS
NON-METAL PERCHLORATES

NITROGEN TRICHLORIDE Cl_3N

1. Mellor, 1940, Vol. 8, 598–604; 1967, Vol. 8, Suppl. 2.2, 411
2. Sidgwick, 1950, 705
3. Brauer, 1963, Vol. 1, 479
4. *ABCM Quart. Safety Summ.*, 1946, **17**, 17
5. Schlessinger, G. C., *Chem. Eng. News,* 1966, **44**(33), 46
6. Kovacic, P. *et al., Org. Synth.*, 1968, **48**, 4

Contact above 0°C of excess chlorine or a chlorinating agent with aqueous ammonia, ammonium salts or a compound containing a hydrolysable amino group, or electrolysis of ammonium chloride solution produces the endothermic explosive nitrogen trichloride as a water-insoluble yellow oil [1,2,4]. It is usually prepared [3] in solution in a solvent, and such solutions in chloroform are reported as stable up to 18% concentration. However, an 18% solution in dibutyl ether exploded on cooling in a refrigerator, and a 12% solution prepared without cooling had vigorously decomposed [5]. In absence of other materials, explosion of the trichloride may be initiated in a wide variety of ways. The solid frozen *in vacuo* in liquid nitrogen explodes on thawing, and the liquid on heating to 60 or 95°C. Exposure to impact, light or ultrasonic irradiation will cause (or sensitise) detonation [1]. The preparation and synthetic use of solutions of nitrogen trichloride in dichloromethane is described in detail [6].

See CHLORINE, Cl_2 : Nitrogen compounds
PHOSPHORUS PENTACHLORIDE, Cl_5: Urea

Initiators
Mellor, 1940, Vol. 8, 601–604; 1967, Vol. 8, Suppl. 2.2, 412
A wide variety of solids, liquids and gases will initiate the violent and often explosive decomposition of nitrogen trichloride. These include concentrated ammonia, arsenic, dinitrogen tetraoxide (above –40°C, with more than 25% solutions of trichloride in chloroform), hydrogen sulphide, hydrogen trisulphide, nitrogen oxide, organic matter (including grease from the fingers), ozone, phosphine, phosphorus (solid, or as carbon disulphide solution), potassium cyanide (solid or aqueous solution), potassium hydroxide solution and selenium.
See other N-HALOGEN COMPOUNDS

NITROSYLRUTHENIUM TRICHLORIDE **Cl_3NORu**

Sidgwick, 1950, 1486
It decomposes violently at 440°C.
See other NITROSO COMPOUNDS

1,3,5-TRICHLOROTRITHIA-1,3,5-TRIAZINE (THIAZYL CHLORIDE) **$Cl_3N_3S_3$**

Ammonia,
Silver nitrate
Goehring, 1957, 67
Thiazyl chloride, treated with aqueous ammonia and then silver nitrate, gives a compound AgN_5S_3 (unknown structure) which is shock-sensitive and explodes violently.

TIN AZIDE TRICHLORIDE **Cl_3N_3Sn**

Explosive solid.
See METAL AZIDE HALIDES

TITANIUM AZIDE TRICHLORIDE Cl_3N_3Ti

Explosive solid.

See METAL AZIDE HALIDES

PHOSPHORYL CHLORIDE Cl_3OP

Carbon disulphide

ABCM Quart. Safety Summ., 1964, **35**, 24

Disposal of a benzene solution of phosphoryl chloride into a waste drum containing carbon disulphide (and other solvents) caused an instantaneous reaction, with evolution of (probably) hydrogen chloride.

N,N-Dimethylformamide,
2,5-Dimethylpyrrole

MCA Case History No. 1460

Poor stirring during formylation of 2,5-dimethylpyrrole with the preformed complex of dimethylformamide with phosphoryl chloride caused eruption of flask contents. Reaction of the complex with local excess of the pyrrole may have been involved.

2,6-Dimethylpyridine *N*-oxide

1. Kato, T. *et al., J. Pharm. Soc. Japan,* 1951, **71**, 217
2. Anon., *Chem. Eng. News,* 1965, **43**(47), 40
3. Evans, R. F. *et al., J. Org. Chem.*, 1962, **27**, 1333

Interaction of the reagents in absence of diluent, according to a published procedure [1], caused an explosion [2]. The use of a chlorinated solvent as diluent prevented explosion, confirming an earlier report [3].

Dimethyl sulphoxide

See DIMETHYL SULPHOXIDE, C_2H_6OS: Acyl halides

Ferrocene-1,1-dicarboxylic acid

Marvel, C. S., *Chem. Eng. News,* 1962, **40**(3), 55

An explosion occurred immediately after pouring and capping of the chloride recovered from preparation of ferrocene-1,1-dicarbonyl chloride. The storage bottle contained phosphoryl chloride recovered from similar preparations and which had been stored for about 3 months. No explanation was offered.

Water
MCA SD-26, 1968
MCA Case History No. 1274
Unpublished observations
The hazards arising from interaction of phosphoryl chloride and water are due to there often being a considerable delay in onset of the exothermic hydrolysis reaction, which may proceed with enough vigour to generate steam and liberate hydrogen chloride gas. Conditions tending to favour delayed or violent reaction include limited quantities of water and/or ice for hydrolysis, lack of stirring, cold or frozen phosphoryl chloride, and reaction in closed or virtually closed containers. A layer of the dense and cold liquid may survive for several minutes under water before violent, almost instantaneous hydrolysis occurs, particularly when disturbed. The Case History describes a violent explosion which occurred when water was added to a drum containing some phosphoryl chloride which had been stored below its freezing point, 2°C.

Zinc
Mellor, 1940, Vol. 8, 1025
Zinc dust ignites in contact with a little phosphoryl chloride, and subsequent addition of water liberates phosphine which ignites.
See other NON-METAL HALIDES

ANTIMONY TRICHLORIDE OXIDE — Cl_3OSb

Bromine trifluoride
See BROMINE TRIFLUORIDE, BrF_3: Antimony trichloride oxide

VANADIUM TRICHLORIDE OXIDE — Cl_3OV

Rubidium
See RUBIDIUM, Rb: Vanadium trichloride oxide

VANADYL TRIPERCHLORATE $Cl_3O_{13}V$

Organic solvents

Schmeisser, M., *Angew. Chem.*, 1955, **67**, 499

A powerful oxidant, igniting organic solvents on contact, which explodes violently above 80°C.

See other METAL OXOHALOGENATES
OXIDANTS

PHOSPHORUS TRICHLORIDE Cl_3P

Acetic acid

1. Coghill, R. D., *J. Amer. Chem. Soc.*, 1938, **60**, 88
2. Peacocke, T. A., *School Sci. Rev.*, 1962, **44**(152), 217

Use of a free flame instead of a heating bath to distil acetyl chloride produced from the two reactants caused the residual phosphonic acid to violently decompose to give spontaneously flammable phosphine [1]. Two later explosions after reflux but before distillation from a water-bath may have been due to ingress of air into the cooling flask and combustion of traces of phosphine [2].

Dimethyl sulphoxide

See DIMETHYL SULPHOXIDE, C_2H_6OS: Acyl halides

Hexafluoroisopropylideneaminolithium

See HEXAFLUOROISOPROPYLIDENEAMINOLITHIUM, C_3F_6LiN: Non-metal halides

Hydroxylamine

See HYDROXYLAMINE, H_3NO: Phosphorus chlorides

Metals

Mellor, 1940, Vol. 8, 1006; 1941, Vol. 2, 470

Potassium ignites in phosphorus trichloride, while molten sodium explodes on contact.

Oxidants

See CHROMYL CHLORIDE, Cl_2CrO_2: Phosphorus, etc.
FLUORINE, F_2: Phosphorus halides
NITRIC ACID, HNO_3: Phosphorus halides

SODIUM PEROXIDE, Na_2O_2: Non-metal halides
LEAD(II) OXIDE, O_2Pb: Non-metal halides
SELENIUM DIOXIDE, O_2Se: Phosphorus trichloride

Water
MCA SD-27, 1972
Reaction with water is exothermic and immediately violent (unlike phosphoryl chloride), and is accompanied by liberation of some diphosphane which ignites.
See other NON-METAL HALIDES

THIOPHOSPHORYL CHLORIDE Cl_3PS

Moeller, T. *et al., Inorg. Synth.,* 1954, **4**, 72
During the preparation from phosphorus trichloride and sulphur, the quantity and quality of the aluminium chloride catalyst is critical to prevent the exothermic reaction going out of control.

Pentaerythritol
MCA Case History No. 1315
On two occasions violent explosions occurred after heating of equimolar proportions of the reagents for 4 h at 160°C according to a literature method had been discontinued.
See other NON-METAL HALIDES

RHODIUM(III) CHLORIDE Cl_3Rh

Pentacarbonyliron,
Zinc
See PENTACARBONYLIRON, C_5FeO_5: Transition metal halides

RUTHENIUM(III) CHLORIDE Cl_3Ru

Pentacarbonyliron,
Zinc
See PENTACARBONYLIRON, C_5FeO_5: Transition metal halides

ANTIMONY TRICHLORIDE Cl_3Sb

MCA SD-66, 1957

TITANIUM TRICHLORIDE Cl_3Ti

Air,
Water
1. Lerner, R. W., *Ind. Eng. Chem.*, 1961, **53**(12), 56A
2. Ingraham, T. K. *et al., Inorg. Synth.*, 1960, **6**, 56

It reacts vigorously with air and/or water (vapour or liquid) and adequate handling precautions are necessary [1]. The finely divided powder is pyrophoric in air [2].
See other METAL HALIDES

VANADIUM TRICHLORIDE Cl_3V

Methylmagnesium iodide
Cotton, F. A., *Chem. Rev.*, 1955, **55**, 560
Reaction of vanadium trichloride and other halides with Grignard reagents is almost explosively violent under a variety of conditions.
See other METAL HALIDES

CAESIUM TETRAPERCHLORATOIODATE Cl_4CsIO_{16}

Christe, K. O. *et al., Inorg. Chem.*, 1972, **11**, 683
Though stable at ambient temperature, samples exploded under laser irradiation at low temperatures.
See other NON-METAL PERCHLORATES

GERMANIUM TETRACHLORIDE Cl_4Ge

Water

Mellor, 1941, Vol. 7, 270

Interaction is very exothermic, accompanied by crackling if the chloride is dropped into water.

See other METAL HALIDES

MOLYBDENUM DIAZIDE TETRACHLORIDE Cl_4MoN_6

Highly explosive.

See METAL AZIDE HALIDES

VANADIUM AZIDE TETRACHLORIDE Cl_4N_3V

Explosive solid.

See METAL AZIDE HALIDES

RHENIUM TETRACHLORIDE OXIDE Cl_4ORe

Ammonia

Sidgwick, 1950, 1302

Interaction with gaseous or liquid ammonia is violent.

See related METAL HALIDES

TITANIUM TETRAPERCHLORATE $Cl_4O_{16}Ti$

Diethyl ether,
or Formamide,
or *N,N*-Dimethylformamide

Laran, R. J., US Pat. 3 157 464, 1964

Insensitive to heat or shock, this powerful oxidant explodes on contact

with diethyl ether, and ignites with formamide or its dimethyl derivative.
See other METAL OXOHALOGENATES

TETRACHLORODIPHOSPHANE Cl_4P_2

Besson, A. *et al., Compt. Rend.*, 1910, **150**, 102
Oxidises rapidly in air, sometimes igniting.
See other NON-METAL HALIDES

LEAD TETRACHLORIDE Cl_4Pb

Potassium
See Potassium, K: Metal halides

Sulphuric acid
Friedrich, H., *Ber.*, 1893, **26**, 1434
It explodes on warming with diluted sulphuric acid or on attempted distillation at 105°C from the concentrated acid in a stream of chlorine.
See other METAL HALIDES

TETRACHLOROSILANE Cl_4Si

Dimethyl sulphoxide
See DIMETHYL SULPHOXIDE, C_2H_6OS: Acyl halides, etc.

TIN TETRACHLORIDE Cl_4Sn

Alkyl nitrates
See ALKYL NITRATES: Lewis acids

Ethylene oxide
See ETHYLENE OXIDE, C_2H_4O: Contaminants

TELLURIUM TETRACHLORIDE Cl_4Te

Ammonia
Mellor, 1943, Vol. 11, 58
Interaction with liquid ammonia at $-15°C$ forms tellurium nitride (?), which explodes at 200°C.
See other NON-METAL HALIDES

TITANIUM TETRACHLORIDE Cl_4Ti

Hydrogen fluoride
See HYDROGEN FLUORIDE, FH: Metal chlorides

ZIRCONIUM TETRACHLORIDE Cl_4Zr

Hydrogen fluoride
See HYDROGEN FLUORIDE, FH: Metal chlorides

Lithium,
Nitrogen
See LITHIUM, Li: Metal chlorides, etc.

MOLYBDENUM AZIDE PENTACHLORIDE Cl_5MoN_3

Extremely explosive.
See METAL AZIDE HALIDES

URANIUM AZIDE PENTACHLORIDE Cl_5N_3U

Explosive solid.
See METAL AZIDE HALIDES

TUNGSTEN AZIDE PENTACHLORIDE Cl_5N_3W

Extremely explosive.
See METAL AZIDE HALIDES

PHOSPHORUS PENTACHLORIDE Cl_5P

Aluminium
See ALUMINIUM, Al: Phosphorus pentachloride

Chlorine dioxide,
Chlorine
Mellor, 1941, Vol. 2, 281; 1940, Vol. 8, 1013
Contact between phosphorus pentachloride and a mixture of chlorine and chlorine dioxide (previously considered to be dichlorine trioxide) usually causes explosion, possibly due to formation of the more sensitive chlorine monoxide.

Diphosphorus trioxide
Mellor, 1940, Vol. 8, 898
Interaction is rather violent at ambient temperature.

Fluorine
See FLUORINE, F_2: Phosphorus halides

Hydroxylamine
See HYDROXYLAMINE, H_3NO: Phosphorus chlorides

Magnesium oxide
Mellor, 1940, Vol. 8, 1016
A heated mixture incandesces brilliantly.

3′-Methyl-2-nitrobenzanilide
Partridge, M. W. private comm., 1968
The residue from interaction of the chloride and anilide in benzene and removal of solvent and phosphoryl chloride *in vacuo* exploded violently on admission of air.

Nitrobenzene
Unpublished observations
A solution of phosphorus pentachloride in nitrobenzene is stable at 110°C but begins to decompose with accelerating violence above 120°C, with evolution of nitrous fumes.

Sodium
See SODIUM, Na: Non-metal halides

Urea
Anon., *Angew. Chem. (Nachr.)*, 1960, 8, 33
A dry mixture exploded after heating, probably owing to formation of nitrogen trichloride. Other chlorinating agents will react similarly with nitrogenous materials.
See CHLORINE, Cl_2: Nitrogen compounds
NITROGEN TRICHLORIDE, Cl_3N

Water
Mellor, 1940, Vol. 8, 1012
Interaction with water in limited quantities is violent, and the hydrolysis products may themselves react violently with more water.
See PHOSPHORYL CHLORIDE, Cl_3OP: Water
See other NON-METAL HALIDES

ANTIMONY PENTACHLORIDE Cl_5Sb

Oxygen difluoride
See OXYGEN DIFLUORIDE, F_2O: Halogens, etc.

Phosphonium iodide
See PHOSPHONIUM IODIDE, H_4IP: Oxidants

AMMONIUM HEXACHLOROPLATINATE(2−) $Cl_6H_8N_2Pt$

Potassium hydroxide,
Combustible material

Mellor, 1942, Vol. 16, 336
Boiling ammonium chloroplatinate with alkali gives a product which, after drying, will explode violently on heating alone to 205°C or with combustible materials.
See other PLATINUM COMPOUNDS

BIS(TRICHLOROPHOSPHORANYLIDENE)SULPHAMIDE $Cl_6N_2O_2P_2S$

Cellulose,
or Water
Vandi, A. *et al., Inorg. Synth.*, 1966, **8**, 119
It is extremely reactive with water or alcohol and causes filter paper to ignite.

BIS-TRIPERCHLORATOSILICON OXIDE $Cl_6O_{25}Si_2$

Schmeisser, M. *Angew. Chem.*, 1955, **67**, 499
This solid decomposition product of silicon tetraperchlorate was so explosive, even in small amounts, that work was discontinued.
See other NON-METAL PERCHLORATES

HEXACHLORODISILANE Cl_6Si_2

See

Chlorine
See CHLORINE, Cl_2: Hexachlorodisilane

OCTACHLOROTRISILANE Cl_8Si_3

See HALOSILANES

TETRAZIRCONIUM TETRAOXIDE HYDROGEN NONAPERCHLORATE $Cl_9HO_{40}Zr_4$

Mellor, 1946, Vol. 2, 403
The salt, 'zirconyl perchlorate', formulated as $4ZrO(ClO_4)_2 \cdot HClO_4$, explodes if heated rapidly.
See related METAL OXOHALOGENATES

DECACHLOROTETRASILANE $Cl_{10}Si_4$

See HALOSILANES

DODECACHLOROPENTASILANE $Cl_{12}Si_5$

See HALOSILANES

COBALT Co

Acetylene
See ACETYLENE, C_2H_2: Cobalt

Hydrazinium nitrate
See HYDRAZINIUM NITRATE, $H_5N_3O_3$

Oxidants
See BROMINE PENTAFLUORIDE, BrF_5: Acids, etc.
NITRYL FLUORIDE, FNO_2: Metals
AMMONIUM NITRATE, $H_4N_2O_3$: Metals

COBALT TRIFLUORIDE CoF_3

Hydrocarbons,
or Water
Priest, H. F., *Inorg. Synth.*, 1950, **3**, 176
It reacts violently with hydrocarbons or water, and finds use as a fluorinating agent.

Silicon
Mellor, 1956, Vol. 2, Suppl. 1, 64
A gently warmed mixture reacts exothermically, attaining red heat.
See other METAL HALIDES

COBALT(III) AMIDE CoH_6N_3

Schmitz-Dumont, O. *et al.*, *Z. Anorg. Chem.*, 1941, **284**, 175
Powdered material is sometimes pyrophoric.
See other *N*-METAL DERIVATIVES

DIAMMINENITRATOCOBALT(II) NITRATE $CoH_6N_4O_6$

McPherson, G. L. *et al.*, *Inorg. Chem.*, 1971, **10**, 1574
A sample of the molten salt exploded at 200°C.
See other AMMINEMETAL OXOSALTS

TRIAMMINETRINITROCOBALT(III) $CoH_9N_6O_6$

Explodes at 305°C, medium impact-sensitivity.
See AMMINEMETAL OXOSALTS

TRIHYDRAZINECOBALT(II) NITRATE $CoH_{12}N_8O_6$

Franzen, H. *et al.*, *Z. Anorg. Chem*, 1908, **60**, 247, 274
It is explosive.
See AMMINEMETAL OXOSALTS

AMMONIUM HEXANITROCOBALTATE(3–) $CoH_{12}N_9O_{12}$

Explodes at 230°C, medium impact-sensitivity.
See AMMINEMETAL OXOSALTS

PENTAAMMINENITRATOCOBALT(III) NITRITE $CoH_{15}N_8O_7$

Explodes at 310°C, medium impact-sensitivity.
See AMMINEMETAL OXOSALTS

HEXAAMMINECOBALT(III) IODATE $CoH_{18}I_3N_6O_9$

Explodes at 360°C, low impact-sensitivity.
See AMMINEMETAL OXOSALTS

HEXAAMMINECOBALT(III) PERMANGANATE $CoH_{18}Mn_3N_6O_{12}$

Mellor, 1942, Vol. 12, 336
Explodes on impact or heating.
See other AMMINEMETAL OXOSALTS

HEXAAMMINECOBALT(III) NITRATE $CoH_{18}N_9O_9$

Explodes at 295°C, medium impact-sensitivity.
See AMMINEMETAL OXOSALTS

HEXAHYDROXYLAMINECOBALT(III) NITRATE $CoH_{18}N_9O_{15}$

Werner, A. *et al., Ber.,* 1905, **38**, 897
It usually exploded during preparation or handling.
See other AMMINEMETAL OXOSALTS

POTASSIUM TRIAZIDOCOBALTATE(1−) $CoKN_9$

Fritzer, H. P. *et al., Angew. Chem. (Intern. Ed.),* 1971, **10**, 829
The complex azide is highly explosive and must be handled with extreme care. The analogous potassium and caesium derivatives of zinc and nickel azides deflagrate strongly in a flame and some are shock-sensitive.
See other METAL AZIDES

POTASSIUM HEXANITROCOBALTATE(3–) $CoK_3N_6O_{12}$

1. Broughton, D. B. *et al., Anal. Chem.*, 1947, **19**, 72
2. Horowitz, O., *Anal. Chem.*, 1948, **20**, 89
3. Tomlinson, W. R. *et al., J. Amer. Chem. Soc.*, 1949, **71**, 375

Evaporation by heating a filtrate from precipitation of potassium cobaltinitrite caused it to turn purple and explode violently [1]. This was attributed to interaction of nitrite, nitrate, acetic acid and residual cobalt with formation of fulminic and methylnitrolic acids or their cobalt salts, all of which are explosive [2]. Mixtures containing nitrates, nitrites and organic material are potentially dangerous, especially in presence of acidic materials and heavy metals. A later publication confirms the suggestion of formation of nitro- or nitrito-cobaltate(III) [3].

COBALT NITRIDE CoN

Schmitz-Dumont, O., *Angew. Chem.*, 1955, **67**, 231
Pyrophoric powder.
See other *N*-METAL DERIVATIVES

COBALT(II) NITRATE CoN_2O_6

Carbon
Crowther, J. R., private comm., 1970
Charcoal impregnated with the nitrate exploded lightly during sieving. Possibly a dust or black powder explosion.
See other METAL OXONON-METALLATES

COBALT(II) OXIDE CoO

Hydrogen peroxide
See HYDROGEN PEROXIDE, H_2O_2: Metals, etc.

HEXAAMMINECOBALT(3+) HEXANITROCOBALTATE(3–)

$Co_2H_{18}N_{12}O_{12}$

Unstable compound, low impact-sensitivity.
See AMMINEMETAL OXOSALTS

DICOBALT TRIOXIDE

Co_2O_3

Hydrogen peroxide
See HYDROGEN PEROXIDE, H_2O_2: Metals, etc.

Nitroalkanes
See NITROALKANES: Metal oxides

CHROMIUM

Cr

Sidgwick, 1950, 1013
Evaporation of mercury from chromium amalgam leaves pyrophoric chromium.

Oxidants
See BROMINE PENTAFLUORIDE, BrF_5: Acids, etc.
AMMONIUM NITRATE, $H_4N_2O_3$: Metals
NITROGEN OXIDE, NO: Metals
SULPHUR DIOXIDE, O_2S: Metals
See other METALS

CHROMIC ACID

CrH_2O_4

Acetone
MCA Case History No. 1583
During glass cleaning operations, acetone splashed into a beaker containing traces of potassium dichromate–sulphuric acid mixture and the solvent ignited. Alcohols behave similarly.
See other OXIDANTS

TRIAMMINEDIPEROXOCHROMIUM(IV) $CrH_9N_3O_4$

1. Hughes, R. G. *et al., Inorg. Chem.*, 1968, **7**, 882
2. Kauffman, G. B., *Inorg. Synth.*, 1966, 8, 133

It must be handled with care because it may explode or become incandescent on sudden heating or shock. Heating at 20°C/min caused a violent explosion at *ca.* 120°C [1]. Preparative and handling precautions have been detailed [2].

See other AMMINECHROMIUM PEROXOCOMPLEXES

AMMONIUM TETRAPEROXOCHROMATE(3–) $CrH_{12}N_3O_8$

Mellor, 1943, Vol. 11, 356

Explodes at 50°C, on impact, or in contact with sulphuric acid.

See other PEROXOACID SALTS

HEXAAMMINECHROMIUM(III) NITRATE $CrH_{18}N_9O_9$

Moderately impact-sensitive, explodes at 265°C.

See AMMINEMETAL OXOSALTS

POTASSIUM TETRAPEROXOCHROMATE(3–) CrK_3O_8

Mellor, 1943, Vol. 11, 356

Not sensitive to impact, but explodes at 178°C, or in contact with sulphuric acid. The impure salt is less stable, and explosive.

See other PEROXOACID SALTS

LITHIUM CHROMATE $CrLiO_4$

Zirconium

de Boer, J. H. *et al., Z. Anorg. Chem.*, 1930, **191**, 113

During reduction of the chromate to lithium at 450–600°C, a considerable excess of zirconium must be used to avoid explosions.

See other METAL OXOSALTS

CHROMYL NITRATE CrN_2O_8

Organic materials
1. Schmeisser, M., *Angew. Chem.*, 1955, **67**, 495
2. Harris, A. D. *et al.*, *Inorg. Synth.*, 1967, **9**, 87
Many hydrocarbons and organic solvents ignite on contact with this powerful oxidant and nitrating agent which reacts like fuming nitric acid in contact with paper, rubber or wood.
See other METAL OXONON-METALLATES

CHROMYL AZIDE CrN_6O_2

Preparative hazard.
See CHROMYL CHLORIDE, Cl_2CrO_2: Sodium azide
See other METAL AZIDES

SODIUM TETRAPEROXOCHROMATE(3–) $CrNa_3O_8$

Mellor, 1943, Vol. 11, 356
Explodes at 115°C.
See other PEROXOACID SALTS

CHROMIUM TRIOXIDE CrO_3

Baker, W., *Chem. & Ind.*, 1965, 280
Presence of nitrates in chromium trioxide may cause oxidation reactions to accelerate out of control, possibly owing to formation of chromyl nitrate. Samples of the oxide should be tested by melting before use, and those evolving oxides of nitrogen should be discarded.
See CHROMYL NITRATE, CrN_2O_2

Acetic acid
BCISC Quart. Safety Summ., 1966, **37**, 30
An explosion occurred during initial heating up of a large volume of glacial acetic acid being treated with chromium troxide. This was attributed to violent interaction of solid chromium trioxide and liquid acetic acid on a hot, exposed steam coil, and subsequent initiation of an

explosive mixture of acetic acid vapour and air. The risk has been obviated by using a solution of dichromate in sulphuric acid as oxidiser, in place of chromium trioxide. The sulphuric acid is essential, as the solid dichromate moist with acetic acid, obtained by evaporating an acetic acid solution to near-dryness, will explode.

See Butyric acid, below

Acetic anhydride

1. Dawber, J. G., *Chem. & Ind.*, 1964, 973
2. Bretherick, L., ibid., 1196
3. Baker, W., *Chem. & Ind.*, 1956, 280

A literature method for preparation of chromyl acetate by interaction of chromium trioxide and acetic anhydride was modified by omission of cooling and agitation. The warm mixture exploded violently on being moved. [1]. A later publication emphasised the need for cooling, and summarised several such previous occurrences [2]. An earlier reference attributes the cause of chromium trioxide–acetic anhydride oxidation mixtures going out of control to presence of nitric acid or nitrates in the chromium trioxide. The latter can readily be checked for freedom from nitrate by melting. If oxides of nitrogen are evolved, the sample should be discarded [3].

Acetic anhydride,
Tetrahydronaphthalene

Peak, D. A., *Chem. & Ind.*, 1949, 14

Use of an anhydride solution of the trioxide to prepare tetralone caused a vigorous fire. This was attributed to use of the more hygroscopic granular trioxide, which is less preferable than the flake type.

Acetone

Delhez, R., *Chem. & Ind.*, 1956, 931

The use of chromium trioxide to purify acetone is hazardous, ignition on contact occurring at ambient temperature. Methanol behaves similarly when used to reduce the trioxide in preparing hexaaquachromium(III) sulphate.

Alcohols

1. Newth, F. H. *et al.*, *Chem. & Ind.*, 1964, 1482
2. Neumann, H., *Chem. Eng. News*, 1970, **48**(28), 4

When methanol was used to rinse a pestle and mortar which had been used to grind coarse chromium trioxide, immediate ignition occurred, due to vigorous oxidation of the solvent. The same occurred with ethanol, 2-propanol, butanol and cyclohexanol. Water is a suitable cleaning agent [1]. For oxidation of *sec*- alcohols in dimethylformamide, the oxide must be finely divided, as lumps cause violent local reaction on addition to the solution [2].

See *N,N*-Dimethylformamide, below

Alkali metals
Mellor, 1943, Vol. 11, 237
Sodium or potassium reacts with incandescence.

Ammonia
Mellor, 1943, Vol. 11, 233
Incandescence on contact.

Arsenic
Mellor, 1943, Vol. 11, 234
Reaction incandesces.

Bromine pentafluoride
See BROMINE PENTAFLUORIDE, BrF_5: Acids, etc.

Butyric acid
Wilson, R. D., *Chem. & Ind.*, 1957, 758
A mixture of chromium trioxide and butyric acid became incandescent on heating to 100°C.
See Acetic acid, above

N,N-Dimethylformamide
Neumann, H., *Chem. Eng. News*, 1970, 48(28), 4
During oxidation of a *sec*-alcohol to ketone in cold DMF solution, addition of solid trioxide caused ignition. Addition of lumps of trioxide was later found to cause local ignition on addition to ice-cooled DMF under nitrogen.

Glycerol
Pieters, 1957, 30

Interaction is violent; the mixture may ignite owing to oxidation of the trihydric alcohol, which is viscous and unable to dissipate the heat of oxidation.

See GLYCERCOL, $C_3H_3O_3$: Solid oxidants

Hydrogen sulphide

Mellor, 1943, Vol. 11, 232

Contact with the heated oxide causes incandescence.

Peroxyformic acid

See PEROXYFORMIC ACID, CH_2O_3: Metals, etc.

Phosphorus

Moissan, H., *Ann. Chim. Phys.* [6], 1885, **5**, 435

Phosphorus and the molten trioxide react explosively.

Potassium hexacyanoferrate(3−)

BCISC Quart. Safety Summ., 1965, **36**(144), 55

Mixtures of the ferrate ('ferricyanide') and chromium trioxide explode and inflame when heated above 196°C. Friction alone is sufficient to violently ignite the mixture when ground with silver sand.

Pyridine

1. Dauben, W. G. *et al., J. Org. Chem.*, 1969, **34**, 3587
2. Poos, G. I. *et al., J. Amer. Chem. Soc.*, 1953, **75**, 427
3. Collins, J. C. *et al., Org. Synth.*, 1972, **52**, 7
4. Ratcliffe, R. *et al., J. Org. Chem.*, 1970, **35**, 4001
5. Stensio, K. E., *Acta Chem. Scand.*, 1971, **25**, 1125
6. *MCA Case History No. 1284*

During preparation of the trioxide–pyridine complex (a powerful oxidant) lack of really efficient stirring led to violent flash fires as the oxide was added to the pyridine at −15 to −18°C. Reversed addition of pyridine to the oxide is extremely dangerous [1], ignition usually occurring [2]. A more recent preparation specifies temperature limits of 10–20°C to avoid an excess of unreacted trioxide [3]. A safe method of preparing the complex in solution has been described [4], and preparation and use of solutions of the isolated complex in dichloromethane [3] or acetic acid [5] have been detailed. The Case History

gives further information on preparation of the complex. Solution of the oxide is not smooth; it first swells, then suddenly dissolves in pyridine with heat evolution. This hazard may be eliminated without loss of yield by dissolving chromium trioxide in an equal volume of water before adding it to 10 volumes of pyridine. Pulverising chromium trioxide before use is not recommended, as this increases its rate of reaction with organic compounds to a hazardous level.
See OXODIPEROXODIPYRIDINECHROMIUM(VI), $C_{10}H_{10}CrN_2O_5$

Selenium
Mellor, 1943, Vol. 11, 233
Interaction is violent.

Sodium
See SODIUM, Na: Metal oxides

Sulphur
Mellor, 1943, Vol. 11, 232
Ignition on warming.
See other OXIDANTS

LEAD CHROMATE **CrO_4Pb**

Iron (3+) hexacyanoferrate(4−)
Anon., *Chem. Processing,* 1967, **30**, 118
During grinding operations, the intimate mixture was ignited by a spark, and burned fiercely.
See other METAL OXOMETALLATES

CHROMIUM(II) SULPHIDE **CrS**

Fluorine
See FLUORINE, F_2: Sulphides

AMMONIUM DICHROMATE **$Cr_2H_8N_2O_7$**

1. Mellor (MIC), 1961, 872
2. *MCA SD-45, 1952*

Thermal decomposition of the salt is initiated by locally heating to 190°C, and flame spreads rapidly through the mass which, if confined, will explode [1].
See other OXOSALTS OF NITROGENOUS BASES

AMMONIUM PENTAPEROXODICHROMATE(2–) $Cr_2H_8N_2O_{12}$

Brauer, 1965, Vol. 2, 1392
Explodes at 50°C.
See other PEROXOACID SALTS

POTASSIUM PENTAPEROXODICHROMATE(2–) $Cr_2K_2O_{12}$

Mellor, 1943, Vol. 11, 357
Sidgwick, 1950, 1007
The powdered salt explodes above 0°C.
See other PEROXOACID SALTS

POTASSIUM DICHROMATE $Cr_2K_2O_7$

Hydroxylamine
See HYDROXYLAMINE, H_3NO: Oxidants

SODIUM DICHROMATE $Cr_2Na_2O_7$

Acetic anhydride
Marszalek, G., private comm., 1973
Addition of the dihydrated salt to acetic anhydride caused an exothermic reaction which accelerated to explosion. Presence of acetic acid (including that produced by hydrolysis of the anhydride by the hydrate water) has a delaying effect on the onset of violent reaction, which occurs when the proportion of anhydride to acid (after hydrolysis) exceeds 0.37:1, with an initial temperature above 35°C. Mixtures of dichromate (30 g) with mixtures of anhydride–acid (70 g, to give ratios of 2:1, 1:1, 0.37:1) originally at 40°C accelerated out of control after 18, 43 and 120 min, to 160, 155 and 115°C, respectively.

Ethanol,
Sulphuric acid
Annable, E. H., *School Sci. Rev.*, 1951, **32**(117), 249
During preparation of acetic acid by acid dichromate oxidation of ethanol according to a published procedure, minor explosions occurred on two occasions after refluxing had been discontinued.

Hydroxylamine
See HYDROXYLAMINE, H_3NO: Oxidants

Sulphuric acid
Bradshaw, J. R., *Process Biochem.*, 1970, **5**(11), 19
The well-known 'chromic acid' mixture for cleaning laboratory glassware is a powerful oxidant by design, and contact with large amounts of tarry residues in vessels should be avoided as it may lead to a violent reaction. Further, if solvents are first used roughly to clean glassware before acid treatment, traces of readily oxidisable solvents must be removed before adding the acid mixture. In many cases treatment with a properly formulated detergent will ensure adequate cleanliness and avoid possible hazard.
See CHROMIC ACID, CrH_2O_4 : Acetone
See also CHROMIUM TRIOXIDE, CrO_3 : Acetic acid

Sulphuric acid,
Trinitrotoluene
Clarke, H. T. *et al.*, *Org. Synth.*, 1941, Coll. Vol. 1, 543–544
During oxidation of TNT in sulphuric acid to trinitrobenzoic acid, stirring of the viscous reaction mixture must be very effective to prevent added portions of solid dichromate causing local ignition.
See other METAL OXOMETALLATES

DICHROMIUM TRIOXIDE Cr_2O_3

Chlorine trifluoride
See CHLORINE TRIFLUORIDE, ClF_3: Metals, etc.

Dirubidium acetylide
See DIRUBIDIUM ACETYLIDE, C_2Rb_2: Metal oxides

Lithium
See LITHIUM Li: Metal oxides

Nitroalkanes
See NITROALKANES: Metal oxides

CAESIUM Cs

Air,
or Oxygen,
or Water
Mellor, 1941, Vol. 2, 468; 1963, Vol. 2, Suppl. 2.2, 2291, 2328
Ignites immediately in air or oxygen, and in contact with cold water, the evolved hydrogen ignites.
See other METALS

CAESIUM FLUORIDE CsF

Benzenediazonium tetrafluoroborate,
Difluoroamine
See DIFLUOROAMINE, F_2HN: Benzendiammonium tetrafluoroborate, etc.

CAESIUM AMIDE CsH_2N

Water
Mellor, 1940, Vol. 8, 256
Interaction is incandescent in presence of air.
See other *N*-METAL DERIVATIVES

CAESIUM TRIOXIDE ('OZONATE') CsO_3

Water
Whaley, T. P. *et al., J. Amer. Chem. Soc.,* 1951, **73**, 79
Reaction of caesium or potassium 'ozonates' with water or aqueous acids is violent, producing oxygen and flashes of light.
See other METAL OXIDES

DICAESIUM OXIDE Cs_2O

Halogens,
or Non-metal oxides
Mellor, 1941, Vol. 2, 487
Above 150–200°C, incandescence occurs with fluorine, chlorine. In presence of moisture, contact at ambient temperature with carbon monoxide or dioxide causes ignition, while dry sulphur dioxide causes incandescence on heating.
See other METAL OXIDES

DICAESIUM SELENIDE Cs_2Se

Sidgwick, 1950, 92
When warm, it ignites in air.
See related METAL SULPHIDES

TRICAESIUM NITRIDE Cs_3N

Mellor, 1940, Vol. 8, 99
Burns in air, and is readily attacked by chlorine, phosphorus or sulphur.
See other *N*-METAL DERIVATIVES

COPPER Cu

1-Bromo-2-propyne
See 1-BROMO-2-PROPYNE, C_3H_3Br: Metals

Ethylene oxide
See ETHYLENE OXIDE, C_2H_4O

Lead azide
See LEAD(II) AZIDE, N_6Pb: Copper

Oxidants

See CHLORINE TRIFLUORIDE, ClF_3 : Metals, etc.
CHLORINE, Cl_2 : Metals
AMMONIUM PERCHLORATE, ClH_4NO_4 : Copper
FLUORINE, F_2 : Metals
SULPHURIC ACID, H_2O_4S: Copper
HYDROGEN SULPHIDE, H_2S: Metals
AMMONIUM NITRATE, $H_4N_2O_3$: Metals
HYDRAZINIUM NITRATE, $H_5N_3O_3$
POTASSIUM DIOXIDE, KO_2 : Metals

COPPER–ZINC COUPLE Cu–Zn

Diiodomethane,
Ether
Foote, C. S., private comm., 1965
Lack of cooling during preparation of the Simmonds–Smith organozinc reagent caused the reaction to erupt. Possible pyrophoric nature of organozinc compounds and the presence of ether presents a severe fire hazard.
See other ALLOYS

COPPER(I) HYDRIDE CuH

Neunhöffer, O. *et al., J. Prakt. Chem.* [2], 1935, **144**, 63
This impure, unstable material may decompose explosively on heating.

Halogens
Mellor, 1941, Vol. 3, 73
Ignition on contact with fluorine, chlorine or bromine.
See other METAL HYDRIDES

COPPER(II) PHOSPHINATE $CuH_2O_4P_2$

Mellor, 1940, Vol. 8, 883
The solid suddenly explodes at about 90°C.
See other METAL OXONON-METALLATES

TETRAAMMINECOPPER(II) NITRITE $CuH_{12}N_6O_4$

Mellor, 1940, Vol. 8, 480

The salt is nearly as shock-sensitive as picric acid. When pure it does not explode on heating but traces of nitrate cause explosive decomposition.

See other AMMINEMETAL OXOSALTS

TETRAAMMINECOPPER(II) NITRATE $CuH_{12}N_6O_6$

Explodes at 330°C, high impact-sensitivity.

See AMMINEMETAL OXOSALTS

TETRAAMMINECOPPER(II) AZIDE $CuH_{12}N_{10}$

Mellor, 1940, Vol. 8, 348; 1967, Vol. 8, Suppl. 2, 26

Explosive on heating or impact.

See other METAL AZIDES

LITHIUM HEXAAZIDOCUPRATE(4−) $CuLi_4N_{18}$

Urbanski, 1967, Vol. 3, 185

It is, like copper(II) azide, an exceptionally powerful initiating detonator.

See other METAL AZIDES

COPPER(II) NITRATE CuN_2O_6

Acetic anhydride

See ACETIC ANHYDRIDE, $C_4H_6O_3$: Metal salts

See also METAL ACETYLIDES

COPPER(I) AZIDE CuN_3

1. Mellor, 1940, Vol. 8, 348; 1967, Vol. 8, Suppl. 2, 42–50
2. Urbanski, 1967, Vol. 3, 185

One of the more explosive metal azides, it decomposes at 205°C [1] and is very highly impact-sensitive.

See other METAL AZIDES

COPPER(II) AZIDE CuN_6

1. Mellor, 1940, Vol. 8, 348; 1967, Vol. 8, Suppl. 2.2, 42–50
2. Urbanski, 1967, Vol. 3, 185

Very explosive, even when moist. Loosening the solid from filter paper caused frictional initiation. Explosion initiated by impact is very violent, and spontaneous explosion has also been recorded [1]. It is also an exceptionally powerful initiator.

See other METAL AZIDES

COPPER(II) OXIDE CuO

Dirubidium acetylide

See DIRUBIDIUM ACETYLIDE, C_2Rb_2: Metal oxides

Hydrogen

Read, C. W. W., *School Sci. Rev.*, 1941, **22**(87), 340

Reduction of the heated oxide in a combustion tube by passage of hydrogen caused a violent explosion. (The hydrogen may have been contaminated with air.)

Hydrogen sulphide

See HYDROGEN SULPHIDE, H_2S: Metal oxides
HYDROGEN TRISULPHIDE, H_2S_3: Metal oxides

Metals

Browne, T. E. W., *School Sci. Rev.*, 1967, **48**(166), 921

An attempted thermite reaction with aluminium powder and copper(II) oxide in place of diiron trioxide caused a violent explosion. An anonymous comment suggests that a greater reaction rate and exothermic effect were involved and adds that attempted use of disilver oxide would be even more violent.

See ALUMINIUM, Al: Metal oxides, etc.
POTASSIUM, K: Metal oxides
MAGNESIUM, Mg: Metal oxides
SODIUM, Na: Metal oxides

Phospham

See PHOSPHAM, HN_2P: Oxidants

Phthalic anhydride

'Leaflet No.5', Inst. of Chem., London, 1940

A mixture of the anhydride and anhydrous oxide exploded violently on heating.

See other METAL OXIDES

COPPER(II) SULPHATE CuO_4S

Hydroxylamine

See HYDROXYLAMINE, H_3NO: Copper(II) sulphate

COPPER MONOPHOSPHIDE CuP

See TRICOPPER DIPHOSPHIDE, Cu_3P_2

COPPER DIPHOSPHIDE CuP_2

See TRICOPPER DIPHOSPHIDE, Cu_3P_2

COPPER(II) SULPHIDE CuS

Chlorates

Mellor, 1956, Vol. 2, Suppl. 1, 584

The copper sulphide explodes in contact with magnesium, zinc, or cadmium chlorates, or with a concentrated solution of chloric acid.

See METAL HALOGENATES

See other METAL SULPHIDES

COPPER(I) OXIDE Cu_2O

Peroxyformic acid

See PEROXYFORMIC ACID, CH_2O_3 : Metals, etc.

COPPER(I) NITRIDE Cu_3N

Mellor, 1940, Vol. 8, 100

May explode on heating in air.

Nitric acid

See NITRIC ACID, HNO_3 : Copper(I) nitride

See other *N*-METAL DERIVATIVES

TRICOPPER DIPHOSPHIDE Cu_3P_2

Oxidants

Mellor, 1940, Vol. 8, 839

The powdered phosphide burns vigorously in chlorine. Mixtures with potassium chlorate explode on impact, and with potassium nitrate, on heating. The monophosphide, CuP, and diphosphide, CuP_2, behave similarly.

See also POTASSIUM CHLORATE, $ClKO_3$: Metal phosphides

See other METAL NON-METALLIDES

DEUTERIUM D_2

Gas above −249°C; E.L., 5–75%

DEUTERIUM OXIDE D_2O

Pentafluorophenyllithium

See PENTAFLUOROPHENYLLITHIUM, C_6F_5Li: Deuterium oxide

EUROPIUM(II) SULPHIDE **EuS**

Sidgwick, 1950, 454
Pyrophoric in air.
See other METAL SULPHIDES

HYDROGEN FLUORIDE **FH**

Keen, M. J. *et al., Chem. & Ind.,* 1957, 805
Braker, 1971, 305
Handling precautions for the gas or anhydrous liquid are detailed [1]. A polythene condenser for disposal of hydrogen fluoride is described [2].

Bismuthic acid
See BISMUTHIC ACID, $BiHO_3$: Hydrofluoric acid

Mercury(II) oxide,
Organic materials
Ormston, J., *School Sci. Rev.,* 1944, **26**(98), 32
During fluorination of organic materials by passing hydrogen fluoride into a vigorously stirred suspension of the oxide (to transiently form mercury difluoride, a powerful fluorinator), it is essential to use adequate and effective cooling below 0°C to prevent loss of control of the reaction system.

Oxides
Mellor, 1956, Vol. 2, Suppl. 1, 122; 1939, Vol. 9, 101
Arsenic trioxide and calcium oxide incandesce in contact with liquid hydrogen fluoride.

N-Phenylazopiperidine
See *N*-PHENYLAZOPIPERIDINE, $C_{11}H_{15}N_3$: Hydrofluoric acid

Phosphorus(V) oxide
See PHOSPHORUS(V) OXIDE, O_5P_2: Hydrogen fluoride

Potassium tetrafluorosilicate(2–)
Mellor, 1956, Vol. 2, Suppl. 1, 121
Contact with liquid hydrogen fluoride causes violent evolution of silicon tetrafluoride. (The same is probably true of metal silicides and other silicon compounds generally.)
See DIALUMINIUM OCTAVANADIUM TRIDECASILICIDE, $Al_2Si_{13}V_8$

Sodium
See SODIUM, Na: Acids

FLUOROAMINE FH_2N

Hoffman, C. J. *et al.*, *Chem. Rev.*, 1962, **62**, 7
The impure material is very explosive.
See other *N*-HALOGEN COMPOUNDS

FLUOROSILANE FH_3Si

Azidogermane
See AZIDOGERMANE, GeH_3N_3: Fluorosilane

AMMONIUM FLUORIDE FH_4N

Chlorine trifluoride
See CHLORINE TRIFLUORIDE, ClF_3: Ammonium fluoride

POTASSIUM FLUORIDE HYDROGEN PEROXIDATE $FK\ H_2O_2$

See CRYSTALLINE HYDROGEN PEROXIDATES

NITROSYL FLUORIDE FNO

Haloalkene (unspecified)
MCA Case History No. 928
Interaction of a mixture in a pressure vessel at –78°C caused it to rupture when moved from the cooling bath.

Metals,
or Non-metals
Schmutzler, R., *Angew. Chem. (Intern. Ed.)*, 1968, **7**, 442
Antimony, bismuth, arsenic, boron, red phosphorus and silicon all react with nitrosyl fluoride with incandescence.

Oxygen difluoride
Ruff, O. *et al., Z. Anorg. Chem.*, 1932, **208**, 293
Explosion on mixing, even at low temperatures.

Sodium
Mellor, 1940, Vol. 8, 612
Interaction is incandescent.
See other N-HALOGEN COMPOUNDS
OXIDANTS

NITRYL FLUORIDE FNO_2

Metals
Aynsley, E. E. *et al., J. Chem. Soc.*, 1954, 1122
When nitryl fluoride is passed at ambient temperature over molybdenum, potassium, sodium, thorium, uranium or zirconium, glowing or white incandescence occurs. Mild warming is needed to initiate similar reactions of aluminium, cadmium, cobalt, iron, nickel, titanium, tungsten, vanadium or zinc, and 200–300°C for lithium or manganese.

Non-metals
Aynsley, E. E. *et al., J. Chem. Soc.*, 1954, 1122
Boron and red phosphorus glow in the fluoride at ambient temperature, while hydrogen explodes at 200–300°C. Carbon and sulphur are also attacked.
See other OXIDANTS

FLUORINE NITRATE FNO_3

Gases
Hoffman, C. J., *Chem. Rev.*, 1964, **64**, 94
Immediate ignition in the gas phase occurs with ammonia, dinitrogen oxide or hydrogen sulphide.

Organic materials
Brauer, 1963, 189
The very powerful liquid oxidant explodes when vigorously shaken, or immediately on contact with alcohol, ether, aniline or grease. It is also sensitive in the vapour or solid state.
See also CHLORINE NITRATE
See other OXIDANTS

NITRYL HYPOFLUORITE FNO_3

1. Schmutzler, R., *Angew. Chem. (Intern. Ed.)*, 1968, **8**, 454
2. Engelbrecht, A., *Monatsh.*, 1964, **95**, 633

It is a toxic colourless gas which is dangerously explosive in the gaseous, liquid and solid states [1]. It is produced during electrolysis of nitrogenous compounds in hydrogen fluoride [2].
See FLUORINE, F_2 : Nitric acid
See other HYPOHALITES

SULPHUR OXIDE (*N*-FLUOROSULPHONYL)IMIDE FNO_3S_2

Roesky, H. W., *Angew. Chem. (Intern. Ed.)*, 1967, **6**, 711
It reacts explosively with water at ambient temperature, but smoothly at −20°C.

FLUORINE AZIDE FN_3

Bauer, S. H., *J. Amer. Chem. Soc.*, 1947, **69**, 3104
This unstable material usually explodes on vaporisation (at −82°C)
See HALOGEN AZIDES

FLUOROTHIOPHOSPHORYL DIAZIDE FN_6PS

O'Neill, S. R. *et al., Inorg. Chem.*, 1972, **11**, 1630
The explosion of the glassy material at –183°C was attributed to crystallisation of the glass.
See other ACYL AZIDES

FLUORINE F_2

1. Kirk-Othmer, 1966, Vol. 9, 506
2. Gall, J.F. *et al., Ind. Eng. Chem.*, 1947, **39**, 262
3. Landau, R. *et al.*, ibid., 262; Turnbull, S. G. *et al.*, ibid., 286
4. Long, G., 'Apparatus for the Disposal of Fluorine on a Laboratory Scale', Harwell, AERE, 1956
5. Gordon, J. *et al., Ind. Eng. Chem.*, 1960, **52**(5), 63A

Fluorine is the most electronegative and reactive element known, reacting, often violently, with most of the other elements and their compounds. Handling hazards and disposal of fluorine on laboratory and plant scales are adequately described [1–5].

Acetylene
See ACETYLENE, C_2H_2: Halogens

Ammonia
Mellor, 1940, Vol. 8, 216
Ammonia ignites in contact with fluorine, and in admixture explodes. Aqueous ammonia ignites or explodes on contact.

Caesium heptafluoropropoxide
Gumprecht, W. H., *Chem. Eng. News,* 1965, **43**(9), 36
MCA Case History No. 1045
Fluorination of caesium heptafluoropropoxide at –40°C with nitrogen-diluted fluorine exploded violently after 10 h. This may have been caused by ingress of moisture, formation of some pentafluoropropionyl derivative, and conversion of this to pentafluoropropionyl hypofluorite, known to be explosive if suitably initiated. Other possible explosive intermediates are peroxides or peresters.
See PENTAFLUOROPROPIONYL HYPOFLUORITE, $C_3F_6O_2$

Covalent halides
Mellor, 1940, Vol. 2, 12; 1956, Vol. 2, Suppl. 1, 64; 1940, Vol. 8, 995, 1003, 1013
Chromyl chloride at high concentration ignites in fluorine, while phosphorus pentachloride, trichloride and trifluoride ignite on contact.
See also Hydrogen fluoride, etc., below

Cyanoguanidine
Rausch, D. A. *et al., J. Org. Chem.*, 1968, **33**, 2522
The products, perfluoro-1-aminomethylguanidine and perfluoro-*N*-aminomethyltriaminomethane, and by-products of the reaction of fluorine and cyanoguanidine, are extremely explosive in gas, liquid and solid states.
See also Sodium dicyanamide, below

Halocarbons
Mellor, 1956, Vol. 2, Suppl. 1
Schmidt, 1967, 82
The violent or explosive reactions which carbon tetrachloride, chloroform, etc., exhibit on direct local contact with gaseous fluorine, can be moderated by suitable dilution, catalysis, and diffused contact.

Halogens,
or Dicyanogen
Mellor, 1940, Vol. 2, 12
Sidgwick, 1950, 1148
While bromine, iodine and dicyanogen all ignite in fluorine at ambient temperature, a mixture of chlorine and fluorine (containing essential moisture) needs sparking before ignition occurs, though an explosion immediately follows.

Hexalithium disilicide
Mellor, 1940, Vol. 6, 169
Incandesces on warming in fluorine.

Hydrocarbons
Mellor, 1940, Vol. 2, 11; 1956, Vol. 2, Suppl. 1, 198
Sidgwick, 1950, 1117
Schmidt, 1967, 82
Violent explosions occur when fluorine directly contacts liquid hydrocarbons, even at $-210°C$ with anthracene or turpentine, or solid

methane at −190°C with liquid fluorine. Many lubricants ignite in fluorine. Contact under carefully controlled conditions with dilution and catalysis can now be effected smoothly.

Hydrofluoric acid
See Hydrogen halides, below

Hydrogen
1. Mellor, 1940, Vol. 2, 11; 1956, Vol. 2, Suppl. 1, 55
2. Sidgwick, 1950, 1102
3. Kirshenbaum, 1956, 46

The violently explosive reactions which sometimes occur when the two elements come into contact under conditions ranging from solid fluorine and liquid hydrogen at −252°C to the mixed gases at ambient temperature [1] are caused by the catalytic effects of impurities or the physical nature of the walls of the containing vessel [2]. Even in absence of such impurities, spontaneous explosions still occur in the range of 75–90 mol % fluorine in the gas phase at −183°C [3].

Hydrogen fluoride,
Seleninyl fluoride
Seppelt, K., *Angew. Chem. (Intern. Ed.)*, 1972, **11**, 630
Preparation of pentafluoroorthoselenic acid from the above reagents in an autoclave above ambient temperature caused occasional explosions. A safer alternative preparation is described.
See also Covalent halides, above

Hydrogen halides
Mellor, 1940, Vol. 2, 12
Hydrogen chloride, bromide and iodide ignite in contact with fluorine, and the concentrated aqueous solutions, including that of hydrogen fluoride, also produce flame.

Hydrogen sulphide
Mellor, 1947, Vol. 10, 133
Ignition on contact.

Ice
UK Scientific Mission Report 68/79, Washington, UKSM, 1968
Mixtures of liquid fluorine and ice are highly impact-sensitive, with a

power comparable to that of TNT. Contact of moist air or water with liquid fluorine can thus be very hazardous.

See Water, below

Metal acetylides and carbides

Mellor, 1946, Vol. 5, 849, 885, 890–891

Mono- and di-caesium, -lithium and -rubidium acetylides, both tungsten carbides, and zirconium dicarbide, all ignite in cold fluorine, while uranium dicarbide ignites in warm fluorine.

Metal hydrides

Mellor, 1940, Vol. 2, 12, 483; 1956, Vol. 2, Suppl. 1, 56; 1941, Vol. 3, 73

Copper, potassium and sodium hydrides all ignite on contact at ambient temperature.

Metal oxides

Mellor, 1940, Vol. 2, 11–13, 487; 1941, Vol. 3, 663; 1942, Vol. 13, 715; 1936, Vol. 15, 380, 399

Oxides of the alkali and alkaline earth metals and nickel monoxide incandesce in cold fluorine, and iron monoxide when warmed. Nickel dioxide also burns in fluorine.

See also Miscellaneous materials, below

Metals

Mellor, 1940, Vol. 2, 13, 469; 1941, Vol. 3, 638; 1940, Vol, 4, 267, 476; 1946, Vol. 5, 421; 1939, Vol. 9, 379, 891; 1943, Vol. 11, 513, 730; 1942, Vol. 12, 344; 1942, Vol. 15, 675

Schmidt, 1967, 78–80

Kirk-Othmer, 1966, Vol. 9, 507–508

Vigour of reaction is greatly influenced by state of subdivision of the metals involved. Massive calcium, moist magnesium, manganese powder, molybenum powder, potassium, sodium, rubidium and antimony all ignite in cold fluorine gas. Warm tantalum powder or cold thallium ignites on contact with fluorine. Fine copper wire (as wool) ignites at 121°C, and osmium and tin begin to burn at 100°C, while iron powder (100-mesh, but not 20-mesh) ignites in liquid fluorine. Titanium will ignite if impacted under the liquid at –188°C and has ignited in presence of catalysts in the gas at –80°C, but in all cases the fluoride film prevents further propagation. Tungsten, and uranium powders, ignite in the gas without heating, while zinc ignites at about 100°C. Molybdenum, tungsten and Monel wires ignited in atmospheric fluorine at 205, 283 and 396°C (averaged values), respectively. Generally, strongly electropositive metals, or those forming volatile fluorides, are attacked the most vigorously.

Metal salts
Mellor, 1940, Vol. 2, 13; 1956, Vol. 2, Suppl. 1, 63
Schmidt, 1967, 83–84
Calcium, lead, basic lead, lithium and sodium carbonates all ignite and burn fiercely in contact with fluorine. Chlorides and cyanides are vigorously attacked by cold fluorine, including lead difluoride and thallium(I) chloride, both of which become molten. Mercury dicyanide ignites in fluorine when warmed gently, and silver cyanide reacts explosively when cold. Sodium metasilicate ignites in fluorine.

Miscellaneous materials
1. Mellor, 1940, Vol. 2, 13
2. Schmidt, 1967, 84, 110
3. Lafferty, R. H. *et al., Chem. Eng. News*, 1948, **26**, 3336

Town gas ignites in contact with gaseous fluorine as does a mixture of lead oxide and glycerol (used as a jointing compound) [1]. Spillage tests involving action of liquid fluorine alone or as a 30% solution in liquid oxygen caused asphalt and crushed limestone to ignite, and coke and charcoal to burn, the latter brilliantly, while liquid JP4 fuel produced violent explosions and a large fireball. Rich soil also burned with a bright flame [2]. Immersion of various glove materials in liquid fluorine was examined. Cotton exploded violently and Neoprene slightly with ignition, and leather charred but did not ignite [3].

Nitric acid
1. Cady, G. H., *J. Amer. Chem. Soc.*, 1934, **56**, 2635
2. Schmutzler, R., *Angew. Chem. (Intern. Ed.),* 1968, **8**, 453

Interaction of fluorine with either the concentrated or very dilute acid caused explosions, while use of 4N acid did not [1]. Later and safer methods of preparing nitryl hypofluorite are summarised [2].

Nitrogenous bases
Hoffman, C. J., *Chem. Rev.*, 1962, **62**, 12

Aniline, dimethylamine and pyridine incandesce on contact with fluorine.
See Ammonia, above
Cyanoguanidine, above

Non-metal oxides

1. Mellor, 1940, Vol. 2, 11; 1939, Vol. 9, 101
2. Hoffman, C. J. *et al., Chem. Rev.*, 1962, **62**, 10

Diarsenic trioxide reacts violently and nitrogen oxide ignites in excess fluorine. Bubbles of sulphur dioxide explode separately on contacting fluorine, while addition of the latter to sulphur dioxide causes an explosion at a certain concentration [1]. Reaction of fluorine with dinitrogen tetraoxide usually causes ignition [2]. Interaction with carbon monoxide and oxygen may be explosive.
See BISFLUOROFORMYL PEROXIDE, $C_2F_2O_4$

Non-metals

Mellor, 1940, Vol. 2, 11, 12; 1956, Vol. 2, Suppl. 1, 60; 1946, Vol. 5, 785, 822; 1940, Vol. 6, 161; 1939, Vol. 9, 34; 1943, Vol. 11, 26
Schmidt, 1967, 52, 107

Boron, phosphorus (yellow or red), selenium, tellurium and sulphur all ignite in contact with fluorine at ambient temperature, silicon attaining a temperature above 1400°C. The reactivity shown by various forms of carbon (charcoal, lampblack, soot), all of which ignite and burn vigorously in fluorine, has been reported to be due to presence of various impurities, moisture and hydrocarbons. Carefully purified carbon (massive graphite) is inert to fluorine at ambient or slightly elevated temperatures for a short period, but may then react explosively.

Phosphorus halides
See Covalent halides, above

Polymeric materials,
Oxygen
Schmidt, 1967, 87

Various polymeric materials were tested statically with both gaseous and liquefied mixtures of fluorine and oxygen containing from 50 to 100% of the former. The materials which burned or reacted violently were: phenol-formaldehyde resins (Bakelite); polyacrylonitrilebutadiene (Buna N); polyamides (Nylon); polychloroprene (Neoprene); polyethylene; poly-trifluoropropylmethyl siloxane (LS 63); polyvinylchloride-vinyl acetate

(Tygan); polyvinylidene fluoride-hexafluoropropylene, (Viton); polyurethane foam. Under dynamic conditions of flow and pressure, the more resistant materials which burned were:
chlorinated polyethylenes; polymethylmethacrylate (Perspex); polytetrafluoroethylene (Teflon).

Polytetrafluoroethylene
Appelman, E. H., *Inorg. Chem.*, 1969, **8**, 223
Teflon tubing, when used to conduct fluorine into a reaction mixture, sometimes catches fire. Combustion stops when the fluorine flow is shut off.

Sodium acetate
Mellor, 1956, Vol. 2, Suppl. 1, 56
Application of fluorine to aqueous sodium acetate solution causes an explosion, involving formation of diacetyl peroxide.
See DIACETYL PEROXIDE, $C_4H_6O_4$

Sodium bromate
Appelman, E. H., *Inorg. Chem.*, 1969, **8**, 223
The oxidation of alkaline bromate to perbromate is not smooth and small explosions may occur in the vapour above the solution. The reaction should not be run unattended.

Sodium dicyanamide
Rausch, D. A. *et al., J. Org. Chem.*, 1968, **33**, 2522
The product, perfluoro-*N*-cyanodiaminomethane, and many of the byproducts from interaction of fluorine and sodium dicyanamide, are extremely explosive in gaseous, liquid and solid states.

Stainless steel
Stewart, J. W., *Proc. 7th Intern. Conf. Low Temp. Phys., Toronto, 1960,* 671
During the study of phase transitions of solidified gases at high pressures, solid fluorine reacted explosively with apparatus made out of stainless steel.

Sulphides
Mellor, 1940, Vol. 2, 11, 13; Vol. 6, 110; 1939, Vol. 9, 522; 1947, Vol. 10, 133; 1943, Vol. 11, 430

Diantimony trisulphide, carbon disulphide vapour, chromium(II) sulphide and hydrogen sulphide all ignite in contact with fluorine at ambient temperature, the solids becoming incandescent.

Water
Mellor, 1940, Vol. 2, 11
Schmidt, 1967, 53, 119
Treatment of liquid air (containing condensed atmospheric moisture) with fluorine gives a potentially explosive precipitate, thought to be fluorine hydrate. Contact of liquid fluorine with a bulk of water caused violent explosions. Ice tends to react explosively with fluorine gas after an indeterminate induction period.
See other OXIDANTS

DIFLUOROAMINE F_2HN

1. Stevens, T. E., *J. Org. Chem.*, 1968, **33**, 2664, 2671
2. Petry, R. C. *et al.*, *J. Org. Chem.*, 1967, **32**, 4034
3. Rosenfeld, D. D. *et al.*, *J. Org. Chem.*, 1968, **33**, 2521
4. Martin, K. J., *J. Amer. Chem. Soc.*, 1965, **87**, 395
5. *MCA Case History No. 768*

It is a dangerous explosive and must be handled with skill and care and appropriate precautions [1,2]. Explosions have occurred when it was condensed at −196°C or allowed to melt [5], and a glass bulb containing the gas exploded violently when accidentally dropped [5].

Benzenediazonium tetrafluoroborate,
Caesium fluoride
Baum, K., *J. Org. Chem.*, 1968, **33**, 4333
Use of caesium fluoride as base to effect condensation caused an explosion in absence of solvent. Pyridine or potassium fluoride and use of methylene chloride gave satisfactory results.
See other N-HALOGEN COMPOUNDS

DIFLUORODIAZENE F_2N_2

Hydrogen
Kuhn, L. P. *et al.*, *Inorg. Chem.*, 1970, **9**, 60
Explosive interaction occurs above 90°C.
See other N-HALOGEN COMPOUNDS

PHOSPHORUS AZIDE DIFLUORIDE F_2N_3P

1. O'Neill, S. R., *et al., Inorg. Chem.*, 1972, **11**, 1630
2. Lines, E. L., *et al.*, ibid. 2270

It is photolytically and thermally unstable and has exploded at 25°C. It is also explosively sensitive to sudden changes in pressure, as in expansion into a vacuum or surging during boiling [1]. It also ignites in air [2].
See other NON-METAL AZIDES

PHOSPHORUS AZIDE DIFLUORIDE-BORANE $F_2N_3P \cdot BH_3$

Lines, E. L. *et al., Inorg. Chem.*, 1972, **11**, 2270
The liquid exploded violently during transfer operations.
See other NON-METAL AZIDES

OXYGEN DIFLUORIDE F_2O

Adsorbents

1. Metz, F. I., *Chem. Eng. News*, 1965, **43**(7), 41
2. Streng, A. G., *Chem. Eng. News*, **43**(12), 5
3. Streng, A. G., *Chem. Rev.*, 1963, **63**, 611

Mixtures of silica gel and the liquid difluoride sealed in tubes at 334 mbar exploded above −196°C, presence of moisture rendering the mixture shock-sensitive at this temperature [1]. Reaction of oxygen difluoride with silica, alumina, molecular sieve or similar surface active solids is exothermic, and under appropriate conditions may be explosive [2]. A quartz fibre can be ignited in the difluoride [3].

Combustible gases
Streng, A. G., *Chem. Rev.*, 1963, **63**, 612
Mixtures with carbon monoxide, hydrogen and methane are stable at ambient temperatures, but explode violently on sparking. Hydrogen sulphide explodes with oxygen difluoride at ambient temperature and, though interaction is smooth at −78°C under reduced pressure, the white solid produced exploded violently when cooling was stopped.

Phosphorus(V) oxide
Streng, A. G., *Chem. Rev.*, 1963, **63**, 611
Ignition occurs spontaneously on contact.

Halogens
or Metal halides
Streng, A. G., *Chem. Rev.*, 1963, **63**, 611
Mixtures with chlorine, bromine or iodine explode on warming. A mixture with chlorine passed through a copper tube at 300°C exploded with variable intensity. Aluminium trichloride explodes in the difluoride, and antimony pentachloride lightly at 150°C.

Metals
Streng. A. G., *Chem. Rev.*, 1963, **63**, 611
Finely divided platinum-group metals react on gentle warming and coarser materials at higher temperatures; aluminium, barium, cadmium, magnesium, strontium, zinc and zirconium evolving light. Lithium, potassium and sodium incandesce brilliantly at 400°C, while tungsten explodes.

Nitrogen oxide
Streng, A. G., *Chem. Rev.*, 1963, **63**, 612
Gaseous mixtures may explode on sparking. The mixed gases slowly react to give a mixture (NO, NOF), which, if liquefied by cooling, will explode on warming.

Nitrosyl fluoride
Hoffman, C. J. *et al.*, *Chem. Rev.*, 1962, **62**, 9
A solid mixture explodes on melting, and the gaseous components ignite on mixing.
See Nitrogen oxide, above

Non-metals
1. Sidgwick, 1950, 1136
2. Streng, A. G., *Chem. Rev.*, 1963, **63**, 611

Pressure of the gas must be limited during concentration by contact with cooled charcoal to avoid violent explosions [1]. Red phosphorus ignites when gently warmed, and powdered boron and silicon generate sparks on heating in the difluoride [2].

Water
Streng, A. G., *Chem. Rev.*, 1963, **63**, 610
Presence of water or water vapour in oxygen difluoride is

dangerous, the mixture (even when diluted with oxygen) exploding violently on spark initiation, especially at 100°C (i.e. with steam).

See Adsorbents, above
OXYGEN FLUORIDES

DIOXYGEN DIFLUORIDE F_2O_2

Various materials

Streng, A. G., *Chem. Rev.*, 1963, **63**, 615

Though not shock-sensitive, it is of limited thermal stability, decomposing below its b.p., –57°C, and explosively in contact with fluorided platinum at –113°C. It is a very powerful oxidant and reacts vigorously or violently with many materials at cryogenic temperatures. It explodes with methane at –194°C, with ice at –140°C, with solid ethanol at below –130°C and with acetone–solid carbon dioxide at –78°C. Ignition or explosion may occur with chlorine, phosphorus trifluoride, sulphur tetrafluoride or tetrafluoroethylene, in the range –130 to –190°C. Even a 2% solution in hydrogen fluoride ignites solid benzene at –78°C.

See OXYGEN FLUORIDES

TRIOXYGEN DIFLUORIDE

Various materials

Streng, A. G., *Chem. Rev.*, 1963, **63**, 619

Though thermally rather unstable, decomposing above its m.p., –190°C, it appears not to be inherently explosive. It is, however, an extremely potent oxidiser and contact with oxidisable materials causes ignition or explosions, even at –183°C. At this temperature, single drops added to solid hydrazine or liquid methane cause violent explosions, while solid ammonia, bromine, charcoal, iodine, red phosphorus and sulphur react with ignition and/or mild explosion. It is also extremely effective at initiating ignition of combustible materials in liquid oxygen, even at 0.1% concentration, whereas mixtures of ozone and fluorine in oxygen are ineffective. This effect has been examined for use in hypergolic rocket propellant systems. Tetryl detonates spontaneously on contact with the difluoride.

See OXYGEN FLUORIDES

FLUORINE FLUOROSULPHATE F_2O_3S

Cady, G. H., *Chem. Eng. News,* 1966, **44**(8), 40; *Inorg. Synth.,* 1968, **11**,155

The crude fluorosulphate, produced as by-product in preparation of peroxodisulphuryl difluoride, was distilled into a cooled steel cylinder and, on warming to room temperature, the cylinder exploded. It decomposes at 200°C, but not explosively.

See related HYPOHALITES

DISULPHURYL DIFLUORIDE $F_2O_5S_2$

Ethanol

Hayek, E., *Monatsh.,* 1951, **82**, 942

Violent reaction on mixing at room temperature.

See other NON-METAL HALIDES

PEROXODISULPHURYL DIFLUORIDE $F_2O_6S_2$

Shreeve, J. M. *et al., Inorg. Synth.*, 1963, **7**, 124

It ignites organic materials immediately on contact.

See also FLUORINE FLUOROSULPHATE, F_2O_3S

See other OXIDANTS

LEAD DIFLUORIDE F_2Pb

Fluorine

See FLUORINE, F_2: Metal salts

XENON DIFLUORIDE F_3HSi

See XENON TRIOXIDE, O_3Xe

TRIFLUOROSILANE F_2Xe

Gas above –80°C, flammable

See other HALOSILANES

IODINE DIOXYGEN TRIFLUORIDE F_3IO_2

Engelbrecht, A. *et al., Angew. Chem. (Intern. Ed.)*, 1969, **8**, 769
It ignites on contact with flammable organic materials.
See other HALOGEN OXIDES

MANGANESE TRIFLUORIDE F_3Mn

Glass
Mellor, 1942, Vol. 12, 344
When heated in contact it attacks glass violently, silicon tetrafluoride being evolved.
See other METAL HALIDES

NITROGEN TRIFLUORIDE F_3N

Charcoal
Massonne, J. *et al., Angew. Chem. (Intern. Ed.)*, 1966, **5**, 317
Adsorption of the fluoride on to activated granular charcoal at −100°C caused an explosion, attributed to the heat of adsorption not being dissipated and causing decomposition to nitrogen and carbon tetrafluoride. No reaction occurs at +100°C in a flow system, but incandescence occurs at 150°C.

Hydrogen-containing compounds
1. Ruff, O., *Z. Angew. Chem.*, 1929, **42**, 807
2. Hoffman, C. J. *et al., Chem. Rev.*, 1962, **62**, 4

Sparking of mixtures with ammonia or hydrogen causes violent explosions, and with steam, feeble ones [1]. Mixtures with ethylene, methane and hydrogen sulphide (also carbon monoxide) explode on sparking [2].

Tetrafluorohydrazine
MCA Case History No. 683
A crude mixture of the two compounds, kept for 3 days in a stainless steel cylinder, exploded violently during valve manipulation.
See other *N*-HALOGEN COMPOUNDS

PHOSPHORUS TRIFLUORIDE F_3P

Dioxygen difluoride
See DIOXYGEN DIFLUORIDE, F_2O_2

Fluorine
See FLUORINE, F_2 : Covalent halides

Hexafluoroisopropylideneaminolithium
See HEXAFLUOROISOPROPYLIDENEAMINOLITHIUM, C_3F_6LiN: Non-metal halides

THIOPHOSPHORYL FLUORIDE F_3PS

Air,
or Sodium
Mellor, 1940, Vol. 8, 1072–1073
In contact with air, the fluoride ignites or explodes, depending on contact conditions. Heated sodium ignites in the gas.

PALLADIUM TRIFLUORIDE F_3Pd

Hydrogen
Sidgwick, 1950, 1574
Contact with hydrogen causes the unheated fluoride to be reduced incandescently.
See other METAL HALIDES

TETRAFLUOROHYDRAZINE F_4N_2

Gas above –73°C; flammable and explosive
Logothetis, A. L., *J. Org. Chem.*, 1966, **31**, 3686, 3689
General precautions recommended in use of tetrafluorohydrazine and derived reaction products include: reactions on as small a scale as possible and behind a barricade; adequate shielding during work-up

of products because explosions may occur; storage of tetrafluorohydrazine at –80°C under 1–2 bar pressure in previously fluorinated Monel or stainless steel cylinders with Monel valves; distillation of volatile (difluoroamino) products in presence of an inert halocarbon oil to prevent explosions in dry distilling vessels.

Air
Martin, K. J., *J. Amer. Soc.*, 1965, **87**, 394
This explodes on contact with air or combustible vapours.

Alkenyl nitrates
Reed, S. F. *et al., J. Org. Chem.*, 1972, **37**, 3329
The products of interaction of tetrafluorohydrazine and alkenyl nitrates, bis(difluoroamino)alkyl nitrates, are heat- and impact-sensitive explosives.
See DIFLUOROAMINO COMPOUNDS

Hydrocarbons
Petry, R. C. *et al., J. Org. Chem.*, 1967, **32**, 4034
Mixtures are potentially highly explosive, approaching the energy of hydrocarbon–oxygen systems.

Hydrogen
Kuhn, L. P. *et al., Inorg. Chem.*, 1970, **9**, 602
Explosive interaction is rather unpredictable, the initiation temperature required (20–80°C) depending on the condition of the vessel walls.

Nitrogen trifluoride
See NITROGEN TRIFLUORIDE, F_3N: Tetrafluorohydrazine

Organic materials
Reed, S. F., *J. Org. Chem.*, 1968, **33**, 2634
Mixtures with organic materials in presence of air constitute explosion hazards. Appropriate precautions are essential.

Ozone
Sessa, P. A. *et al., Inorg. Chem.*, 1971, **10**, 2067
When tetrafluorohydrazine was pyrolysed at 310°C to

generate NF_2 radicals and the mixture contacted liquid ozone at −196°C, a violent explosion occurred.
See other N-HALOGEN COMPOUNDS

PLATINUM TETRAFLUORIDE **F_4Pt**

Water
Sidgwick, 1950, 1614
Interaction is violent.
See other METAL HALIDES

SULPHUR TETRAFLUORIDE **F_4S**

Dioxygen difluoride
See DIOXYGEN DIFLUORIDE, F_2O_2

SELENIUM TETRAFLUORIDE **F_4Se**

Water
Aynsley, E. E. *et al., J. Chem. Soc.*, 1952, 1231
Violent interaction.
See other NON-METAL HALIDES

XENON TETRAFLUORIDE **F_4Xe**

1. Shieh, J. C. *et al., J. Org. Chem.*, 1970, **35**, 4022
2. Malm, J. G. *et al., Inorg. Synth.*, 1966, **8**, 254
3. Chernick, C. L., *J. Chem. Educ.*, 1966, **43**, 619
4. Holloway, J. H., *Talanta,* 1967, **14**, 871
5. Falconer, E. E. *et al., J. Inorg. Nucl. Chem.*, 1967, **29**, 1380

Moisture converts it to highly shock-sensitive xenon oxides [1]. Necessary precautions are given [2–5].
See XENON TRIOXIDE, O_3Xe

PENTAFLUOROORTHOSELENIC ACID F_5HOSe

Preparative hazard.

See FLUORINE, F_2: Hydrogen fluoride
: Seleninyl fluoride

IODINE PENTAFLUORIDE F_5I

Benzene
Ruff, O. *et al., Z. Anorg. Chem.*, 1931, **201**, 245
Interaction becomes violent above 50°C.

Calcium carbide,
or Potassium hydride
Booth, H. S. *et al., Chem. Rev.*, 1947, **41**, 425
Both incandesce on contact (the carbide when warmed).

Dimethyl sulphoxide
Lawless, E. M., *Chem. Eng. News,* 1969, **47**(13), 8, 109
Unmoderated reaction with the sulphoxide is violent, and in the presence of diluents reaction may be delayed and become explosively violent. Although small-scale reactions were uneventful, reactions involving about 0.15 g mol of the pentafluoride and sulphoxide in presence of trichlorofluoromethane or tetrahydrothiophene-1, 1-dioxide as diluents caused delayed and violent explosions. Silver difluoride and other fluorinating agents also react violently with the sulphoxide.
See DIMETHYL SULPHOXIDE, C_2H_6OS: Acyl halides, etc.

Limonene,
Tetrafluoroethylene
See TETRAFLUOROETHYLENE, C_2F_4: Iodine pentafluoride, etc.

Metals,
or Non-metals
1. Sidgwick, 1950, 1159

2. Mellor, 1940, Vol. 2, 114
3. Booth, H. S. *et al., Chem. Rev.*, 1947, **41**, 424–425

Contact with boron, silicon, red phosphorus, sulphur, or arsenic, antimony or bismuth usually causes incandescence [1]. Solid potassium or molten sodium explodes with the pentafluoride, and aluminium foil ignites on prolonged contact [2]. Molybdenum and tungsten incandesce when warmed [3].

Potassium hydride

See Calcium carbide, above

Water

Ruff, O. *et al., Z. Anorg. Chem.*, 1931, **201**, 245

Reaction with water or water-containing materials is violent.

See other INTERHALOGENS

POTASSIUM HEXAFLUOROSILICATE(2–) **F_6K_2Si**

Hydrogen fluoride

See HYDROGEN FLUORIDE, FH: Potassium tetrafluorosilicate(2–)

PENTAFLUOROSULPHUR HYPOFLUORITE **F_6OS**

Ruff, J. K., *Inorg. Synth.*, 1968, **11**, 137

Considered to be potentially explosive.

See other HYPOHALITES

PENTAFLUOROSULPHUR HYPOFLUORITE **F_6OSe**

1. Mitra, G. *et al., J. Amer. Chem. Soc.*, 1959, **81**, 2646
2. Smith, J. E. *et al., Inorg. Chem.*, 1970, **9**, 1442

A cooled sample exploded when allowed to warm rapidly [1], but this may have been due to impurities, as it did not happen in the later work [2].

See other HYPOHALITES

SULPHUR HEXAFLUORIDE F_6S

Disilane
See DISILANE, H_6Si_2: Sulphur hexafluoride

URANIUM HEXAFLUORIDE F_6U

Aromatic hydrocarbons,
or Hydroxy compounds
Sidgwick, 1950, 1072
Interaction with benzene, toluene or xylene is very vigorous, with separation of carbon, and violent with ethanol or water.
See other METAL HALIDES

XENON HEXAFLUORIDE F_6Xe

Water
See XENON TRIOXIDE, O_3Xe

IODINE HEPTAFLUORIDE F_7I

Carbon,
or Combustible gases
Booth, H. S. *et al., Chem. Rev.*, 1947, **41**, 428
Activated carbon ignites immediately in the gas, mixtures with methane ignite, and with carbon monoxide ignite on warming, while those with hydrogen explode on heating or sparking.

Metals
Booth, H. S. *et al., Chem. Rev.*, 1947, **41**, 427
Interaction with barium, potassium and sodium is immediate, accompanied by evolution of heat and light. Aluminium, magnesium and tin are passivated on contact, but on heating react similarly to the former metals.

Organic solvents

Booth, H. S. *et al., Chem. Rev.*, 1947, **41**, 428

Benzene, light petroleum, ethanol and ether ignite in contact with the gas, while the exotherm with acetic acid, acetone or ethyl acetate caused rapid boiling. General organic materials (cellulose, grease, oils) ignite if excess heptafluoride is present.

See other INTERHALOGENS

XENON BIS(PENTAFLUOROORTHOSELENATE) $F_{10}O_2Se_2Xe$

Oxidisable materials

Seppelt, K., *Angew. Chem. (Intern. Ed.)*, 1972, **11**, 724

Interaction is explosive.

See other XENON COMPOUNDS

XENON(II) PENTAFLUOROORTHOTELLURATE $F_{10}O_2Te_2Xe$

Organic solvents

Sladky, F., *Angew. Chem. (Intern. Ed.)*, 1969, 8, 523

Explosive or very vigorous reactions occur on contact with acetone, benzene or ethanol.

See other XENON COMPOUNDS

IRON Fe

Air,

Water

1. Brimelow, H. C., private comm., 1972
2. Unpublished observations, 1949

o-Nitrophenylpyruvic acid was reduced to oxindole using iron pindust–ferrous sulphate in water. The iron oxide residues, after filtering and washing with chloroform, rapidly heated on exposure to air and shattered the Buchner funnel [1]. Previously, rapid heating effects had been observed on sucking air through the metal oxide residue from hot filtration of aqueous liquor from reduction of a nitro compound with reduced iron powder [2].

Disodium acetylide

See DISODIUM ACETYLIDE, C_2Na_2: Metals

Halogens or Interhalogens

See BROMINE PENTAFLUORIDE, BrF_5: Acids, etc.
CHLORINE TRIFLUORIDE, ClF_3: Metals
CHLORINE, Cl_2: Metals
FLUORINE, F_2: Metals

Oxidants

See Halogens, above
PEROXYFORMIC ACID, CH_2O_3: Metals
NITRYL FLUORIDE, FNO_2: Metals
HYDROGEN PEROXIDE, H_2O_2: Metals
AMMONIUM NITRATE, $H_4N_2O_3$: Metals
AMMONIUM PEROXODISULPHATE, $H_8N_2O_8S_2$: Iron
DINITROGEN TETRAOXIDE, N_2O_4: Metals

Polystyrene

Unpublished observation, 1971

Iron flake powder and polystyrene beads had been blended in a high-speed mixer. The mixture ignited and burned rapidly when discharged into a polythene bag. Rapid oxidation of the finely divided metal and/or static discharge may have initiated the fire. No ignition occurred when the iron powder was surface-coated with stearic acid.

See other METALS

IRON–SILICON Fe–Si

Anon., *Chem. Trade J.*, 1956, **139**, 1180

Ferrosilicon containing from 30 to 75% of silicon is hazardous, particularly when finely divided, and must be kept in a moisture-tight drum. In contact with water, the impurities present (arsenide, carbide, phosphide) evolve poisonous arsine, combustible acetylene, and spontaneously flammable phosphine.

AMMONIUM IRON(III) SULPHATE $FeH_4NO_8S_2$

Sulphuric acid

See SULPHURIC ACID, H_2O_4S: Ammonium iron(III) sulphate

IRON(II) IODIDE FeI_2

Alkali metals

See POTASSIUM, K: Metal halides
SODIUM, Na: Metal halides

POTASSIUM PEROXOFERRATE(2–) FeK_2O_5

Alone,
or Non-metals,
or Sulphuric acid

Goralevich, D. K., *J. Russ. Phys. Chem. Soc.*, 1926, **58**, 1155

Explodes on heating or impact, or in contact with charcoal, phosphorus, sulphur or sulphuric acid.

See other PEROXOACID SALTS

IRON(II) OXIDE FeO

Air,
or Sulphur dioxide

Mellor, 1941, Vol. 13, 715

The oxide (prepared at 300°C) incandesces when heated in sulphur dioxide, and burns in air above 200°C. The finely divided oxide prepared by reduction may be pyrophoric in air at ambient temperature.

Oxidants

See NITRIC ACID, HNO_3: Iron(II) oxide
HYDROGEN PEROXIDE, H_2O_2: Metals, etc.

See other METAL OXIDES

IRON(II) SULPHIDE FeS

1. Mellor, 1942, Vol. 14, 157
2. Anon., *Chem. Age*, 1939, **40**, 267

The moist sulphide readily oxidises in air exothermically, and may reach incandescence. Grinding in a mortar hastens this [1]. The impure sulphide formed when steel equipment is used with materials containing hydrogen sulphide or volatile sulphur compounds is pyrophoric, and has caused many fires and explosions when such equipment is opened without effective purging. Various methods of purging are discussed [2].

Lithium,
See LITHIUM, Li: Metal oxides, etc.
See other METAL SULPHIDES

IRON DISULPHIDE **FeS_2**

Anon., *Angew. Chem. (Nachr.),* 1954, **2**, 219
Finely powdered pyrites, especially in presence of water, will rapidly heat spontaneously and ignite, particularly in contact with combustible materials. Inert gas blanketing will prevent this.
See other METAL SULPHIDES

DIIRON TRIOXIDE **Fe_2O_3**

Aluminium
Mellor, 1946, Vol. 5, 217
An intimately powdered mixture, usually ignited by magnesium ribbon, reacts with an intense exotherm to give molten iron and has been used commercially as 'thermite' for welding purposes. Incendive particles have been produced by this reaction on impact between aluminium and rusty iron.
See Calcium disilicide, below
LIGHT ALLOYS
ALUMINIUM, Al: Metal oxides, etc.

Aluminium,
Propene
Batty, G. F., private comm., 1972
Use of a rusty iron tool on an aluminium compressor piston caused incendive sparks which ignited residual propene–air mixture in the cylinder.
See Aluminium, above

Calcium disilicide
Berger, E., *Compt. Rend.*, 1920, **170**, 29
The mixture ('silicon thermite') attains a very high temperature when heated, producing molten iron, similar to the normal thermite.
See Aluminium, above

Carbon monoxide
Othen, C. W., *School Sci. Rev.*, 1964, **45**(156), 459
The reason for a previously reported explosion during reduction of iron oxide with carbon monoxide is given as the formation of pentacarbonyliron at temperatures between 0 and 150°C. Suitable heating arrangements and precautions will eliminate this hazard.
See PENTACARBONYLIRON, C_5FeO_5

Ethylene oxide
See ETHYLENE OXIDE, C_2H_4O: Contaminants

Hydrogen peroxide
See HYDROGEN PEROXIDE, H_2O_2: Metals, etc.

Magnesium
See MAGNESIUM, Mg: Metal oxides

Metal acetylides
See CALCIUM ACETYLIDE, C_2Ca: Iron(III) chloride, etc.
DICAESIUM ACETYLIDE, C_2Cs_2: Diiron trioxide
DIRUBIDIUM ACETYLIDE, C_2Rb_2: Metal oxides
See other METAL OXIDES

TRIIRON TETRAOXIDE Fe_3O_4

Hydrogen trisulphide
See HYDROGEN TRISULPHIDE, H_2S_3: Metal oxides

GALLIUM Ga

Halogens

Walker, H. L., *School Sci. Rev.*, 1956, **37**(132), 196

The metal reacts with cold chlorine strongly exothermically, and the compact metal with bromine even at $-33^\circ C$, reaction being violent at ambient temperature.

LITHIUM TETRAHYDROGALLATE GaH_4Li

Gaylord, 1956, 26

Though of lower stability than the analogous aluminate, its reactivity is generally similar to that of the latter.

See other COMPLEX HYDRIDES

SODIUM TETRAHYDROGALLATE GaH_4Na

Mackay, 1966, 169

It is explosively hydrolysed by water.

See other COMPLEX HYDRIDES

GERMANIUM Ge

Halogens

Mellor, 1941, Vol. 7, 260

The powdered metal ignites in chlorine, and lumps will ignite on heating in chlorine or bromine.

Oxidants

Mellor, 1941, Vol. 7, 260–261

The powdered metal reacts violently with nitric acid, and mixtures with potassium chlorate or nitrate explode on heating.

See POTASSIUM HYDROXIDE, HKO: Germanium

See other METALS

GERMANIUM MONOHYDRIDE $(GeH)_n$

Jolly, W. L. *et al., Inorg. Synth.*, 1963, **7**, 39
The solid polymeric hydride sometimes decomposes explosively into its elements on exposure to air.
See other METAL HYDRIDES

GERMANIUM(II) IMIDE GeHN

Oxygen
Johnson, O. H., *Chem. Rev.*, 1952, **51**, 449
On exposure to air it reacts violently, and in oxygen incandescence occurs.
See other *N*-METAL DERIVATIVES

AZIDOGERMANE GeH_3N_3

Fluorosilane
Anon., *Angew. Chem. (Nachr.),* 1970, **18**, 27
An attempt to prepare azidosilane by interaction of azidogermane and fluorosilane exploded.
See related METAL AZIDES

GERMANE GeH_4

Gas above –90°C; flammable
Brauer, 1963, Vol. 1, 715
Germane and its higher homologues decompose in air, often igniting.

Bromine
See BROMINE, Br_2: Non-metal hydrides
See other METAL HYDRIDES

SODIUM GERMANIDE $GeNa$

Air,
or Water
Johnson, O. H., *Chem. Rev.*, 1952, **51**, 452
The binary alloy is pyrophoric and may ignite in contact with water, as do other alkali metal germanides.
See other ALLOYS

GERMANIUM(II) SULPHIDE GeS

Potassium nitrate
See POTASSIUM NITRATE, KNO_3: Metal sulphides

DIGERMANE Ge_2H_6

1. Mellor, 1941, Vol. 7, 264
2. Brauer, 1963, 715

May ignite in air, particularly if air is admitted suddenly into the gas at reduced pressure.

TRIGERMANE Ge_3H_8

Brauer, 1963, 715
Air-sensitive, may ignite.
See other METAL HYDRIDES

DIMERCURY IMIDE OXIDE HHg_2NO

Sidgwick, 1950, 318
The anhydride of Millon's base explodes if touched or heated to 130°C.
See MERCURY, Hg: Ammonia
See other *N*-METAL DERIVATIVES

HYDRIODIC ACID HI

Muir, G. D., private comm., 1968
During preparation of hydriodic acid by distillation of phosphorus and wet iodine, the condenser became blocked with by-product phosphonium iodide, and an explosion, possibly also involving phosphine, occurred. There is also a purification hazard.
See PHOSPHORUS, P: Hydriodic acid

Metals
See MAGNESIUM, Mg: Hydrogen iodide
POTASSIUM, K: Hydrogen iodide

Oxidants
See ETHYL HYDROPEROXIDE, $C_2H_6O_2$: Hydriodic acid
PERCHLORIC ACID, $ClHO_4$: Iodides
POTASSIUM CHLORATE, $ClKO_3$: Hydrogen iodide

IODIC ACID HIO_3

Non-metals
1. Mellor, 1946, Vol. 5, 15
2. Partington, 1967, 813
Interaction with boron below 40°C is vigorous, attaining incandescence [1]. Charcoal, phosphorus and sulphur deflagrate on heating [2].
See METAL HALOGENATES
See other OXIDANTS
OXOHALOGEN ACIDS

PERIODIC ACID HIO_4

Dimethyl sulphoxide
Rowe, J. J. M. *et al., J. Amer. Chem. Soc.*, 1968, **90**, 1924
Although 1.5 M solutions of periodic acid in dimethyl

sulphoxide explode after a few minutes, 0.15 M solutions appear stable.

See DIMETHYL SULPHOXIDE, C_2H_6OS: Metal oxosalts
: Perchloric acid

See other OXIDANTS
OXOHALOGEN ACIDS

DIIODOAMINE HI_2N

Mellor, 1940, Vol. 8, 607

Explosive, formed on prolonged contact of nitrogen triiodide with water.

See NITROGEN TRIIODIDE-AMMONIA, $I_3N \cdot H_3N$

See other *N*-HALOGEN COMPOUNDS

POTASSIUM HYDRIDE HK

Air

See POTASSIUM HEXAHYDROALUMINATE (3–), AlH_6K_3

Oxidants

See FLUORINE, F_2: Metal hydrides
OXYGEN (Gas), O_2: Metal hydrides

POTASSIUM HYDROXIDE HKO

Acids

MCA Case History No. 920

Incautious addition of acetic acid to a vessel contaminated with potassium hydroxide caused eruption of the acid.

Ammonium hexachloroplatinate(2–)

See AMMONIUM HEXACHLOROPLATINATE (2–), $Cl_6H_8N_2Pt$:
Potassium hydroxide

Chlorine dioxide

See CHLORINE DIOXIDE, ClO_2: Potassium hydroxide

Germanium
Partington, 1967, 181
Germanium is oxidised by the fused hydroxide with incandescence.

Hyponitrous acid
See HYPONITROUS ACID, $H_2N_2O_2$

Maleic anhydride
See MALEIC ANHYDRIDE, $C_4H_2O_3$: Cations, etc.

Nitroalkanes
See NITROALKANES: Inorganic bases
NITROMETHANE, CH_3NO_2: Acids, etc.

Nitrobenzene
See NITROBENZENE, $C_6H_5NO_2$: Potassium hydroxide, etc.

Nitrogen trichloride
See NITROGEN TRICHLORIDE, Cl_3N: Initiators

Potassium peroxodisulphate
See POTASSIUM PEROXODISULPHATE, $K_2O_8S_2$: Potassium hydroxide

2,2,3,3-Tetrafluoropropanol
See 2,2,3,3-TETRAFLUOROPROPANOL, $C_3H_4F_4O$: Potassium hydroxide

Tetrahydrofuran
See TETRAHYDROFURAN, C_4H_8O: Alkalies

Thorium dicarbide
See THORIUM DICARBIDE, C_2Th: Non-metals, etc.

2,4,6-Trinitrotoluene
See 2,4,6-TRINITROTOLUENE, $C_7H_3N_3O_6$: Added impurities

POTASSIUM PEROXOMONOSULPHATE HKO_5S

Castrantas, 1965, 5

Melts with decomposition at 100°C; forms explosive mixtures with as little as 1% of organic matter.
See other PEROXOACID SALTS

LITHIUM HYDRIDE — HLi

Oxygen
See OXYGEN (Gas), O_2: Metal hydrides
OXYGEN (Liquid), O_2: Lithium hydride

PERMANGANIC ACID — $HMnO_4$

Organic materials
1. Frigerio, N. A., *J. Amer. Chem. Soc.,* 1969, **91**, 6201
2. von Schwartz, 1918, 327

The crystalline acid and its dihydrate are very unstable, often exploding at about 3 and 18°C, respectively, but they may be stored virtually unchanged at −75°C. The anhydrous solid ignited explosively every organic compound with which it came into contact except mono-, di- or tri-chloromethanes [1]. The solution of permanganic acid (or its explosive anhydride, dimanganese heptoxide) produced by interaction of permanganates and sulphuric acid, will explode on contact with benzene, carbon disulphide, diethyl ether, ethanol, flammable gases, petroleum or other organic substances [2].
See other OXIDANTS

NITROUS ACID — HNO_2

A semicarbazone,
Silver nitrate
Mitchell, J. J., *Chem. Eng. News,* 1956, **34**, 4704
Use of nitrous acid to liberate a free keto-acid from its semicarbazone caused formation of hydrogen azide which was co-extracted into ether with the product. Addition of silver nitrate to precipitate the silver salt of the acid also precipitated silver azide, which later exploded on scraping from a sintered disc. The possibility of formation of free

hydrogen azide from interaction of nitrous acid and hydrazine or hydroxylamine derivatives is stressed.

Phosphine
See PHOSPHINE, H_3P: Air

Phosphorus trichloride
Mellor, 1940, Vol. 8, 1004
The trichloride explodes with nitrous (or nitric) acid.
See other OXIDANTS

NITRIC ACID HNO_3

MCA SD-5, 1961
The oxidising power (and hazard potential) of nitric acid increases progressively with increase in strength from the concentrated acid (70% wt. HNO_3) through fuming acids (above 85% wt.) to the anhydrous 100% acid. The presence of dissolved oxides of nitrogen in the red fuming grades of acid further enhances the potency of the oxidant.

Acetic acid,
Sodium hexahydroxyplatinate(IV)
1. Davidson, J. M. *et al., Chem. & Ind.*, 1966, 306
2. Malerbi, B. W., *Chem. & Ind.*, 1970, 796

During preparation of diacetatoplatinum(II) by alternative procedures, the hexahydroxyplatinate in mixed nitric–acetic acids was evaporated to a syrup and several explosions were experienced [1], possibly due to formation of acetyl nitrate. On one occasion a brown solid was isolated and dried, but subsequently exploded with great violence when touched with a glass rod. The material was thought to be a mixture of platinum-(IV) acetate–nitrate species [2].

Acetic anhydride
1. Brown, T. A. *et al., Chem. Brit.*, 1967, **3**, 504
2. Dubar, J. *et al., Compt. Rend.*, 1968C, **266**, 1114
3. Dingle, L. E. *et al., Chem. Brit.*, 1968, **4**, 136
4. *MCA Case History No. 103*

Mixtures containing between 50 and 85% wt. of fuming nitric acid are detonable and very sensitive to initiation by friction or shock (possibly owing to formation of acetyl nitrate or tetranitromethane). For preparation of mixtures outside these limits, the order of mixing is important (below 50%, acid into anhydride; above 85%, vice versa; and all below 10°C) [1]. Similar information is also presented diagrammatically [2].

Mixtures containing less than 50% of nitric acid are also dangerous in that addition of small amounts of water (or water-containing mineral acids) readily initiates an uncontrollable exothermic fume-off, which will evaporate most of the liquid present. Equimolar mixtures of 38% nitric acid with acetic anhydride can be detonated at room temperature after a few hours' ageing [3]. Accidental contact of the two materials caused a violent explosion [4].

Acetone,
Acetic acid
1. Frant, M. S., *Chem. Eng. News,* 1960, **38**(43), 56
2. Secunda, W. J., *Chem. Eng. News,* 1960, **38**(46), 5

A mixture of equal parts of nitric acid, acetone and 75% acetic acid, used to etch nickel, will explode 1½–6 h after mixing if kept in a closed bottle. The presence of the diluted acetic acid would probably slow the known violent oxidation of acetone by nitric acid [1]. Alternatively, the formation of tetranitromethane and subsequent oxidation of acetone is suspected [2].

Acetone,
Sulphuric acid
Fawcett, H. H., *Ind. Eng. Chem.,* 1959, **51**, 89A

Acetone is oxidised violently by mixed nitric–sulphuric acids, and if the mixture is confined in a narrow-mouthed vessel, it may be ejected or explode.

Acetonitrile
Andrussow, L., *Chim. Ind.,* 1961, **86**, 542

Mixtures of fuming nitric acid and acetonitrile are high explosives.

Acrylonitrile
See ACRYLONITRILE, C_3H_3N: Acids

Alcohols
1. Fawcett, H. H., *Chem. Eng. News,* 1949, 27, 1396
2. Unpublished observations, 1956

3. *MCA Case History No. 1152*
4. Potter, C. R., *Chem. & Ind.*, 1971, 501
5. Spengler, G. *et al.*, *Brennst. Chem.*, 1965, **46**, 117
6. Long, L. A. private comm., 1972

A 15% solution of nitric acid in ethanol was used to etch a bismuth crystal. After removing the metal, the mixture decomposed vigorously. Mixtures or nitric acid and alcohols ('Nital') are quite unstable when the concentration of acid is above 10%, and mixtures containing over 5% of acid should not be stored [1]. The use of a little alcohol and excess nitric acid to clean sintered glassware (by 'nitric acid fizzing') is not recommended. At best it is a completely unpredictable approximation to a nitric acid–alcohol rocket propulsion system. At worst, if heavy metals are present, fulminates capable of detonating the mixture may be formed [2]. Chromic acid mixture is less hazardous for such cleaning operations. The Case History describes a violent explosion caused by addition of concentrated acid to a tank car contaminated with a little alcohol [3]. During oxidation of cyclohexanol to the 1,2-dione by an established process, a violent explosion occurred. Two intermediates are possible suspects [4]. Furfuryl alcohol is hypergolic with high-strength nitric acid [5] and methanol has been used as a propellant fuel. It also readily forms the explosive ester, methyl nitrate [6].

2-Aminothiazole,
Sulphuric acid
Silver, L., *Chem. Eng. Progr.*, 1967, **63**(8), 43
Nitration of 2-aminothiazole with nitric/sulphuric acids was normally effected by mixing the reactants at low temperature, heating to 90°C during 30 min and then positively cooling. When positive cooling was omitted, a violent explosion occurred. Experiment showed that this was due to a slow exothermic reaction accelerating out of control under the adiabatic conditions. *N*-Nitroamines were not involved.

Ammonia
Mellor, 1940, Vol. 8, 219
A jet of ammonia will ignite in nitric acid vapour.

Anilinium nitrate
Andrussow, L., *Chim. Ind.*, 1961, **86**, 542
Although aniline may be hypergolic with nitric acid (below), anilinium nitrate dissolves unchanged in 98% nitric acid and can be stored for long

periods, though the solution has high-explosive properties.
See Cyclohexylamine, below

Aromatic amines
1. Kit and Evered, 1960, 239, 242
2. *Aniline,* 85, Allied Chemical Corp., New York, 1964
3. Miller, R. O., *Tech. Note No. 3884,* 1–32, Washington, Nat. Advisory Comm. Aeronaut., 1956
4. Spengler, G. *et al., Brennst. Chem.*, 1965, **46**, 117

Many aromatic amines (aniline, *N*-ethylaniline, *o*-toluidine, xylidine, etc., and their mixtures) are hypergolic with red fuming nitric acid [1]. When the amines are dissolved in triethylamine, ignition occurs at −60°C and below [2]. Addition of a mixture of aniline, dimethylaniline, xylidine and pentacarbonyliron renders hydrocarbons hypergolic with concentrated nitric acid [3]. Although aniline is not hypergolic with 96% nitric acid, presence of sulphuric acid (5% or above) renders it so. Presence of dinitrogen tetraoxide further reduces ignition delay [4].
See Anilinium nitrate, above

Arsine-borontribromide
See ARSINE-BORONTRIBROMIDE, $AsH_3 \cdot BBr_3$: Oxidants

Benzo[b]thiophene derivatives
Brown, I. *et al., Chem. & Ind.*, 1962, 982
During nitration of several derivatives, diazotisation and oxidation occurred to produce internal diazonium phenolate derivatives. 5-Acetylamino-3-bromo-benzo[*b*]thiophene unexpectedly underwent hydrolysis, diazotisation and oxidation to the explosive compound below.
See 3-BROMO-2,7-DINITRO-5-BENZO[*b*]THIOPHENEDIAZONIUM-4-OLATE, $C_8HBrN_4O_5S$
See 4-Chloro-2-nitroaniline, below

Bromine pentafluoride
See BROMINE PENTAFLUORIDE, BrF_5: Acids, etc.

Butanethiol
McCullough, F. *et al., Proc. Fifth Combustion Symp.*, 181, New York, Reinhold, 1955
Technical material (containing 28% of propane- and 7% of pentanethiols) is hypergolic with 96% nitric acid.

Cellulose
MCA SD-5, 1961
Cellulose may be converted to the highly flammable nitrate ester on contact with the vapour of nitric acid, as well as the liquid itself.

4-Chloro-2-nitroaniline
1. Elderfield, R. C. *et al., J. Org. Chem.*, 1946, **11**, 820
2. *MCA Case History No. 1489*

The literature procedure for preparation of 4-chloro-2,6-dinitroaniline [1], involving direct nitration in 65% nitric acid, was modified by increasing the reaction temperature to 60°C 1 h after holding at 30–35°C as originally specified. This procedure was satisfactory on the bench scale, and was scaled up into a 900 litre reactor. After the temperature had reached 30°C, heating was discontinued, but the temperature continued to rise to 100–110°C and decomposition set in with copious evolution of nitrous fumes and production of a very shock-sensitive explosive solid. This was identified as 4-chloro-2,5-dinitrobenzenediazonium-6-olate produced by hydrolysis of a nitro group in the expected product by the diluted nitric acid at high temperature, diazotisation of the free amino group by the nitrous acid produced in the hydrolysis (or by the nitrous fumes), and introduction of a further nitro group under the prevailing reaction conditions. It is recommended that primary aromatic amines should be protected by acetylation before nitration, to avoid the possibility of accidental diazotisation [2].
See Benzo[*b*]thiophene derivative, above
See other ARENEDIAZONIUMOLATES

Crotonaldehyde
Andrussow, L., *Chim. Ind.,* 1961, **86**, 542
Crotonaldehyde is hypergolic with concentrated nitric acid, ignition delay being 1 ms.
See ROCKET PROPELLANTS

Copper(I) nitride
Mellor, 1940, Vol. 8, 100
Interaction with concentrated acid is very violent.

Cyclohexylamine
Andrussow, L., *Chim. Ind.,* 1961, **86**, 542
Although cyclohexylamine has been used as a fuel with nitric acid in

rocket motors, cyclohexylammonium nitrate dissolves unchanged in fuming nitric acid to give a solution stable for long periods.

See Aromatic amines, above
ROCKET PROPELLANTS

1,2-Diaminoethanebis-trimethylgold

See 1,2-DIAMINOETHANEBIS-TRIMETHYLGOLD, $C_8H_{26}Au_2N_2$

Dichloromethane

Andrussow, L., *Chim. Ind.,* 1961, **86**, 542

Dichloromethane dissolves endothermically in concentrated nitric acid to give a detonable solution.

Diethyl ether

Foote, C. S., private comm., 1965

Addition of ether to a nitration mixture (*o*-bromotoluene and conc. nitric acid) diluted with an equal volume of water in a separating funnel caused a low-order explosion. This was attributed to oxidation of the ether (possibly containing alcohol) by the acid. Addition of more water before adding ether was recommended.

1,1-Dimethylhydrazine,
Organic compounds

Spengler, G. *et al., Brennst. Chem.*, 1965, **46**, 117

Contact of nitric acid (or dinitrogen tetraoxide) with dimethylhydrazine is hypergolic and well described in rocket technology. While hydrocarbons and several other classes of organic compounds are not hypergolic with these oxidisers, addition of a proportion of dimethylhydrazine to a wide range of hydrocarbons, alcohols, amines, esters and heterocyclic compounds renders them hypergolic in contact with nitric acid or dinitrogen tetraoxide.

Diphenyldistibene

See DIPHENYLDISTIBENE, $C_{12}H_{10}Sb_2$: Air, etc.

Diphenyltin

Krause, E. *et al., Ber.*, 1920, **53**, 177

Ignition occurs on contact with fuming nitric acid.

Disodium phenylorthophosphate
Muir, G. D., private comm., 1968
Concentration of the nitration product of the phosphate ester caused a violent explosion. Picric acid derivatives may have been involved.

Divinyl ether
Andrussow, L., *Chim. Ind.*, 1961, **86**, 542
Divinyl ether is hypergolic with concentrated nitric acid, ignition delay being 1 ms.
See ROCKET PROPELLANTS

5-Ethyl-2-methylpyridine
1. Frank, R. L., *Chem. Eng. News,* 1952, **30**, 3348
2. Rubinstein, H. *et al., J. Chem. Eng. Data*, 1967, **12**, 149
Following a patented procedure, the two reactants were being heated together at 145°C/14.5 bar to produce 2,5-pyridinedicarboxylic acid. The temperature and pressure rose to 160°C/43.5 bar, and the autoclave was vented and cooled, but 90 s later, a violent explosion occurred, although both rupture discs (105 and 411 bar) had gone. General precautions are discussed [1]. The later publication disclosed that up to 20% of 5(1,1-dinitroethyl)2-methylpyridine, probably an explosive compound, is produced in this reaction. However, in presence of added water, no instability was seen in a series of reactions at temperatures up to 160°C and pressure to 102 bar [2].

Fluorine
See FLUORINE, F_2: Nitric acid

Hexalithium disilicide
See HEXALITHIUM DISILICIDE, Li_6Si_2: Acids

2,2,4,4,6,6-Hexamethyltrithiane
Baumann, E. *et al., Ber.*, 1889, **22**, 2596
Interaction of 'tri(thioacetone)' with the concentrated acid is explosively violent.
See Thioaldehydes, etc. below

Hydrazine
Andrussow, L., *Chim. Ind.*, 1961, **86**, 542

Hydrazine is hypergolic in contact with concentrated nitric acid.
See ROCKET PROPELLANTS

Hydrocarbons

1. Andrussow, L., *Chim. Ind.*, 1961, **86**, 542
2. Wilson, P. J. *et al., Chem. Rev.*, 1944, **34**, 8
3. Sykes, W. G. *et al., Chem. Eng. Progr.*, 1963, **59**(1), 70–71
4. Mason, C. M. *et al., J. Chem. Eng. Data*, 1965, **10**, 173
5. Urbanski, 1961, Vol. 1, 140
6. *Proc. Symp. Chem. Process Hazards Plant Design, Manchester, I.Ch.E., 1960,* 37–41
7. Powell, G. *et al., Org. Synth.*, 1943, Coll. Vol. 2, 450
8. Wilms, H. *et al., Angew. Chem.*, 1962, **74**, 465

Dienes and acetylene derivatives are hypergolic in contact with concentrated nitric acid, ignition delay being 1 ms [1].
See Phenylacetylene, below
ROCKET PROPELLANTS

Cyclopentadiene reacts explosively with fuming nitric acid [2].
See also SULPHURIC ACID, H_2O_4S: Cyclopentadiene

Burning fuel oil and other petroleum products detonate immediately on contact with concentrated nitric acid [3]. Very high sensitivity to detonation is shown by mixtures with benzene close to the stoicheiometric proportions of *ca.* 84% acid [4]. Lack of proper control in nitration of toluene with mixed acids may lead to runaway or explosive reaction. A contributory factor is the oxidative formation, and subsequent nitration and decomposition, of nitrocresols [5]. Oxidation of *p*-xylene with nitric acid under pressure in manufacture of terephthalic acid carries explosion hazards in the autoclaves and condensing systems [6]. During nitration of mesitylene in acetic acid–anhydride solution, fuming nitric acid must be added slowly to the cooled mixture to prevent temperature exceeding 20°C, when an explosive reaction may occur [7]. During oxidation of mesitylene with nitric acid in an autoclave at 115°C to give 3,5-dimethylbenzoic acid, a violent explosion occurred. This was attributed to local overheating, formation of 1,3,5-tri(nitromethyl) benzene and violent decomposition of the latter. Smaller-scale preparations with better temperature control were uneventful [8].

Hydrogen iodide,
or Hydrogen selanide,
or Hydrogen sulphide
Hofmann, A. W., *Ber.*, 1870, 3, 660
Ignition occurs on contact of fuming nitric acid with excess hydrogen iodide, and on contact of the sulphide or selenide with the acid.
See Non-metal hydrides, below

Hydrogen peroxide,
Ketones
See HYDROGEN PEROXIDE, H_2O_2: Ketones, etc.

Hydrogen peroxide,
Mercury(II) oxide
See HYDROGEN PEROXIDE, H_2O_2: Mercury(II) oxide, etc.

Ion exchange resins
1. Barghusen, J. *et al., Reactor Fuel Processing*, 1964, 7, 297
2. McBride, J. A. *et al., 3rd Geneva Conf. on Peaceful Uses of Atomic Energy,* A/CONF. 28 28/P/278, 1965

Several cases of interaction between anion exchange resins and nitric acid causing rapid release of energy or explosion have occurred [1]. The cause has been attributed to oxidative degradation of the organic resin matrix and/or nitration of the latter. Suggested precautions include control of temperature, acid concentration and contact time. Presence of heavy ions (Pu) or oxidising agents (dichromates) tends to accelerate the decomposition [2].

Iron(II) oxide
Mellor, 1941, Vol. 13, 716
The finely divided (pyrophoric) oxide incandesces with nitric acid.

Lactic acid,
Hydrofluoric acid
Bubar, S. F. *et al., J. Chem. Educ.*, 1966, 43, A956
Mixtures of the three acids,used as metal polishing solutions, are unstable and should not be stored. Lactic and nitric acids react autocatalytically after a quiescent period, producing a temperature of about 90°C and vigorous gas evolution after 12 h. Prepare freshly, discard after use and handle carefully.

Metal acetylides
Mellor, 1946, Vol. 5, 848
Dicaesium and dirubidium acetylides explode in contact with nitric acid, and the sodium and potassium analogues probably react violently.

Metal hexacyanoferrates (3–) or (4–)
Sidgwick, 1950, 1344
The action of 30% nitric acid on hexacyanoferrates (3–) or (4–) to produce pentacyanonitrosoferrates(2–) ('nitroprussides') is violent.

Metals
Mellor, 1940, Vol. 2, 470; Vol. 4, 270, 483; 1941, Vol. 7, 260; 1940, Vol. 9, 627; 1942, Vol. 12, 32, 188
Bismuth powder glows red hot in contact with fuming nitric acid, while the molten metal (271°C) explodes in contact with the concentrated acid. Powdered germanium reacts violently with concentrated acid, and lithium ignites. Manganese powder incandesces and explodes feebly with nitric acid, and sodium ignites in contact with nitric acid of density above 1·056. Titanium alloys form an explosive deposit with fuming nitric acid. Although uranium powder reacts vigorously with red fuming nitric acid, under some conditions explosive deposits may be formed. Addition of concentrated nitric acid to molten zinc (419°C) causes it to incandesce. Magnesium burns brilliantly in nitric acid vapour.

Metal salicylates
Belcher, R., private comm., 1968
Metal salicylates are occasionally incorporated in mixtures of 'unknowns' for qualitative inorganic analysis. During the conventional group separation, organic radicals are removed by evaporation with nitric acid. When salicylates are present, this can lead to the formation of trinitrophenol through nitration and decarboxylation. This may react with any heavy metal ions present to form unstable explosive picrates, if the evaporation is taken to dryness. An alternative scheme of analysis obviates this danger.

Metal thiocyanate
MCA Case History No. 853
When the (unspecified) thiocyanate solution was pumped through a 80 mm pipeline containing nitric acid, a violent explosion occurred.

This was later confirmed experimentally and attributed to oxidation of the thiocyanate solution by nitric acid.

4-Methylcyclohexanone
Dye, W. T., *Chem. Eng. News,* 1959, **37**, 48
Oxidation of 4-methylcyclohexanone by addition to nitric acid at about 75°C caused a detonation to occur. These conditions had been used previously to oxidise the corresponding alcohol, but, although the ketone is apparently an intermediate in the oxidation of the alcohol, the former requires a much higher temperature to start and maintain the reaction. An OTS report, PB73591, mentions a similar violent reaction with cyclohexanone.
See Acetone, above

Nitroaromatics
Urbanski, 1967, Vol. 3, 290
A series of mixtures of nitric acid with one or more of mono- and di-nitrobenzenes, di- and tri-nitrotoluenes have been shown to possess high explosive properties.
See Nitrobenzene, below

Nitrobenzene,
Nitric acid,
Water
1. Anon., *J. R. Inst. Chem.*, 1960, **84**, 451
2. Van Dolah, R. W., *Loss. Prev.,* 1969(3), 32
A plant explosion involved a mixture of nitrobenzene, nitric acid and a substantial quantity of water. Detonation occurred with a speed and power comparable to TNT. This was unexpected in view of the presence of water in the mixture [1]. The later reference deals with a detailed practical and theoretical study of this system and determination of the detonability limits and shock sensitivity. The limits of detonability coincided with the limits of miscibility over a wide portion of the ternary composition diagram. In absence of water, very high sensitivity (similar to that of glyceryl trinitrate) occurred between 50 and 80% nitric acid, the stoicheiometric proportion being 73%.
See Benzene, above
Nitroaromatics, above

Nitromethane
Olah, G. A. *et al., Org. Synth.*, 1967, **47**, 60
Mixtures are extremely explosive.
See NITROMETHANE, CH_3NO_2: Acids

Non-metal hydrides
1. Mellor, 1946, Vol. 5, 36; 1940, Vol. 6, 814; 1939, Vol. 9, 56, 397
2. Hofmann, A. W., *Ber.*, 1870, **3**, 658
Arsine, phosphine and tetraborane(10) are all oxidised explosively by fuming nitric acid, while stibine behaves similarly with the concentrated acid [1]. Phosphine, hydrogen sulphide and selenide all ignite when the fuming acid is dripped into the gas [2].
See Phosphine derivatives, below

Non-metals
Mellor, 1946, Vol. 5, 16; 1940, Vol. 8, 787, 845
Boron (finely divided forms) reacts violently with concentrated acid and may attain incandescence. The vapour of phosphorus, heated in nitric acid in presence of air, may ignite. Boron phosphide ignites with the concentrated acid.

Organic matter
Bowen, H. J. M., private comm., 1968
When 16M (70%) nitric acid was poured down a sink without diluting water, interaction with (unspecified) organic matter in the trap caused a delayed explosion.

Organic matter
Perchloric acid.
See PERCHLORIC ACID, $ClHO_4$: Nitric acid, etc.

Organic matter,
Sulphuric acid
ABCM Quart. Safety Summ., 1934, **5**, 17
Use of the mixed concentrated acids to dissolve an organic residue caused a violent explosion. Nitric acid is a very powerful and rapid oxidant and may form unstable fulminic acid or polynitro compounds under these conditions.

Phenylacetylene,
1,1-Dimethylhydrazine
Spengler, G. *et al., Sci. Tech. Aerospace Rep.*, 2(17), 2392, Washington, NASA, 1964
Phenylacetylene does not itself ignite on contact with nitric acid but addition of 1,1-dimethylhydrazine causes it to become hypergolic.
See Hydrocarbons, above

Phosphine derivatives
von Schwartz, 1918, 325
Mellor, 1947, Vol. 8, 827, 1041
Graham, T., *Trans. R. Soc. Edinburgh*, 1835, **13**, 88
Phosphine ignites in concentrated nitric acid and addition of warm fuming nitric acid to phosphine causes explosion. Phosphonium iodide ignites with nitric acid, and ethylphosphine explodes with fuming acid. Tris(iodomercuri)phosphine is violently decomposed by nitric acid or aqua regia.

Phosphorus halides
Mellor, 1947, Vol. 8, 827, 1004, 1038
Tetraphosphorus iodide ignites in contact with cold concentrated nitric acid. Phosphorus trichloride explodes with nitric (or nitrous) acid.

Phthalic anhydride,
Sulphuric acid
1. Tyman, J. H. P. *et al., Chem. & Ind.*, 1972, 664
2. Bentley, R. K., ibid., 767
3. Bretherick, L., ibid., 790

Attempts to follow a published method for nitration of phthalic anhydride in sulphuric acid at 80–100°C with fuming nitric acid caused an eruptive decomposition to occur after 2 h delay [1]. The hazard can be eliminated by use of a smaller excess of nitrating acid at 55–65°C [2]. Possible causes of the delayed eruption are suggested as acyl nitrates [3].
See Sulphuric acid, Terephthalic acid, below

Polydibromosilane
See POLYDIBROMOSILANE, $(Br_2Si)_n$: Oxidants

Polyalkenes
Marsh, J. R., *Chem. & Ind.*, 1968, 1718
Fuming nitric acid had seeped past the protective polytetrafluoroethylene liner inside the polyethylene or polypropylene screw cap and attacked the latter, causing pressure build-up in the glass bottle.

Pyrotechol
Andrussow, L., *Chim. Ind.*, 1961, **86**, 542
The phenol is hypergolic with concentrated nitric acid, with a 1 ms ignition delay.
See ROCKET PROPELLANTS

Reducants
A variety of reducants ignite or explode with nitric acid.
See HYDROGEN IODIDE, HI: Nitric acid
HYDROGEN SULPHIDE, H_2S: Nitric acid
POTASSIUM PHOSPHINATE, H_2KO_2P: Nitric acid
HYDRAZINE, H_4N_2 : Nitric acid
Sulphur dioxide, below

Sulphur dioxide
Coleman, G. H. *et al., Inorg. Synth.*, 1939, **1**, 55
Presence of dinitrogen tetraoxide appears to be essential to catalyse smooth formation of nitrosylsulphuric acid. In its absence, reaction may be delayed and then proceed explosively.

Sulphur halides
Mellor, 1947, Vol. 10, 646
Interaction with sulphur dichloride or dibromide is violently effervescent, hydrogen chloride or bromide being evolved.

Sulphuric acid
BCISC Quart. Safety Summ., 1964, **35**, 3
The gland of a centrifugal pump being used to pump nitrating acid (nitric–sulphuric acids, 1:3) exploded after 10 min use. This was attributed to nitration of the gland packing, followed by frictional detonation. Inert shaft sealing material is advocated.

Sulphuric acid,
Terephthalic acid
Withers, C. V., *Chem. & Ind.*, 1972, 821; private comm., 1972
During nitration of the acid with fuming nitric acid in oleum, a delayed exotherm increased the temperature after 2 h from 100°C to 160°C, causing eruption of the contents. At 120°C the delay was 1 h and at 130°C 30 min.
See Phthalic anhydride, above

Thioaldehydes,
or Thioketones
Campaigne, E., *Chem. Rev.*, 1946, **39**, 57
Nitric acid generally reacts too violently with thials or thiones for the reactions to be of synthetic interest.
See Hexamethyltrithiane, above

Thiophene
1. Meyer, V., *Ber.*, 1883, **16**, 1472
2. Babasinian, V. S., *Org. Synth.*, Coll. Vol. 2, 467
Interaction of thiophene with fuming nitric acid is very violent if uncontrolled, extensive oxidation occurring [1]. Use of a diluent and close control of temperature is necessary for preparation of nitrothiophene [2].

Tricadmium diphosphide
Juza, R. *et al.*, *Z. Anorg. Chem.*, 1956, **283**, 230
Reaction with concentrated acid is explosive.

Triethylgallium monoetherate
See TRIETHYLGALLIUM, $C_6H_{15}Ga$

Trimagnesium diphosphide
Mellor, 1940, Vol. 8, 842
Oxidation proceeds with incandescence.

2,4,6-Trimethyltrioxane
Muir, G. D., private comm., 1968
Oxidation of 'paraldehyde' to glyoxal by action of nitric acid is subject to an induction period, and the reaction may become violent if addition of trioxane is too fast. Presence of nitrous acid eliminates the induction period.

Uranium disulphide
Sidgwick, 1950, 1081
Interaction is violent.
See other OXIDANTS

PEROXONITRIC ACID **HNO_4**

Schwarz, R., *Z. Anorg. Chem.*, 1948, **256**, 3
The pure material, prepared at –80°C, decomposes explosively at –30°C. Solutions in acetic acid or water of below the limiting concentration (corresponding to a stoicheiometric mixture of 70% aqueous nitric acid and 100% hydrogen peroxide) are stable, while those above the limit decompose autocatalytically, eventually exploding.
See other PEROXOACIDS

NITROSYLSULPHURIC ACID **HNO_5S**

Preparative hazard.
See NITRIC ACID, HNO_3: Sulphur dioxide

Dinitroaniline
BCISC Quart. Safety Summ., 1970, **41**, 28
During plant-scale diazotisation of a dinitroaniline hydrochloride, local increase in temperature, due to high concentration of reaction mixture, caused a violent explosion.
See also NITRIC ACID, HNO_3: 4-Chloro 2-nitroaniline

LEAD IMIDE **HNPb**

Mellor, 1940, Vol. 8, 265
It explodes on heating or in contact with water or dilute acids.
See other *N*-METAL DERIVATIVES

PHOSPHAM HN_2P

Hydrogen sulphide
Mellor, 1940, Vol. 8, 270
The solid produced by interaction of phospham and hydrogen sulphide at red heat is probably a trimeric triphosphatriazine derivative such as phospham. The solid ignites in slightly warm air or in dinitrogen tetraoxide, and is violently oxidised by nitric acid.

Oxidants
Mellor, 1940, Vol. 8, 269–270
Interaction with copper(II) or mercury(II) oxides proceeds incandescently. Mixtures with a chlorate or nitrate explode on heating. Phospham ignites in dinitrogen tetraoxide.

HYDROGEN AZIDE (HYDRAZOIC ACID) HN_3

1. Smith, 1966, Vol. 2, 214
2. Bowden, F. P. *et al., Endeavour*, 1962, **21**, 121
3. Audreith, L. F. *et al., Inorg. Synth.*, 1939, **1**, 77
4. Kemp, M. D., *J. Chem. Educ.*, 1960, **37**, 142
5. Birkofer, L. *et al., Org. Synth.*, 1970, **50**, 109

Hydrogen azide is quite safe in dilute solution, but is violently explosive and of variable sensitivity in the concentrated (17–50%) or pure states. Wherever possible a low-boiling solvent (ether, pentane) should be added to its solutions to prevent inadvertent concentration by evaporation and recondensation. If this is not possible, no unwetted part of apparatus containing its solutions should be kept appreciably below the boiling point (35°C) to prevent condensation of the pure acid. The pure acid has often been isolated by distillation, but appears to undergo rapid sensitisation on standing, so that, after an hour, faint vibrations or speech are enough to initiate detonation [1]. The solid acid is also very unstable [2]. Preparative procedures have been detailed [3]. It is readily formed on contact of hydrazine or its salts with nitrous acid or its salts. A safe procedure for the preparation of virtually anhydrous hydrogen azide is described [4]. Trimethylsilyl azide serves as a safe and stable substitute for hydrogen azide in many cases [5].

See METHYL AZIDE, CH_3N_3
p-CHLOROPHENYL ISOCYANATE, C_7H_4ClNO

Heavy metals

1. Napier, D. H., private comm., 1972
2. Cowley, B. R. *et al., Chem. & Ind.*, 1973, 444

Great care is necessary to prevent formation of explosive heavy metal azides from unsuspected contact of hydrogen azide with heavy metals. Interaction of hydrazine and nitrite salts in a copper drainage system caused formation and explosion of copper azide [1]. Use of a brass water-pump and vacuum gauge during removal of excess hydrogen azide under vacuum formed deposits which exploded when the pump and gauge were handled later [2].

See other NON-METAL AZIDES

SODIUM HYDRIDE HNa

Acetylene

Mellor, 1941, Vol. 2, 483

Dry acetylene does not react with sodium hydride below 42°C, but in presence of moisture reaction is vigorous even at −60°C.

Air

Plesek, J. *et al., Sodium Hydride* (Eng. Transl., Jones, G.) 5,8, London, Iliffe Books, 1968

Commercial sodium hydride may contain traces of sodium which render it spontaneously flammable in moist air, or air enriched by carbon dioxide. The very finely divided dry powder ignites in dry air. Dispersions of the hydride in oil are safe to handle. All normal extinguishers are unsuitable for solid sodium hydride fires. Powdered sand, ashes or sodium chloride are suitable.

Dimethyl sulphoxide

See DIMETHYL SULPHOXIDE, C_2H_6OS : Sodium hydride

Ethyl 2,2,3-trifluoropropionate

Bagnall, R. D., private comm., 1972

The ester decomposes violently in presence of sodium hydride, probably owing to hydride-induced elimination of hydrogen fluoride and subsequent exothermic polymerisation.

Glycerol
Unpublished observations, 1956
Exothermic interaction of granular hydride with undiluted (viscous) glycerol with inadequate stirring caused charring to occur. Dilution with tetrahydrofuran to reduce viscosity and improve mixing prevented local overheating during formation of monosodium glyceroxide.

Halogens
Mellor, 1940, Vol. 2, 483
Interaction with chlorine or fluorine is incandescent at ambient temperature, and with iodine at 100°C.

Oxygen
See OXYGEN, O_2: Metal hydrides

Sulphur
See SULPHUR, S: Sodium hydride

Sulphur dioxide
Moissan, H., *Compt. Rend.*, 1902, **135**, 647
Sulphur dioxide reacts explosively in contact with sodium hydride unless diluted with hydrogen.

Water
1. Anon., *J. Chem. Educ.*, 1967, **44**, 321
2. *MCA Case History No. 1587*

Addition of sodium hydride to a damp reactor which had not been purged with an inert gas caused evolution of hydrogen, and a violent explosion. Solid dispersions of the hydride in mineral oil are more easily and safely handled[1]. When an unprotected polythene bag containing the hydride was moved, some of the powder leaked from a hole, contacted moisture and immediately ignited. Such materials should be kept in tightly closed metal containers in an isolated, dry location[2].
See other METAL HYDRIDES

SODIUM HYDROXIDE — HNaO

Chloroform,
Methanol
See CHLOROFORM, $CHCl_3$: Sodium hydroxide, etc.

4-Chloro-2-methylphenol

See 4-CHLORO-2-METHYLPHENOL, C_7H_7ClO: Sodium hydroxide

Cinnamaldehyde

See CINNAMALDEHYDE, C_9H_8O: Sodium hydroxide

Cyanogen azide

See CYANOGEN AZIDE, CN_4: Sodium hydroxide

Diborane

See DIBORANE, B_2H_6: Octanal oxime, etc.

Maleic anhydride

See MALEIC ANHYDRIDE, $C_4H_2O_3$: Cations, etc.

4-Methyl-2-nitrophenol

See 4-METHYL-2-NITROPHENOL, $C_7H_7NO_3$: Sodium hydroxide

3-Methyl-2-penten-4-yn-1-ol

See 3-METHYL-2-PENTEN-4-YN-1-OL, C_6H_8O: Sodium hydroxide

1,2,4,5-Tetrachlorobenzene

See 1,2,4,5-TETRACHLOROBENZENE, $C_6H_2Cl_4$: Sodium hydroxide

1,1,1-Trichloroethanol

See 1,1,1-TRICHLOROETHANOL, $C_2H_3Cl_3O$: Sodium hydroxide

Trichloronitromethane

See TRICHLORONITROMETHANE, CCl_3NO_3: Sodium hydroxide

Water

MCA SD-9, 1968

Haz. Chem. Data, 1969, 199

Contact with water in limited amounts is very exothermic, and boiling or ignition of adjacent combustibles may be caused.

Zinc

See ZINC, Zn: Sodium hydroxide

Zirconium

See ZIRCONIUM, Zr: Oxygen-containing compounds

SODIUM HYDROGENSULPHATE $HNaO_4S$

Calcium hypochlorite
See CALCIUM HYPOCHLORITE, $CaCl_2O_2$: Sodium hydrogensulphate

'SOLID PHOSPHORUS HYDRIDE' HP_2

Mellor, 1940, Vol. 8, 851
Sidgwick, 1950, 730
This material (possibly phosphine adsorbed on phosphorus and produced by decomposition of diphosphane in light) ignites in air, on impact, or on sudden heating to 100°C.
See other NON-METAL HYDRIDES

RUBIDIUM HYDRIDE HRb

Acetylene
Mellor, 1941, Vol. 2, 483
In presence of moisture interaction of the hydride and acetylene is vigorous at –60°C. In dry acetylene reaction only occurs above 42°C.

Oxygen
See OXYGEN (Gas), O_2 : Metal hydrides

Water
Mellor, 1963, Vol. 2, Suppl. 2.2, 2187
Interaction with water is too violent to permit of safe use of the hydride as a drying agent. When dispersed as a solid solution in a metal halide, it can be used as a drying or reducing agent.
See other METAL HYDRIDES

SILICON MONOHYDRIDE $(HSi)_n$

Alkali
Stock, G. *et al., Angew, Chem.*, 1956, **68**, 213
The polymeric hydride is relatively stable to water, but reacts violently with aqueous alkali, evolving hydrogen.
See other NON-METAL HYDRIDES

HYDROGEN H_2

Gas above –253°C; E.L., 4.1–74.2%

Air,
Catalysts
Mellor, 1942, Vol. 1, 325; 1937, Vol. 16, 146
Catalytically active platinum and similar metals containing adsorbed oxygen or hydrogen will heat and cause ignition in contact with hydrogen or air, respectively, Nitrogen-purging before exposure to atmosphere will eliminate the possibility.

Air,
Various vapours
Schumacher, H. J., *Angew. Chem.*, 1951, **63**, 560–561
The effects of the presence of 44 gaseous or volatile materials upon the upper explosion limits of hydrogen—air mixtures have been tabulated.

Halogens,
or Interhalogens
See BROMINE TRIFLUORIDE, BrF_3: Hydrogen-containing materials
BROMINE, Br_2: Hydrogen
CHLORINE TRIFLUORIDE, ClF_3: Hydrogen-containing materials
CHLORINE, Cl_2: Hydrogen
FLUORINE, F_2: Hydrogen
IODINE HEPTAFLUORIDE, F_7I: Carbon, etc.

o-Nitroanisole
See *o*-NITROANISOLE, $C_7H_7NO_2$: Hydrogen

Oxidants
See FLUORINE PERCHLORATE, $ClFO_4$: Hydrogen
CHLORINE DIOXIDE, ClO_2: Hydrogen
DICHLORINE OXIDE, Cl_2O: Oxidisable materials
COPPER(II) OXIDE, CuO: Hydrogen
NITRYL FLUORIDE, FNO_2: Non-metals
DIFLUORODIAZENE, F_2N_2: Hydrogen
NITROGEN OXIDE, NO: Hydrogen, etc.
DINITROGEN TETRAOXIDE, N_2O_4: Hydrogen, etc.
PALLADIUM(II) OXIDE, OPd: Hydrogen
OXYGEN (Gas), O_2: Hydrogen

Palladium trifluoride
See PALLADIUM TRIFLUORIDE, F_3Pd: Hydrogen

LIQUID HYDROGEN H_2

Air

1. Kit and Evered, 1960, 123
2. Report *UCRL-3072,* Univ. of California, Berkeley, 1955
3. Weintraub, A. A. *et al., Health Phys.*, 1962, 8, 11

The main precaution necessary for use of liquid hydrogen is to prevent air leaking into the system, where it will be condensed and solidified. Fracture of a crystal of solid air or oxygen could produce a spark to initiate explosion [1]. Procedures for the safe handling of liquid hydrogen in the laboratory [2] and in liquid hydrogen bubble chambers [3] have been detailed.

POTASSIUM AMIDE H_2KN

Brandsma, 1971, 20–21

It has similar properties to the much more widely investigated sodium amide, but may be expected on general grounds to be more violently reactive than the former. The frequent fires or explosions observed during work-up of reaction mixtures involving the amide were attributed to presence of unreacted (oxide-coated) particles of potassium in the amide solution in ammonia. A safe filtration technique is described.

Potassium nitrite

See POTASSIUM NITRATE KNO_2: Potassium amide

Water

Mellor, 1940, Vol. 8, 255

Interaction is violent and ignition may occur, even in contact with humid air. Old samples may explode with a considerable delay after contact with liquid water.

See other N-METAL DERIVATIVES

POTASSIUM AMIDOSULPHATE H_2KNO_3S

Metal nitrates or nitrites

See METAL AMIDOSULPHATES

POTASSIUM PHOSPHINATE ('HYPOPHOSPHITE') H_2KO_2P

Nitric acid

Mellor, 1940, Vol. 8, 882

The salt burns (owing to evolution of phosphine) when heated in air, and explodes when evaporated with nitric acid.

See other REDUCANTS

MAGNESIUM HYDRIDE H_2Mg

Oxygen

See OXYGEN (Gas), O_2: Metal hydrides

SODIUM AMIDE H_2NNa

Air

1. Bergstrom, F. W. *et al., Chem. Rev.*, 1933, **12**, 61; Brauer, 1963, 467
2. Krüger, G. R. *et al., Inorg. Synth.*, 1966, **8**, 15
3. Shreve, R. N. *et al., Ind. Eng. Chem.*, 1940, **32**, 173
4. Sandor, S., *Munkavédelem*, 1960, **6**(4–6), 20

It frequently ignites or explodes on heating or grinding in air, particularly if previously exposed to air or moisture to produce degradation products (possibly peroxidic) [1]. Only one explosion not involving exposure to air has been recorded, during pulverisation [2]. The following oxidation products, all liable to explode, have been identified [3]:

Sodium hyponitrite, $N_2Na_2O_2$

Sodium trioxodinitrate (2–), $N_2Na_2O_3$

Sodium tetraoxodinitrate (2–), $N_2Na_2O_4$

Sodium pentaoxodinitrate (2–), $N_2Na_2O_5$

Sodium hexaoxodinitrate (2–), $N_2Na_2O_6$

Oxidants

See POTASSIUM CHLORATE, $ClKO_3$: Sodium amide
SODIUM NITRITE, $NNaO_2$: Sodium amide
DINITROGEN TETRAOXIDE, N_2O_4: Sodium amide

Water

1. Mellor, 1941, Vol. 2, 255
2. Personal experience

Fresh material behaves like sodium, hissing, forming a diminishing mobile

globule, and often finally exploding [1]. Old, degraded (yellow) samples may be immersed in water for appreciable periods with little action, and then explode very violently. Disposal by controlled burning is safer [2].
See other N-METAL DERIVATIVES

SODIUM HYDROXYLAMIDE H_2NNaO

See HYDROXYLAMINE, H_3NO: Metals

SODIUM AMIDOSULPHATE H_2NNaO_3S

Metal nitrates,
or Metal nitrites
See METAL AMIDOSULPHATES

HYPONITROUS ACID $H_2N_2O_2$

Sidgwick, 1950, 693
Mellor, 1940, Vol. 8, 407
An extraordinarily explosive solid, of which the sodium salt also explodes on heating to 260°C. An attempt to prepare the acid by treating its silver salt with hydrogen sulphide caused explosive decomposition. Contact with solid potassium hydroxide caused ignition.
See LEAD HYPONITRITE, N_2O_2Pb
See other REDUCANTS

NITRIC AMIDE (NITRAMIDE) $H_2N_2O_2$

Canis, C., *Rev. Chim. Minerale,* 1964, **1**, 521
Nitramide is quite unstable and various reactions in which it is formed are violent. Attempts to prepare it by interaction of various nitrates and sulphamates showed that the reactions became explosive at specific temperatures.

Alkalies
Thiele, J. *et al., Ber.,* 1894, **27**, 1909
A drop of concentrated alkali added to solid nitramide causes a flame and explosive decomposition.

Sulphuric acid
Urbanski, 1967, Vol. 3, 16
Nitramide decomposes explosively on contact with concentrated sulphuric acid.
See other N-NITRO COMPOUNDS

AMIDOSULPHURYL AZIDE $H_2N_4O_2S$

Shozda, R. J. *et al., J. Org. Chem.*, 1967, **32**, 2876
It is a low-melting explosive solid, as shock-sensitive as glyceryl nitrate.
See other ACYL AZIDES

SODIUM PHOSPHINATE ('HYPOPHOSPHITE') H_2NaO_2P

Mellor, 1940, Vol. 8, 881
Evaporation of aqueous solutions by heating may cause an explosion, phosphine being evolved.

Oxidants
1. Costa, R. L., *Chem. Eng. News,* 1947, **25**, 3177
2. Mellor, 1940, Vol. 8, 881

Evaporation of a moist mixture of sodium phosphinate and a trace of sodium chlorate by slow heating caused a violent explosion. It was concluded that, once started, the decomposition of the phosphinate proceeds spontaneously [1]. Similar reactions have been reported with nitrates instead of chlorates. Such mixtures had previously been proposed as explosives [2].
See PERCHLORIC ACID, $ClHO_4$: Sodium phosphinate
See other REDUCANTS

SODIUM DIHYDROGENPHOSPHIDE H_2NaP

Albers, H. *et al., Ber.*, 1943, **76**, 23
It ignites in air.
See related NON-METAL HYDRIDES

OXOSILANE H_2OSi

Kautsky, K., *Z. Anorg. Chem.*, 1921, **117**, 209
It ignites in air.
See related NON-METAL HYDRIDES

HYDROGEN PEROXIDE H_2O_2

1. Shanley, E. S. *et al., Ind. Eng. Chem.*, 1947, **39**, 1536
2. Naistat, S. S. *et al., Chem. Eng. Progr.*, 1961, **57**(8), 76
3. Kirk-Othmer, 1966, Vol. 11, 407; *MCA SD-53,* 1969
4. Anon., *Fire Prot. Ass. J.*, 1954, 215
5. Smith, I. C. P., private comm., 1973
6. *MCA Case History No. 1121*
7. Campbell, G. A. *et al., Proc. 4th IChE Symp. on Chem. Proc. Hazards,* 1971, 37–43

The hazards attendant upon use of concentrated hydrogen peroxide solutions have been reviewed [1–3]. Salient points include:

Release of enough energy during catalytic decomposition of 65% peroxide to evaporate all water present and formed, and subsequent liability of ignition of combustible materials.

Most cellulosic materials contain enough catalyst to cause spontaneous ignition with 90% peroxide.

Contamination of concentrated peroxide causes possibility of explosion. Readily oxidisable materials, or alkaline substances containing heavy metals, may react violently.

Soluble fuels (acetone, ethanol, glycerol) will detonate on admixture with peroxide of over 30% concentration, the violence increasing with concentration.

Handling systems must exclude fittings of iron, brass, copper, Monel, and screwed joints caulked with red lead.

Concentrated peroxide may decompose violently in contact with iron, copper, chromium, and most other metals or their salts, and dust (which frequently contains rust). Absolute cleanliness, suitable equipment (PVC, butyl or Neoprene rubber) and personal protection are essential for safe handling [4]. During concentration under vacuum of aqueous [5] or of aqueous–alcoholic [6] solutions of hydrogen peroxide, violent explosions occurred when the concentration was sufficiently high (probably above 90%) [3]. Detonation of hydrogen peroxide vapour has been studied experimentally [7].

Acetal,
Acetic acid
Ashley, J. M. *et al., Chem. & Ind.*, 1957, 702
An organic sulphide containing an acetal group in the molecule had been oxidised to the sulphone with 30% hydrogen peroxide in acetic acid. After the residue had been concentrated by vacuum distillation at 50–60°C, the residue exploded during handling. This was attributed to formation of the peroxide of the acetal (formally a *gem*-diether) or of the aldehyde formed by hydrolysis, but peracetic acid may also have been involved.

Acetic acid
Grundmann, C. *et al., Ber.*, 1939, **69**, 1755
During preparation of peracetic acid, the temperature should not be too low to prevent reaction as the reagents are mixed, because reaction may begin later with explosive violence.
See Oxygenated compounds, below
PEROXYACETIC ACID, $C_2H_4O_3$

Acetone,
Other reagents
1. Anon., *Angew. Chem. (Nachr.)*, 1970, **18**, 3
2. *MCA Case History No. 233*
3. *MCA Case History No. 223*
4. Stirling, C. J. M., *Chem. Brit.*, 1969, **5**, 36
5. Seidl, H., *Angew. Chem.* (*Intern. Ed.*), 1964, **3**, 640
6. Treibs, W., ibid., 802

Acetone and hydrogen peroxide readily form explosive dimeric and trimeric peroxides, particularly during evaporation of the mixture. Many explosions have occurred during work-up of peroxide reactions run in acetone, including partial hydrolysis of a nitrile [1] and oxidation of 2,2′-thiodiethanol [2] and of an unspecified material [3]. The reaction mixture from oxidation of a sulphide with hydrogen peroxide in acetone exploded violently during vacuum evaporation at 90°C. On another occasion oxidation of a sulphide in acetone in presence of molybdate catalyst proceeded with explosive violence. A general warning about use of acetone as solvent for peroxide oxidations is given [4]. During the isolation of 1-tetralone, produced by oxidation of tetralin with hydrogen peroxide in acetone, a violent explosion occurred which was attributed to acetone peroxide [5]. The originator of the method later gave detailed instructions for a safe procedure, which include the exclusion of mineral acids, even in traces [6].

See Oxygenated compounds, below
KETONE PEROXIDES

Alcohols

1. *MCA SD-53*, 1969
2. Spengler, G. *et al.*, *Brennst. Chem.*, 1965, **46**, 117

Homogeneous mixtures of concentrated peroxide and alcohols or other peroxide-miscible organic liquids are capable of detonation by shock or heat [1]. Furfuryl alcohol ignites in contact with 85% peroxide within 1 s [2].

See Oxygenated compounds, below

Alcohols,
Sulphuric acid

Hedaya, E. *et al.*, *Chem. Eng. News*, 1967, **45**(43), 73

During conversion of alcohols to hydroperoxides, the order of mixing the reagents is important. Addition of concentrated acid to mixtures of an alcohol and concentrated peroxide almost inevitably leads to explosion, particularly if the mixture is inhomogeneous and the alcohol is a solid.

See *tert*-Butanol, below
2-Methyl-1-phenyl-2-propanol, below

tert-Butanol,
Sulphuric acid

Schenach, T. A., *Chem. Eng. News*, 1973, **51**(6), 39

Preparation of di-*tert*-butyl peroxide by addition of *tert*-butanol to 50% hydrogen peroxide: 78% sulphuric acid mixtures (1:2 by wt.) is a dangerously deceptive procedure. On the small scale and with adequate cooling capacity it may be possible to prevent the initial stage (exothermic formation of *tert*-butyl hydroperoxide) getting out of control and initiating violent or explosive decomposition of the peroxide–peroxomonosulphuric acid mixture. This hazard diminishes as the reaction proceeds with consumption of hydrogen peroxide and dilution by the water of reaction. On the plant scale several severe explosions have occurred, preceded only by a gradual temperature increase, during attempted process development work.

See Alcohols, above

Carbon
1. Mellor, 1939, Vol. 1, 936–938
2. Schumb, 1955, 402, 478
The violent decomposition observed on adding charcoal to concentrated hydrogen peroxide is mainly due to catalysis by metallic impurities present and the surface of the charcoal, rather than to direct oxidation of the carbon [1]. Charcoal mixed with a trace of manganese dioxide ignites immediately on contact with concentrated peroxide [2].

Carboxylic acids
Admixture produces peroxyacids, some of which are unstable and explosive.
See PEROXYACIDS

Iron(II) sulphate,
2-Methylpyridine,
Sulphuric acid
Mond Div., ICI, private comm., 1969
Addition of 30% peroxide and sulphuric acid to 2-methylpyridine and iron(II) sulphate caused a sudden exotherm, followed by a vapour phase explosion and ignition. Lack of stirring is thought to have caused localised overheating, vaporisation of the base, and ignition in the possibly oxygen-enriched atmosphere.

Ketones,
Nitric acid
Bjorklund, G. H. *et al., Trans. R. Soc. Can.*, 1950(Sect. III), **44**, 25
Unless the temperature and concentrations of reagents were carefully controlled, mixtures of hydrogen peroxide, nitric acid and acetone overheated and exploded violently. Under controlled conditions, the explosive dimeric or trimeric acetone peroxides were produced. Butanone-2 and pentanone-3 gave shock- and heat-sensitive oily peroxides. Cyclopentanone reacts vigorously, producing a solid which soon produces a series of explosions if left in contact with the undiluted reaction liquor. The isolated trimeric peroxide is very sensitive to shock, slight friction or rapid heating, and explodes very violently. Cyclohexanone and 3-methylcyclohexanone gave oily, rather explosive peroxides.
See KETONE PEROXIDES

Mercury(II) oxide,
Nitric acid
Mellor, 1940, Vol. 4, 781
Although red mercury oxide usually vigorously decomposes hydrogen peroxide, the presence of traces of nitric acid inhibits decomposition and promotes the formation of red mercury(II) peroxide. This explodes on impact or friction, even when wet, if the mercury oxide was finely divided.

Metals,
or Metal oxides,
or Metal salts
1. Mellor, 1939, Vol. 1, 936–944
2. Schumb, 1955, 480
3. Kit and Evered, 1960, 136
4. 'Hydrogen Peroxide Data Manual', Laporte Chemicals Ltd., Luton, 1960

The noble metals are all very active catalysts, particularly when finely divided, for the decomposition of hydrogen peroxide, silver being used for this purpose in peroxide-powered rocket motors. Gold and the platinum group metals behave similarly [2,4]. Addition of platinum black to concentrated peroxide solution may cause an explosion, and powdered magnesium and iron, promoted by traces of manganese dioxide, ignite on contact [1].

Oxides of cobalt, iron (especially rust), lead (also the hydroxide), manganese, mercury and nickel are also very active and the parent metals and their alloys must be rigorously excluded from peroxide handling systems [1,4]. Soluble derivatives of many other metals, particularly under alkaline conditions, will also catalyse the exothermic decomposition, even at low concentrations [4]. Calcium permanganate has been used, either as a solid or in concentrated solution, to ignite peroxide rocket motors [3]. The last reference gives comprehensive data on all aspects of handling and use of concentrated peroxide.
See also Carbon, above

2-Methyl-1-phenyl-2-propanol,
Sulphuric acid
1. Winstein, S. *et al.*, *J. Amer. Chem. Soc.*, 1967, **89**, 1661
2. Hedaya, E. *et al.*, *Chem. Eng. News*, 1967, **45**(43), 73
3. Hiatt, R. R. *et al.*, *J. Org. Chem.*, 1963, **28**, 1893

Directions given [1] for preparation of 2-methyl-1-phenyl-2-propyl hydroperoxide by adding sulphuric acid to a mixture of the alcohol and 90% hydrogen peroxide are wrong [2] and will lead to explosion [3]. The acidified peroxide (30–50% solution is strong enough) is preferably added to the alcohol, with suitable cooling and precautions [2].

See Alcohols, above
Oxygenated compounds, below

Nitric acid,
Thiourea
Bjorklund, G. H. *et al., Trans. R. Soc. Can.*, 1950 (Sect.III), **44**, 28
The solid peroxide produced by action of hydrogen peroxide and nitric acid on thiourea in acetic acid solution decomposed violently on drying in air with evolution of sulphur dioxide and free sulphur. (The solid may have been a peroxidate of thiourea dioxide.)

See CRYSTALLINE HYDROGEN PEROXIDATES

Nitrogenous bases
1. Stone, F. S. *et al., J. Chem. Phys.*, 1952, **20**, 1339
2. Urbanski, 1967, Vol. 3, 306
3. *Haz. Chem. Data,* 1969, 220

Ammonia dissolved in 99.6% peroxide gave an unstable solution which exploded violently [1]. In the absence of catalysts, concentrated peroxide does not react immediately with hydrazine hydrate. This induction period has caused a number of explosions and accidents due to sudden reaction of accumulated materials [2]. 1,1-Dimethylhydrazine is hypergolic with high-test peroxide [3].

Organic compounds
Hutton, E., *Chem. Brit.,* 1969, **5**, 287
Although under certain circumstances mixtures of hydrogen peroxide and organic compounds are capable of developing more explosive power than an equivalent weight of TNT, in many cases interaction can be effected safely and well under control by applying well-established procedures, to which several references are given.

Oxygenated compounds,
Water
Monger, J. M. *et al., J. Chem. Eng. Data,* 1961, **6**(1), 23
The explosion limits have been determined for liquid systems

containing hydrogen peroxide, water and acetaldehyde, acetic acid, acetone, ethanol, formaldehyde, formic acid, methanol, 2-propanol or propionaldehyde, under various types of initiation.
See Alcohols, above

Phosphorus
Anon., *J. R. Inst. Chem.*, 1957, **81**, 473
If yellow or red phosphorus is incompletely immersed while undergoing oxidation in hydrogen peroxide solutions, heating at the air–solution interface can ignite the phosphorus and lead to a violent reaction. Such behaviour has been observed with hydrogen peroxide solutions above 30% (110 volume) concentration.

Phosphorus(V) oxide
Toennies, G., *J. Amer. Chem. Soc.*, 1937, **59**, 555
The extremely violent interaction of phosphorus(V) oxide and concentrated hydrogen peroxide to give permonophosphoric acid may be moderated by using acetonitrile as a diluent.

Tin(II) chloride
Vickery, R. C. *et al.*, *Chem. & Ind.*, 1949, 657
Interaction is strongly exothermic, even in solution. Addition of peroxide solutions of greater than 3% wt. strength causes a violent reaction.

Vinyl acetate
ABCM Quart. Safety Summ., 1948, **19**, 18
Vinyl acetate had been hydroxylated by treatment with excess hydrogen peroxide in presence of osmium tetraoxide catalyst. An explosion occurred while excess vinyl acetate and solvent were being removed by vacuum distillation. This was attributed to the presence of peracetic acid, formed by interaction of excess hydrogen peroxide and acetic acid produced by hydrolysis of the vinyl acetate.
See other OXIDANTS

ZINC HYDROXIDE H_2O_2Zn

Chlorinated rubber
See CHLORINATED RUBBER: Metal oxides, hydroxides

MCA SD-20, 1963
Safe handling procedures for sulphuric acid and fuming sulphuric acid (oleum) are detailed.

Acetone,
Nitric acid
See NITRIC ACID, HNO_3: Acetone, etc.

Acetonitrile,
Sulphur trioxide
See ACETONITRILE, C_2H_3N: Sulphuric acid

Acrylonitrile
See ACRYLONITRILE, C_3H_3N: Acids

Alkyl nitrates
See ALKYL NITRATES: Lewis acids

Ammonium iron(III) sulphate dodecahydrate
Clark, R. E. D., private comm., 1973
A few dense crystals heated with sulphuric acid exploded, owing to the exotherm in contact with water liberated as the crystals disintegrated.

Aniline,
Glycerol,
Nitrobenzene
See QUINOLINE, C_9H_7N

Bromine pentafluoride
See BROMINE PENTAFLUORIDE, BrF_5: Acids, etc.

Copper,
Campbell, D. A., *School Sci. Rev.*, 1939, **20**(80), 631

The generation of sulphur dioxide by reduction of sulphuric acid with copper is considered too dangerous for a school experiment.

2-Cyano-2-propanol
1. *Occupancy Fire Record*, FR 57-5, 5, Boston, NFPA, 1957
2. Kirk-Othmer, 1967, Vol. 13, 333

Addition of sulphuric acid to the cyano-alcohol caused a vigorous reaction which pressure-ruptured the vessel [1]. This seems likely to have been due to insufficient cooling to prevent dehydration of the alcohol to methacrylonitrile and lack of inhibitors to prevent exothermic polymerisation of the nitrile [2].

Cyclopentadiene
Wilson, P. J. *et al., Chem. Rev.*, 1944, **34**, 8

It reacts violently with charring, or explodes in contact with concentrated sulphuric acid.

See also NITRIC ACID, HNO_3: Hydrocarbons

Cyclopentanone oxime
Brookes, F. R., private comm., 1968

Heating the oxime with 85% sulphuric acid to effect the Beckmann rearrangement caused eruption of the stirred flask contents. Benzenesulphonyl chloride in alkali was a less vigorous reagent.

See BUTYRALDOXIME, C_4H_9NO
ETHYL FORMYLPROPIONATE OXIME, $C_6H_{11}NO_3$

Dimethoxydinitroanthraquinone
See *mixo*-DIMETHOXYDINITROANTHRAQUINONE, $C_{16}H_{10}N_2O_8$: Sulphuric acid

Hexalithium disilicide
See HEXALITHIUM DISILICIDE, Li_6Si_6: Acids

Metal acetylides or carbides
1. Mellor, 1946, Vol. 5, 849
2. *MCA SD-20*, 1963

Monocaesium and monorubidium acetylides ignite with concentrated sulphuric acid [1]. Other carbides are hazardous in contact [2].

Metal chlorates

See METAL CHLORATES: Acids
METAL HALOGENATES: Metals and oxidisable derivatives, etc.

Metal perchlorates

See METAL PERCHLORATES: Sulphuric acid

Nitroaryl bases and derivatives

1. Hodgson, J. F., *Chem. & Ind.*, 1968, 1399; private comm., 1973
2. Poshkus, A. C. *et al.*, *J. Appl. Polymer Sci.*, 1970, **14**, 2049–2052

A series of *o*- and *p*-nitroaniline derivatives and analogues when heated with sulphuric acid to above 200°C undergo, after an induction period, a vigorous reaction. This is accompanied by gas evolution which produces up to a 150-fold increase in volume of a solid foam, and is rapid enough to be potentially hazardous if confined. *o*-Nitroaniline reacts almost explosively [1] and *p*-nitroaniline, *p*-nitroacetanilide, aminonitrodiphenyls, -naphthalenes and various derivatives [2], as well as some nitro-*N*-heterocycles [1,2], also react vigorously. *p*-Nitroanilinium sulphate and 4-nitroaniline-2-sulphonic acid and its salts also generate foams when heated without sulphuric acid. The mechanism is not clear, but involves generation of a polymeric matrix foamed by sulphur dioxide and water eliminated during the reaction [1].

See also DIETHYL SULPHATE, $C_4H_{10}O_4S$: 2,7-Dinitro-9-phenylphenanthridine

N-Nitromethylamine

See *N*-NITROMETHYLAMINE, $CH_4N_2O_2$: Sulphuric acid

Nitramide

See NITRIC AMIDE, $H_2N_2O_2$: Sulphuric acid

Nitric acid,
Organic matter

See NITRIC ACID, HNO_3: Organic matter, etc.

Nitric acid,
Toluene

See NITRIC ACID, HNO_3: Hydrocarbons (reference 5)

Nitrobenzene

See NITROBENZENE, $C_6H_5NO_2$: Sulphuric acid

m-Nitrobenzenesulphonic acid

See *m*-NITROBENZENESULPHONIC ACID, $C_6H_5NO_5S$: Sulphuric acid

Nitromethane

See NITROMETHANE, CH_3NO_2: Acids

p-Nitrotoluene

See *p*-NITROTOLUENE, $C_7H_7NO_2$: Sulphuric acid

Permanganates

Interaction produces the powerful oxidant, permanganic acid.

See PERMANGANIC ACID, $HMnO_4$: Organic materials
POTASSIUM PERMANGANATE, $KMnO_4$: Sulphuric acid

Phosphorus

Mellor, 1940, Vol. 8, 786

White phosphorus ignites in contact with boiling sulphuric acid or its vapour.

Phosphorus trioxide

See PHOSPHORUS(III) OXIDE, O_3P_2: Sulphuric acid

Potassium

See POTASSIUM, K: Sulphuric acid

Potassium *tert*-butoxide

See POTASSIUM *tert*-BUTOXIDE, C_4H_9KO: Acids

Silver peroxochromate

See SILVER PEROXOCHROMATE, $AgCrO_5$: Sulphuric acid

Sodium

See SODIUM, Na: Acids

Sodium carbonate
See SODIUM CARBONATE, CNa_2O_3: Sulphuric acid

Trimercury dinitride
See TRIMERCURY DINITRIDE, Hg_3N_2

1,3,5-Trinitrosohexahydro-1,3,5-triazine
See 1,3,5-TRINITROSOHEXAHYDROTRIAZINE, $C_3H_6N_6O_3$: Sulphuric acid

Water
Mellor, 1947, Vol. 10, 405–408
Dilution of sulphuric acid with water is vigorously exothermic, and must be effected by adding acid to water to avoid local boiling. Mixtures of sulphuric acid and excess snow form powerful freezing mixtures. Fuming sulphuric acid (containing sulphur trioxide) reacts violently with water.
See also *m*-NITROBENZENESULPHONIC ACID, $C_6H_5NO_5S$: Sulphuric acid
See other OXIDANTS

PEROXOMONOSULPHURIC ACID **H_2O_5S**

1. Edwards, J. O., *Chem. Eng. News,* 1955, **33**, 3336
2. Brauer, 1963, Vol. 1, 389

A small sample had been prepared from chlorosulphuric acid and 90% hydrogen peroxide; the required acid phase was separated and stored at 0°C overnight. After warming slightly, it exploded [1]. The handling of large amounts is dangerous owing to possibility of local overheating (e.g. contact with moisture) and explosive decomposition [2].

Acetone
1. *MCA Case History No. 662*
2. Bayer, A., *Ber.,* 1900, **33**, 858

Accidental addition of a little acetone to the residue from wet ashing a polymer with mixed nitric–sulphuric acids and hydrogen peroxide caused a violent explosion [1]. The peroxo acid would be produced under these conditions, and is known to react with acetone to produce the explosive acetone peroxide [2].

Alcohols
Toennies, G., *J. Amer. Chem. Soc.*, 1937, **59**, 552
Contact of the acid with secondary and tertiary alcohols, even with cooling, may lead to violent explosions.

Aromatic compounds
Sidgwick, 1950, 939
Mixtures with aniline, benzene, phenol, etc., explode.

Catalysts
Mellor, 1946, Vol. 2, 483–484
The 92% acid is decomposed explosively on contact with massive or finely divided platinum, manganese dioxide or silver. Neutralised solutions of the acid also froth violently on treatment with silver nitrate, lead or manganese dioxides.

Fibres
Ahrle, H., *Z. Angew. Chem.*, 1909, **23**, 1713
Wool and cellulose are rapidly carbonised by the 92% acid, while cotton ignites after a short delay.
See other OXIDANTS

PEROXODISULPHURIC ACID $H_2O_8S_2$

Organic liquids
1. D'Ans, J. *et al.*, *Ber.*, 1910, **43**, 1880; *Z. Anorg. Chem.*, 1911, **73**, 1911
2. Sidgwick, 1950, 938

A very powerful oxidant; uncontrolled contact with aniline, benzene, ethanol, ether, nitrobenzene or phenol may cause explosion [1]. Alkanes are slowly carbonised [2].
See other OXIDANTS

HYDROGEN SULPHIDE H_2S

Gas above −62°C; E.L., 4.3–46%
MCA SD-36, 1968

p-Bromobenzenediazonium chloride
See DIAZONIUM SULPHIDES

Metal oxides
Mellor, 1947, Vol. 10, 129, 141
Hydrogen sulphide is rapidly oxidised, and may ignite, in contact with a range of metal oxides, including barium dioxide, chromium trioxide, copper oxide, lead dioxide, manganese dioxide, nickel oxide, silver mono- and dioxides, sodium peroxide, thallium trioxide. In the presence of air, contact with mixtures of calcium or barium oxide with mercury or nickel oxide may cause vivid incandescence or explosion.
See Rust, below

Metals
Mellor, 1947, Vol. 10, 140; 1943, Vol. 11, 731
A mixture with air passed over copper powder may attain red heat. Finely divided tungsten glows red-hot in a stream of hydrogen sulphide.

Nitrogen trichloride
See NITROGEN TRICHLORIDE, Cl_3N: Initiators

Oxidants
Interaction of hydrogen sulphide with a variety of oxidants may be violent if uncontrolled.
See Metal oxides, above
Rust, below
HEPTASILVER NITRATE OCTAOXIDE, Ag_7NO_{11}
BROMINE PENTAFLUORIDE, BrF_5: Hydrogen-containing materials
PERCHLORYL FLUORIDE, $ClFO_3$: Hydrocarbons, etc.
CHLORINE TRIFLUORIDE, ClF_3: Hydrogen-containing materials
DICHLORINE OXIDE, Cl_2O: Oxidisable materials
CHROMIUM TRIOXIDE, CrO_3: Hydrogen sulphide
FLUORINE, F_2: Hydrogen sulphide
OXYGEN DIFLUORIDE, F_2O: Combustible gases
NITRIC ACID, HNO_3: Hydrogen iodide, etc.
SODIUM PEROXIDE, Na_2O_2: Hydrogen sulphide
LEAD(IV) OXIDE, O_2Pb: Hydrogen sulphides

Rust

Mee, A. J., *School Sci. Rev.*, 1940, 22(85), 95

Hydrogen sulphide may ignite if passed through rusty iron pipes.

See Metal oxides, above

Soda-lime

1. Mellor, 1947, Vol. 10, 140
2. Bretherick, L., *Chem. & Ind.*, 1971, 1042

Interaction is exothermic, and if air is present, incandescence may occur with freshly prepared granular material. Admixture with oxygen causes a violent explosion [1]. Soda-lime, used to absorb hydrogen sulphide, will subsequently react with atmospheric oxygen and carbon dioxide with a sufficient exotherm in contact with moist paper (in a laboratory waste bin) to cause ignition [2]. Spent material should be saturated with water before separate disposal.

See other NON-METAL HYDRIDES

HYDROGEN DISULPHIDE H_2S_2

Fl.P., below 22°C

HYDROGEN TRISULPHIDE H_2S_3

Metal oxides

Mellor, 1947, Vol. 10, 159

Contact with copper oxide, lead mono- or dioxides, mercury(II) oxide, tin dioxide and triiron tetraoxide causes violent decomposition and ignition.

Nitrogen trichloride

See NITROGEN TRICHLORIDE, Cl_3N: Initiators

Pentyl alcohol

Mellor, 1947, Vol. 10, 158

Interaction is explosively violent.

Potassium permanganate
See POTASSIUM PERMANGANATE, $KMnO_4$: Hydrogen trisulphide
See other NON-METAL SULPHIDES

HYDROGEN SELENIDE **H_2Se**

Gas above –42° C; flammable

POLSILYLENE **$(H_2Si)_n$**
Brauer, 1963, Vol, 1, 682
The dry solid ignites in air.
See other NON-METAL HYDRIDES

ZINC HYDRIDE **H_2Zn**

Barbaras, G. D. *et al., J. Amer. Chem. Soc.*, 1951, **73**, 4587
Fresh samples reacted slowly with air, but aged and partly decomposed samples (containing finely divided zinc) may ignite in air.
See other METAL HYDRIDES

'ZIRCONIUM HYDRIDE' **H_2Zr**

Mellor, 1941, Vol. 7, 114
Sidgwick, 1950, 633
The product of sorbing hydrogen on to hot zirconium powder when heated in air burns with incandescence and mild explosions.
See other METAL HYDRIDES

LANTHANUM TRIDHYDRIDE **H_3La**

Mackay, 1966, 66
Ignites in air.
See other METAL HYDRIDES

AMMONIA H_3N

Gas above −33°C; E.L., 16–25%

1. *MCA SD-8, 1960*
2. *ABCM Quart. Safety Summ.*, 1950, **21**, 1

Although there is a high lower explosive limit in air and ignition is not easy, there is a long history of violent explosions in refrigeration practice, where ammonia was previously widely used [1].

Mixtures of ammonia and air lying within the explosive limits can occur above aqueous solutions of certain strengths. Welding operations on a vessel containing aqueous ammonia caused a violent explosion [2].

Boron halides

Sidgwick, 1950, 380

The boron halides react violently with ammonia.

Chlorine azide

See CHLORINE AZIDE, ClN_3: Ammonia

Ethylene oxide

See ETHYLENE OXIDE, C_2H_4O: Ammonia

Gold(III) chloride

See GOLD(III) CHLORIDE, $AuCl_3$: Ammonia, etc.

Halogens,
or Interhalogens

Ammonia either reacts violently, or produces explosive products, with all four halogens and some of the interhalogens.

See BROMINE, Br_2: Ammonia
CHLORINE, Cl_2: Nitrogen compounds
FLUORINE, F_2: Hydrides
IODINE, I_2: Ammonia
BROMINE PENTAFLUORIDE, BrF_5: Hydrogen-containing materials
CHLORINE TRIFLUORIDE, ClF_3: Hydrogen-containing materials

Mercury

See MERCURY, Hg: Ammonia

Nitrogen trichloride

See NITROGEN TRICHLORIDE, Cl_3N: Initiators

Oxidants

See POTASSIUM CHLORATE, $ClKO_3$: Ammonia
NITRYL CHLORIDE, $ClNO_2$: Inorganic materials
CHROMYL CHLORIDE, Cl_2CrO_2: Ammonia
DICHLORINE OXIDE, Cl_2O: Oxidisable materials
CHROMIUM TRIOXIDE, CrO_3: Ammonia
TRIOXYGEN DIFLUORIDE, F_2O_3
NITRIC ACID, HNO_3: Ammonia
HYDROGEN PEROXIDE, H_2O_2: Nitrogenous bases
AMMONIUM PEROXODISULPHATE, $H_8N_2O_8S_2$: Ammonia, etc.
NITROGEN OXIDE, NO: Phosphine, etc.
DINITROGEN TETRAOXIDE, N_2O_4: Ammonia

Oxygen,
Platinum

Nealy, D., *School Sci. Rev.*, 1935, **16**(63), 410; Williams, E. J. *et al.*, ibid.
In school demonstrations of oxidation of ammonia to nitric acid over platinum catalysts, substitution of oxygen for air causes fairly vigorous explosions to occur.

Silver compounds

See SILVER CHLORIDE, AgCl: Ammonia
SILVER NITRATE, $AgNO_3$: Ammonia
SILVER OXIDE, Ag_2O: Ammonia, etc.

Stibine

See STIBINE, H_3Sb: Ammonia

Tellurium tetrachloride

See TELLURIUM TETRACHLORIDE, Cl_4Te: Ammonia

Tetramethylammonium amide

See TETRAMETHYLAMMONIUM AMIDE, $C_4H_{14}N_2$: Ammonia

Sulphinyl chloride

See SULPHINYL CHLORIDE, Cl_2OS: Ammonia

Thiotrithiazyl chloride

See THIOTRITHIAZYL CHLORIDE, ClN_3S_4: Ammonia
See other NON-METAL HYDRIDES

HYDROXYLAMINE

H_3NO

1. Ashford, J. S., private comm., 1967
2. Brauer, 1963, Vol. 1, 502
3. Vervalin, C. H., *Hydrocarbon Proc. Petr. Ref.*, 1963, **42**(2), 174

The material was being prepared by distilling a mixture of hydroxylamine hydrochloride and sodium hydroxide in methanol under reduced pressure A violent explosion occurred towards the end of the distillation [1], probably due to an increase in pressure to above 53 mbar. It explodes when heated under atmospheric pressure [2]. Traces of hydroxylamine remaining after reaction with acetonitrile to form acetamide- oxime caused an explosion during evaporation of solvent. Traces can be removed by treatment with diacetyl monoxime and ammoniacal nickel sulphate, forming nickel dimethylglyoxime [3].

Barium oxide
See Oxidants, below

Carbonyl compound,
Pyridine
Anon., *Chem. Processing (Chicago)*, 1963, **26**(24), 30
During preparation of an unspecified oxime the carbonyl compound, pyridine, hydroxylamine hydrochloride and sodium acetate were heated in a stainless steel autoclave. At 90°C sudden reaction caused a pressure increase to 340 bar, when the bursting disc failed. The reaction had been previously run uneventfully on one-tenth scale in a glass-lined autoclave.

Copper(II) sulphate
Mellor, 1940, Vol. 8, 292
Hydroxylamine ignites with the anhydrous salt.

Metals
Mellor, 1940, Vol. 8, 290
Sodium ignites in contact with hydroxylamine alone, but reacts smoothly in ether solution to give sodium hydroxylamide which may be pyrophoric in air. Calcium reacts to give the bis-hydroxylamide which explodes at 180°C. Finely divided zinc either ignites or explodes when warmed in contact with hydroxylamine.

Oxidants
Mellor, 1941, Vol. 3, 670; 1940, Vol. 8, 287–294
Brauer, 1963, Vol. 1, 502
Hydroxylamine is a powerful reducant, particularly when anhydrous, and if exposed to air on a fibrous extended surface (filter paper), it rapidly heats by aerial oxidation. It explodes in contact with air above 70°C. Barium dioxide will ignite aqueous hydroxylamine, while the solid ignites in contact with dry barium mono- and dioxides, lead dioxide, potassium permanganate, but with chlorates, bromates and perchlorates only when moistened with sulphuric acid. Contact of the anhydrous base with potassium or sodium dichromate is violently explosive, but less so with ammonium dichromate or chromium trioxide. Ignition occurs in gaseous chlorine, and vigorous oxidation occurs with hypochlorites.

Phosphorus chlorides
Mellor, 1940, Vol. 8, 290
The base ignites in contact with phosphorus tri- or pentachloride.
See HYDROXYLAMINE SALTS
See other REDUCANTS

AMIDOSULPHURIC ACID **H_3NO_3S**

Metal nitrates or nitrites.
See METAL AMIDOSULPHATES

SODIUM HYDRAZIDE **H_3N_2Na**

Air,
or Alcohol,
or Water
1. Mellor, 1940, Vol. 8, 317
2. Kauffmann, T., *Angew. Chem. (Intern. Ed.)*, 1964, **3**, 342
Contact with traces of air, alcohol or water causes a violent explosion [1], as does heating to 100°C [2].
See other N-METAL DERIVATIVES

AZIDOSILANE H_3N_3Si

Preparative hazard.
See AZIDOGERMANE, GeH_3N_3: Fluorosilane
See other NON-METAL AZIDES

PHOSPHINIC ('HYPOPHOSPHOROUS') ACID H_3O_2P

Mercury(II) oxide
Mellor, 1940, Vol. 4, 778
The redox reaction is explosive.
See other REDUCANTS

PHOSPHOROUS ACID H_3O_3P

See PHOSPHORUS TRICHLORIDE, Cl_3P: Acetic acid

ORTHOPHOSPHORIC ACID H_3O_4P

MCA SD-70, 1958
Although it is not an oxidant, it is a strong acid which, because of its availability in high concentration (90% soln. contains 50 geq./1), can develop a large exotherm on neutralisation (or dilution). Handling precautions are detailed.

Nitromethane
See NITROMETHANE, CH_3NO_2: Acids

PEROXOMONOPHOSPHORIC ACID H_3O_5P

Preparative hazard.
See HYDROGEN PEROXIDE, H_2O_2: Phosphorus(V) oxide

Organic matter
Castrantas, 1965, 5
The 80% solution causes ignition when dropped into organic matter.
See other PEROXOACIDS

PHOSPHINE H_3P

Gas above –87°C; E.L., ~1– %

Air
1. Sidgwick, 1950, 729
2. Mellor, 1940, Vol. 8, 811, 814
3. McKay, H. A. C., *Chem. & Ind.*, 1964, 1978

Pure phosphine does not spontaneously ignite in air below 150°C unless it is thoroughly dried, when it ignites in cold air. The presence of traces (0.2%) of diphosphane in phosphine as normally prepared causes it to ignite spontaneously in air, even at below –15°C [1]. Pure phosphine is rendered flammable by a trace of dinitrogen trioxide, nitrous acid, or similar oxidant [2].

Phosphine, generated by action of water on calcium phosphide, was dried by passage through towers packed with the latter. Soon after refilling the generator (but not the towers) and starting purging with argon, a violent explosion occurred. This was attributed to the air, displace from the generator by argon, reacting explosively with dry phosphine present in the drying towers, possibly catalysed by the orange-yellow polyphosphine which forms on the surface of calcium phosphide. Fresh calcium phosphide in generator and drying towers, with separate purging, is recommended [3].

See Oxygen, below

Dichlorine oxide

See DICHLORINE OXIDE, Cl_2O: Oxidisable materials

Halogens

Mellor, 1940, Vol. 8, 812

Ignition occurs on contact with chlorine or bromine or their aqueous solutions.

Metal nitrates

Mellor, 1941, Vol. 3, 471; 1940, Vol. 4, 993

Passage of phosphine into concentrated silver nitrate solution causes ignition or explosion, depending on the gas rate. Mercury(II) nitrate solution gives a complex phosphide, explosive when dry.

Nitric acid
See NITRIC ACID, HNO_3: Non-metal hydrides

Nitrogen trichloride
See NITROGEN TRICHLORIDE, Cl_3N: Initiators

Oxygen
Fischer, E. O. *et al., Angew. Chem. (Intern. Ed.),* 1968, 7, 136
Even small amount of oxygen present in phosphine gives explosive mixture, in which autoignition occurs at low pressures.
See Air, above
See other NON-METAL HYDRIDES

PLUTONIUM(III) HYDRIDE **H_3Pu**

Stout, E. L., *Chem. Eng. News,* 1958, **36**(8), 64
The hydride, and the metal, when finely divided are spontaneously flammable, and burning causes a specially dangerous contamination problem, in view of the radioactivity and toxic hazards.
See other METAL HYDRIDES

STIBINE **H_3Sb**

Gas above −17°C; flammable
Mellor, 1939, Vol. 9, 394
During evaporation of liquid stibine at −17°C, a relatively weak and isothermal explosive decomposition may occur. Gaseous stibine at ambient temperature may propagate an explosion from a hot spot on the retaining vessel, and autocatalytically decomposes, sometimes explosively, at 200°C.

Ammonia
Mellor, 1939, Vol. 9, 397
A heated mixture explodes.

Oxidants
See CHLORINE, Cl_2: Non-metal hydrides
NITRIC ACID, HNO_3: Non-metal hydrides
OZONE, O_3: Stibine
See other NON-METAL HYDRIDES

AZIDOSILANE H_3SiN_3

Preparative hazard.
See AZIDOGERMANE, GeH_3N_3: Fluorosilane

URANIUM(III) HYDRIDE H_3U

Stout, E. L., *Chem. Eng. News,* 1958, **36**(8), 64
The dry powdered hydride readily ignites in air.
See other METAL HYDRIDES

AMMONIUM IODIDE H_4IN

Bromine trifluoride
See BROMINE TRIFLUORIDE, BrF_3: Ammonium halides

AMMONIUM IODATE H_4INO_3

ABCM Quart. Safety Summ., 1955, **26**, 24
Violent decomposition occurred on touching with a scoop. A similar batch of material contained less than 100 p.p.m. of periodate.
See related METAL OXOHALOGENATES

AMMONIUM PERIODATE H_4INO_4

1. Anon., *Chem. Eng. News*, 1951, **29**, 1770
2. Mellor, 1946, Vol. 2, 408

It exploded while being transferred by scooping [1]. The sensitivity towards heat, but not to abrasive impact, was known previously [2].
See related METAL OXOMETALLATES

PHOSPHONIUM IODIDE H_4IP

Oxidants
Mellor, 1947, Vol. 8, 827
Bromates, chlorates or iodates ignite in contact with phosphonium

iodide at ambient temperature if dry, or in presence of acid to generate bromic acid, etc. Ignition also occurs with nitric acid, and reaction with dry silver nitrate is very exothermic. Interaction with antimony pentachloride at ambient temperature proceeds explosively.
See other REDUCANTS

AMMONIUM PERMANGANATE **H_4MnNO_4**

Mellor, 1942, Vol. 12, 302
Dry material is friction-sensitive, and explodes at 60°C.
See related METAL OXOMETALLATES

AMMONIUM HYDROXIDE **H_4NO**

Nitromethane
See NITROMETHANE, CH_3NO_2: Acids, etc.

HYDRAZINE **H_4N_2**

(Fl.P., 38°C); E.L., 4.7–100%; A.I.T., 23°C on rusty surface, 132°C on iron, 156°C on stainless steel.

Alone
Troyan, J. E., *Ind. Eng. Chem.*, 1953, **45**, 2608
Hydrazine vapour is exceptionally hazardous in that once it is ignited it will continue to burn by exothermic decomposition in complete absence of air or other oxidant.

Air
Day, A. C. *et al.*, *Org. Synth.*, 1970, **50**, 4
Distillation of anhydrous hydrazine (prepared by dehydrating hydrazine hydrate with solid sodium hydroxide) must be carried out under nitrogen to avoid the possibility of an explosion if air is present.
See Barium oxide, below

Barium oxide,
or Calcium oxide

1. Gmelin, 1935, Syst. 4, 318
2. *ABCM Quart. Safety Summ.*, 1950, **21**, 18

The residue from dehydrating hydrazine with barium oxide slowly decomposes exothermically in daylight and finally explodes [1]. Dehydration of 95% hydrazine by boiling with calcium oxide under nitrogen caused violent explosions on two occasions. It is concluded that use of calcium or barium oxides for this purpose is potentially dangerous, and that boiling under reduced pressure may be advisable to lower the liquid temperature. General precautions in handling hydrazine are discussed [2].

Metal catalysts

1. Mellor, 1940, Vol. 8, 317; 1967, Vol. 8, Suppl. 2.2, 83
2. Audrieth, L. F. *et al.*, *Ind. Eng. Chem.*, 1951, **43**, 1774
3. Troyan, J. E., *Ind. Eng. Chem.*, 1953, **45**, 2608

In contact with metallic catalysts (platinum black, Raney nickel) hydrazine is catalytically decomposed, yielding ammonia, nitrogen and hydrogen. With concentrated hydrazine the reaction may be violent, and the ammonia and hydrogen evolved could be ignited by particles of dry catalyst [1]. Measures to prevent autoxidation of hydrazine catalysed by traces of copper are discussed, including displacement of air from transportation containers with nitrogen [2]. Catalytic decomposition is effected by iron oxide, molybdenum-stabilised stainless steel, molybdenum and its oxides, and finely divided solids. Handling procedures are discussed [3].

Metal salts

Mellor, 1941, Vol. 7, 430; 1940, Vol. 8, 318; 1967, Vol. 8, Suppl. 2.2, 88

Several of its complexes with metallic salts are unstable, including those of basic cadmium perchlorate (highly explosive), cadmium nitrite (explosive), copper chlorate (explodes on drying without heat), manganese nitrate (ignites at 150°C), mercury(I) and (II) chlorides and nitrates (all are explosive) and tin(II) chloride (explodes on heating).

See also AMMINEMETAL OXOSALTS

Oxidants

1. *Haz. Chem. Data,* 1969, 127
2. Mellor, 1941, Vol. 3, 137; 1940, Vol. 8, 313–319; 1943, Vol. 11, 234
3. *ASESB Op. Incid. Rep. No. 105*
4. Troyan, J. E., *Ind. Eng. Chem.,* 1953, **45**, 2608

Hydrazine is a powerful, endothermic reducing agent, and its interaction with oxidants may be expected to be violent if unmoderated, as in rocket propulsion systems (often hypergolic). Mixtures of hydrazine vapour and air are explosive over a very wide range (4.7–100%) and ignition temperatures can be very low (24°C on a rusty iron surface). When mixed with, or spilt on, highly porous materials (asbestos, cloth, earth, wood), autoxidation in air may proceed fast enough to cause ignition. Contact with hydrogen peroxide or nitric acid may cause ignition with concentrated reactants, and dinitrogen tetraoxide is hypergolic [1]. Contact with chromate salts or chromium trioxide causes explosive decomposition. Copper or lead oxides cause a vigorous decomposition, while dropping hydrazine hydrate on to mercury oxide can cause an explosion. This may be due to known oxidation of hydrazine to hydrogen azide by two-electron transfer, and formation of explosive mercury azide [2]. Contact of hydrazine with (explosive) *N*,2,4,6-tetranitro-*N*-methylaniline caused ignition [3]. Contact with iron oxide, chlorates or peroxides may lead to a violent reaction [4].

See *N*-HALOIMIDES
ROCKET PROPELLANTS
SILVER OXIDE, Ag_2O: Ammonia, etc.
TRIOXYGEN DIFLUORIDE, F_2O_3
POTASSIUM PEROXODISULPHATE, $K_2O_8S_2$: Hydrazine, etc.

Potassium

Gmelin, 1935, Syst. 4, 318

Explosive interaction.

See also SODIUM HYDRAZIDE, H_3N_2Na

Sodium

See HYDRAZINE, H_4N_2: Sodium
See other NON-METAL HYDRIDES
REDUCANTS

AMMONIUM NITRITE $H_4N_2O_2$

See NITRITE SALTS OF NITROGENOUS BASES

1. Federoff, 1960, 35
2. Urbanski, 1965, Vol. 2, 460
3. Popper, H., *Chem. Eng.*, 1962, **70**, 91
4. Sykes, W. G. *et al., Chem. Eng. Progr.*, 1963, **59**(1), 66
5. Mellor, 1964, Vol. 8, Suppl. 1, 543
6. Croysdale, L. G. *et al., Chem. Eng. Progr.*, 1965, **61**(1), 76
7. *MCA Case History No. 873*

The decomposition, fire and explosion hazards of this salt of negative oxygen balance have been extensively reviewed [1–4]. In the absence of impurities, it is difficult, but not impossible, to cause ammonium nitrate to detonate. Use of explosives to break up the caked double salt, ammonium nitrate: sulphate (2:1), caused a 4.5 Mkg dump to detonate [4] (even though some 45% of inert ammonium sulphate was present in the salt). The effect of various impurities and additions on the thermal stability of solid ammonium nitrate has been widely studied [1,2,5].

A few incidents involving explosive decomposition of aqueous solutions of the salt during transfer operations (or in presence of oil) have been recorded [6]. Flame-cutting a mild steel pipe blocked by the solid (impure) salt caused it to explode [7].

See Metals, below
Organic fuels, below

Acetic acid
von Schwartz, 1918, 322
Concentrated mixtures ignite on warming.

Ammonium sulphate,
Potassium
Staudinger, H., *Z. Elektrochem.*, 1925, **31**, 549
Ammonium nitrate containing the sulphate readily explodes on contact with potassium or its alloy with sodium.

Alkali metals
Mellor, 1964, Vol. 8, Suppl. 1, 546
Sodium progressively reduces the nitrate, eventually forming a yellow explosive solid, probably sodium hyponitrite.

Metals

Mellor, 1964, Vol. 8, Suppl. 1, 543–546

Many of the following powdered metals reacted violently or explosively with fused ammonium nitrate below 200°C: aluminium, antimony, bismuth, cadmium, chromium, cobalt, copper, iron, lead, magnesium, manganese, nickel, tin, zinc; also brass and stainless steel. Sodium reacts to form the yellow explosive compound sodium hyponitrite and presence of potassium sensitises the nitrate to shock.

See Ammonium sulphate etc., above
ALUMINIUM, Al: Ammonium nitrate
POTASSIUM, K: Nitrogen-containing explosives

Non-metals

Mellor, 1941, Vol. 2, 841–842

Glowing charcoal explodes in contact with the solid salt, and phosphorus ignites on the fused salt.

See Sulphur, below
PHOSPHORUS, P: Nitrates

Organic fuels

1. Sykes, W. G. *et al., Chem. Eng. Progr.*, 1963, **59**(1), 66
2. Urbanski, 1965, Vol. 2, 461

In a review of ammonium nitrate explosion hazards several incidents involving mixtures of this negative oxygen balance material with various organic materials were analysed [1]. In general, fire incidents involving ammonium nitrate admixed with 1% of wax, oil or stearates (as anti-caking additives) tended more towards explosion than the pure salt, although the degree of confinement is also important [1,2]. The ease of detonation of mixtures is much greater when 2–4% of oil is present, and such mixtures have been used as explosives[2].

Potassium nitrite

Mellor, Vol. 2, 1946, 842

Contact of solid nitrate with fused nitrite causes incandescence.

See NITRITE SALTS OF NITROGENOUS BASES

Potassium permanganate

See POTASSIUM PERMANGANATE, $KMnO_4$: Ammonium nitrate

Sulphur

1. Mason, C. M. *et al., J. Agric. Food Chem.*, 1967, **15**, 954
2. Prugh, R. W., *Chem. Eng. Progr.*, 1967, **63**(11), 53–55

The fire risks of nitrate–sulphur mixtures have been discussed [1] and explosion risks assessed by differential thermal analysis [2].

Urea

Croysdale, L. C. *et al., Chem. Eng. Progr.*, 1965, **61**(1), 72

Concentrated solutions of ammonium nitrate and urea exploded during large-scale mixing operations. Although the cause was not established, hazards associated with these operations are discussed in relation to the circumstances.

See other OXIDANTS
OXOSALTS OF NITROGENOUS BASES

HYDROXYLAMMONIUM NITRATE $H_4N_2O_4$

See HYDROXYLAMMONIUM SALTS
OXOSALTS OF NITROGENOUS BASES

AMMONIUM AZIDE H_4N_4

1. Mellor, 1940, Vol. 8, 344
2. Obenland, C. O. *et al., Inorg. Synth.*, 1966, 8, 53

It explodes on rapid heating [1] and contains ~93% of nitrogen. Preparative precautions have been detailed [2].

See other NON-METAL AZIDES

SODIUM HEXAHYDROXOPLATINATE(2–) $H_4Na_2O_6Pt$

Acetic acid,
Nitric acid

See NITRIC ACID HNO_3: Acetic acid, etc.

See other PLATINUM COMPOUNDS

'SODIUM PERPYROPHOSPHATE' $H_4Na_4O_{11}P_2$

See SODIUM PYROPHOSPHATE HYDROGEN PEROXIDATE, $Na_4O_7P_2 \cdot 2H_2O_2$

OXODISILANE H_4OSi_2

Kautsky, K., *Z. Anorg. Chem.*, 1921, **117**, 209
Ignites in air.
See related NON-METAL HYDRIDES

ZIRCONIUM HYDROXIDE H_4O_4Zr

Zirconium
See ZIRCONIUM, Zr: Oxygen-containing compounds

DIPHOSPHANE H_4P_2

Mellor, 1940, Vol. 8, 829
Ignites in air, and will cause other flammable gases to ignite in air when present at 0.2% v/v concentration.
See also 'PHOSPHORUS HYDRIDE', HP_2
PHOSPHINE, H_3P
See other NON-METAL HYDRIDES

SILANE H_4Si

Alone,
or Covalent halides,
or Halogens
1. Mellor, 1940, Vol. 6, 220–221
2. Braker, 1971, 505

Pure material is said not to ignite in air unless the temperature be increased or the pressure reduced. Presence of other hydrides as impurities causes ignition always to occur on air contact [1]. However,

99.95% pure material, even at concentrations down to 1% in hydrogen and/or nitrogen, ignites in contact with air unless emerging at very high gas velocity, when mixtures with up to 10% silane content may not ignite [2]. Silane burns in contact with bromine, chlorine or covalent chlorides (carbonyl chloride, antimony pentachloride, tin(IV) chloride, etc.)[1]. Extreme caution is necessary when handling silane in systems with halogenated compounds, as a trace of free halogen may cause violent explosions[2].
See other NON-METAL HYDRIDES

THORIUM HYDRIDE H_4Th

1. Mellor, 1941, Vol. 7, 207
2. Stout, E. L., *Chem. Eng. News,* 1958, **36**(8), 64

Thorium hydride explodes on heating in air[1], and the powdered hydride readily ignites on handling in air[2].
See other METAL HYDRIDES

URANIUM(IV) HYDRIDE H_4U

Oxygen
See OXYGEN (Gas), O_2 : Metal hydrides

ORTHOPERIODIC ACID H_5IO_6

Azo-pigment,
Perchloric acid
See PERCHLORIC ACID, $ClHO_4$: Azo-pigment, etc.

HYDRAZINIUM SALTS $H_5N_2^+$

Metal nitrites
See HYDROGEN AZIDE, HN_3
NITRITE SALTS OF NITROGENOUS BASES

DIAMIDOPHOSPHOROUS ACID H_5N_2OP

Water
Mellor, 1940, Vol. 8, 704
Violent reaction with incandescence.

HYDRAZINIUM NITRITE $H_5N_3O_2$

See NITRITE SALTS OF NITROGENOUS BASES

HYDRAZINIUM NITRATE $H_5N_3O_3$

Alone,
or Metals,
or Metal acetylides, nitrides, oxides or sulphides
Mellor, 1940, Vol. 8, 327; 1967, Vol. 8, Suppl. 2.2, 84, 96
It is an explosive, less stable than ammonium nitrate, and has been studied in detail. Stable on slow heating to 300°C, it decomposes explosively on rapid heating or under confinement. Presence of zinc, copper, most other metals and their nitrides, oxides or sulphides causes flaming decomposition above the melting point (70°C). Commercial cobalt (cubes) causes an explosion also.
See other OXOSALTS OF NITROGENOUS BASES

HYDRAZINIUM AZIDE H_5N_5

It contains ~94% of nitrogen.
See HYDRAZINIUM SALTS
See other NON-METAL AZIDES

HYDROXYLAMMONIUM PHOSPHINATE H_6NO_3P

Mellor, 1940, Vol. 8, 880
The salt detonates above its melting point, 92°C.
See HYDROXYLAMMONIUM SALTS

AMMONIUM AMIDOSULPHATE ('SULPHAMATE') $H_6N_2O_3S$

1. Hunt, J. K., *Chem. Eng. News*, 1952, **30**, 707
2. Rogers, M. G., private comm., 1973

A 60% solution of ammonium sulphamate (pH above 4.5) will not undergo rapid hydrolysis below 200°C. Addition of acid (to pH 2) causes a runaway exothermic hydrolysis to set in at 130°C. Superheating and vigorous boiling can occur under appropriate physical conditions [1]. The use of urea–formaldehyde resins as temporary binders in the firing of refractories and ceramics at high temperatures can lead to the formation of substantial deposits of ammonium sulphamate in the cooler parts of kilns, should a fuel oil containing appreciable quantities of sulphur be used for firing. Ammonia from decomposition of the resin combines with sulphur trioxide to form ammonium sulphamate which accumulates as either a solidified mass in flues or a white deposit on walls, causing corrosion and handling problems [2].

See OXOSALTS OF NITROGENOUS BASES

HYDRAZINIUM HYDROGENSELENATE $H_6N_2O_4Se$

Meyer, J. *et al., Ber.*, 1928, **61**, 1839

The salt explodes in contact with a hot glass rod.

See other OXOSALTS OF NITROGENOUS BASES

cis-DIAMMINEDINITROPLATINUM(II) $H_6N_4O_4Pt$

Brauer, 1965, Vol. 2, 1560

Explosive decomposition at 200°C.

See other PLATINUM COMPOUNDS

DIAMMINEPALLADIUM(II) NITRATE $H_6N_4O_6Pd$

White, J. H., private comm., 1965

There is a danger of explosion if the nitrate is dried.

See other AMMINEMETAL OXOSALTS

ZINC DIHYDRAZIDE H_6N_4Zn

Mellor, 1940, Vol. 8, 315
It explodes at 70°C.
See other N-METAL DERIVATIVES

2,4,6-TRISILATRIOXANE ('TRIPROSILOXANE') $H_6O_3Si_3$

Air,
or Chlorine
Mellor, 1940, Vol. 6, 234
The solid polymer (approximating to the trimer) ignites in air or chlorine.
See related NON-METAL HYDRIDES

DISILANE H_6Si_2

Bromine
See BROMINE, Br_2: Non-metal hydrides

Non-metal halides,
or Oxygen
Mellor, 1940, Vol. 6, 220–224
It explodes on contact with carbon tetrachloride or sulphur hexafluoride and contact with chloroform causes incandescence. Disilane ignites spontaneously in air, even when pure, and ingress of air or oxygen into a volume of disilane causes explosion.
See SILANES
See other NON-METAL HYDRIDES

AMMINEPENTAHYDROXOPLATINUM H_8NO_5Pt

Jacobsen, I., *Compt. Rend.*, 1909, **149**, 575
Explodes fairly violently above 250°C, as does the pyridine analogue.
See other PLATINUM COMPOUNDS

AMMONIUM SULPHATE $H_8N_2O_4S$

Potassium chlorate
See POTASSIUM CHLORATE, $ClKO_3$: Ammonia, etc.

HYDROXYLAMMONIUM SULPHATE $H_8N_2O_6S$

See HYDROXYLAMMONIUM SALTS

AMMONIUM PEROXODISULPHATE $H_8N_2O_8S_2$

Aluminium,
Water
Pieters, 1957, 30
A mixture including the powdered metal may explode.

Ammonia,
Silver salts
Mellor, 1947, Vol. 10, 466
In concentrated solutions the silver-catalysed oxidation of ammonia to nitrogen may be very violent.

Iron
Mellor, 1947, Vol. 10, 470
Iron exposed to the action of a slightly acid concentrated solution of ammonium peroxodisulphate dissolves violently.

Sodium peroxide
See SODIUM PEROXIDE, Na_2O_2: Ammonium peroxodisulphate

Zinc,
Ammonia
Report of H.M. Inspector of Explosives, 33, London, HMSO, 1950
The salt exploded during drying, but no cause was determined. However, the reputed explosive character of tetraamminezinc peroxodisulphate, possibly formed from interaction of the ammonium salt, galvanised iron and ammonia, was mentioned as a possible cause.
See other PEROXOACID SALTS

AMMONIUM SULPHIDE H_8N_2S

Zinc
Chemiearbeit, 1960, **12**(4), 29
A closed zinc container filled with a concentrated solution of ammonium sulphide exploded, owing to liberation of hydrogen sulphide and hydrogen, accompanying the formation of zinc ammonium sulphide.

AMMONIUM TETRANITROPLATINATE(II) $H_8N_6O_8Pt$

Nilson, L. F., *J. Prakt. Chem.*, [2], 1877, **16**, 249
Decomposes explosively on heating.
See other PLATINUM COMPOUNDS

TRISILANE H_8Si_3

Air,
or Carbon tetrachloride
Mellor, 1940, Vol. 6, 224
Ignites and explodes in air or oxygen, and reacts vigorously with carbon tetrachloride.
See SILANES
See other NON-METAL HYDRIDES

TRISILYLAMINE H_9NSi_3

Mellor, 1940, Vol. 8, 262
The liquid ignites in air.
See related NON-METAL HYDRIDES

TRIAMMINENTITRATOPLATINUM(II) NITRATE $H_9N_5O_6Pt$

Mellor, 1942, Vol. 16, 409
Decomposes violently on heating.
See other AMMINEMETAL OXOSALTS
PLATINUM COMPOUNDS

TETRASILANE $H_{10}Si_4$

Air,
or Carbon tetrachloride
Mellor, 1940, Vol. 6, 224
Ignites and explodes in air or oxygen, and reacts vigorously with carbon tetrachloride.
See SILANES
See other NON-METAL HYDRIDES

TETRAAMMINEZINC PEROXODISULPHATE $H_{12}N_4O_8S_2Zn$

Barbieri, G. A. *et al., Z. Anorg. Chem.*, 1911, **71**, 347
Explodes on heating or impact.
See AMMONIUM PEROXODISULPHATE, $H_8N_2O_8S_2$: Zinc
See other AMMINEMETAL OXOSALTS

TETRAAMMINEPALLADIUM(II) NITRATE $H_{12}N_6O_6Pd$

White, J. H., private comm., 1965
Evaporation of a solution of the salt used for plating gave a moist residue which ignited and burned violently. Reclamation of palladium from such solutions by direct reduction is recommended.
See other AMMINEMETAL OXOSALTS

TRIHYDRAZINENICKEL(II) NITRATE $H_{12}N_8NiO_6$

Ellern, H. *et al., J. Chem. Educ.*, 1955, **32**, 24
A dry sample exploded violently and a moist sample spontaneously deflagrated.
See other AMMINEMETAL OXOSALTS

TETRAAMMINEHYDROXONITRATOPLATINUM(IV) NITRATE $H_{13}N_7O_{10}Pt$

Mellor, 1942, Vol. 16, 411
Explodes violently on heating.
See other AMMINEMETAL OXOSALTS
PLATINUM COMPOUNDS

TETRAAMMINELITHIUM DIHYDROGENPHOSPHIDE $H_{14}LiN_4P$

Legoux, C., *Compt. Rend.*, 1938, **207**, 634
Reacts vigorously with water, evolving phosphine and some ammonia, which may ignite.
See related NON-METAL HYDRIDES

UNDECAAMMINETETRARUTHENIUM DODECAOXIDE $H_{33}N_4O_{12}Ru_4$

Preparative hazard.
See RUTHENIUM TETRAOXIDE, O_4Ru: Ammonia

HAFNIUM Hf

Alone,
or Non-metals,
or Oxidants
MCA SD-92, 1966
Although the massive metal is relatively inert, when powdered it becomes very reactive. The dry powder may react explosively at elevated temperatures with nitrogen, phosphorus, oxygen, sulphur and other non-metals. The halogens react similarly, and in contact with hot concentrated nitric acid or other oxidants it may explode (often after a delay with nitric acid). The powder is pyrophoric and readily ignitable by friction, heat or static sparks, and if dry burns fiercely. Presence of water (5–10%) slightly reduces the ease of ignition, but combustion of the damp powder proceeds explosively. A minimum of 25% water is necessary to reduce handling hazards to a minimum. Full handling precautions are detailed.
See PYROPHORIC METALS

Acetylenic compounds

See DISODIUM ACETYLIDE, C_2Na_2: Metals
1-BROMO-2-PROPYNE, C_3H_3Br: Metals
2-BUTYNE-1,4-DIOL, $C_4H_6O_2$

Ammonia

Sampey, J. J., *Chem. Eng. News,* 1947, **25**, 2138; *J. Chem. Educ.*, 1967, **44**, A324

A mercury manometer used with ammonia became blocked by deposition of a grey-brown solid, which exploded during attempts to remove it mechanically or on heating. The solid appeared to be a dehydration product of Millon's base and was freely soluble in sodium thiosulphate solution. This method of cleaning is probably safer than others, but the use of mercury manometers with ammonia should be avoided as intrinsically unsafe.

See DIMERCURY IMIDE OXIDE, HHg_2NO

Boron diiodophosphide

See BORON DIIODOPHOSPHIDE, BI_2P: Metals

Ethylene oxide

See ETHYLENE OXIDE, C_2H_4O

Metals

Bretherick, L., *Lab. Pract.*, 1973, **22**, 533

The high mobility and tendency to dispersion exhibited by mercury, and the ease with which it forms alloys (amalgamates) with many laboratory and electrical contact metals, can cause severe corrosion problems in laboratories. A trap is described to contain mercury accidentally ejected by overpressuring of mercury manometers and similar items.

See also ALUMINIUM, Al: Mercury
POTASSIUM, K: Mercury
LITHIUM, Li: Mercury
SODIUM, Na: Mercury
RUBIDIUM, Rb: Mercury

Methyl azide

See METHYL AZIDE, CH_3N_3: Mercury

Methylsilane,
Oxygen
See METHYLSILANE, CH_6Si: Mercury, etc.

Oxidants
See BROMINE, Br_2: Metals
PEROXYFORMIC ACID, CH_2O_3: Metals
CHLORINE DIOXIDE, ClO_2: Mercury
NITRIC ACID, HNO_3: Ethanol

Tetracarbonylnickel,
Oxygen
See TETRACARBONYLNICKEL, C_4NiO_4: Mercury, etc.
See other METALS

ZINC AMALGAM Hg–Zn

Air
Brimelow, H. C., private comm., 1972
Amalgamated zinc residues isolated from Clemmensen reduction of an alkyl aryl ketone in glacial acetic acid were pyrophoric, and had to be immediately dumped into water after filtration to prevent ignition.
See other ALLOYS

MERCURY(II) IODIDE HgI_2

Chlorine trifluoride
See CHLORINE TRIFLUORIDE, ClF_3: Metals, etc.

MERCURY(I) NITRATE $HgNO_3$

Phosphorus
See PHOSPHORUS, P: Nitrates

MERCURY(II) NITRATE HgN_2O_6

Acetylene
Mellor, 1940, Vol. 4, 993
Contact of acetylene with the nitrate solution gives mercury acetylide, an explosive sensitive to heat, friction or contact with sulphuric acid.

Ethanol
Mellor, 1940, Vol. 4, 993
Addition of aqueous nitrate solution to ethanol gives mercury fulminate, $C_2HgN_2O_2$.

Phosphine
Mellor, 1940, Vol. 4, 993
Phosphine reacts with an aqueous salt solution to give a complex nitrate-phosphide, which explodes when dry on heating or impact.

Petroleum hydrocarbons
1. Mixer, R. Y., *Chem. Eng. News,* 1948, **26**, 2434
2. Ball, J., ibid., 3300

Gas oil was stirred vigorously with the finely divided solid nitrate to complete the removal of sulphur compounds. After the mixture had congealed, preventing further stirring, a violent reaction set in, which reached incandescence, accompanied by vigorous evolution of nitrous fumes. This was attributed to the self-catalysed nitrating action of mercury nitrate in a semi-solid environment unable to lose heat effectively. A similar occurrence was observed when crushed nitrate was just covered with cracked naphtha [1].

The second publication reveals that this type of hazard was known in cracked distillates containing high proportions of unsaturates and aromatics, when allowed to stand in prolonged contact with mercury(II) nitrate. The hazard may be avoided by using several small portions of the salt sequentially and working on 100 g portions of hydrocarbons [2].
See other METAL OXONON-METALLATES

MERCURY(II) AZIDE

1. Mellor, 1940, Vol. 8, 351; Mellor 1967, Vol. 8, Suppl. 2, 43, 50; Barton, A. F. M. *et al., Chem. Rev.,* 1973, **73**, 138

2. Heathcock, C. H., *Angew. Chem. (Intern. Ed.)*, 1969, 8, 134

The very high friction-sensitivity, particularly of large crystals, and brisance on explosion are to be expected from the thermodynamic properties of the salt. Its great sensitivity, even under water, renders it unsuitable as a practical detonator. Spontaneous explosions during intercrystalline transformations have been observed, or on crystallisation from hot water [1]. A safe method of preparing solutions in aqueous tetrahydrofuran for synthetic purposes is available [2].

See other METAL AZIDES

MERCURY(II) OXIDE HgO

Under appropriate conditions, it can function as a powerful oxidant and/or catalyst, owing to the tendency to dissociate.

Acetyl nitrate

See ACETYL NITRATE, $C_2H_3NO_4$: Mercury(II) oxide

Butadiene,
Ethanol,
Iodine

See 2-ETHOXY-1-IODO-3-BUTENE, $C_6H_{11}IO$

Chlorine,
Hydrocarbons

See METHANE, CH_4: Chlorine
ETHYLENE, C_2H_4: Chlorine

Disulphur dichloride

See DISULPHUR DICHLORIDE, Cl_2S_2: Mercury(II) oxide

Hydrogen peroxide

See HYDROGEN PEROXIDE, H_2O_2: Mercury(II) oxide
Metals, etc.

Hydrogen trisulphide

See HYDROGEN TRISULPHIDE, H_2S_3: Metal oxides

Metals
Staudinger, H., *Z. Elektrochem.*, 1925, **31**, 549
Mellor, 1940, Vol. 4, 272
Mixtures of the red or yellow oxides with sodium–potassium alloy explode violently on impact, the yellow (more finely particulate) oxide giving the more sensitive mixture. Mixtures with magnesium or potassium may explode on heating.

Methanethiol
Klason, P., *Ber.*, 1887, **20**, 3410
Interaction is rather violent in absence of diluent.

Non-metals
Mellor, 1940, Vol. 4, 777–778
Mixtures with phosphorus explode on impact or on boiling with water. A mixture with sulphur exploded on heating.

Phospham
See PHOSPHAM, HN_2P: Oxidants

Reducants
Mellor, 1940, Vol. 4, 778; Vol. 8, 318
Hydrazine hydrate and phosphinic acid both explosively reduce the oxide when dropped on to it.
See other METAL OXIDES

MERCURY PEROXIDE HgO_2

See HYDROGEN PEROXIDE, H_2O_2: Mercury(II) oxide

MERCURY(II) SULPHIDE HgS

Oxidants
Mellor, 1940, Vol. 2, 242; 1941, Vol. 3, 377; 1940, Vol. 4, 952
The sulphide causes dichlorine oxide to explode and incandesces in chlorine. Grinding with silver oxide ignites the mixture.
See other METAL SULPHIDES

MERCURY(I) THIONITROSYLATE $(Hg_2N_2S_2)_n$

Goehring, M. *et al., Z. Anorg. Chem.*, 1956, **285**, 70
Explodes on heating in a flame.

MERCURY(I) AZIDE Hg_2N_6

Mellor, 1940, Vol. 8, 351; 1967, Vol. 8, Suppl. 2, 25, 50
It is less sensitive and a less powerful explosive than silver or lead azides. It explodes on heating in air to above 270°C, or at 140°C *in vacuo.*

See other METAL AZIDES

'MERCURY(I) OXIDE' Hg_2O

Sidgwick, 1950, 292
This material is known to be an intimate mixture of mercury(II) oxide and metallic mercury.

Alkali metals
Mellor, 1940, Vol. 4, 771
Interaction with molten potassium or sodium is accompanied by a brilliant light and light explosion.

Chlorine,
Ethylene
See ETHYLENE, C_2H_4: Chlorine

Hydrogen peroxide
Antropov, V., *J. Prakt. Chem.*, 1908, **77**, 316
Contact causes explosive decomposition.
See HYDROGEN PEROXIDE, H_2O_2: Metals, etc.

Non-metals
Mellor, 1940, Vol. 4, 771
Mixtures with phosphorus explode on impact and that with sulphur ignites on frictional initiation.

TRIS(IODOMERCURI)PHOSPHINE Hg_3I_3P

Nitric acid

See NITRIC ACID, HNO_3 : Phosphine derivatives

TRIMERCURY DINITRIDE Hg_3N_2

Alone,
or Sulphuric acid

1. Mellor, 1940, Vol. 8, 108
2. Fischer, F. *et al.*, *Ber.*, 1910, **43**, 1469

It explodes on friction, impact, heating, or in contact with sulphuric acid [1]. A sample at below −40°C exploded when disturbed [2].

See other *N*-METAL DERIVATIVES

TRIMERCURY TETRAPHOSPHIDE Hg_3P_4

Oxidants

Mellor, 1940, Vol. 8, 844

It ignites when warmed in air, or cold on contact with chlorine. A mixture with potassium chlorate explodes on impact.

See POTASSIUM CHLORATE, $ClKO_3$: Metal phosphides

See other METAL NON-METALLIDES

POTASSIUM IODIDE IK

Diazonium salts

See DIAZONIUM TRIIODIDES

Diisopropyl peroxydicarbonate

See DIISOPROPYL PEROXYDICARBONATE, $C_8H_{14}O_6$

Oxidants

See BROMINE PENTAFLUORIDE, BrF_5 : Acids, etc.
TRIFLUOROACETYL HYPOFLUORITE, $C_2F_4O_2$
FLUORINE PERCHLORATE, $ClFO_4$: Potassium iodide
CHLORINE TRIFLUORIDE, ClF_3 : Metals, etc.

POTASSIUM IODATE IKO_3

Metals and oxidisable derivatives

See METAL HALOGENATES: Metals and oxidisable derivatives
PHOSPHORUS, P: Metal halogenates

POTASSIUM PERIODATE IKO_4

Ammonium perchlorate

See AMMONIUM PERCHLORATE, ClH_4NO_4: Impurities

IODINE AZIDE IN_3

Hantzsch, A., *Ber.*, 1900, **33**, 525

The isolated solid is a very shock- and friction-sensitive explosive.

See HALOGEN AZIDES

Sulphur-containing alkenes

Hassner, A. *et al.*, *J. Org. Chem.*, 1968, **33**, 2690

Interaction is accompanied by violent decomposition of the azide.

SODIUM IODIDE INa

Oxidants

See PERCHLORYL FLUORIDE, $ClFO_3$: Calcium acetylide, etc.
PERCHLORIC ACID, $ClHO_4$: Iodides

SODIUM IODATE $INaO_3$

Potassium

See POTASSIUM, K: Oxidants

TETRAPHOSPHORUS IODIDE IP_4

Nitric acid

See NITRIC ACID, HNO_3: Phosphorus halides

Ammonia

Mellor, 1940, Vol. 8, 605; 1967, Vol. 8, Suppl. 2.2, 416

Ammonia solutions react with iodine (or potassium iodide) to produce highly explosive addition compounds between nitrogen triiodide and ammonia.

See NITROGEN TRIIODIDE-AMMONIA, $I_3N{\cdot}H_3N$

Butadiene,
Ethanol,
Mercuric oxide

See 2-ETHOXY-1-IODO-3-BUTENE, $C_6H_{11}IO$

Dicaesium oxide

See DICAESIUM OXIDE, Cs_2O: Halogens

Ethanol,
Phosphorus

1. Read, C. W. W., *School Sci. Rev.*, 1940, **21**(83), 967
2. Llowarch, D., *School Sci. Rev.*, 1956, **37**(133), 434

Interaction of ethanol, phosphorus and iodine to form iodoethane is considered too dangerous for a school experiment [1]. A safer modification has now become available [2].

See PHOSPHORUS, P: Halogens

Formamide,
Pyridine,
Sulphur trioxide

See FORMAMIDE, CH_3NO: Iodine, etc.

Halogens,
or Interhalogens

See BROMINE TRIFLUORIDE, BrF_3: Halogens
BROMINE PENTAFLUORIDE, BrF_5: Acids, etc.
CHLORINE TRIFLUORIDE, ClF_3: Iodine
FLUORINE, F_2: Iodine

Metal acetylides or carbides

Several react very exothermally with iodine.

See BARIUM ACETYLIDE, C_2Ba: Halogens
CALCIUM ACETYLIDE, C_2Ca: Halogens
DICAESIUM ACETYLIDE, C_2Cs_2: Halogens
CUPROUS ACETYLIDE, C_2Cu: Halogens
DILITHIUM ACETYLIDE, C_2Li_2: Halogens
DIRUBIDIUM ACETYLIDE, C_2Rb_2: Halogens
STRONTIUM ACETYLIDE, C_2Sr: Halogens
ZIRCONIUM DICARBIDE, C_2Zr: Halogens

Metals

Mellor, 1941, Vol. 2, 469; 1963, Vol. 2, Suppl. 2.2, 1563

A mixture of potassium and iodine explodes weakly on impact, while potassium ignites in contact with molten iodine.

See ALUMINIUM, Al: Oxidants
ALUMINIUM–TITANIUM ALLOYS, Al–Ti: Oxidants
HAFNIUM, Hf

Oxygen difluoride

See OXYGEN DIFLUORIDE, F_2O: Halogens

Silver azide

See SILVER AZIDE, AgN_3: Halogens

Trioxygen difluoride

See TRIOXYGEN DIFLUORIDE, F_2O_3
See other HALOGENS

IODINE(V) OXIDE I_2O_5

Bromine pentafluoride

See BROMINE PENTAFLUORIDE, BrF_5: Acids, etc.

Non-metals

Mellor, 1941, Vol. 2, 295

Iodine pentaoxide reacts explosively with warm carbon, sulphur, resin, sugar or powdered, easily combustible elements.

See other HALOGEN OXIDES

NITROGEN TRIIODIDE-SILVER AMIDE $I_3N \cdot AgH_2N$

Mellor, 1967, Vol. 8, Suppl. 2, 418
The dry complex may explode.
See other *N*-HALOGEN COMPOUNDS
N-METAL DERIVATIVES

NITROGEN TRIIODIDE-AMMONIA $I_3N \cdot H_3N$

Alone,
or Halogens,
or Oxidants,
or Concentrated acids
Mellor, 1940, Vol. 8, 607; 1967, Vol. 8, Suppl. 2, 418
When dry it is an extremely sensitive, unstable detonator capable of initiation by virtually any form of energy (light, heat, sound, nuclear radiation or mechanical vibration), even at sub-zero temperatures, and occasionally even with moisture present. It may be handled cautiously when wet, but heavy friction will still initiate it. It explodes in boiling water and is decomposed by cold water to explosive diiodamine. It explodes, possibly owing to heat-initiation, on contact with virtually any concentrated acid, with chlorine or bromine, ozone or hydrogen peroxide solution.
See other *N*-HALOGEN COMPOUNDS

INDIUM In

Acetonitrile,
Dinitrogen tetraoxide
See DINITROGEN TETRAOXIDE, N_2O_4: Acetonitrile, etc.

Mercury(II) bromide
See MERCURY(II) BROMIDE, Br_2Hg: Indium

Sulphur
See SULPHUR, S: Metals

IRIDIUM

Ir

The finely divided catalytic metal may be pyrophoric.

See ZINC, Zn: Catalytic metals
See other HYDROGENATION CATALYSTS

Interhalogens

See BROMINE PENTAFLUORIDE, BrF_5: Acids, etc.
CHLORINE TRIFLUORIDE, ClF_3: Metals

IRIDIUM(IV) OXIDE

IrO_2

Peroxyformic acid

See PEROXYFORMIC ACID, CH_2O_3: Metals, etc.

POTASSIUM

K

1. Gilbert, H. N., *Chem. Eng. News,* 1948, **26**, 2604
2. Johnson, W. S. *et al., Org. Synth.*, 1963, Coll. Vol. 4, 134–135

A review of the comparative properties of sodium and potassium, the latter invariably being the more hazardous. Potassium readily reacts with carbon monoxide to form an explosive carbonyl, while sodium does so only under exceptional circumstances[1]. Detailed laboratory procedures for safe handling of potassium have been published [2].

See POTASSIUM CARBONATE, CK_2O_3: Carbon

Acetylene

See ACETYLENE, C_2H_2: Potassium

Air

It is convenient to consider interaction of potassium and air under two headings.

Rapid oxidation

Mellor, 1941, Vol. 2, 468; 1963, Vol. 2, Suppl. 2.2, 1559

Reaction with air or oxygen in complete absence of moisture does not occur, even on heating to boiling point. However, in contact with normally moist air, oxidation may become so fast that melting or ignition occurs, particularly if pressure is applied locally to cause melting and exposure of a fresh surface, as when potassium is pressed through a die to form potassium wire.

Slow oxidation

1. Mellor, 1941, Vol. 2, 493
2. Gilbert, H. N., *Chem. Eng. News*, 1948, **26**, 2604
3. Short, J. F., *Chem. & Ind.*, 1964, 2132; Brazier, A. D., *Chem. & Ind.*, 1965, 220; Balfour, A. E., ibid., 353; Bil, M. S., ibid., 812; Cole, R. J., ibid., 944
4. Brandsma, 1971, 10, 21

Metallic potassium on prolonged exposure to air forms a coating of yellow potassium superoxide (KO_2) under which is a layer of potassium monoxide in contact with the metal. Normal contact of potassium superoxide with potassium causes ignition to occur [1], but if the layer of superoxide is driven into the underlying potassium by dry metal-cutting operations or a hammer blow, very violent explosions occur [2,3]. It has also been observed that potassium-cutting operations under oil cause flashes of light or fire to occur [3]. Fresh potassium should be stored under dry xylene in airtight containers to prevent oxidation [1]. Old stocks, where the coating is orange or yellow, should not be cut but destroyed by burning on an open coke fire, or by addition of *tert*-butanol to small portions under xylene in a hood. It is dangerous to use methyl or ethyl alcohol (either dry or wet) as a replacement for *tert*-butanol. Recommended procedures include cutting and handling the metal with forceps under dry xylene and disposal of scraps in xylene by addition of *tert*-butanol [2,3]. Even the latter may react violently [4].

See also POTASSIUM–SODIUM ALLOY, K–Na

Alcohols

See Air (slow oxidation), above

Carbon

1. Mellor, 1963, Vol. 2, Suppl. 2.2, 1566
2. Mellor, D. P., *Chem. & Ind.*, 1965, 723
3. *Alkali Metals*, 1957, 169

Reaction of various forms of carbon (soot, graphite or activated charcoal) is exothermic and vigorous at elevated temperatures, and if the carbon is finely divided and air is present, ignition may occur leading to explosion, possibly due to interaction with the potassium superoxide which would be formed [1]. Explosions caused by attempts to extinguish potassium fires with graphite powder have been so attributed [2]. Potassium cannot be produced by electrolysis of potassium chloride because of interaction of the metal with graphite electrodes and formation of explosive carbonyl-potassium [3].

Ethylene oxide

See ETHYLENE OXIDE, C_2H_4O: Contaminants

Halocarbons

1. Staudinger, H., *Z. Angew. Chem.*, 1922, **35**, 657; *Z. Elektrochem.*, 1925, **31**, 549
2. Lenze, F. *et al.*, *Chem. Ztg.*, 1932, **56**, 921
3. Rampino, L. D., *Chem. Eng. News*, 1958, **36**(32), 62

Although apparently stable on contact, mixtures of potassium (or its alloys) with a wide range of halocarbons are shock-sensitive and may explode with great violence on light impact. Mono-, di-, tri- and pentachloro-ethane, bromoform, dibromo- and diiodo-methane are among those investigated. Sensitivity increases generally with the degree of substitution, and potassium–sodium alloy gives extremely sensitive mixtures. The mixture with carbon tetrachloride is 150–200 times as shock-sensitive as mercury fulminate, and a mixture of potassium with bromoform was exploded by a door slamming nearby. Mixtures with tetra- and pentachloro-ethane will often explode spontaneously after a short delay during which a voluminous solid separates out [1,2]. When heated together, potassium and tetrachloroethylene exploded at 97–99°C, except when the metal had been very recently freed of its usual oxide film. Sodium did not explode under the same conditions [3].

See CHLOROMETHANE, CH_3Cl: Metals

Halogens,
or Interhalogens

Mellor, 1941, Vol. 2, 114, 469; 1963, Vol. 2, Suppl. 2.2, 1563

Potassium ignites in fluorine and in dry chlorine (unlike sodium). In bromine vapour it incandesces, and explodes violently in liquid bromine. Mixtures with iodine incandesce on heating, and explode weakly on impact. Potassium reacts explosively with molten iodine bromide and iodine, and a mixture with the former is shock-sensitive and explodes strongly. Molten potassium reacts explosively with iodine pentafluoride.

See IODINE CHLORIDE, ClI: Metals
IODINE HEPTAFLUORIDE, F_7I: Metals

Hydrazine

See HYDRAZINE, H_4N_2: Potassium

Hydrogen iodide

Cueilleron, J., *Bull. Soc. Chim. Fr.*, 1945, **12**, 88

Impact causes a mixture of potassium and anhydrous hydrogen iodide to explode very violently.

Maleic anhydride

See MALEIC ANHYDRIDE, $C_4H_2O_3$: Cations, etc.

Mercury

Mellor, 1941, Vol. 2, 469

Interaction to form amalgams is vigorously exothermic and may become violent if too much potassium is added at once.

Metal halides

Staudinger, H., *Z. Elektrochem.*, 1925, **31**, 549
Cueilleron, J., *Bull. Soc. Chim. Fr.*, 1945, **12**, 88
Mellor, 1963, Vol. 2, Suppl. 2.2, 1571

Mixtures of potassium with metal halides are sensitive to mechanical shock, and the ensuing explosions have been classified. Very violent explosions occurred with calcium bromide, iron(III) bromide or chloride, iron(II) bromide or iodide, or cobalt(II) chloride.
Strong explosions occurred with silver fluoride, all four mercury(II) halides, copper(I) chloride, bromide or iodide, copper(II) chloride and bromide, and ammonium tetrachlorocuprate; zinc and cadmium chlorides, bromides, iodides; aluminium fluoride, chloride, bromide, thallium(I) bromide; tin(II) or (IV) chloride, tin(IV) iodide (with sulphur), arsenic trichloride and triiodide, antimony and bismuth

trichlorides, tribromides and triiodides; vanadium(V) chloride chromium(IV) chloride, manganese(II) and iron(II) chlorides, and nickel chloride, bromide and iodide. Weak explosions occurred with ammonium bromide or iodide, silver chloride or iodide, strontium iodide, thallium(I) chloride; potassium tetrachlorocuprate, zinc, cadmium and nickel fluorides, chromium(III) fluoride and manganese(II) bromide or iodide. Vanadyl chloride reacts violently with potassium at 100°C, and explosions have occurred with copper and lead chloride oxides or lead(IV) chloride.

Metal oxides

1. Mellor, 1963, Vol. 2, Suppl. 2.2, 1571
2. Mellor, 1941, Vol. 3, 138
3. Mellor, 1940, Vol. 4, 770, 779
4. Mellor, 1941, Vol. 7, 401
5. Mellor, 1943, Vol. 11, 237, 542

Lead peroxide reacts explosively [1] and copper(II) oxide incandescently [2] with warm potassium. Mercury(II) and (I) oxides react with molten potassium with incandescence and explosion [3]. Tin(IV) oxide is reduced incandescently on warming [4] and molybdenum(III) oxide on heating [5]. Potassium superoxide explodes on intimate contact with potassium metal.

See Air (slow oxidation), above
'MERCURY(I) OXIDE', Hg_2O

Nitrogen-containing explosives

Staudinger, H., *Z. Elektrochem.*, 1925, **31**, 549

Nitro or nitrate explosives, normally shock-insensitive, are rendered extremely sensitive by addition of traces of potassium or potassium–sodium alloy. Ammonium nitrate, and nitrate–sulphate mixtures, picric acid, and even nitrobenzene respond in this way.

Non-metal halides

1. Mellor, 1963, Vol. 2, Suppl. 2.2, 1564–1568
2. Mellor, 1940, Vol. 8, 1006
3. Mellor, 1947, Vol. 10, 642–646, 908, 912

Diselenium dichloride and seleninyl chloride both explode on addition of potassium [1,3], while the metal ignites in contact with phosphorus trichloride vapour or liquid [2]. Mixtures of potassium with sulphur dichloride or dibromide, phosphorus trichloride or tribromide, and with phosgene are shock-sensitive, usually exploding violently on impact. Potassium also explodes violently on heating with disulphur dichloride, and with sulphur dichloride or seleninyl bromide without heating [3].

Non-metal oxides

1. Gilbert, H. N., *Chem. Eng. News,* 1948, **26**, 2604
2. Mellor, 1941, Vol. 2, 241
3. Mellor, 1940, Vol. 8, 436, 544, 554, 945

Mixtures of potassium and solid carbon dioxide are shock-sensitive and explode violently on impact, and carbon monoxide readily reacts to form explosive carbonylpotassium [1]. Dichlorine oxide explodes on contact with potassium [2]. Potassium ignites in dinitrogen tetra- or pentaoxide at ambient temperature, and incandesces when warmed with phosphorus(V) or nitrogen oxides [3].

Oxalyl dihalides

Staudinger, H., *Z. Angew. Chem.*, 1922, **35**, 657; *Ber.*, 1913, **46**, 1426

In absence of mechanical disturbance, potassium or potassium–sodium alloy appears to be stable in contact with oxalyl dibromide or dichloride, but the mixtures are very shock-sensitive and explode very violently.

See Halocarbons, above

Oxidants

Mellor, 1963, Vol. 2, Suppl. 2.2, 1571

The potential for violence of interaction between the powerful reducing agent potassium and oxidant classes has been well described. Other miscellaneous oxygen-containing substances which react violently or explosively include sodium and silver iodates, lead sulphate and boric acid.

See Air, above
Metal oxides, above
Non-metal oxides, above
Selenium, below
CHLORINE TRIFLUORIDE, ClF_3: Metals
DICHLORINE OXIDE, Cl_2O: Oxidisable materials
CHROMIUM TRIOXIDE, CrO_3: Alkali metals
NITRYL FLUORIDE, FNO_2: Metals
AMMONIUM NITRATE, $H_4N_2O_3$: Ammonium sulphate, etc.

Selenium,
or Tellurium
Mellor, 1947, Vol. 10, 767; 1943, Vol. 11, 40
Interaction of selenium and potassium is exothermic and ignition occurs, or, with excess potassium, a mild explosion. Tellurium and potassium become incandescent when warmed in hydrogen atmosphere to prevent aerial oxidation.

Sulphuric acid
Kirk-Othmer, 1961, Vol. 16, 362
Interaction is explosive.

Water
1. Mellor, 1941, Vol. 2, 469; 1963, Vol. 2, Suppl.2.2, 1560
2. Angus, L. H., *School Sci. Rev*., 1950, **31**(115), 402
Interaction is violently exothermic, and the heat evolved with water at 20°C is enough to ignite evolved hydrogen. Larger pieces of potassium invariably explode in water and scatter burning potassium particles over a wide area. Aqueous alcohols should not be used for waste metal disposal [1]. Small pieces of potassium will also explode with a restricted amount of water [2].
See Air (slow oxidation), above
ALKALI METALS

POTASSIUM–SODIUM ALLOY **K–Na**

Health & Safety Information, **251**, Washington, USAEC, 1967
This describes appropriate precautions in handling potassium–sodium alloy, used as a liquid coolant. It is generally more hazardous than either of the component metals, because alloys in the range 50–85% wt. potassium are liquid at ambient temperature and can therefore come into more intimate contact with reagents than the solid metals. Most of the entries under Potassium, and some of those under Sodium, may be expected to apply to their alloys, depending on the composition.

Air,
Mellor, 1961, Vol. 2, Suppl. 2.1, 562
Anon., *Fire Prot. Assoc. J*., 1965, **68**, 140
The alloy usually ignites on exposure to air, with which it reacts in any

case to form potassium dioxide, a very powerful oxidant. Fires may be extinguished with dried sodium carbonate, calcium carbonate, sand or resin-coated sodium chloride. Graphite is not suitable since violent reaction with the superoxide is possible. Carbon dioxide or halocarbon extinguishers must not be used.

See Carbon dioxide, below
Halocarbons, below
POTASSIUM, K: Air (slow oxidation)
POTASSIUM DIOXIDE, KO_2: Carbon

Carbon dioxide,
or Carbon disulphide
Staudinger, H., *Z. Elektrochem.*, 1925, **31**, 549; *Z. Angew. Chem.*, 1922, **35**, 657
Mixtures of the alloy and solid carbon dioxide are powerful explosives, some 40 times more sensitive to shock than mercury fulminate. Carbon disulphide behaves similarly.

Halocarbons
1. Staudinger, H., *Z. Elektrochem.*, 1925, **31**, 550
2. *Inform. Exch. Bull.*, Lawrence Radiation Lab., 1961, **1**, 2
3. Kimura, T. *et al.*, *J. Chem. Educ.*, 1973, **50**, A85

The liquid alloy gives mixtures with halocarbons even more shock-sensitive than those with potassium [1]. Highly chlorinated methane derivatives are more reactive than those of ethane, often exploding spontaneously after a delay [1]. Contact of 1,1,1-trichloroethane with a trapped alloy residue in a valve caused an explosion [2]. It is to be expected that chloro-fluorocarbons will also form hazardous mixtures in view of their reactivity with barium. A Teflon-coated magnetic bar used to stir the alloy under propane atmosphere ignited when the speed was increased and generated enough heat to melt the glass. Triboelectric initiation was postulated [1].
See other METALS: Halocarbons

Metal oxides
Staudinger, H., *Z. Elektrochem.*, 1925, **31**, 551
Mixtures of the alloy with silver and mercury oxides are shock-sensitive powerful explosives. The red form of mercury(I) oxide is 40 times, and the yellow form 120 times, as shock-sensitive as mercury fulminate.

Nitrogen-containing explosives
See POTASSIUM, K: Nitrogen-containing explosives
See other ALLOYS

POTASSIUM PERMANGANATE **$KMnO_4$**

Acetic acid,
or Acetic anhydride
von Schwartz, 1918, 34
Cooling is necessary to prevent possible explosion from contact of potassium permanganate (or the calcium or sodium salts) with acetic acid or its anhydride.

Aluminium carbide
Mellor, 1946, Vol. 5, 872
Incandescence on warming.

Ammonia,
Sulphuric acid
Mellor, 1941, Vol. 1, 907
Ammonia is oxidised with incandescence in contact with the permanganic acid formed in the mixture.

Ammonium nitrate
Urbanski, 1965, Vol. 2, 491
Admixture of 0.5% of potassium permanganate with an ammonium nitrate explosive caused an explosion 7 h later. This was due to formation and exothermic decomposition of ammonium permanganate, leading to ignition.

Ammonium perchlorate
See AMMONIUM PERCHLORATE, ClH_4NO_4: Impurities

Antimony,
or Arsenic
Mellor, 1942, Vol. 12, 322
Antimony ignites on grinding in a mortar with the solid permanganate, while arsenic explodes.

Glycerol
'Leaflet No. 5', London, Inst. of Chem., April, 1940
Contact of glycerol with solid potassium permanganate caused a vigorous fire.

Hydrochloric acid
Curry, J. C., *School Sci. Rev.*, 1965, **46**(160), 770
During preparation of chlorine by addition of the concentrated acid to solid permanganate, a sharp explosion occurred on one occasion. (Contamination of the acid with sulphuric acid could have produced permanganic acid.)

Hydrogen peroxide
Anon., *J. Pharm. Chim.*, 1927, **6**, 410
Contact of hydrogen peroxide solution from a broken bottle with pervious packages of permanganate caused a violent reaction and fire.
See also HYDROGEN PEROXIDE, H_2O_2: Metals, etc. (reference 3)

Hydrogen trisulphide
Mellor, 1947, Vol. 10, 159
Contact with solid permanganate ignites the liquid sulphide.

Hydroxylamine
See HYDROXYLAMINE, H_3NO: Oxidants

Non-metals
Mellor, 1942, Vol. 12, 319–323
A mixture of carbon and potassium permanganate is not friction-sensitive, but burns vigorously on heating. Mixtures with phosphorus or sulphur react explosively on grinding and heating, respectively.

Sulphuric acid,
Water
1. Archer, J. R., *Chem. Eng. News*, 1948, **26**, 205
2. *ABCM Quart. Safety Summ.*, 1946, **17**, 2
3. Stevens, G. C., private comm., 1971

Addition of concentrated sulphuric acid to the slightly damp permanganate caused an explosion. This was attributed to formation of permanganic acid, dehydration to dimanganese heptoxide and explosion of the latter, caused by heat liberated from interaction of sulphuric acid and moisture [1]. A similar incident was reported previously, when a solution of potassium permanganate in sulphuric acid, prepared as a cleaning agent, exploded violently [2]. There is, however, a reputedly safe procedure for the final ultra-cleaning of glassware in cases where the adsorbed chromium film left by chromic acid cleaning would be intolerable. This involves treatment of the glassware with concentrated sulphuric acid (10 ml) to which one or two small crystals (not more) of permanganate have been added. If the solution changes to a brown colour, it must immediately be discarded into excess water [3].

Titanium
Mellor, 1941, Vol. 7, 20
A mixture of powdered metal and oxidant explodes on heating.

Wood
Anon., *Chem. Ztg.*, 1927, **51**, 221
Dunn, B. W., Amer. Rail Assoc., Bur. Explosives, *Report No. 10*, 1917
Contact between the solid material and wood, either in presence of moisture or of mechanical friction, may cause a fire.
See other METAL OXOMETALLATES
OXIDANTS

POTASSIUM NITRITE **KNO_2**

Ammonium salts
Mellor, 1941, Vol. 2, 702
Addition of ammonium sulphate to the fused nitrate causes effervescence and ignition.
See NITRITE SALTS OF NITROGENOUS BASES

Boron
Mellor, 1946, Vol. 5, 16
Addition of boron to the fused nitrite causes violent decomposition.

Potassium amide
Bergstrom, F. W. *et al., Chem. Rev.,* 1933, **12**, 64
Heating a mixture of the solids under vacuum causes a vigorous explosion.
See other METAL OXONON-METALLATES

Calcium disilicide
Berger, F., *Compt. Rend.*, 1920, **170**, 1492
A mixture of potassium (or sodium) nitrate and calcium silicide (60:40) is a readily ignited primer which burns at a very high temperature. It is capable of initiating many high-temperature reactions.

Metals
Mellor, 1941, Vol. 7, 20, 116, 261; 1939, Vol. 9, 382
Mixtures of potassium nitrate and powdered titanium, antimony or germanium explode on heating, and with zirconium at fusion temperature of the mixture.

Metal sulphides
1. Mellor, 1939, Vol. 9, 270, 524
2. Mellor, 1941, Vol. 3, 745
3. Mellor, 1941, Vol. 7, 91, 274
4. Mellor, 1943, Vol. 11, 647

Mixtures of potassium nitrate with antimony trisulphide [1], barium or calcium sulphides [2], germanium monosulphide or titanium disulphide [3], all explode on heating. The mixture with arsenic disulphide is detonable, and addition of sulphur gives a pyrotechnic composition [3]. Mixtures with molybdenum disulphide are also detonable [4].

Non-metals
1. *MCA Case History No. 1334*
2. Mellor, 1946, Vol. 5, 16
3. Mellor, 1941, Vol. 2, 820, 825; 1963, Vol. 2, Suppl. 2.2, 1939
4. Mellor, 1940, Vol. 8, 788
5. Mellor, 1939, Vol. 9, 35

A pyrotechnic blend of a finely divided mixture with boron ignited and exploded when the aluminium container was dropped [1]. Boron is not attacked at below 400°C, but is at fusion temperature or at lower temperatures if decomposition products (nitrites) are present [2]. Contact of powdered carbon with the fused nitrate causes vigorous combustion and a mixture explodes on heating. Gunpowder is the oldest explosive known and contains potassium nitrate, charcoal and sulphur, the latter to reduce ignition temperature and increase the speed of combustion [3]. Mixtures of white phosphorus and potassium nitrate explode on percussion and a mixture of red phosphorus and potassium nitrate reacts vigorously on heating [4]. Mixtures of potassium nitrate with arsenic explode vigorously on ignition [5].

Phosphides
Mellor, 1940, Vol. 8, 839, 845
Boron phosphide ignites in molten nitrates; mixtures of the nitrate with tricopper diphosphide explode on heating, and that with copper monophosphide on impact.

Reducants
Mellor, 1941, Vol. 2, 820
Mixtures of potassium nitrate with sodium phosphinate and sodium thiosulphate are explosive, the former being rather powerful.

Sodium acetate
Pieters, 1957, 30
Mixtures may be explosive.

Thorium dicarbide
See THORIUM DICARBIDE, C_2Th: Non-metals, etc.
See other METAL OXONON-METALLATES
OXIDANTS

POTASSIUM NITRIDOOSMATE(1−) **KNO_3Os**

Clifford, A. F. *et al., Inorg. Synth.,* 1960, **6**, 205
Heating above 180°C at atmospheric pressure causes it to explode.
See other *N*-METAL DERIVATIVES

POTASSIUM AZIDE **KN_3**

Mellor, 1940, Vol. 8, 347
Insensitive to shock. On heating progressively it melts, then

decomposes evolving nitrogen, and the residue ignites with a feeble explosion.

Carbon disulphide
See CARBON DISULPHIDE, CS_2: Metal azides

Manganese dioxide
See MANGANESE DIOXIDE, MnO_2: Potassium azide

Sulphur dioxide
Mellor, 1940, Vol. 8, 347
The salt explodes at 120°C when heated in liquid sulphur dioxide.
See other METAL AZIDES

POTASSIUM AZIDOSULPHATE KN_3O_3S

Mellor, 1940, Vol. 8, 314
Explodes on heating.
See other ACYL AZIDES

POTASSIUM AZIDODISULPHATE $KN_3O_6S_2$

Alone,
or Water
Mellor, 1967, Vol. 8, Suppl. 2.2, 36
On keeping or slow heating, the salt produces explosive disulphuryl azide. The salt reacts explosively with water.
See DISULPHURYL AZIDE, $N_6O\ S_2$
See other ACYL AZIDES

POTASSIUM DIOXIDE (SUPEROXIDE) KO_2

Alkali Metals, 1957, 174
Gilbert, H. N., *Chem. Eng. News,* 1948, **26**, 2604
This powerful oxidant is stable when pure, but impact-sensitive when formed as a layer on metallic potassium.
See POTASSIUM, K: Air (slow oxidation)

Carbon
Mellor, D. P., *Chem. & Ind.,* 1965, 723
Residues from extinguishing small sodium–potassium alloy fires with graphite were accumulated in an airtight drum. Later, burning alloy fell into the drum and caused a violent explosion. This was attributed to formation of potassium superoxide during storage and explosive reaction of the latter with graphite, initiated by the burning alloy. Graphite is not a suitable extinguisher for potassium or alloy fires.

Diselenium dichloride
Mellor, 1947, Vol. 10, 897
Interaction is very violent.

Ethanol
Health & Safety Information, **251**, Washington, USAEC, 1967
Disposal of a piece of sodium–potassium alloy under argon in a glove box by addition of alcohol caused violent gas evolution which burnt a glove and produced a flame. A piece of highly oxidised potassium exploded when dropped into alcohol. Both incidents were attributed to violent interaction of potassium superoxide and ethanol.

Hydrocarbons
Health & Safety Information, **251,** Washington, USAEC, 1967
Residues of sodium–potassium alloy in metal containers were covered with oil prior to later disposal. When a lid was removed later, a violent explosion occurred. This was attributed to frictional initiation of the mixture of potassium superoxide (formed on long standing), and oil.

Metals
Mellor, 1941, Vol. 2, 493
Oxidation of arsenic, antimony, copper, potassium, tin or zinc proceeds with incandescence. Impaction of the superoxide into potassium causes explosion.
See POTASSIUM, K: Air (slow oxidation)
See related METAL PEROXIDES

POTASSIUM TRIOXIDE KO_3

See CAESIUM TRIOXIDE, CsO_3

POTASSIUM TETRAPEROXOMOLYBDATE(2–) K_2MoO_8

Jahr, K. F., *FIAT Rev. of German Science: Inorg. Chem.*, Part III, 1948, 170
It is explosive.
See other PEROXOACID SALTS

POTASSIUM NITROSODISULPHATE(2–) $K_2NO_7S_2$

1. Zimmer, H. *et al., Chem. Rev.*, 1971, **71**, 230, 243
2. Wehrli, P. A. *et al., Org. Synth.*, 1972, **52**, 86

The observed instability of the solid radical on storage, ranging from slow decomposition to violent explosion (probably depending on the degree of confinement), depends largely upon the degree of contamination by nitrite ion. Storage conditions to enhance stability are detailed [1]. Synthetic use of solutions of the salt has been preferred, and the storage of any isolated salt as an alkaline slurry is recommended [2].
See other NITROSO COMPOUNDS

POTASSIUM DINITROSOSULPHITE $K_2N_2O_5S$

Weitz, E. *et al., Ber.* 1933, **66**, 1718
Explodes on heating.
See other NITROSO COMPOUNDS

POTASSIUM AZODISULPHONATE(2–) $K_2N_2O_6S_2$

Konrad, E. *et al., Ber.* 1926, **59**, 135
There is a possibility of explosion during vacuum desiccation of the ether-damp solid.
See related AZO COMPOUNDS

POTASSIUM SULPHUR DIIMIDE(2–) K_2N_2S

Goehring, 1957, 35, 69
Ignites in air, reacts violently and may explode with traces of water.

POTASSIUM THALLIUM(I)AMIDE(2–) AMMONIATE $K_2NTl \cdot NH_3$

Alone,
or Acids,
or Water
Mellor, 1946, Vol. 5, 262
The amide solvated with 4, 2 or $1\frac{1}{3}$ mol of ammonia, explodes violently when subjected to shock, heat, dilute acids or water.
See other N-METAL DERIVATIVES

POTASSIUM DIPEROXOORTHOVANADATE(2–) K_2O_6V

Mellor, 1939, Vol. 9, 795
Emeléus, 1960, 425
Explodes on heating.
See other PEROXOACID SALTS

POTASSIUM PEROXODISULPHATE(2–) $K_2O_8S_2$

Hydrazine salts,
Alkali
Mellor, 1947, Vol. 10, 466
Addition of alkali to the mixed salts liberates hydrazine which is vigorously oxidised to nitrogen gas.

Potassium hydroxide
BCISC Quart. Safety Summ., 1965, **36**, 41
Surface contamination of 2 kg of the dry salt with as little as two flakes of moist potassium hydroxide caused a vigorous, self-sustaining fire, which was extinguished with water, but not by carbon dioxide or dry powder extinguishers.

Water
Castrantas, 1965, 5
The salt liberates oxygen rapidly above 100°C when dry, but at only 50°C when wet.
See other PEROXOACID SALTS

POTASSIUM TETRAPEROXOTUNGSTATE(2–) **K_2O_8W**

Mellor, 1943, Vol. 11, 836
Explodes on friction, or rapid heating to 80°C.
See other PEROXOACID SALTS

POTASSIUM SULPHIDE **K_2S**

Nitrogen oxide
See NITROGEN OXIDE, NO: Potassium sulphide

POTASSIUM NITRIDE(3–) **K_3N**

Mellor, 1940, Vol. 8, 99
Usually ignites in air.

Phosphorus
See PHOSPHORUS, P: Potassium nitride

Sulphur
See SULPHUR, S: Potassium nitride
See other *N*-METAL DERIVATIVES

POTASSIUM HEXAOXOXENONATE(4–)-XENON TRIOXIDE
$K_4O_6Xe \cdot 2O_3Xe$

1. Grosse, A. V., *Chem. Eng. News*, 1964, **42**(29), 42
2. Spittler, T. M. *et al., J. Amer. Chem. Soc.*, 1966, **88**, 2942
3. Appelman, E. H. *et al., J. Amer. Chem. Soc.*, 1964, **86**, 2141

The complex salt is very sensitive to mechanical shock and explodes violently [1,2], even when wet [3].
See other XENON COMPOUNDS

LANTHANUM — La

Nitric acid
Mellor, 1946, Vol. 5, 603
Oxidation is violent.

Phosphorus
See PHOSPHORUS, P: Metals
See other METALS

DILANTHANUM TRIOXIDE — La_2O_3

Chlorine trifluoride
See CHLORINE TRIFLUORIDE, ClF_3: Metals, etc.

Water
Sidgwick, 1950, 441
Interaction is vigorously exothermic, accompanied by hissing, as for calcium oxide.
See CALCIUM OXIDE, CaO: Water
See other METAL OXIDES

LITHIUM — Li

Atmospheric gases
1. Mellor, 1941, Vol. 2, 468; 1961, Vol. 2, Suppl. 2.1, 71
2. *Haz. Chem. Data,* 1969, 141

Finely divided metal may ignite in air at ambient temperature, and massive metal at temperatures above the m.p. (180°C), especially if oxide or nitride is present. Since lithium will burn in oxygen, nitrogen or carbon dioxide, and then reacts with sand, sodium carbonate, etc., it is difficult to extinguish once alight [1]. Molten lithium is very reactive, and will attack concrete and refractory materials. Use of normal fire

extinguishers (containing water, foam, carbon dioxide, halocarbons, dry powders) will either accelerate combustion or cause explosion. Powdered graphite, lithium or potassium chlorides or zirconium silicate are suitable extinguishants [2].
See Metal chlorides, below

Bromine pentafluoride
See BROMINE PENTAFLUORIDE BrF_5: Acids, etc.

Diazomethane
See DIAZOMETHANE, CH_2N_2: Alkali metals

Halocarbons
1. Staudinger, H., *Z. Elektrochem.*, 1925, **31**, 549
2. *Pot. Incid. Rep.*, 1968, **39**, ASESB, Washington
3. Pittwell, L. R., *J. R. Inst. Chem.*, 1959, **80**, 552

Mixtures of lithium shavings and several halocarbon derivatives are impact-sensitive and will explode, sometimes violently [1,2]. Such materials include: bromoform, carbon tetra-bromide, -chloride and -iodide, chloroform, dichloro- and diiodo-methane, fluorotrichloromethane, tetrachloroethylene, trichloroethylene and 1,1,2-trichlorotrifluoroethane. In an operational incident, shearing samples off a lithium billet immersed in carbon tetrachloride caused an explosion and continuing combustion of the immersed metal [3].
See Poly-1,1-difluoroethylene-hexafluoropropylene, below
See other HALOCARBONS: Metals

Halogens
1. Staudinger, H., *Z. Elektrochem.*, 1925, **31**, 549
2. Mellor, 1961, Vol. 2, Suppl. 2.1, 82

Mixtures of lithium and bromine are unreactive unless subject to heavy impact, when explosion occurs [1]. Lithium and iodine react above 200°C with a large exotherm [2].

Mercury
Smith, G. McP. *et al.*, *J. Amer. Chem. Soc.*, 1909, **31**, 799
Alexander, J. *et al.*, *J. Chem. Educ.*, 1970, **47**, 277

Interaction to form lithium amalgam is violently exothermic and may be explosive if large pieces of lithium are used [1]. An improved technique, using *p*-cymene as inerting diluent is described in the later reference [2].

Metal chlorides,
Nitrogen
BCISC Quart. Safety Summ., 1969, **40**, 16
Accidental contamination of lithium strip with anhydrous chromium trichloride or zirconium tetrachloride caused it to ignite and burn vigorously in the nitrogen atmosphere in a glove box.
See Atmospheric gases, above

Metal oxides and chalcogenides
1. Mellor, 1961, Vol. 2, Suppl. 2.1, 81–82
2. *Alkali Metals,* 1957, 11
Lithium is used to reduce metallic oxides in metallurgical operations, and the reactions, after initiation at moderate temperatures, are violently exothermic and rapid. Chromium(III) oxide reacts at 180°C, reaching 965°C; similarly molybdenum trioxide (180 to 1400°C), niobium pentoxide (320 to 490°C), titanium dioxide (200–400 to 1400°C), tungsten trioxide (200 to 1030°C), vanadium pentoxide (394 to 768°C); also iron(II) sulphide, (260 to 945°C) and manganese telluride, (230 to 600°C)[1]. Residual mixtures from lithium production cells containing lithium and rust sometimes ignite when left as thin layers exposed to air [2].

Nitric acid
Accident & Fire Prevention Information, Washington, USAEC, March, 1954
Lithium ignited on contact with nitric acid and the reaction became violent, ejecting burning lithium.

Nitryl fluoride
See NITRYL FLUORIDE, FNO_2: Metals

Non-metal oxides
Mellor, 1961, Vol. 2, Suppl. 2.1, 74, 84, 88
Although carbon dioxide reacts slowly with lithium at ambient temperature, the molten metal will burn vigorously in the gas, which cannot be used as an extinguisher on lithium fires. Carbon monoxide reacts in liquid ammonia to give the carbonyl, which reacts explosively with water or air. Lithium rapidly attacks silica or glass at 250°C.

Platinum
See PLATINUM; Pt: Lithium

Poly-1,1-difluoroethylene-hexafluoropropylene (Viton)
Markowitz, M. M. *et al., Chem. Eng. News,* 1961, **39**(32), 4
Mixtures of finely divided metal and shredded polymer ignited in air on contact with water or on heating to 369°C, or at 354°C under argon.
See Halocarbons, above

Sodium carbonate,
or Sodium chloride
Mellor, 1961, Vol. 2, Suppl. 2.1, 25
Sodium carbonate and chloride are unsuitable to use as extinguishants for lithium fires, since burning lithium will liberate the more reactive sodium in contact with them.

Sulphur
Mellor, 1961, Vol. 2, Suppl. 2.1, 75
Interaction when either is molten is very violent and, even in presence of an inert diluent, the reaction begins explosively. Reaction of sulphur with lithium dissolved in liquid ammonia at –33°C is also very vigorous.

Trifluoromethylhypofluorite
See TRIFLUOROMETHYL HYPOFLUORITE, CF_4O: Lithium

Water
Mellor, 1961, Vol. 2, Suppl. 2.1, 72
Reaction with cold water is of moderate vigour, but violent with hot water, and the liberated hydrogen may ignite.
See ALKALI METALS
See other METALS

LITHIUM–TIN ALLOYS Li–Sn

Alkali Metals, 1957, 13
Lithium–tin alloys are pyrophoric.
See other ALLOYS

LITHIUM SODIUM NITROXYLATE $LiNNaO_2$

Mellor, 1961, Vol. 2, Suppl. 2.1, 78
Decomposes violently at 130°C.
See other METAL OXONON-METALLATES
N-METAL DERIVATIVES

LITHIUM NITRATE $LiNO_3$

Propene,
Sulphur dioxide
See PROPENE, C_3H_6: Lithium nitrate, etc.

LITHIUM AZIDE LiN_3

Mellor, 1940, Vol. 8, 345
Insensitive to shock, the moist or dry salt decomposes explosively at 115–298°C depending on the rate of heating.

Alkyl nitrates,
Dimethylformamide
Polansky, O. E., *Angew. Chem. (Intern. Ed.)*, 1971, **10**, 412
Although these reaction mixtures are stable at 25°C during preparation of *tert*-alkyl azides, above 200°C the mixtures are shock-sensitive and highly explosive.
See other METAL AZIDES

LITHIUM NITRIDE Li_3N

'Data Sheet TD 121', Exton, Foote Mineral Co., 1965
The finely divided powder may ignite if sprayed into moist air.
See other N-METAL DERIVATIVES

HEXALITHIUM DISILICIDE Li_6Si_2

Acids
Mellor, 1940, Vol. 6, 170
The silicide incandesces in conc. hydrochloric acid, and with dil. acid evolves silicon hydrides which ignite. It explodes with nitric acid and incandesces when floated on sulphuric acid.

Halogens
Mellor, 1940, Vol. 6, 169
When warmed, it ignites in fluorine, but higher temperatures are required with chlorine, bromine and iodine.

Non-metals
Mellor, 1940, Vol. 6, 169
Interaction with phosphorus, selenium or tellurium causes incandescence.

Water
Mellor, 1940, Vol. 6, 170
Reaction with water is very violent, and the silicon hydrides evolved ignite.
See other METAL NON-METALLIDES

MAGNESIUM Mg

Air
The rapid and brilliant combustion of magnesium powder, initiated in air at 520°C, was used formerly for photographic flash illumination. Very finely divided magnesium powder dispersed in air is a serious dust explosion hazard.
See DUST EXPLOSIONS
MAGNESIUM FIRES

Beryllium fluoride
Walker, H. L., *School Sci. Rev.*, 1954, **35**(127), 348
Interaction is violently exothermic.

Boron diiodophosphide
See BORONDIIODOPHOSPHIDE, BI_2P: Metals

Ethylene oxide

See ETHYLENE OXIDE, C_2H_4O

Halocarbons

1. Clogston, C. C., *Bull. Res. Underwriters Lab.*, 1945, **34**, 5
2. *Pot. Incid. Rep.*, **39**, ASESB, Washington, 1968
3. Mond Div. ICI, private comm., 1968

Powdered metal reacts vigorously and may explode on contact with chloromethane, chloroform or carbon tetrachloride, or mixtures of these [1]. Mixtures of powdered magnesium with carbon tetrachloride or trichloroethylene will flash on heavy impact [2]. Violent decomposition with evolution of hydrogen chloride can occur when 1,1,1-trichloroethane comes into contact with magnesium or its alloys with aluminium [3].

See Polytetrafluoroethylene, below
BROMOMETHANE, CH_3Br: Metals

Halogens,
or Interhalogens

Mellor, 1940, Vol. 4, 267

It ignites if moist and burns violently in fluorine, and ignites in moist or warm chlorine. It burns not very readily in bromine vapour and may ignite if finely divided on heating in iodine vapour.

See CHLORINE TRIFLUORIDE, ClF_3
IODINE HEPTAFLUORIDE, F_7I

Hydrogen iodide

Mellor, 1940, Vol. 2, 206

Contact causes momentary ignition.

Metal cyanides

Mellor, 1940, Vol. 4, 271

Magnesium reacts with incandescence on heating with cadmium, cobalt, copper, lead, nickel and zinc cyanides. With gold and mercury cyanide, the cyanogen released by thermal decomposition of these salts reacts explosively with magnesium.

Metal oxides

1. Mellor, 1941, Vol. 3, 138, 378; 1940, Vol. 4, 272; 1941, Vol. 7, 401
2. Stout, E. L., *Chem. Eng. News*, 1958, **3**(8), 64

Magnesium will violently reduce metal oxides on heating, similarly to aluminium powder in the 'thermite' reaction. Beryllium, cadmium, copper, mercury, molybdenum, tin and zinc oxides are all reduced explosively on heating. Silver oxide reacts with explosive violence when heated with magnesium powder in a sealed tube [1]. Interaction of molten magnesium and iron oxide scale is violent[2].

Metal oxosalts
Mellor, 1940, Vol. 4, 272; Vol. 8, Suppl. 2.1, 545
Pieters, 1957, 30
Interaction with fused ammonium nitrate or with metal nitrates, phosphates or sulphates may be explosively violent. Lithium and sodium carbonates also react vigorously.

See SILVER NITRATE, $AgNO_3$: Magnesium
POTASSIUM PERCHLORATE, $ClKO_4$: Metal powders

Methanol
Personal experience
Vogel, 1957, 169
The reaction of magnesium and methanol to form magnesium methoxide and used to prepare dry methanol is very vigorous, but often subject to a lengthy induction period. Sufficient methanol must be present to absorb the often violent exotherm which sometimes occurs.

Oxidants
Mellor, 1940, Vol. 2, 90; Vol. 4, 270–272
Magnesium is a powerful reducing agent and undergoes violent or explosive reactions with a variety of oxidants, particularly when powdered.

See Halogens, or Interhalogens, above
Metal oxides, above
Metal oxosalts, above
Sulphur, below
POTASSIUM CHLORATE, $ClKO_3$: Metals
POTASSIUM PERCHLORATE, $ClKO_4$: Powdered metals
NITRIC ACID, HNO_3: Metals
HYDROGEN PEROXIDE, H_2O_2: Metals
AMMONIUM NITRATE, $H_4N_2O_3$: Metals
DINITROGEN TETRAOXIDE, N_2O_4: Metals
SODIUM PEROXIDE, Na_2O_2: Metals
OXYGEN (Liquid), O_2: Metals
LEAD DIOXIDE, O_2Pb: Metals

Polytetrafluoroethylene
Anon., *Indust. Res.*, 1968(9),15
Finely divided magnesium and Teflon are described as a hazardous combination of materials.
See Halocarbons, above

Potassium carbonate
Winckler, C., *Ber.*, 1890, **23**, 44
The mixture of magnesium and potassium carbonate recommended by Castellana as a safer substitute for molten sodium in the Lassaigne test can itself be hazardous, as an equimolar mixture gives an explosive substance (probably carbonylpotassium) on heating.

Silicon dioxide
Barker, W. B., *School Sci. Rev.*, 1938, **20**(77), 150
Heating a mixture of powdered magnesium and silica (later found not to be absolutely dry) caused a violent explosion rather than the vigorous interaction expected.

Sulphur,
or Tellurium
Brauer, 1963, Vol. 1, 913
Hutton, K., *School Sci. Rev.*, 1950, **31**(114), 265
Interaction may be very violent or explosive at elevated temperatures.
See SULPHUR, S: Metals
See other METALS
REDUCANTS

MAGNESIUM NITRATE MgN_2O_6

N,N-Dimethylformamide
'DMF Brochure', Billingham, ICI Ltd., 1965
Magnesium nitrate has been reported to undergo spontaneous decomposition in DMF. Although this effect has not been observed with other nitrates, such reactions should be treated with care.
See other METAL OXONON-METALLATES

MAGNESIUM OXIDE MgO

Interhalogens

See BROMINE PENTAFLUORIDE, BrF_5: Acids, etc.
CHLORINE TRIFLUORIDE, ClF_3: Metals, etc.

Phosphorus pentachloride

See PHOSPHORUS PENTACHLORIDE, Cl_5P: Magnesium oxide

MAGNESIUM SULPHATE MgO_4S

Ethoxyethynyl alcohols

See ETHOXYETHYNYL ALCOHOLS

TRIMAGNESIUM DIPHOSPHIDE Mg_3P_2

Oxidants

Mellor, 1940, Vol. 8, 842

It ignites on heating in chlorine, or in bromine and iodine vapours at higher temperatures. Reaction with nitric acid causes incandescence.

Water

Mellor, 1940, Vol. 8, 842

Phosphine is evolved and may ignite.

See other METAL NON-METALLIDES

MANGANESE Mn

Oxidants

1. *Occup. Hazards,* 1966, 28(11), 44
2. Mellor, 1942, Vol. 12, 186–188

The finely divided metal is pyrophoric, and a mixture of manganese and aluminium dusts accidentally released into air from a filter bag exploded violently [1]. Powdered manganese ignites and becomes incandescent in fluorine or on warming in chlorine. Contact with concentrated hydrogen peroxide causes violent decomposition and/or ignition, and with nitric acid incandescence and a feeble explosion were observed.

Contact with dinitrogen tetraoxide caused ignition [2].
See BROMINE PENTAFLUORIDE, BrF_5: Acids, etc.
NITRYL FLUORIDE, FNO_2: Metals
AMMONIUM NITRATE, $H_4N_2O_3$: Metals

Phosphorus
See PHOSPHORUS, P: Metals

Sulphur dioxide
Mellor, 1942, Vol. 12, 187
Pyrophoric manganese burns brilliantly on warming in sulphur dioxide.
See other METALS

SODIUM PERMANGANATE $MnNaO_4$

Acetic acid,
or Acetic anhydride
See POTASSIUM PERMANGANATE, $KMnO_4$: Acetic acid

MANGANESE(II) OXIDE MnO

Hydrogen peroxide
See HYDROGEN PEROXIDE, H_2O_2: Metals, etc.

MANGANESE(IV) OXIDE MnO_2

Aluminium
Sidgwick, 1950, 1265
Interaction on heating is very violent.
See ALUMINIUM, Metal oxides

Dirubidium acetylide
See DIRUBIDIUM ACETYLIDE, C_2Rb_2: Metal oxides

Hydrogen sulphide
See HYDROGEN SULPHIDE, H_2S: Metal oxides

Oxidants
1. Mellor, 1942, Vol. 12, 254
2. Mellor, 1941, Vol. 1, 936
3. Ahrle, H., *Z. Angew. Chem.*, 1909, 22, 1713
4. von Schwartz, 1918, 323

Action of chlorine trifluoride causes incandescence [1]. Manganese dioxide catalytically decomposes powerful oxidising agents, often violently. Dropped into concentrated hydrogen peroxide the powdered oxide may cause explosion [2]. Either the massive or powdered oxide explosively decomposes 92% peroxomonosulphuric acid [3], and mixtures with chlorates (heated to generate oxygen) may react with explosive violence [4].

See PEROXYFORMIC ACID, CH_2O_3: Metals, etc.

Potassium azide
Mellor, 1940, Vol. 8, 347
On gentle warming, interaction is violent.

See other METAL OXIDES
OXIDANTS

MANGANESE(II) SULPHIDE — MnS

Mellor, 1942, Vol. 12, 394
The vacuum-dried red sulphide becomes red-hot on exposure to air.

See other METAL SULPHIDES

MANGANESE(II) TELLURIDE — MnTe

Lithium
See LITHIUM, Li: Metal oxides, etc.

MANGANESE(VII) OXIDE — Mn_2O_7

Mellor, 1942, Vol. 12, 293

It explodes at between 40 and 70°C, or on friction or impact, sensitivity being as great as that of mercury fulminate.
See POTASSIUM PERMANGANATE, $KMnO_4$: Sulphuric acid

Organic material
1. Patterson, A. M., *Chem. Eng. News*, 1948, **26**, 711
2. Mellor, 1940, Vol. 12, 293
A sample of the heptoxide exploded in contact with the grease on a stopcock [1] and explosion or ignition had been noted with various solvents, oils, fats and fibres [2].
See other OXIDANTS

TRIMANGANESE DIPHOSPHIDE Mn_3P_2

Chlorine
See CHLORINE, Cl_2: Phosphorus compounds

MOLYBDENUM Mo

Oxidants

See BROMINE TRIFLUORIDE, BrF_3: Halogens, etc.
BROMINE PENTAFLUORIDE, BrF_5: Acids, etc.
CHLORINE TRIFLUORIDE, ClF_3: Metals
NITRYL FLUORIDE, FNO_2: Metals
FLUORINE, F_2: Metals
IODINE PENTAFLUORIDE, F_5I: Metals
LEAD DIOXIDE, O_2Pb: Metals

SODIUM TETRAPEROXOMOLYBDATE(2−) $MoNa_2O_8$

Castrantas, 1965, 5
Decomposes explosively under vacuum.
See other PEROXOACID SALTS

MOLYBDENUM(VI) OXIDE MoO_3

Interhalogens

See BROMINE PENTAFLUORIDE, BrF_5: Acids, etc.
CHLORINE TRIFLUORIDE, ClF_3: Metals, etc.

Metals

Mellor, 1943, Vol. 11, 542

Reduction of the oxide by heated sodium or potassium proceeds with incandescence, and explosion occurs in contact with molten magnesium.

See LITHIUM Li: Metal oxides

See other METAL OXIDES

MOLYBDENUM(IV) SULPHIDE MoS_2

Potassium nitrate

See POTASSIUM NITRATE, KNO_3: Metal sulphides

SODIUM NITRITE $NNaO_2$

Aminoguanidine salts

Urbanski, 1967, Vol. 3, 207

Interaction, without addition of acid, produces tetrazolylguanyltriazene ('tetrazene'), a primary explosive of equal sensitivity to mercury(II) azide, but more readily initiated.

Ammonium salts

1. von Schwartz, 1918, 299
2. Mellor, 1967, Vol. 8, Suppl. 2, 388
3. *RoSPA Occup. Safety and Health Suppl.*, 1972, 2(10), 3

Heating a mixture of an ammonium salt with a nitrite salt causes a violent explosion on melting [1], due to formation and decomposition of ammonium nitrite. Salts of other nitrogenous bases behave similarly. Mixtures of ammonium chloride and sodium nitrite are used as commercial explosives [2]. Accidental contact of traces of ammonium nitrate with sodium nitrite residues caused wooden decking on a lorry to ignite [3].

See Wood, below
NITRITE SALTS OF NITROGENOUS BASES

Butadiene
Beer, R. N., private comm., 1972
Sodium nitrite solution is used to inhibit 'popcorn' polymerisation of butadiene in processing plants. If concentrated nitrite solutions (5%) are used, a black sludge is produced which, when dry, will ignite and burn when heated to 150°C, even in absence of air.
See also NITROGEN OXIDE, NO: Dienes

Metal cyanides
1. von Schwartz, 1918, 299; Mellor, 1940, Vol. 8, 478
2. Greenwood, P. H. S., *J. and Proc. R. Inst. Chem.*, 1947, 137
3. Elson, C. H. R., ibid., 19
4. Eiter, K. *et al.*, Austrian Pat. 176 784, 1953

Mixtures of sodium (or other) nitrites and various cyanides explode on heating, including potassium cyanide [1] or hexacyanoferrate (3–) or sodium pentacyanonitrosylferrate(2–), potassium hexacyanoferrate (4–) [3], or mercury(II) cyanide [4]. Such mixtures have been proposed as explosives, initiable by heating or detonator [4].

Phthalic acid,
Phthalic anhydride
Hawes, B. V. W., *J. R. Inst. Chem.*, 1955, **79**, 668
Mixtures of sodium nitrite and phthalic acid or anhydride explode violently on heating. A nitrite ester may have been produced.
See ACYL NITRITES

Reducants
See Sodium disulphite, below
Sodium thiosulphate, below
SODIUM NITROXYLATE(2–), NNa_2O_2

Sodium amide
Bergstrom, F. W. *et al.*, *Chem. Rev.*, 1933, **12**, 64
Addition of solid nitrate to the molten amide causes immediate gas evolution, followed by a violent explosion.

Sodium disulphite
MCA Case History No. 183
Large-scale addition of solid disulphite to an unstirred and too-concentrated solution of nitrite caused a vigorous exothermic reaction.

Sodium thiocyanate
Mellor, 1940, Vol. 8, 478
A mixture explodes on heating.

Sodium thiosulphate
Mellor, 1940, Vol. 8, 478; 1947, Vol. 10, 501
There is no interaction between solutions, but evaporation of the mixture gives a residue which explodes on heating. The mixed solids behave similarly.

Urea
Bucci, F., *Ann.Chim. Rome,* 1951, **41**, 587
Fusion of urea (2 mol) with sodium (or potassium) nitrite (1 mol) to give high yields of the cyanate must be carried out exactly as described to avoid the risk of explosion.

Wood
ABCM Quart. Safety Summ., 1944, **15**, 30
Wooden staging, which had become impregnated over a number of years with sodium nitrite, became accidentally ignited and burned as fiercely as if impregnated with potassium chlorate. Although the effect of impregnating cellulosic material with sodium nitrate is well known, that due to nitrite was unexpected.
See Ammonium salts, above
See other METAL OXONON-METALLATES
OXIDANTS

SODIUM NITRATE $NNaO_3$

Acetic anhydride
See ACETIC ANHYDRIDE, $C_4H_6O_3$: Metal salts

Aluminium,
or Aluminium oxide
Anon, *Fire,* 1935, 28, 30
Mixtures of the nitrate with the powdered metal or oxide are explosive.

Antimony
Mellor, 1939, Vol. 9, 382
Powdered antimony explodes when heated with an alkali metal nitrate.

Barium thiocyanate
Pieters, 1957, 30
Mixtures may explode.

Boron phosphide
Mellor, 1940, Vol. 8, 845
Deflagration occurs in molten alkali nitrates.

Fibrous material
1. Mellor, 1961, Vol. 2, Suppl. 2.1, 1244
2. *ABCM Quart, Safety Summ.*, 1944, **15**, 30
Fibrous organic material (jute storage bags) is oxidised in contact with sodium nitrate above 160°C, and will ignite at or below 220°C [2]. Wood and similar cellulosic material are rendered highly combustible by nitrate impregnation [2].

Metal cyanides
See MOLTEN SALT BATHS

Non-metals
Mellor, 1941, Vol. 2, 820
Contact of powdered charcoal with the molten nitrate, or of the solid nitrate with glowing charcoal, causes vigorous combustion of the carbon. Mixtures with charcoal and sulphur have been used as black powder.
See POTASSIUM NITRATE, KNO_3: Non-metals

Sodium
Mellor, 1961, Vol. 2, Suppl. 2.1, 518
Interaction of sodium nitrate and sodium alone, or dissolved in liquid ammonia, eventually gives a yellow explosive compound.
See SODIUM NITROXYLATE(2–), NNa_2O_2

Sodium phosphinate
Mellor, 1941, Vol. 2, 820
Rüst, 1947, 337
A mixture exploded violently on warming.

Sodium thiosulphate
Mellor, 1941, Vol. 2, 820
A mixture is explosive when heated.

Wood

See Fibrous material, above

See other METAL OXONON-METALLATES
OXIDANTS

SODIUM NITROXYLATE(2–) NNa_2O_2

Mellor, 1963, Vol. 2, Suppl. 2.2, 1566

The solid is very reactive towards air, moisture and carbon dioxide, and tends to explode readily, also decomposing violently on heating. It is produced from sodium nitrite or nitrate by electrolytic reduction, or action of sodium alone or in liquid ammonia solution.

See other *N*-METAL DERIVATIVES

SODIUM NITRIDE(3–) NNa_3

Fischer, F. *et al., Ber.*, 1910, **43**, 1468

It decomposes explosively on gentle warming.

See other *N*-METAL DERIVATIVES

NITROGEN OXIDE ('NITRIC OXIDE') NO

UK Sci. Mission Rep. 68/79, Washington, UKSM, 1968

Liquid nitrogen oxide and other cryogenic oxidisers (ozone, fluorine in presence of water) are very sensitive to detonation in absence of fuel and can be initiated as readily as glyceryl nitrate.

Carbon disulphide

Winderlich, R., *J. Chem. Educ.*, 1950, **27**, 669

A demonstration of combustion of carbon disulphide in nitrogen oxide exploded violently.

Dichlorine oxide

See DICHLORINE OXIDE, Cl_2O: Nitrogen oxide

Dienes,
Oxygen
1. Haseba. S. *et al., Chem. Eng. Progr.*, 1966, **62**(4), 92
2. Schuftan, P. M., *Chem. Eng. Progr.*, **62**(7), 8
Violent explosions which occurred at −100 to −180°C in ammonia synthesis gas units were traced to the formation of explosive addition products between dienes and oxides of nitrogen, produced from interaction of nitrogen oxide and oxygen. Laboratory experiments showed that the addition products from 1,3-butadiene or cyclopentadiene formed rapidly at about −150°C, and ignited or exploded on warming to −35 to −15°C. The unconjugated propadiene, and alkenes, or acetylene reacted slowly and the products did not ignite until +30 to 50°C [1]. This type of derivative ('pseudo-nitrosite') was formerly used (Wallach) to characterise terpene hydrocarbons. Further comments were made later [2].
See also SODIUM NITRITE, $NNaO_2$: Butadiene

Fluorine
See FLUORINE, F_2 : Oxides

Hydrogen,
Oxygen
Chanmugam, J. *et al., Nature,* 1952, **170**, 1067
Pre-addition of nitrogen oxide (or nitrosyl chloride as its precursor) to stoicheiometric hydrogen–oxygen mixtures at 240 mbar/360°C will cause immediate ignition under a variety of circumstances.

Metal acetylides,
Metal carbides
Mellor, 1946, Vol. 5, 848, 891
Dirubidium acetylide ignites on heating and uranium dicarbide incandesces in the gas at 370°C.

Metals
Mellor, 1940, Vol. 8, 436; 1943, Vol. 11, 162; 1942, Vol. 12, 32
Pyrophoric chromium attains incandescence in the oxide, while calcium, potassium and uranium need heating before ignition occurs, when combustion is brilliant.

Pentacarbonyliron
Manchot, W. *et al.*, *Ann.*, 1929, **470**, 275
Rapid heating to above 50°C in an autoclave caused an explosive reaction.

Nitrogen trichloride
See NITROGEN TRICHLORIDE, Cl_3N: Initiators

Non-metals
Mellor, 1940, Vol. 8, 109, 433, 435
Amorphous (not crystalline) boron reacts with brilliant flashes at ambient temperature, and charcoal and phosphorus continue to burn more brilliantly than in air.

Perchloryl fluoride
See PERCHLORYL FLUORIDE, $ClFO_3$: Hydrocarbons, etc.

Phosphine,
Oxygen
Mellor, 1940, Vol. 8, 435
Addition of oxygen to a mixture of phosphine and nitrogen oxide causes ignition.
See also PHOSPHINE, H_3P: Oxygen

Potassium sulphide
Mellor, 1940, Vol. 8, 434
The pyrophoric sulphide ignites in the oxide.
See other NON-METAL OXIDES
OXIDANTS

RUBIDIUM NITRIDE(3–) **NRb**

Mellor, 1940, Vol. 8, 99
The alkali nitrides burn in air.
See other *N*-METAL DERIVATIVES

ANTIMONY(III) NITRIDE NSb

Alone,
or Water

1. Fischer, F. *et al.*, *Ber.*, 1910, **43**, 1471
2. Franklin, E. C., *J. Amer. Chem. Soc.*, 1905, **27**, 850

Explosive decomposition occurs on warming *in vacuo* [1] and impure material explodes mildly on heating in air, or on contact with water or dilute acids [2].

See other N-METAL DERIVATIVES

THALLIUM(I) NITRIDE NTl_3

Alone,
or Acids,
or Water

Mellor, 1946, Vol. 5, 262

The nitride explodes violently on exposure to shock, heat, water or dilute acids.

See other N-METAL DERIVATIVES

NITROGEN (Gas) N_2

Lithium

See LITHIUM, Li: Atmospheric gases
Metal chlorides, etc.

NITROGEN (Liquid)

BCISC Quart. Safety Summ., 1964, **35**, 25

Although liquid nitrogen is inherently safer than liquid oxygen as a coolant, its ability to condense liquid oxygen out of the atmosphere can create hazards. A distillation tube containing a little solvent was cooled in liquid nitrogen while being sealed off in a blowpipe flame. A few minutes later the tube exploded violently, probably owing to high internal pressure caused by evaporation of liquid oxygen which had condensed into the tube during sealing. It is possible that the atmosphere close to the blow pipe was oxygen-enriched.

Open vessels which contain organic materials, or which are to be hermetically sealed, should not be cooled in liquid nitrogen, but in a coolant at a higher temperature. Liquid nitrogen should normally be used only for cooling evacuated or closed vessels where extreme cooling is necessary, and removed from around such vessels before opening them to atmosphere.

SODIUM HYPONITRITE $N_2Na_2O_2$

See HYPONITROUS ACID, $H_2N_2O_2$
SODIUM AMIDE, H_2NNa: Air

SODIUM TRIOXODINITRATE(2–) $N_2Na_2O_3$

See SODIUM AMIDE, H_2NNa: Air

SODIUM TETRAOXODINITRATE(2–) $N_2Na_2O_4$

See SODIUM AMIDE, H_2NNa: Air

SODIUM PENTAOXODINITRATE(2–) $N_2Na_2O_5$

See SODIUM AMIDE, H_2NNa: Air

SODIUM HEXAOXODINITRATE(2–) $N_2Na_2O_6$

See SODIUM AMIDE, H_2NNa: Air

DINITROGEN OXIDE ('NITROUS OXIDE') N_2O

Boron
Mellor, 1947, Vol. 8, 109
Amorphous boron (but not crystalline) ignites on heating in the dry oxide.

Phosphine
Thénard, J., *Compt. Rend.*, 1844, **18**, 652
A mixture with excess oxide can be exploded by sparking.

Tin oxide
See TIN(II) OXIDE, OSn: Non-metal oxides

Tungsten carbides
See TUNGSTEN CARBIDE, CW: Nitrogen oxides
DITUNGSTEN CARBIDE, CW_2: Oxidants
See other NON-METAL OXIDES

LEAD HYPONITRITE N_2O_2Pb

Partington, J. R. *et al.*, *J. Chem. Soc.*, 1932, **135**, 2589
The lead salt decomposes with explosive violence at 150–160°C and should not be dried by heating. The salt prepared from lead acetate explodes more violently than that from the nitrate.
See HYPONITROUS ACID, $H_2N_2O_2$
See other METAL OXONON-METALLATES

DINITROGEN TRIOXIDE N_2O_3

Phosphine
See PHOSPHINE, H_3P: Air

Phosphorus
See PHOSPHORUS, P: Non-metal oxides

DINITROGEN TETRAOXIDE (NITROGEN DIOXIDE) N_2O_4

Acetonitrile,
Indium
Addison, C. C. *et al.*, *Chem. & Ind.*, 1958, 1004
Shaking a slow-reacting mixture caused detonation, attributed to indium-catalysed oxidation of acetonitrile.

Alcohols
Daniels, F., *Chem. Eng. News,* 1955, **33**, 2372
A violent explosion occurred during the ready interaction to produce alkyl nitrates.

Ammonia
Mellor, 1940, Vol. 8, 541
Liquid ammonia reacts explosively with the solid tetraoxide at −80°C, while aqueous ammonia reacts vigorously with the gas at ambient temperature.

Barium oxide
Mellor, 1940, Vol. 8, 545
In contact with the gas at 200°C the oxide suddenly reacts, reaches red heat and fuses.

Carbon disulphide
Mellor, 1940, Vol. 8, 543
Mixtures proposed for use as explosives are stable up to 200°C, but detonable by mercury fulminate.

Carbonylmetals
Cloyd, 1965, 74
Combination is hypergolic.

Cellulose,
Magnesium perchlorate
See MAGNESIUM PERCHLORATE, Cl_2O_8Mg: Cellulose, etc.

Cycloalkenes,
Oxygen
Lachowicz, D. R. *et al.*, US Pat. 3 621 050, 1971
Contact of cycloalkenes with a mixture of dinitrogen tetraoxide and excess oxygen at temperatures of 0°C or below produces nitroperoxonitrates of the general formula $-CHNO_2-CH(O_2NO_2)-$ which appear to be unstable at temperatures above 0°C, owing to the presence of the peroxonitrate grouping.
See Hydrocarbons, below

Dimethyl sulphoxide
See DIMETHYL SULPHOXIDE, C_2H_6OS: Dinitrogen tetraoxide

Halocarbons

1. Turley, R. E., *Chem. Eng. News*, 1964, **42**(47), 53
2. Benson, S. W., *Chem. Eng. News*, 1964, **42**(51), 4
3. Shanley, E. S., *Chem. Eng. News*, 1964, **42**(52), 5

Mixtures of the tetraoxide with dichloromethane, chloroform, carbon tetrachloride, 1,2-dichloroethane, trichloroethylene and tetrachloroethylene are explosive when subjected to shock of 25 g TNT equivalent or less [1]. Mixtures of the tetraoxide with trichloroethylene react violently on heating to 150°C [2]. Partially fluorinated chloroalkanes were more stable to shock. Theoretical aspects are discussed in the later references [2,3].

See also URANIUM, U: Nitric acid

Heterocyclic bases

Mellor, 1940, Vol. 8, 543

Pyridine and quinoline are violently attacked by the liquid oxide.

Hydrazine derivatives

Cloyd, 1965, 74

Combination with hydrazine, methylhydrazine, 1,1-dimethylhydrazine or mixtures is hypergolic and used in rocketry.

See ROCKET PROPELLANTS

Hydrogen,
Oxygen

Lewis, B., *Chem. Rev.*, 1932, **10**, 60

The presence of small amounts of the oxide in non-explosive mixtures of hydrogen and oxygen renders them explosive.

Hydrocarbons

1. Mellor, 1967, Vol. 8, Suppl. 2.2, 264
2. Fierz, H. E., *J. Soc. Chem. Ind.*, 1922, **41**, 114R
3. *MCA Case History No. 128*
4. Folecki, J. *et al.*, *Chem. & Ind.*, 1967, 1424
5. Cloyd, 1965, 74
6. Urbanski, 1967, Vol. 3, 289

During attempted separation by low temperature distillation of an accidental mixture of light petroleum and the tetraoxide, a large bulk of material awaiting distillation became heated by unusual atmospheric conditions to 50°C and exploded violently [2].

Erroneous addition of liquid instead of gaseous tetraoxide to hot cyclohexane caused an explosion [3]. During kinetic studies, one sample of a 1:1 molar solution of tetraoxide in hexane exploded during (normally slow) decomposition at 28°C [4]. A mixture of tetraoxide and toluene exploded, possibly initiated by unsaturated impurities [1]. Cyclopentadiene is hypergolic with the tetraoxide [5]. These incidents are understandable because of their similarity to rocket propellant systems and liquid mixtures previously used as bomb fillings [6].

See Cycloalkenes, above

Metal acetylides,
Metal carbides
Mellor, 1946, Vol. 5, 849
Caesium acetylide ignites at 100°C in the gas.

See TUNGSTEN CARBIDE, CW: Nitrogen oxides
DITUNGSTEN CARBIDE, CW_2: Oxidants

Metals
Mellor, 1940, Vol. 8, 544–545; 1942, Vol. 13, 342
Reduced iron, potassium and pyrophoric manganese all ignite in the gas at ambient temperature. Magnesium filings burn vigorously when heated in the gas.

See ALUMINIUM, Al: Oxidants

Nitrobenzene
Urbanski, 1967, Vol. 3, 288
Mixtures were formerly used as liquid high explosives, with addition of carbon disulphide to reduce the freezing point. High sensitivity to mechanical stimuli was disadvantageous.

Nitrogen trichloride

See NITROGEN TRICHLORIDE, Cl_3N: Initiators

Organic compounds
Riebsomer, J. L., *Chem. Rev.*, 1945, **36**, 158
In a review of the interaction of the tetraoxide with organic compounds, attention is drawn to the possibility of formation of unstable or explosive products.

Ozone
See OZONE, O_3: Dinitrogen tetraoxide

Phospham
See PHOSPHAM, HN_2P: Oxidants

Phosphorus
See PHOSPHORUS, P: Non-metal oxides

Sodium amide
Beck, G., *Z. Anorg. Chem.*, 1937, **233**, 158
Interaction with the oxide in carbon tetrachloride is vigorous, producing sparks.

Triethylamine
Davenport, D. A. *et al.*, *J. Amer. Chem. Soc.*, 1953, **75**, 4175
The complex, containing excess oxide over amine, exploded at below 0°C when free of solvent.
See Triethylammonium nitrate, below

Triethylammonium nitrate
Addison, C. C. *et al.*, *Chem. & Ind.*, 1953, 1315
The two ingredients form an addition complex with diethyl ether, which exploded violently after partial desiccation: an ether-free complex is also unstable.
See Triethylamine, above
See other NON-METAL OXIDES
OXIDANTS

DINITROGEN PENTAOXIDE N_2O_5

Dichloromethane
See NITRATING AGENTS

Disodium acetylide
See DISODIUM ACETYLIDE, C_2Na_2: Oxidants

Metals
Mellor, 1940, Vol. 8, 554
Potassium and sodium burn brilliantly in the gas, while mercury and arsenic are vigorously oxidised.

Naphthalene
Mellor, 1940, Vol. 8, 554
Naphthalene explodes and other organic materials react vigorously with the pentaoxide.

Sulphur dichloride,
or Sulphuryl chloride
Schmeisser, M., *Angew. Chem.*, 1955, **67**, 495, 499
Interaction is explosively violent.
See other NON-METAL OXIDES
OXIDANTS

LEAD(II) NITRATE N_2O_6Pb

Potassium acetate
Mee, A. J., *School Sci. Rev.*, 1940, **22**(86), 95
A heated mixture exploded violently.
See other METAL OXONON-METALLATES

ZINC NITRATE N_2O_6Zn

Carbon
Mellor, 1940, Vol. 4, 655
When the nitrate is sprinkled on to hot carbon, an explosion occurs.
See other METAL OXONON-METALLATES

TIN(II) NITRATE OXIDE $N_2O_7Sn_2$

Alone,
or Organic dust
1. Mellor, 1941, Vol. 7, 481
2. Anon., *J. Soc. Chem. Ind.*, 1922, **41**, 424R

The basic nitrate burns brilliantly when dry, and explodes on shock, friction or heating to above 100°C [1]. It was also involved, either as oxidant or initiator, in a flour mill explosion [2].
See related METAL OXONON-METALLATES

URANYL NITRATE N_2O_8U

Cellulose
Clinton, T. G., *J. R. Inst. Chem.*, 1958, **82**, 633
The analytical use of cellulose to absorb uranyl nitrate solution prior to ignition has caused explosions during ignition, due to formation of cellulose nitrate. An alternative method is described.
See related METAL OXONON-METALLATES

TRILEAD DINITRIDE N_2Pb_3

Fischer, F. *et al., Ber.*, 1910, **43**, 1470
Very unstable, it decomposes explosively during vacuum degassing.
See other N-METAL DERIVATIVES

DISULPHUR DINITRIDE N_2S_2

Brauer, 1963, Vol. 1, 410
Explosive, initiated by shock, friction, pressure, or temperatures above 30°C.
See other NON-METAL SULPHIDES

TETRASULPHUR DINITRIDE N_2S_4

Brauer, 1963, Vol. 1, 408
Explosively decomposes to its elements at 100°C.
See other NON-METAL SULPHIDES

SODIUM AZIDE N_3Na

1. Mellor, 1940, Vol. 8, 345
2. Lambert, B., *School Sci. Rev.*, 1927, 8(31), 218

Insensitive to impact, it decomposes, sometimes explosively, above the m.p. [1], particularly if heated rapidly [2].

Barium carbonate
Henneburg, G. O. *et al., Can. J. Res.,* 1950, **28B**, 345
Interaction to form cyanide ion requires careful control of temperature at 630°C to prevent explosions.

Bromine
See BROMINE, Nr_2: Metal azides

Carbon disulphide
See CARBON DISULPHIDE, CS_2: Metal azides

Chromyl chloride
See CHROMYL CHLORIDE, Cl_2CrO_2: Sodium azide

Sulphuric acid
Ross, F. F., *Water & Waste Treatment,* 1964, **9**, 528; private comm., 1966
One of the reagents required for the determination of dissolved oxygen in polluted water is a solution of sodium azide in 50% sulphuric acid. It is important that the diluted acid should be quite cold before adding the azide, since hydrogen azide boils at 36°C and is explosive in the condensed liquid state.

Water
Anon., *Angew. Chem.,* 1952, **64**, 169
Addition of water to sodium azide which had been strongly heated caused a violent reaction. This was attributed to formation of metallic sodium or sodium nitride in the azide.
See other METAL AZIDES

SODIUM AZIDOSULPHATE N_3NaO_3S

Shozda, R. J. *et al., J. Org. Chem.,* 1967, **32**, 2876
It is a weak explosive with variable sensitivity to mechanical shock and heating.
See other ACYL AZIDES

VANADIUM TRINITRATE OXIDE $N_3O_{10}V$

1. Schmeisser, M., *Angew. Chem.*, 1955, **67**, 495
2. Harris, A. D. *et al.*, *Inorg. Synth.*, 1967, **9**, 87

Many hydrocarbons and organic solvents ignite on contact with this powerful oxidant and nitrating agent [1], which reacts like fuming nitric acid with paper, rubber or wood [2].

See related METAL OXONON-METALLATES
See other OXIDANTS

THALLIUM(I) AZIDE N_3Tl

Mellor, 1940, Vol. 8, 352

A relatively stable azide, it can be exploded on fairly heavy impact, or by heating at 350–400°C.

See other METAL AZIDES

NITROSYL AZIDE N_4O

Lucien, H. W. *J. Amer. Chem. Soc.*, 1958, **80**, 4458

Explosions were experienced on several occasions during preparation of nitrosyl azide by various methods. It decomposes even at –50°C.

See other NON-METAL AZIDES

THIOTRITHIAZYL NITRATE $N_4O_3S_4$

Goehring, 1957, 74

Explodes on friction or impact.

PLUTONIUM(IV) NITRATE N_4O_8Pu

MCA Case History No. 1498

Polythene bottles are not suitable for long-term storage of plutonium nitrate solution as stress cracks appeared in bases of several 10 litre bottles during 6 months' storage. Short-term storage and improved venting are recommended.

See other METAL OXONON-METALLATES

TETRASULPHUR TETRANITRIDE N_4S_4

1. Mellor, 1940, Vol. 8, 625; Brauer, 1963, Vol. 1, 406
2. Villema-Blanco, M. *et al., Inorg. Synth.,* 1967, **9**, 101

The endothermic nitride is susceptible to explosive decomposition on friction, shock or heating above 100°C. Explosion is violent if initiated by a detonator [1]. Sensitivity towards heat and shock increases with purity. Preparative precautions have been detailed [2].
See other NON-METAL SULPHIDES

TETRASELENIUM TETRANITRIDE N_4Se_4

Alone,
or Halogens and derivatives
Mellor, 1947, Vol. 10, 789
The dry material explodes on slight compression, or on heating at 130–230°C. Contact with bromine, chorine or a little fuming hydrochloric acid also causes explosion.
See related N-METAL DERIVATIVES

TRITELLURIUM TETRANITRIDE N_4Te_3

Fischer, F. *et al., Ber.,* 1910, **43**, 1472
Two forms are described, one black which explodes on impact, and one yellow which explodes at 200°C.
See related N-METAL DERIVATIVES

TRITHORIUM TETRANITRIDE N_4Th_3

Air,
or Oxygen
Mellor, 1940, Vol. 8, 122
It burns incandescently on heating in air, and very vividly in oxygen.
See other N-METAL DERIVATIVES

SULPHURYL DIAZIDE N_6O_2S

Curtius, T. *et al., Ber.*, 1922, **55**, 1571
It explodes violently when heated and often spontaneously at ambient temperature.
See other ACYL AZIDES

DISULPHURYL DIAZIDE $N_6O_5S_2$

Alone,
or Alkali
Lehmann, H. A. *et al., Z. Anorg. Chem.*, 1957, **293**, 314
The azide decomposes explosively below 80°C and should only be stored in 1 g quantities. In contact with dilute alkali at 0°C an explosive deposit is produced.
See other ACYL AZIDES

LEAD(II) AZIDE N_6Pb

1. Mellor, 1940, Vol. 8, 353; 1967, Vol. 8, Suppl. 2, 21, 50
2. Taylor, G. W. C. *et al., J. Crystal Growth,* 1968, **3**, 391
3. Barton, A. F. M. *et al., Chem. Rev.,* 1973, **73**, 138

As a widely used detonator, its properties have been studied in great detail. Although quantitatively inferior to mercury fulminate as a detonator, it has proved to be more reliable in service. The pure compound occurs in two crystalline forms, one of which appears to be much more sensitive to initiation [1]. Factors which suppressed spontaneous explosions of lead azide during crystallisation were vigorous agitation and use of hydrophilic colloids [2]. These aspects have been recently reviewed [3].

Calcium stearate
MCA Case History No. 949
An explosion occurred during blending and screening operations on a mixture of lead azide and 0.5% of calcium stearate. If free stearic acid were present as impurity in the calcium salt, free hydrogen azide may have been involved.

Copper,
or Zinc
Federoff, 1960, A532, 551
Lead azide, on prolonged contact with copper, zinc or their alloys, forms traces of the extremely sensitive copper or zinc azides which may initiate explosion of the whole mass of azide.
See other METAL AZIDES

STRONTIUM AZIDE N_6Sr

See BARIUM AZIDE, BaN_6
CALCIUM AZIDE, CaN_6

PHOSPHORUS TRIAZIDE N_9P

Lines, E. L. *et al., Inorg. Chem.,* 1972, **11**, 2270
It is highly explosive.
See other NON-METAL AZIDES

LEAD(IV) AZIDE $N_{12}Pb$

Mellor, 1967, Vol. 8, Suppl. 2, 22
The crystalline product appears less stable than the diazide, spontaneously decomposing, sometimes explosively.
See other METAL AZIDES

SILICON TETRAAZIDE $N_{12}Si$

1. Mellor, 1967, Vol. 8, Suppl. 2, 20
2. Anon., *Angew. Chem. (Nachr.),* 1970, **18**, 27

Spontaneous explosions have been observed [1]. A residue containing the tetraazide, the chlorotriazide and probably the dichlorodiazide, exploded on standing for 2 or 3 days, possibly owing to hydrazoic acid produced by hydrolysis.
See other NON-METAL AZIDES

1,1,3,3,5,5-HEXAAZIDO-2,4,6-TRIAZA-1,3,5-TRIPHOSPHORINE

$N_{21}P_3$

Grundmann, C., *Z. Naturforsch,* 1955, **106**, 116
This viscous liquid (containing 73.5% N) is violently explosive when subjected to shock or friction.
See other NON-METAL AZIDES

SODIUM

Na

1. Hawkes, A. S. *et al., J. Chem. Educ*., 1953, **30**, 467
2. *MCA SD-47,* 1952
3. *Alkali Metals,* 1957

In a discussion of safe methods for laboratory use of sodium, disposal of small quantities (up to 5–10 g) by immersion in isopropanol, which may contain up to 2% of water to increase the rate of reaction, is recommended. Quantities up to 50 g should be burned in a heavy metal dish, using a gas flame [1]. Handling techniques and safety precautions for large-scale operations have also been detailed for this reactive metal [2,3].

Acids
Mellor, 1941, Vol. 2, 469–70
Anhydrous hydrogen chloride, hydrogen fluoride or sulphuric acids react slowly with sodium, while the aqueous solutions react explosively. Nitric acid of density above 1.056 causes ignition of sodium.

Air
Muir, G. D., private comm., 1968
Dispersions of sodium in volatile solvents become pyrophoric if the solvent evaporates round the neck of a flask or bottle. Serum cap closures are safer.

Calcium,
Mixed oxides
See CALCIUM, Ca: Sodium, etc.

Chloroform,
Methanol
See CHLOROFORM, $CHCl_3$: Sodium Methanol

Diazomethane
See DIAZOMETHANE, CH_2N_2: Alkali metals

Diethyl ether
Hey, P., private comm., 1965
While sodium wire was being pressed into ether, the jet blocked. Increasing the pressure to free it caused ignition of the ejected sodium and explosion of the flask of ether. Pressing the sodium into xylene or toluene and subsequent transfer to the ether is recommended.

N,N-Dimethylformamide
'DMF Brochure', Billingham, ICI, 1965
A vigorous reaction occurs on heating DMF with sodium metal.

Fluorinated compounds
Herring, D. E., private comm., 1964
Fusion of fluorinated compounds with sodium for qualitative analysis requires a high temperature for reaction because of their unreactivity. When reaction does occur, there is often an explosion and suitable precautions must be taken.

Halocarbons
1. Staudinger, H., *Z. Elektrochem,* 1925, **31**, 549
2. Lenze, F. *et al., Chem. Ztg.,* 1932, **56**, 921
3. Ward, E. R., *Proc. Chem. Soc.,* 1963, 15
4. *MCA SD-34,* 1949
5. Rayner, P. N. G., private comm., 1968
6. Anon., *Angew. Chem. (Nachr.),* 1970, **18**, 397

Although apparently stable standing in contact, mixtures of sodium with a range of halogenated alkanes (solvents) are metastable and capable of explosion by shock or impact. Carbon tetrachloride [1–3], chloroform, dichloromethane and chloromethane [1,2], tetrachloroethane [1,2,4] have been investigated, among others.

Generally the sensitivity to initiation and the force of the explosion increase with the degree of halogen substitution; the two former are less than in the corresponding systems with potassium or potassium–sodium alloy. Any aliphatic halocarbon (except fully fluorinated alkanes) may be expected to behave in this way. Small portions of sodium and hexachlorocyclopentadiene were mixed in preparation for a sodium fusion test. On shaking a few minutes later, the tube exploded [5]. On the small scale, no reaction occurred in boiling perfluorohexyl iodide at 114°C in contact with metallic sodium. With 140 g of iodide and 7 g of sodium an explosion occurred after 30 min [6].

See Chloroform, above
See IODOMETHANE, CH_3I: Sodium
See other METALS: Halocarbons

Halogen azides
See HALOGEN AZIDES

Halogens,
or Interhalogens

1. Mellor, 1961, Vol. 2, Suppl. 2. 1, 450–452
2. Mellor, 1941, Vol. 2, 114, 469
3. Mellor, 1941, Vol. 2, 92, 469; Staudinger, H., *Z. Elektrochem.*, 1925, **31**, 549
4. Booth, H. S. *et al., Chem. Rev.*, 1947, **41**, 427

Sodium ignites in fluorine gas, but is inert in the liquefied gas [1]. Cold sodium ignites in moist chlorine [2] but may be distilled unchanged in the dry gas [1]. Sodium and liquid bromine appear to be unreactive on prolonged contact, but mixtures may be detonated violently by mechanical shock [3]. Finely divided sodium luminesces in bromine vapour [1]. Iodine bromide or chloride react slowly with sodium, but mixtures will explode under a hammer blow [1]. Interaction of iodine pentafluoride with solid sodium is initially vigorous, but soon slows with film-formation, while that with molten sodium is explosively violent [2]. Sodium reacts immediately and incandescently with iodine heptafluoride [4].

Hydrazine
Mellor, 1940, Vol. 8, 316
Anhydrous hydrazine and sodium react in ether to form sodium hydrazide, which explodes in contact with air. Hydrazine hydrate and sodium react very exothermically, generating hydrogen and ammonia.

Hydroxylamine

See HYDROXYLAMINE, H_3NO: Metals

Mercury

Brauer, 1965, Vol. 2, 1802

Interaction of sodium and mercury to form sodium amalgam is violently exothermic, and moderation of the reaction with an inert liquid or by adding mercury slowly to the sodium is necessary. Even so, temperatures of 400°C may be attained.

Metal halides

1. *Alkali Metals,* 1957, 129
2. Mellor, 1961, Vol. 2, Suppl. 2.1, 494–496
3. Staudinger, H., *Z. Elektrochem.*, 1925, **31**, 549
4. Cueilleron, J., *Bull. Soc. Chim. Fr.*, 1945, **12**, 88

Sodium dispersions will reduce many metal halides exothermically. Iron(III) chloride is reduced fairly smoothly at room temperature or below in presence of 1,2-dimethoxyethane. Nickel, cobalt, lead or cadmium chlorides require higher temperatures to initiate the reaction, the exotherm with cobalt increasing the temperature from 325 to 375°C and causing evaporation of most of the dispersing oil [1]. The finely powdered metals produced by reduction of halides of cadmium, chromium, cobalt, copper, iron, manganese, molybdenum, nickel, silver, tin or zinc with sodium, dispersed in ether or toluene, are all pyrophoric [2]. Interaction of sodium and vanadyl chloride at 180°C is violent [2]. Mixtures of sodium with metal halides are sensitive to mechanical shock [3]; the ensuing explosions have been classified [4]. Very violent explosions occurred with iron(III) chloride or bromide, iron(II) bromide or iodide, and cobalt(II) chloride or bromide. Strong explosions occurred with the various halides of aluminium, antimony, arsenic, bismuth, copper(II), mercury, silver and tin (including a mixture of tin(IV) iodide and sulphur), also vanadium pentachloride and ammonium tetrachlorocuprate. Ammonium, copper(I), cadmium and nickel halides generally gave weak explosions. Most of the alkali and alkaline earth halides were insensitive.

Metal oxides

1. Mellor, 1939, Vol. 9, 649
2. Mellor, 1943, Vol. 11, 237, 542

3. Mellor, 1941, Vol. 3, 138
4. Mellor, 1941, Vol. 7, 658, 401
5. Mellor, 1940, Vol. 4, 770
6. Brewer, L. *et al., Rep. UCRL 1864,* 7, Washington, USAEC, 1952
7. Mellor, 1961, Vol. 2, Suppl. 2.1, 474

Bismuth(III) oxide [1], chromium trioxide [2] and copper(II) oxide [3] are reduced with incandescence on heating with sodium. Finely divided sodium ignites on admixture with fine lead oxide without heating, while the coarse material reacts vigorously with molten sodium [4]. The latter reduces mercury(I) oxide [5] or molybdenum trioxide [2] with incandescence, the former producing a light explosion also. Sodium reduces sodium peroxide vigorously at 500°C [6] and tin(IV) oxide incandescently on gentle heating [4]. Finely dispersed sodium reduces metal oxides on heating at temperatures between 100 and 300°C, producing, e.g. pyrophoric iron, nickel and zinc.

Non-metal halides

1. Mellor, 1940, Vol. 8, 1033
2. Mellor, 1941, Vol. 2, 470; 1940; Vol. 8, 1016
3. Mellor, 1961, Vol. 2, Suppl. 2.1, 455, 460
4. Mellor, 1940, Vol. 8, 1073
5. Cueilleron, J., *Bull. Soc. Chim. Fr.*, 1945, **12**, 88
6. Mellor, 1947, Vol. 10, 912

Sodium floats virtually unchanged on phosphorus tribromide, but added drops of water cause a violent explosion [1]. Molten sodium explodes with phosphorus trichloride, and may ignite or explode with the pentachloride [2]. Diselenium dichloride reacts vigorously with sodium on heating, emitting light and heat [3]. Sodium ignites in sulphinyl chloride vapour at 300°C [3], or in a stream of thiophosphoryl fluoride [4]. The shock-sensitive mixtures of sodium with phosphorus pentachloride, phosphorus tribromide or sulphur dichloride gave very violent explosions on impact, while those with boron tribromide or sulphur dibromide gave strong explosions [5]. Seleninyl bromide reacts explosively [6].

Non-metal oxides

1. Mellor, 1940, Vol. 6, 70; 1961, Vol. 2, Suppl. 2.1, 468
2. Gilbert, H. N., *Chem. Eng. News,* 1948, **26**, 2604
3. Mellor, 1940, Vol. 8, 554, 945; 1961, Vol. 2, Suppl. 2.1, 458

Sodium and carbon dioxide are normally unreactive till red heat is attained [1], but mixtures of the two solids are impact-sensitive and explode violently [2]. Carbon dioxide is unsuitable as an extinguishant for the burning metal alone, as the intensity of combustion is increased by replacing air with carbon dioxide. However, it has been used successfully to extinguish solvent fires where sodium is also present, since it often fails to ignite because of the blanketing effect of solvent vapour. Conversely, addition of kerosene to burning sodium enables the whole to be extinguished with carbon dioxide [1]. Carbon monoxide reacts with sodium in liquid ammonia (but not otherwise) to form sodium carbonyl, which explodes on heating to 90°C [2]. Finely divided silica (sand) will often react with burning sodium, so is not entirely suitable as an extinguishant [1,2]. (Sodium carbonate and chloride are suitably unreactive for this purpose.) Solid sodium is inert to dry liquid or gaseous sulphur dioxide, but molten sodium reacts violently [2]. The moist gas reacts with sodium almost as vigorously as water [1]. Phosphorus pentaoxide becomes incandescent on warming with sodium, which also ignites in dinitrogen pentaoxide [3].

Non-metals

1. Mellor, 1961, Vol. 2, Suppl. 2.1, 466–467
2. Ibid., 454–455; Mellor, 1941, Vol. 2, 469
3. Mellor, 1947, Vol. 10, 766

Explosions have occasionally occurred when carbon powder is in contact with evaporating sodium and air [1]. The violent interaction of ground or heated mixtures of sodium and sulphur may be moderated by the presence of sodium chloride or boiling toluene [2]. Selenium reacts incandescently with sodium when heated [3] and molten tellurium reacts vigorously when poured on to solid sodium [2].

Oxidants

See Halogens or Interhalogens, above
Metal halides, above
Metal oxides, above
Non-metal oxides, above
NITROSYL FLUORIDE, FNO: Sodium
NITRYL FLUORIDE, FNO_2: Metals

AMMONIUM NITRATE, $H_4N_2O_3$: Metals
SODIUM NITRITE, $NNaO_2$: Reducants
SODIUM NITRATE, $NNaO_3$: Sodium

2,2,3,3-Tetrafluoropropanol
See 2,2,3,3-TETRAFLUOROPROPANOL, $C_3H_4F_4O$: Potassium hydroxide, etc.

Water
1. Mellor, 1941, Vol. 2, 469
2. Mellor, 1961, Vol. 2, Suppl. 2.1, 362
3. *MCA Case History No. 456*

Small pieces of sodium in contact with water react vigorously but do not usually ignite the evolved hydrogen unless the water is at above 40°C, or if rapid dissipation of heat is prevented by immobilising the sodium by use of a viscous solution (starch paste) or wet filter paper [1]. In contact with ice, sodium explodes violently [2]. Small, hot particles of sodium (remaining from dissolution of larger pieces) may finally explode as do large lumps of the metal in water [1]. Sodium residues from a Wurtz reaction were treated with alcohol to destroy sodium, but accidental contact with water caused a fire, showing that the alcohol treatment was incomplete. This may have been due to a crust of alcohol-insoluble halide over the residual sodium [3].
See ALKALI METALS
See other METALS

SODIUM–ANTIMONY ALLOY Na–Sb

See NITROSYL TRIBROMIDE, Br_3NO: Sodium–antimony alloy

SODIUM SILICIDE NaSi

Mellor, 1961, Vol. 2, Suppl. 2.1, 564
Ignites in air.
See other METAL NON-METALLIDES

SODIUM OXIDE Na_2O

Phosphorus(V) oxide
See PHOSPHORUS(V) OXIDE, O_5P_2: Inorganic bases

Water
Interaction is likely to be violently exothermic.

SODIUM PEROXIDE Na_2O_2

Allen, C. F. H. *et al., J. Chem. Educ.,* 1942, **19**, 72
Hazards attendant upon use of this powerful oxidant may in many cases be removed by substitution of sodium perborate.

Acetic acid
von Schwartz, 1918, 321
Admixture causes explosion. This may be due to direct oxidation of the acetic acid, or may involve peracetic acid or its sodium salt.

Aluminium,
Aluminium chloride
Anon., *Chem. Eng. News,* 1954, **32**, 258
A mixture of the three slowly reacted, creating a pressure of 122 bar in 41 days, and the residue reacted spontaneously on exposure to air.
See Metals, below

Ammonium peroxodisulphate
Mellor, 1947, Vol. 10, 464
A mixture explodes on being subjected to friction in a mortar, heating above 75°C, or exposure to carbon dioxide or drops of water.

Boron nitride
Mellor, 1940, Vol. 8, 111
Addition of powdered nitride to the molten peroxide causes incandescence.

Calcium acetylide
Mellor, 1941, Vol. 2, 490
A mixture is explosive.

Fibrous materials
von Schwartz, 1918, 328
Kirk-Othmer, 1967, Vol. 14, 749
Contact of the solid with moist cloth, paper or wood often causes ignition.

Hydrogen sulphide
1. Barrs, C. E., *J. R. Inst. Chem.*, 1955, **79**, 43
2. Mellor, 1947, Vol. 10, 129
Solid sodium peroxide causes immediate ignition in contact with gaseous hydrogen sulphide [1]. Barium peroxide and other peroxides behave similarly [2].

Hydroxy compounds
Castrantas, 1965, 4
The exothermic oxidation of ethanol, glycerol, sugar or acetic acid may lead to fire or explosion.
See Water, below

Metals,
Carbon dioxide
Mellor, 1941, Vol. 2, 490; 1961, Vol. 2, Suppl. 2.1, 633; 1946, Vol. 5, 217
Intimate mixtures with aluminium, magnesium or tin powders ignite on exposure to moist air and become incandescent on heating or on moistening with water (magnesium may explode). Exposure of such mixtures to carbon dioxide causes an explosion. (Interaction of the peroxide and carbon dioxide is highly exothermic). Sodium is oxidised vigorously at 500°C.

Non-metal halides
1. Mellor, 1947, Vol. 10, 897; 1961, Vol. 2, Suppl. 2.1, 634
2. Comanducci, E., *Chem. Ztg.*, 1911, **15**, 706
Violent interactions occur with diselenium or disulphur dichlorides, the latter emitting light and heat [1]. The very exothermic reaction with phosphorus trichloride may accelerate to explosion [2].

Non-metals
1. von Schwartz, 1918, 328
2. Mellor, 1941, Vol. 2, 490
3. *ABCM Quart. Safety Summ.*, 1948, **19**, 13
Intimate mixtures with carbon or phosphorus ignite or explode [1]. Other readily oxidisable materials (probably antimony, arsenic, boron sulphur, selenium) also form explosive mixtures [2]. Use of finely powdered carbon, rather than the granular carbon specified for a reagent, mixed with sodium peroxide caused an explosion [3].

Organic liquids,
Water
von Schwartz, 1918, 328
Simultaneous contact of sodium peroxide with water and aniline, or benzene, or diethyl ether, or glycerol, causes ignition. (Equivalent to contact with concentrated hydrogen peroxide.)

Peroxyformic acid
See PEROXYFORMIC ACID, CH_2O_3: Metals, etc.

Silver chloride,
Charcoal
Mellor, 1941, Vol. 3, 401
An intimate mixture ignites after a short delay, but the same is probably true if silver chloride is omitted.
See Non-metals, above

Water
1. *Haz. Chem. Data,* 1969, 201
2. Friend, J. N. *et al.*, *Nature*, 1934, **134**, 778
3. Cheeseman, G. H. *et al.*, ibid., 971
Reaction with water is vigorous, and with large quantities of peroxide it may be explosive. Contact of the peroxide with combustible materials and traces of water may cause ignition [1]. Violent explosions on two occasions during attempted preparation of oxygen were attributed to presence of sodium metal in the peroxide. The former would liberate hydrogen and ignite the detonable mixture [2,3];

SODIUM THIOSULPHATE $Na_2O_3S_2$

Metal nitrates

See POTASSIUM NITRATE, KNO_2: Sodium thiosulphate
SODIUM NITRATE, $NNaO_3$: Sodium thiosulphate

Sodium nitrite

Stevens, H. P., *J. Proc. R. Inst. Chem.*, 1946, 285

A mixture of these reactants will explode violently when most of the water of crystallisation has been driven off by heating.

See other METAL OXONON-METALLATES
REDUCANTS

SODIUM METASILICATE Na_2O_3Si

Fluorine

See FLUORINE, F_2: Metal salts

SODIUM DITHIONITE ('HYDROSULPHITE') $Na_2O_4S_2$

MCA Case History No. 350

A batch decomposed violently during drying in a graining bowl. No explanation was offered but contamination with water and/or an oxidant seems likely.

Sodium chlorite

See SODIUM CHLORITE, $ClNaO_2$: Sodium dithionite

Water

Anon., *Chem. Trade J.*, 1939, **104**, 355

Addition of 10% of water to the solid anhydrous material caused a vigorous exotherm and spontaneous ignition. Bulk material may decompose at 135°C.

See METAL OXONON-METALLATES
REDUCANTS

SODIUM DISULPHITE ('METABISULPHITE') $Na_2O_5S_2$

Sodium nitrite
See SODIUM NITRITE, $NNaO_2$: Sodium disulphite

SODIUM TETRAPEROXOTUNGSTATE(2–) $Na_2O_8W_2$

Mellor, 1943, Vol. 11, 835
Explodes feebly on warming.
See other PEROXOACID SALTS

SODIUM SULPHIDE Na_2S

ABCM Quart. Safety Summ., 1942, **13**, 5
Fused sodium sulphide in small lumps is liable to spontaneous heating, temperatures of up to 120°C being observed after exposure to moisture and oxygen. Packing in hermetically closed containers is essential.
See also SULPHUR BLACK

Carbon
Creevey, J., *Chem. Age*, 1941, **44**, 257
Mixtures of sodium sulphide and finely divided carbon react exothermically, probably owing to co-promotion of aerial oxidation.

Diazonium salts
See DIAZONIUM SULPHIDES

N, N-Dichloromethylamine
See *N,N*-DICHLOROMETHYLAMINE, CH_3Cl_2N
See other METAL SULPHIDES

SODIUM DISULPHIDE Na_2S_2

Diazonium salts
See DIAZONIUM SULPHIDES

SODIUM POLYSULPHIDE NaS_x

Diazonium salts

See DIAZONIUM SULPHIDES

See also SULPHUR BLACK

SODIUM PHOSPHIDE(3–) Na_3P

Water

Mellor, 1940, Vol. 8, 834

Sodium (and potassium) phosphide is decomposed by moist air or water, evolving phosphine, which often ignites.

See other METAL NON-METALLIDES

SODIUM PYROPHOSPHATE HYDROGEN PEROXIDATE $Na_4O_7P_2{\cdot}2H_2O_2$

See CRYSTALLINE HYDROGEN PEROXIDATES

NIOBIUM Nb

Halogens,

or Interhalogens

Mellor, 1939, Vol. 9, 849; 1956, Vol. 2, Suppl. 1, 165

Niobium ignites in cold fluorine, and in chlorine at 205°C, and incandesces in contact with bromine trifluoride.

See other METALS

NIOBIUM(V) OXIDE Nb_2O_5

Lithium

See LITHIUM, Li: Metal oxides

NEODYMIUM Nd

Phosphorus
See PHOSPHORUS, P: Metals

NICKEL Ni

1. Sasse, W. H. F., *Org. Synth.*, 1966, **46**, 5
2. Whaley, T. P., *Inorg. Synth.*, 1957, **5**, 197

Raney nickel catalysts must not be degassed by heating under vacuum, as large amounts of heat and hydrogen may be evolved suddenly and dangerous explosions may be caused [1]. Nickel powder prepared by several alternative methods may be pyrophoric if particles are fine enough [2].

Aluminium
See ALUMINIUM, Al: Nickel

Aluminium trichloride,
Ethylene
See ETHYLENE, C_2H_4: Aluminium trichloride

p-Dioxan
See *p*-DIOXAN, $C_4H_8O_2$: Nickel

Hydrogen
Adkins, H. *et al., J. Amer. Chem. Soc.*, 1948, **70**, 695; *Org. Synth.*, 1955, Coll. Vol. 3, 176
During hydrogenation of an unspecified substrate (possibly *p*-nitrotoluene) at high pressure with the highly active W6 type of Raney nickel catalyst at 150°C, a sudden exotherm caused the initial pressure to rapidly double to 680 bar. This does not happen at 100°C or below. Care is necessary with selection of reaction conditions for highly active catalysts.

Methanol
MCA Case History No. 1225
Ignition occurred when methanol was poured through the open

manhole of a 230 litre reactor containing Raney nickel catalyst. Though the reactor had been thoroughly purged with nitrogen before opening, it was later shown that air was entrained during pouring operations, and this would cause nickel particles round the man-hole to glow. A closed system was recommended.

Non-metals

Mellor, 1942, Vol. 15, 148, 151

On heating, mixtures of powdered nickel with sulphur or selenium react incandescently.

See Sulphur compounds, below

Oxidants

See BROMINE PENTAFLUORIDE, BrF_5 : Acids, etc.
PEROXYFORMIC ACID, CH_2O_3 : Metals
POTASSIUM PERCHLORATE, $ClKO_3$: Metal powders
CHLORINE, Cl_2 : Metals
NITRYL FLUORIDE, FNO_2 : Metals
AMMONIUM NITRATE, $H_4N_2O_3$: Metals

Sulphur compounds

Kornfeld, E. C., *J. Org. Chem.*, 1951, **16**, 137

Raney nickel catalyst, containing appreciable amounts of the sulphide (after use to desulphurise thioamides) is rather pyrophoric.

See HYDROGENATION CATALYSTS
PYROPHORIC CATALYSTS

See other PYROPHORIC METALS

NICKEL(II) OXIDE **NiO**

Fluorine

See FLUORINE, F_2: Metal oxides

Hydrogen peroxide

See HYDROGEN PEROXIDE, H_2O_2: Metals, etc.

Hydrogen sulphide

See HYDROGEN SULPHIDE, H_2S: Metal oxides

NICKEL(IV) OXIDE NiO_2

Fluorine
See FLUORINE, F_2: Metal oxides

DINICKEL TRIOXIDE Ni_2O_3

Hydrogen peroxide
See HYDROGEN PEROXIDE, H_2O_2: Metals, etc.

Nitroalkanes
See NITROALKANES: Metal oxides

LEAD(II) OXIDE OPb

Chlorinated rubber
See CHLORINATED RUBBER: Metal oxides

Chlorine,
Ethylene
See ETHYLENE, C_2H_4: Chlorine

Fluorine,
Glycerol
See FLUORINE, F_2: Miscellaneous materials

Glycerol,
Perchloric acid
See PERCHLORIC ACID, $ClHO_4$: Glycerol, etc.

Hydrogen trisulphide
See HYDROGEN TRISULPHIDE, H_2S_3: Metal oxides

Metal acetylides
Mellor, 1946, Vol. 5, 849

Interaction at 200°C with dirubidium acetylide is explosive, and with dilithium acetylide, incandescent.

Metals
Mellor, 1946, Vol. 5, 217; 1941, Vol. 7, 116, 656–658
Mixtures of the oxide with aluminium powder give a violent or explosive 'thermite' reaction on heating. Finely divided sodium ignites on admixture with the oxide, and a mixture of the latter with zirconium explodes on heating.

Non-metals
Mellor, 1941, Vol. 7, 657
A mixture with boron incandesces on heating, and with silicon the reaction is vigorous. If aluminium is present, the mixture explodes on heating (but the same is true if silicon is absent).

Peroxyformic acid
See PEROXYFORMIC ACID, CH_2O_3 : Metals, etc.

Seleninyl chloride
See SELENINYL CHLORIDE, Cl_2OSe: Metal oxides
See other METAL OXIDES
OXIDANTS

PALLADIUM(II) OXIDE **OPd**

Hydrogen
Sidgwick, 1950, 1558
It is a strong oxidant and glows in contact with hydrogen at ambient temperature.

THORIUM OXIDE SULPHIDE **OSTh**

Mellor, 1941, Vol. 7, 240
Ignites in contact with air.
See related METAL SULPHIDES

SILICON OXIDE OSi

Zintl, E. *et al., Z. Anorg. Chem.*, 1940, **245**, 1
The freshly prepared material is pyrophoric in air.
See other NON-METAL OXIDES

TIN(II) OXIDE OSn

Non-metal oxides
Mellor, 1941, Vol. 7, 388
The oxide ignites in nitrous oxide at 400°C, and incandesces when heated in sulphur dioxide.
See other METAL OXIDES

ZINC OXIDE OZn

Chlorinated rubber
See CHLORINATED RUBBER: Metal oxides

Linseed oil
Anon., *Chem. Trade J.*, 1933, **92**, 278
Slow addition of zinc white (a voluminous oxide containing much air) to cover the surface of linseed oil varnish caused generation of heat and ignition. Lithopone, a denser oxide, did not cause heating.

Magnesium
See MAGNESIUM, Mg: Metal oxides
See other METAL OXIDES

OXYGEN (Gas) O_2

Acetaldehyde
See Acetaldehyde, C_2H_4O: Air, or oxygen

Alcohols
1. Davies, 1961, **80**
2. Redemann, E. G., *J. Amer. Chem. Soc.*, 1942, **64**, 3049

Secondary alcohols are readily autoxidised in contact with oxygen or air, forming ketones and hydrogen peroxide [1]. A partly full bottle of 2-propanol exposed to sunlight for a long period became 0.36M in peroxide and potentially explosive [2].

See HYDROGEN PEROXIDE, H_2O_2: Acetone
: Alcohols

Alkali metals
Mellor, 1941, Vol. 2, 469
Sidgwick, 1950, 65
Reactivity towards air or oxygen increases from lithium to caesium and intensity depends on state of subdivision and on presence or absence of moisture. Lithium normally ignites in air above its melting point, while potassium may ignite after exposure to the atmosphere, unless it is unusually dry. Rubidium and caesium ignite immediately on exposure. It is reported that sodium and potassium may be distilled unchanged under perfectly dried oxygen.

Alkaline earth metals
Mellor, 1941, Vol. 3, 637–638
Finely divided calcium may ignite in air, and the massive metal ignites on heating in air, and burns vigorously at 300°C in oxygen. Strontium and barium behave similarly.

Aluminium–titanium alloys
See ALUMINIUM–TITANIUM ALLOYS, Al–Ti: Oxidants

Biological material,
Ether?
Napier, D. H., private comm., 1972
Biological material in a polythene bag filled with oxygen and being prepared for analytical combustion exploded. Diethyl ether used previously to anaesthetise the experimental animal from which the sample was derived may have still been present and initiation from static on the plastic bag could have been involved.

Carbon disulphide,
Mercury,
Anthracene

ABCM Quart. Safety Summ., 1953, **25**, 2
Shortly after mercury was accidentally introduced into a system containing a solution of anthracene in carbon disulphide under an atmosphere of oxygen, an explosion occurred. Presence of mercury may have catalysed rapid oxidation of carbon disulphide.

Diboron tetrafluoride
See DIBORON TETRAFLUORIDE, B_2F_4 : Oxygen

Ethers
Davies, 1961, 79
Many ethers, either of open-chain (diethyl or diisopropyl ether) or cyclic type (tetrahydrofuran, dioxan), are readily autoxidised on exposure to air in presence of light. The hydroperoxides formed are less volatile than the parent ether and may be concentrated to a dangerous extent if distillation of peroxidised material is attempted.
See AUTOXIDATION
PEROXIDES IN SOLVENTS

Fibrous fabrics
Most fibrous fabrics will sorb oxygen when exposed to concentrations greater than the normal 21%, and this is retained for a long time after excess oxygen is no longer present, greatly increasing the possibility of ignition. Many industrial fire accidents have been caused by this.
See OXYGEN ENRICHMENT

Halocarbons
1. Fawcett, H. H., *Chem. Eng. News,* 1957, **35**(43), 60
2. Weber, U., *Chim. Ind.*, 1963, **90**, 178
3. Bashford, L. A. *et al. J. Chem. Soc.*, 1938, 1964

1,1,1-Trichloroethane exploded after heating under oxygen at 54 bar and 100°C for 3 h [1]. Trichloroethylene, remaining in a pipe after cleaning operations, exploded under 27 bar pressure of oxygen at ambient temperature. It was later found possible to explode stoicheiometric mixtures [2]. Chlorotrifluoroethylene and bromotrifluoroethylene each react explosively with oxygen at ambient temperature [3].
See Tetrafluoroethylene, below
OXYGEN (Liquid): Halocarbons,

Hydrocarbons

1. Davies, 1961, 11
2. Staudinger, H., *Z. Elektrochem.*, 1925, **31**, 549
3. Unpublished observations, 1968

Interaction may be slow or fast, but peroxidic products are always involved, and the rate of formation is highest at a C–H link adjacent to an aromatic ring or double bond [1]. Predictable conditions for explosion are those relevant to the internal combustion engine. Unpredicted conditions include the following cases. During studies on autoxidation of diphenylethylene with oxygen at high pressure and low temperature an explosion occurred [2]. Partial oxidation of a gasoline fraction in an autoclave under oxygen, initially at 22 bar and 100°C, ran wild and exploded; several smaller reactions had proceeded uneventfully [3].

See also TETRACARBONYLNICKEL, C_4NiO_4: Oxygen

Hydrogen

1. Bailey, D. J. C., *Chem. & Ind.*, 1954, 492
2. Cook, M. A. *et al.*, *Chem. Eng. News*, 1956, **34**, 4436
3. Weissveiler, A., *Z. Elektrochem.*, 1936, **42**, 499

When oxygen was passed through a drying tower containing activated alumina previously used to dry hydrogen, explosions occurred. Nitrogen purging between changing gases prevented this [1]. During preparations of stoicheiometric mixtures in a steel mixing tank (at 17–82 bar, with or without 15% argon), several spontaneous explosions occurred during valve manipulation at 30 min after mixing, but not 10 min after mixing. A possible catalytic effect of the steel tank was eliminated by coating it thinly with silicone grease [2]. There is a narrow range of concentrations in which the mixture is supersensitive to initiation [3].

Mercury,
Tetracarbonylnickel

See TETRACARBONYLNICKEL, C_4NiO_4: Mercury, etc.

Metal hydrides
Mellor, 1941, Vol. 2, 483

Mackay, 1966, 30, 67

Sodium hydride ignites in oxygen at 230°C, and finely divided uranium hydride ignites on contact. Lithium, sodium and potassium hydrides react slowly in dry air, while rubidium and caesium hydrides ignite. Reaction is accelerated in moist air, and even finely divided lithium hydride ignites then. Finely divided magnesium hydride, prepared by pyrolysis, ignites immediately in air.

See also LITHIUM HEXAHYDROALUMINATE, AlH_4Li
SODIUM HEXAHYDROALUMINATE, AlH_4Na
POTASSIUM HEXAHYDROALUMINATE(3–), AlH_6K_3
See other COMPLEX HYDRIDES

Non-metal hydrides

Mellor, 1946, Vol. 5, 36

Sidgwick, 1950, 344, 553

Mellor, 1940, Vol. 6, 220–225

The reported ignition of diborane and tetraborane(10) in contact with air or oxygen is due to the presence of traces of silicon hydrides. The lower members of the latter class ignite or explode in air or oxygen, especially at reduced pressure. Phosphine is also sensitive.

See PHOSPHINE, H_3P: Oxygen
ALUMINIUM TETRAHYDROBORATE, AlB_3H_{12}
PENTABORANE(11), B_5H_{11}
DECABORANE(14), $B_{10}H_{14}$

Organic analytical samples

Pack, D. E., *Chem. Eng. News,* 1966, **44**(2), 4

Devereaux, H. D., ibid., (6), 4

Heyssel, R. M. *et al.,* ibid., (10), 4

MacDonald, A. M. G., ibid., (19), 7

Hazards involved in use of the oxygen flask combustion technique are discussed and illustrated with seven examples. Some explosions occurred after completion of combustion, when the oxygen concentration in the flask is still *ca.* 75%. Shielding of the flask, minimally with wire gauze, seems advisable.

Phosphorus trioxide

See PHOSPHORUS(III) OXIDE, O_3P_2: Oxygen

Polymers
1. *MCA Case History No. 395*
2. *MCA Case History No. 1111*
A foam rubber sample, being tested for oxidation resistance under oxygen at 34 bar at 90°C, exploded with extreme violence after 4 days [1]. Use of Neoprene-lined hose in a high-pressure oxygen manifold caused failure and ignition of the burst hose. Possible ignition sources include adiabatic compressive heating, and friction from vibration of metal fibres in a high-velocity stream of oxygen [2].

Rhenium
Mellor, 1942, Vol. 12, 471
The metal ignites in oxygen at 300°C

Tetrafluoroethylene
1. Pajaczkowski, A., *Chem. & Ind.*, 1964, 659
2. Ger. Pat. 773 900, 1971
Accidental admixture of oxygen gas with unstabilised liquid tetrafluoroethylene produced a polymeric peroxide which was powerfully explosive, and sensitive to heat, impact or friction [1]. Removal of oxygen by treatment with pyrophoric copper to prevent explosion of tetrafluoroethylene has been claimed [2].
See other POLYPEROXIDES

Titanium
ABCM Quart. Safety Summ., 1962, **33**, 24
Passage of oxygen through a titanium feed pipe into a titanium autoclave caused a titanium–oxygen fire and explosion at 44 bar. When the surface oxide film is damaged, titanium can ignite at 24 bar under static conditions, and at 3.4 bar under dynamic conditions, with oxygen at ambient temperature.
See also ALUMINIUM–TITANIUM ALLOYS, Al–Ti: Oxidants

Tricalcium diphosphide
See TRICALCIUM DIPHOSPHIDE, Ca_3P_2 : Oxygen
See other NON-METALS
OXIDANTS

OXYGEN (Liquid) O_2

Although the section below refers to the specific hazards observed with a few organic or oxidisable materials with liquid oxygen, it is probable that most organic materials and inorganic reducing agents are highly hazardous with liquid oxygen under appropriate conditions of contact and initiation.

See CRYOGENIC FLUIDS

Acetone

Blau, K., private comm., 1965

Accidental addition of liquid oxygen to vacuum jars containing residual acetone caused a violent explosion. Liquid nitrogen is less hazardous as a trap coolant, but only under controlled conditions.

See NITROGEN (Liquid), N_2

Asphalt

Weber, J., *Chem. Ind.*, 1963, **90**, 178

Mechanical impact upon a road surface on to which liquid oxygen had leaked caused a violent explosion.

Carbon

Kirshenbaum, 1965, 4; Stettbacher, A., *Spreng- und Schiess-stoffe*, 71, Zurich, Racher, 1948

Mixtures of carbon and liquid oxygen have been used as detonable explosives for some time.

Halocarbons

Anon., *Chem. Eng. News*, 1965, **43**(24), 41

Mixture of liquid oxygen with dichloromethane, 1,1,1-trichloroethane, trichloroethylene and 'chlorinated dye penetrants 1 and 2' exploded violently when initiated with a blasting cap. Carbon tetrachloride exploded only mildly, and a partly fluorinated chloroalkane not at all. Trichloroethylene has been used for degreasing metallic parts before use with liquid oxygen, but is not safe.

See OXYGEN (Gas): Halocarbons,

Hydrocarbons

1. Mellor, 1941, **1**, 379
2. Kirshenbaum, 1956, 17, 22

3. *MCA Case History No. 865*

Mixtures of hydrocarbons with liquid oxygen are highly dangerous explosives, not always requiring external initiation. The earliest example must be that of Claudé, who accidentally knocked a lighted candle into a bucket of liquid oxygen in 1903 and suffered from the violent explosion ensuing [1]. Mixtures with liquid methane and benzene are specifically described as explosive [2]. Traces of oil in a liquid oxygen transfer pump caused the explosion described in the Case History [3]. Mixtures with petroleum and absorbent charcoal have been used experimentally as blasting explosives [1]. Addition of a little powdered aluminium to liquid methane–oxygen mixtures increases the explosive power [2].

Liquefied gases

Kirshenbaum, 1956, 17, 30, 34

Mixtures with liquefied carbon monoxide, cyanogen (solidified) and methane are highly explosive.

Lithium hydride

Kirshenbaum, 1956, 44

Mixtures of lithium hydride powder and liquid oxygen are detonable explosives of greater power than TNT.

Metals

1. Anon., *Chem. Eng. News*, 1957, **35**(24), 90
2. Ibid., (31), 10
3. Austin, C. M. *et al.*, *J. Chem. Educ.*, 1959, **36**, 54, 308
4. *MCA Case History No. 824*
5. Kirshenbaum, 1956, 4, 16
6. *MCA Case History No. 988*

A demonstration of combustion of aluminium powder in oxygen exploded violently, probably owing to presence of unevaporated liquid at ignition [1]. Stoicheiometric mixtures of the two are explosives much more powerful than TNT [2]. Further details and comments were given later [3]. An aluminium filter in a high-capacity liquid oxygen transfer line exploded violently, possibly owing to frictional or impact ignition of an aluminium component in contact with an abrasive particle [4]. Powdered magnesium, titanium and zirconium mixed with liquid oxygen are detonable [5]. A nitrogen-pressurised liquid oxygen dispenser made of titanium alloy exploded during nitrogen pressurising. The vessel failed because of reaction of the titanium alloy with liquid oxygen. Titanium is more reactive towards oxygen than either stainless steel or aluminium, and should therefore not be used for oxygen service, either gaseous or liquid [6].

See ALUMINIUM, Al: Oxidants
MAGNESIUM, Mg: Oxidants

1,3,5-Trioxane ('paraformaldehyde')
Kirshenbaum, 1956, 31
Mixtures with liquid oxygen are highly explosive.

Wood (+ charcoal)
1. Weber, U., *Chim. Ind.*, 1963, **90**, 178
2. Lang, A., *Chem. Eng. Progr.*, 1962, **58**(2), 71

Several major accidents during handling of tonnage quantities are described [1]. One involved a violent explosion caused by leakage of liquid oxygen on to metal-encased timber (which had previously been charred) and spark ignition from welding in the enriched atmosphere of an air separation plant [2].

See other OXIDANTS

OSMIUM(IV) OXIDE O_2Os

Sidgwick, 1950, 1493
The amorphous form, prepared by dehydration at low temperature, is a pyrophoric powder.

See other METAL OXIDES

LEAD(IV) OXIDE O_2Pb

Chlorine trifluoride
See CHLORINE TRIFLUORIDE, ClF_3 : Metals, etc.

Hydrogen sulphides
Mellor, 1941, Vol. 7, 689; 1947, Vol. 10, 129, 159
Contact of hydrogen sulphide with dry or moist lead dioxide causes

attainment of red heat and ignition. Contact of hydrogen trisulphide with the dioxide causes violent decomposition and ignition.

Metal acetylides,
Metal carbides
Mellor, 1946, Vol. 5, 850, 872
Interaction with aluminium carbide is incandescent, and with caesium acetylide at 350°C is explosive. Other related compounds may be expected to be oxidised violently.

Metals
Mellor, 1963, Vol. 2, Suppl. 2.2, 1571; 1940, Vol. 4, 272; 1946, Vol. 5, 217; 1941, Vol. 7, 658, 691
Warm potassium reacts explosively, and sodium probably behaves similarly. Magnesium reacts violently and powdered aluminium probably does also, as the monoxide reacts violently. Mixtures of powdered molybdenum or tungsten with the oxide incandesce on heating.

Metal sulphides
Mellor, 1941, Vol. 3, 745
Calcium, strontium and barium sulphides react vigorously on heating.

Nitroalkanes
See NITROALKANES: Metal oxides

Nitrogen compounds
Mellor, 1941, Vol. 7, 637; 1940, Vol. 8, 291
Hydroxylamine ignites in contact with the oxide, while phenylhydrazine immediately reacts vigorously.

Non-metals
Mellor, 1946, Vol. 5, 17; 1941, Vol. 7, 689–690
Boron or yellow phosphorus explodes violently on grinding with the oxide, while red phosphorus ignites. Mixtures with sulphur ignite on grinding or addition of sulphuric acid.

Peroxyformic acid
See PEROXYFORMIC ACID, CH_2O_3: Metals, etc.

Non-metal halides
Mellor, 1941, Vol. 7, 690; 1947, Vol. 10, 676, 909
Warm phosphorus trichloride reacts with incandescence, and seleninyl chloride vigorously. Sulphonyl dichloride may react explosively.

Potassium
See POTASSIUM, K: Metal oxides

Sulphur dioxide
Mellor, 1941, Vol. 7, 689
Interaction is incandescent.
See other METAL OXIDES
OXIDANTS

PLATINUM(IV) OXIDE **O_2Pt**

Acetic acid,
Hydrogen
Adams, R. *et al., Org. Synth.*, 1941, Coll. Vol. 1, 66
Addition of fresh platinum oxide catalyst to a hydrogenation reaction in acetic acid caused an immediate explosion. Several similar incidents, usually involving acetic acid as solvent, are known to the author.
See other METAL OXIDES

SULPHUR DIOXIDE **O_2S**

MCA SD-52, 1953
Preparative hazard.
See SULPHURIC ACID, H_2O_4S: Copper

Halogens,
or Interhalogens
See BROMINE PENTAFLUORIDE, BrF_5: Acids, etc.
CHLORINE TRIFLUORIDE, ClF_3: Metals, etc.
FLUORINE, F_2: Non-metal oxides

Lithium acetylide-ammonia
See LITHIUM ACETYLIDE-AMMONIA, $C_2HLi \cdot H_3$: Gases

Lithium nitrate,
Propene
See PROPENE, C_3H_6 : Lithium nitrate, etc.

Metal acetylides
Mellor, 1946, Vol. 5, 848–849
Monocaesium or monopotassium acetylides, and the ammoniate of monolithium acetylide, all ignite and incandesce in unheated sulphur dioxide. The di-metal (including sodium) salts appear to be less reactive, needing heat before ignition occurs.

Metal oxides
1. Mellor, 1941, Vol. 2, 487
2. Mellor, 1942, Vol. 13, 715
3. Mellor, 1941, Vol. 7, 388

Dicaesium monoxide [1], iron(II) oxide [2] and tin(II) oxide [3] all ignite and incandesce on heating in sulphur dioxide.

Metals
1. Mellor, 1943, Vol. 11, 161
2. Mellor, 1942, Vol. 12, 187
3. Gilbert, H. N., *Chem. Eng. News,* 1948, **26**, 2604; Mellor, 1961, Vol. 2, Suppl. 2.1, 468

Finely divided (pyrophoric) chromium incandesces in sulphur dioxide [1], while pyrophoric manganese burns brilliantly on heating in the gas [2]. Molten sodium reacts violently with the dry gas or liquid, while the moist gas reacts as vigorously as water with cold sodium.

Polymeric tubing
MCA Case History No. 1044
Plastics tubing normally capable of withstanding an internal pressure of 7 bar failed below 2 bar when used to convey gaseous sulphur dioxide.

Potassium chlorate
See POTASSIUM CHLORATE, $ClKO_3$: Sulphur dioxide

Sodium hydride

See SODIUM HYDRIDE, HNa: Sulphur dioxide

See other NON-METAL OXIDES

SELENIUM DIOXIDE O_2Se

Phosphorus trichloride

Mellor, 1940, Vol. 8, 1005

A mixture of the cold components attains red-heat.

See other OXIDANTS

SILICON DIOXIDE O_2Si

Metals

See MAGNESIUM, Mg: Silicon dioxide
SODIUM, Na: Non-metal oxides

Oxygen difluoride

See OXYGEN DIFLUORIDE, F_2O: Adsorbents

Vinyl acetate

See VINYL ACETATE, $C_4H_6O_2$: Desiccants

TIN(IV) OXIDE O_2Sn

Chlorine trifluoride

See CHLORINE TRIFLUORIDE, ClF_3: Metals, etc.

Hydrogen trisulphide

See HYDROGEN TRISULPHIDE, H_2S_3: Metal oxides

Metals

See ALUMINIUM, Al: Metal oxides
POTASSIUM, K: Metal oxides
MAGNESIUM, Mg: Metal oxides
SODIUM, Na: Metal oxides

TITANIUM(IV) OXIDE O_2Ti

Metals

Mellor, 1941, Vol. 7, 10, 44; 1961, Vol. 2, Suppl.2.1, 81

Reduction of the oxide by aluminium, calcium, magnesium, potassium, sodium or zinc is accompanied by more or less incandescence (lithium magnesium and zinc especially).

See other METAL OXIDES

URANIUM(IV) OXIDE O_2U

Mellor, 1942, Vol. 12, 42

The finely divided oxide prepared at low temperature is pyrophoric when heated in air, and burns brilliantly.

See other METAL OXIDES

TUNGSTEN(IV) OXIDE O_2W

Chlorine

See Chlorine, Cl_2: Tungsten dioxide

ZINC PEROXIDE O_2Zn

Alone,

or Metals

Mellor, 1940, Vol. 4, 530

Sidgwick, 1950, 270

The hydrated peroxide (of indefinite composition) explodes at 212°C, and mixtures with aluminium or zinc powders burn brilliantly.

See other OXIDANTS

OZONE O_3

1. Sidgwick, 1950, 860
2. Adley, F. E., *Nucl. Sci. Abs.*, 1962, **17**, 18
3. Clough, P. N. *et al., Chem. & Ind.*, 1966, 1971

4. Gatwood, G. T. *et al., J. Chem. Educ.*, 1969, **46**, A103
5. Streng, A. G., *Explosivstoffe,* 1960, **8**, 225
6. Loomis, J. S. *et al., Nucl. Instr. Methods,* 1962, **15**, 243
7. *Ozone Reactions with Organic Compounds,* ACS 112, Washington, ACS, 1972

Pure solid or liquid materials are highly explosive, and evaporation of a solution of ozone in liquid oxygen causes ozone enrichment and ultimately explosion [1]. Organic liquids and oxidisable materials dropped into liquid ozone will also cause explosion of the ozone [2]. Ozone technology and hazards have been reviewed [5], and a safe process to concentrate ozone by selective adsorption on silica gel at low temperature has been described [3], and safe techniques for generation and handling of ozone have been detailed [4]. Ozone is now commercially available as a dissolved gas in cylinders. Explosive hazards involved in use of liquid nitrogen as coolant where ozone may be incidentally produced during operation of van der Graaff generators, etc. have been discussed [6]. The chemistry of interaction with organic compounds compounds has been recently reviewed [7].

Alkenes
Sidgwick, 1950, 862
Davies, 1961, 97
Interaction of alkenes with ozonised oxygen tends to give several types of products or their polymers, some of which show more pronounced explosive tendencies than others. The cyclic *gem*-diperoxides are more explosive than the true ozonides.
See CYCLIC PEROXIDES

Aromatic compounds
Mellor, 1941, Vol. 1, 911
Benzene, aniline and other aromatic compounds give explosive gelatinous ozonides, among other products, on contact with ozonised oxygen.

Benzene,
Rubber
Morrell, S. H., private comm., 1968
During ozonisation of rubber dissolved in benzene, an explosion occurred. This seems unlikely to have been due to formation of

benzene tri-ozonide (which separates as a gelatinous precipitate after prolonged ozonisation), since the solution remained clear. A rubber ozonide may have been involved.

Bromine
Lewis, B. *et al.*, *J. Amer. Chem. Soc.*, 1931, **53**, 2710
Interaction becomes explosive above 20°C and a minimum critical pressure.
See Hydrogen bromide, below
BROMINE TRIOXIDE, BrO_3

Dicyanogen
Anon., *Chem. Eng. News,* 1957, **35**(21), 22
This hypergolic combination appears to have rocket-propellant capabilities.
See ROCKET PROPELLANTS

Diethyl ether
Mellor, 1941, Vol. 1, 911
Contact with ozonised oxygen produces some of the explosive diethyl peroxide.

Dinitrogen tetraoxide
Pinkus, A. *et al.*, *J. Chim. Phys.*, 1920, **18**, 366
Luminescent reaction is often explosive.

Hydrogen bromide
Lewis, B. *et al.*, *J. Amer. Chem. Soc.*, 1931, **53**, 3565
Interaction is rapid and becomes explosive above a total pressure of ~40 mbar, even at −104°C.
See Bromine, above

4-Hydroxy-4-methyl-1,6-heptadiene
See 4-HYDROXY-4-METHYL-1,6-HEPTADIENE, $C_8H_{14}O$: Ozone

Nitrogen trichloride
See NITROGEN TRICHLORIDE, Cl_3N: Initiators

Stibine
Stock, A. *et al.*, *Ber.*, 1905, **38**, 3837
Passage of oxygen containing 2% ozone through stibine at −90°C caused an explosion. On standing, a suspension of solid stibine in a liquid oxygen solution of ozone eventually exploded, as oxygen evaporated and increased the concentration of ozone and the temperature.

Tetrafluorohydrazine
See TETRAFLUOROHYDRAZINE, F_4N_2: Ozone
See other OXIDANTS

PHOSPHORUS(III) OXIDE O_3P_2

Ammonia
Thorpe, T. E. *et al.*, *J. Chem. Soc.*, 1891, **59**, 1019
Interaction of ammonia and the molten oxide under nitrogen is rather violent, and the mixture ignites. Use of a solvent and cooling controls the reaction to produce phosphorus diamide.

Disulphur dichloride
Mellor, 1940, Vol. 8, 898
Interaction is very violent.

Halogens
Thorpe, T. E. *et al.*, *J. Chem. Soc.*, 1891, **59**, 1019
Mellor, 1940, Vol. 8, 897
The oxide ignites in contact with excess chlorine gas, and reacts violently, usually igniting, with liquid bromine.

Oxygen
Mellor, 1940, Vol. 8, 897–898
Interaction with air or oxygen is rapid and, at slightly elevated temperatures in air or at high concentration of oxygen, ignition is very probable, particularly if the oxide is molten (above 22°C) or distributed as a thin layer. The solid in contact with oxygen at 50–60°C ignites and burns very brilliantly. During distillation of phosphorus trioxide, air must be rigorously excluded to avoid the possibility of explosion.

Phosphorus pentachloride
See PHOSPHORUS PENTACHLORIDE, Cl_5P: Diphosphorus trioxide

Sulphur
1. Thorpe, T. E. *et al., J. Chem. Soc.*, 1891, **59**, 1019
2. Pernert, J. C. *et al., Chem. Eng. News,* 1949, 27, 2143
Interaction of a mixture under inert atmosphere above 160°C to form tetraphosphorus tetrathiohexoxide (or phosphorus oxysulphide) is violent and dangerous on scales of working other than small [1]. A safer procedure involving distillation of phosphorus pentaoxide and pentasulphide is described [2].

Sulphuric acid
Mellor, 1940, Vol. 8, 898
Addition of sulphuric acid to the oxide causes violent oxidation, and ignition if more than 1–2 g is used.

Water
Mellor, 1940, Vol. 8, 897
Reaction with cold water is slow, but with hot water, violent, the evolved phosphine igniting. With more than 2 g of oxide, violent explosions occur.
See other NON-METAL OXIDES

DIPALLADIUM TRIOXIDE O_3Pd_2

Sidgwick, 1950, 1573
If the hydrated oxide is heated to remove water, it incandesces or explodes, giving the monoxide.
See other METAL OXIDES

SULPHUR TRIOXIDE O_3S

Acetonitrile,
Sulphuric acid
See ACETONITRILE, C_2H_3N: Sulphuric acid, etc.

Dimethyl sulphoxide
Seel, F., *Inorg. Synth.*, 1967, **9**, 114
Dissolution of sulphur trioxide in the sulphoxide is very exothermic and must be done slowly with cooling to avoid decomposition.

Dioxan
Sisler, H. H. *et al., Inorg. Synth.*, 1947, **2**, 174
Since the 1:1 addition complex sometimes decomposes violently on storing at ambient temperature, it should only be prepared immediately before use.

Diphenylmercury
See DIPHENYLMERCURY, $C_{12}H_{10}Hg$: Non-metal oxides

Formamide,
Iodine,
Pyridine
See FORMAMIDE, CH_3NO: Iodine, etc.

Metal oxides
Partington, 1967, 706
Mellor, 1941, Vol. 7, 654
The violent interaction causes incandescence with barium and lead oxides.

Nitryl chloride
See NITRYL CHLORIDE, $ClNO_2$: Inorganic materials

Phosphorus
See PHOSPHORUS, P: Non-metal oxides

Water
Mellor, 1947, Vol. 10, 344
Interaction is vigorously exothermic, sometimes explosive, with evolution of light and heat. A mixture of oxide and water in ratio of 4:1 completely vaporises, with simultaneous emission of light.
See other NON-METAL OXIDES

ANTIMONY(III) OXIDE O_3Sb_2

Bromine trifluoride
See BROMINE TRIFLUORIDE, BrF_3: Antimony trichloride oxide

Chlorinated rubber
See CHLORINATED RUBBER: Metal oxides

THALLIUM(III) OXIDE O_3Tl_2

Sulphur,
or Sulphur compounds
Mellor, 1946, Vol. 5, 434
Mixtures of the oxide with sulphur or antimony sulphide explode on grinding in a mortar. Dry hydrogen sulphide ignites, and sometimes feebly explodes, over thallium oxide.
See other METAL OXIDES

VANADIUM(III) OXIDE O_3V_2

Sidgwick, 1950, 825
Ignites on heating in air.
See other METAL OXIDES

TUNGSTEN(VI) OXIDE O_3W

Interhalogens
See BROMINE PENTAFLUORIDE, BrF_5 : Acids, etc.
CHLORINE TRIFLUORIDE, ClF_3 : Metals, etc.
Lithium
See LITHIUM, Li: Metal oxides

XENON TRIOXIDE O_3Xe

1. Malm, J. G. *et ul.*, *Chem. Rev.*, 1965, **65**, 199
2. Chernick, C. L. *et al.*, *J. Chem. Educ.*, 1966, **43**, 619; *Inorg. Synth.*, 1968, **11**, 206, 209
3. Anon., *Chem. Eng. News*, 1963, **41**(13), 45
4. Holloway, J. H., *Talanta*, 1967, **14**, 871–873

It is a powerful explosive [1] produced when xenon tetra- or hexa-fluorides are exposed to moist air and hydrolysed. Some tetrafluoride is usually present in xenon difluoride, so the latter is potentially dangerous. Although safe to handle in small amounts in aqueous solution, great care must be taken to avoid solutions drying out, e.g. around ground stoppers [2]. Full safety precautions have been discussed [2–4].
See other XENON COMPOUNDS

OSMIUM(VIII) OXIDE O_4Os

Hydrogen peroxide
See HYDROGEN PEROXIDE, H_2O_2: Metals, etc.

LEAD SULPHATE O_4PbS

Potassium
See POTASSIUM, K: Oxidants

TRILEAD TETRAOXIDE O_4Pb_3

Peroxyformic acid
See PEROXYFORMIC ACID, CH_2O_3: Metals, etc.

Seleninyl chloride
See SELENINYL CHLORIDE, Cl_2OSe: Metal oxides

2,4,6-Trinitrotoluene
See 2,4,6-TRINITROTOLUENE, $C_7H_3N_3O_6$: Added impurities

RUTHENIUM(VIII) OXIDE O_4Ru

Mellor, 1942, Vol. 15, 520
Brauer, 1965, Vol. 2, 1600
The liquid (which shows no true boiling point) decomposes with explosive violence above 106°C.

Ammonia
Watt, J. *et al.*, *J. Inorg. Nucl. Chem.*, 1965, **27**, 262
Rapid interaction at –70°C/1 mbar causes ignition. Slow interaction produced a solid which exploded at 206°C, probably owing to the formation of $RuO(NH_2)$ and RuN.

Organic materials
Mellor, 1942, Vol. 15, 520
Brauer, 1965, Vol. 2, 1600
The solid material, or its concentrated solutions or vapour, tends to oxidise ethanol, cellulose fibres, etc., explosively.
See other METAL OXIDES
OXIDANTS

PHOSPHORUS(V) OXIDE O_5P_2

MCA SD-28, 1948
Safe handling procedures for this extremely powerful desiccant are detailed.

Formic acid
See FORMIC ACID, CH_2O_2 : Phosphorus pentaoxide

Hydrogen fluoride
Gore, G., *J. Chem. Soc.,* 1869, 22, 368
Interaction is vigorous below 20°C.

Inorganic bases
Mellor, 1940, Vol. 8, 945
Van Wazer, 1958, Vol. 1, 279
Dry mixtures with sodium or calcium oxides do not react in the cold, but interact violently if warmed or moistened, evolving phosphorus(V) oxide vapour. A mixture of the oxide and sodium carbonate may be initiated by strong local heating, when the whole mass will suddenly become hot.

Metals,
Mellor, 1940, Vol. 8, 945
Interaction with warm sodium or potassium is incandescent and explosive with heated calcium.

Oxidants
See BROMINE PENTAFLUORIDE, BrF_5 : Acids, etc.
CHLORINE TRIFLUORIDE, ClF_3 : Metals, etc.
PERCHLORIC ACID, $ClHO_4$: Dehydrating agents
OXYGEN DIFLUORIDE, F_2O: Phosphorus(V) oxide
HYDROGEN PEROXIDE, H_2O_2 : Phosphorus(V) oxide

Water
Mellor, 1940, Vol. 8, 944
MCA SD-28, 1948
Interaction with water is very energetic and highly exothermic. The increase in temperature may be enough to ignite combustible materials if present and in contact.
See other NON-METAL OXIDES

TANTALUM(V) OXIDE O_5Ta_2

Chlorine trifluoride
See CHLORINE TRIFLUORIDE, ClF_3 : Metals, etc.

VANADIUM(V) OXIDE O_5V_2

Chlorine trifluoride
See CHLORINE TRIFLUORIDE, ClF_3 : Metals, etc.

Lithium
See LITHIUM, Li: Metal oxides

Peroxyformic acid
See PEROXYFORMIC ACID, CH_2O_3 : Metals, etc.

DISULPHUR HEPTAOXIDE O_7S_2

Meyer, F. *et al.*, *Ber.*, 1922, **55**, 2923
The crystalline material (of unknown structure) soon explodes on exposure to moist air.
See other NON-METAL OXIDES

TRIURANIUM OCTAOXIDE O_8U_3

Barium oxide
Mellor, 1942, Vol. 12, 49
Interaction below 300°C is vigorous and very exothermic.
See other METAL OXIDES

OSMIUM — Os

Chlorine trifluoride
See CHLORINE TRIFLUORIDE, ClF_3 : Metals

Fluorine
See FLUORINE, F_2 : Metals

Phosphorus
See PHOSPHORUS, P: Metals

PHOSPHORUS — P

A.I.T., 45°C
1. *MCA SD-16*, 1947
2. Anon., *Chem. Trade J.*, 1936, **98**, 522

Handling precautions for the reactive allotrope, white (or yellow) phosphorus, are detailed [1]. Red phosphorus powder being charged into a reaction vessel exploded when the lid of the tin fell in and was struck by the agitator [2].
See Oxygen, below

Alkalies
Mellor, 1940, Vol. 8, 802
Contact of phosphorus with boiling caustic alkalies or with hot calcium hydroxide evolves phosphine, which usually ignites in air.

Chlorosulphuric acid
Heumann, K. *et al., Ber.*, 1882, **15**, 417
White phosphorus begins to reduce the acid at 25–30°C and the vigorous reaction accelerates to explosion. Red phosphorus is similar at a higher initial temperature.

Cyanogen iodide
Mellor, 1940, Vol. 8, 791
Molten phosphorus reacts incandescently.

Halogen azides
See HALOGEN AZIDES

Halogen oxides

See CHLORINE DIOXIDE, ClO_2 : Non-metals
DICHLORINE OXIDE, Cl_2O: Oxidisable materials
OXYGEN DIFLUORIDE, F_2O: Non-metals
TRIOXYGEN DIFLUORIDE, F_2O_3

Halogens,
or Interhalogens

1. Mellor, 1940, Vol. 8, 785; 1956, Vol. 2, Suppl.1, 379
2. Christomanos, A. C., *Z. Anorg. Chem.*, 1904, **41**, 279

Both yellow and red phosphorus ignite on contact with fluorine and chlorine; red ignites in liquid bromine or in a heptane solution of chlorine at 0°C. Yellow phosphorus explodes in liquid bromine or chlorine, and ignites in contact with bromine vapour or solid iodine [1]. Interaction of bromine and white phosphorus in carbon disulphide gives a slimy by-product which explodes violently on heating [2].

See BROMINE TRIFLUORIDE, BrF_3 : Halogens, etc.
BROMINE PENTAFLUORIDE, BrF_5 : Acids, etc.
CHLORINE TRIFLUORIDE, ClF_3 : Metals, etc.
IODINE PENTAFLUORIDE, F_5I: Metals, etc.

Hexalithium disilicide
Mellor, 1940, Vol. 6, 169
Interaction is incandescent.

Hydriodic acid
Blau, K., private comm., 1965
During removal of free iodine from hydriodic acid by distillation from red phosphorus, phosphine was produced. When air was admixed by changing the receiver, an explosion occurred. Omission of distillation, by boiling the reactants in inert atmosphere, and separating the phosphorus by filtration through a sintered glass funnel containing solid carbon dioxide, gave a colourless product.

Hydrogen peroxide

See HYDROGEN PEROXIDE, H_2O_2 : Phosphorus

Magnesium perchlorate
1965 Summary of Serious Accidents, Washington, USAEC, 1966
Phosphorus and magnesium perchlorate exploded violently while being

mixed. It seems likely that this was initiated by interaction of phosphorus and traces of perchloric acid to form magnesium phosphate. The acid may have been present in the magnesium perchlorate, or have been formed by traces of phosphoric acid in the phosphorus. Once formed, the perchloric acid would be rendered anhydrous by the powerful dehyrating action of magnesium perchlorate.

Metal acetylides
Mellor, 1946, Vol. 5, 848–849
Mono-rubidium and -caesium acetylides incandesce with warm phosphorus. Di-lithium and -sodium acetylides burn vigorously in phosphorus vapour; the di-potassium, -rubidium and -caesium derivatives should react with increasing violence.

Metal halogenates
Mellor, 1941, Vol. 2, 310; 1940, Vol. 8, 785–786
Dry finely divided mixtures of red (or white) phosphorus with chlorates, bromates or iodates of barium, calcium, magnesium, potassium, sodium or zinc will readily explode upon initiation by friction, impact or heat. Fires have been caused by accidental contact in the pocket between the red phosphorus in the friction strip on safety match boxes and potassium chlorate tablets. Addition of a little water to a mixture of white or red phosphorus and potassium iodate causes a violent or explosive reaction. Addition of a little of a solution of phosphorus in carbon disulphide to potassium chlorate causes an explosion when the solvent evaporates. The extreme danger of mixtures of red phosphorus (or sulphur) with chlorates was recognised in the U K some 40 years ago when unlicensed preparation of such mixtures was prohibited by Orders in Council.

Metal halides
Mellor, 1939, Vol. 9, 467; 1943, Vol. 11, 234
Phosphorus ignites in contact with antimony pentafluoride and explodes with chromyl chloride in presence of moisture at ambient temperature.

Metal oxides
Mellor, 1941, Vol. 2, 490; Vol. 7, 690; 1940, Vol. 8, 792; 1943, Vol. 11, 234
von Schwartz, 1918, 328
Red phosphorus reacts vigorously on heating with copper oxide or

manganese dioxide; and on grinding or warming with lead monoxide or mercury or silver oxides, ignition may occur. Red phosphorus ignites in contact with lead, sodium or potassium peroxides, while white phosphorus explodes, and also in contact with molten chromium trioxide at 200°C.

Metals
1. Mellor, 1941, Vol. 7, 115; 1940, Vol. 8, 842, 847, 853
2. Van Wazer, 1958, Vol. 1, 159
3. Mellor, 1942, Vol. 15, 696; 1937, Vol. 16, 160

Reaction of beryllium, manganese, thorium or zirconium is incandescent when heated with phosphorus [1] and that of cerium, lanthanum, neodymium and praseodymium is violent above 400°C [2]. Osmium incandesces in phosphorus vapour, and platinum burns vividly below red-heat [3].

Metal sulphates
Berger, E., *Compt. Rend.*, 1920, **170**, 1492
Excess red phosphorus will burn admixed with barium or calcium sulphates if primed at a high temperature with potassium nitrate–calcium silicide mixture.

See POTASSIUM NITRATE, KNO_3 : Calcium silicide

Nitrates
Mellor, 1941, Vol. 3, 470; 1940, Vol. 4, 987; Vol. 8, 788
Yellow phosphorus ignites in molten ammonium nitrate, and mixtures of phosphorus with ammonium, mercury (I) and silver nitrates explode on impact. Red phosphorus is oxidised vigorously when heated with potassium nitrate.

See HEPTASILVER NITRATE OCTAOXIDE, Ag_7NO_{11}

Nitric acid
See NITRIC ACID, HNO_3 : Non-metals

Nitrogen halides
See NITROGEN TRIBROMIDE HEXAAMMONIATE, $Br_3N{\cdot}H_{18}N_6$: Initiators
NITROGEN TRICHLORIDE, Cl_3N: Initiators

Nitrosyl fluoride
See NITROSYL FLUORIDE, FNO: Metals, etc.

Nitryl fluoride
See NITRYL FLUORIDE, FNO_2 : Non-metals

Non-metal halides
Mellor, 1946, Vol. 5, 136; 1940, Vol. 8, 787; 1947, Vol. 10, 906
Either the white or red form incandesces with boron triiodide. Red phosphorus incandesces in seleninyl chloride while white explodes. Red phosphorus reacts vigorously on warming with sulphuryl or disulphuryl chlorides.

Non-metal oxides
Mellor, 1940, Vol. 8, 786–787
Warm or molten white phosphorus burns vigorously in nitrogen oxide, dinitrogen tetroxide or pentaoxide. White phosphorus ignites after some delay in contact with the vapour of sulphur trioxide, but immediately in contact with the liquid if a large piece is used.

Oxygen
Mellor, 1940, Vol. 8, 771
Van Wazer, 1958, **1**, 97
The reactivity of phosphorus with air or oxygen depends on the allotrope of phosphorus involved and the conditions of contact, white (yellow) phosphorus being by far the most reactive. White phosphorus readily ignites in air if warmed, finely divided (e.g. from an evaporating solution) or under conditions where the slow oxidative exotherm cannot be dissipated. Contact with finely divided charcoal or lampblack promotes ignition, probably by the adsorbed oxygen. Contact with amalgamated aluminium also promotes ignition.

Peroxyformic acid
See PEROXYFORMIC ACID, CH_2O_3 : Non-metals

Potassium nitride
Mellor, 1940, Vol. 8, 99
Potassium and other alkali metal nitrides react on heating with phosphorus to give a highly inflammable mixture which evolves ammonia and phosphine with water

Potassium permanganate
Mellor, 1942, Vol. 12, 322

Grinding mixtures of phosphorus and potassium permanganate causes explosion, more violent if also heated.

Sulphur
See SULPHUR, S: Non-metals

Sulphuric acid
See SULPHURIC ACID, H_2O_4S: Phosphorus
See other NON-METALS

PHOSPHORUS(V) SULPHIDE P_2S_5

Air
MCA SD-71, 1958
The general reactivity of the sulphide depends markedly on the physical form, and differences of a factor of 10 may be involved. It is ignitable by friction, sparks or flames, and ignites if heated in dry air close to the m.p., 275–280°C. The dust (200-mesh) forms explosive mixtures in air above a concentration of 0.5% w/v.

Alcohols
1. *MCA Case History No. 227*
2. Ashford, J. S., private comm., 1964

A mixture of the sulphide, ethylene glycol and hexane in a mantle-heated flask spontaneously overheated and exploded at an internal temperature of *ca.* 180°C. It had been intended to maintain the reaction at 60°C, but since alcoholysis of phosphorus pentasulphide is exothermic, presence of the heating mantle prevented dissipation of heat, and the reaction accelerated continuously until explosive decomposition occurred [1]. An incident in similar circumstances involving interaction of the sulphide with 4-methyl-2-pentanol also led to violent eruption of the flask contents [2].

Water
Haz. Chem. Data, 1971, 180
It heats and may ignite in contact with limited amounts of water, hydrogen sulphide also being evolved.
See other NON-METAL SULPHIDES

TRIZINC DIPHOSPHIDE P_2Zn_3

Perchloric acid

See PERCHLORIC ACID, $ClHO_4$: Trizinc diphosphide

TETRAPHOSPHORUS TRISELENIDE P_4Se_3

Mellor, 1947, Vol. 10, 790

Ignites when heated in air.

LEAD Pb

Mézáros, L., *Tetrahedron Lett.,* 1967, 4951

The finely divided lead produced by reduction of the oxide with furfural vapour at 290°C is pyrophoric and chemically reactive.

Disodium acetylide

See DISODIUM ACETYLIDE, C_2Na_2: Metals

See other METALS

PYROPHORIC METALS

Oxidants

See CHLORINE TRIFLUORIDE, ClF_3: Metals, etc.

HYDROGEN PEROXIDE, H_2O_2: Metals

AMMONIUM NITRATE, $H_4N_2O_3$: Metals

ZIRCONIUM–LEAD ALLOYS Pb–Zr

Alexander, P. P., US Pat., 2 611 316, 1952

Zirconium–lead alloys containing 10–70% of the former will pulverise and ignite on impact.

See other ALLOYS

PALLADIUM Pd

Arsenic

See ARSENIC, As: Metals

Carbon

1. Mozingo, R., *Org. Synth.*, 1955, Coll. Vol. 3, 687
2. Rachlin, A. I. *et al., Org. Synth.*, 1971, **51**, 9
3. Fraser, R., private comm., 1973

Palladium-on-carbon catalysts prepared by formaldehyde reduction are less pyrophoric than those reduced with hydrogen [1]. Palladium-on-carbon catalysts become extremely pyrophoric on thorough vacuum drying [2]. Those prepared on high surface area supports (up to 2000 m^2/g) are highly active and will readily cause catalytic ignition of hydrogen/air or solvent/air mixtures. Methanol is notable for easy ignition because of its high volatility [3].

Ozonides

See OZONIDES

Sodium tetrahydroborate

Augustine, 1968, 75

In preparation for the reduction of a nitro compound, the tetrahydroborate solution is added to an aqueous suspension of palladium-on-charcoal catalyst. Addition of catalyst to the tetrahydroborate solution may cause ignition of liberated hydrogen.

Sulphur

See SULPHUR, S: Metals

See other HYDROGENATION CATALYSTS
PYROPHORIC METALS

PLATINUM Pt

1. van Campen, M. G., *Chem. Eng. News,* 1954, **32**, 4698
2. Tilford, C. H., private comm., 1965

Finely divided (catalytic forms) of platinum are hazardous to handle if allowed to dry. Used Adams' hydrogenation catalyst exploded while being sieved in air. It had been well washed, finally with water, and then air-dried [1]. Manipulation of used catalyst on filter paper caused a violent explosion [2]. The explosive oxidation of adsorbed hydrogen is suspected here.

See also ZINC, Zn: Catalytic metals

Acetone,
Nitrosyl chloride
See NITROSYL CHLORIDE, ClNO: Acetone, etc.

Arsenic
Wöhler, L., *Z. Anorg. Chem.*, 1930, **186**, 324
During interaction to give platinum diarsenide in a sealed tube at 270°C, the highly exothermic reaction may explode.

Dioxygen difluoride
See DIOXYGEN DIFLUORIDE, F_2O_2

Ethanol
Elkins, J. S., private comm., 1968
Addition of platinum-black catalyst to ethanol caused ignition. Pre-reduction with hydrogen and/or nitrogen purging of air prevented this.

Hydrazine
See HYDRAZINE, H_4N_2: Catalysts

Hydrogen,
Air
Coleman, J. W., private comm., 1965
A platinised alumina catalyst had been treated with flowing hydrogen at ambient temperature. Purging with air caused an explosion. Inert gas purging is essential after treating catalysts or adsorbents with hydrogen.

Hydrogen peroxide
See HYDROGEN PEROXIDE, H_2O_2: Metals

Lithium
Nash, C. P., *Chem. Eng. News,* 1961, **39**(5), 42
Interaction to form an intermetallic compound sets in violently at 540°±20°C.

Ozonides
See OZONIDES

Peroxomonosulphuric acid
See PEROXOMONOSULPHURIC ACID, H_2O_5S: Catalysts

Phosphorus
See PHOSPHORUS, P: Metals

Selenium,
or Tellurium
Mellor, 1937, Vol. 16, 158
Finely divided or spongy platinum reacts incandescently when heated with selenium or tellurium.
See other HYDROGENATION CATALYSTS
PYROPHORIC METALS

PLUTONIUM **Pu**

Williamson, G. K., *Chem. & Ind.*, 1960, 1384
This pyrophoric metal may be worked under nitrogen, argon or helium (but not carbon dioxide) at temperatures above 600°C.

Carbon tetrachloride
Serious Accid. Series, No. 246, Washington, USAEC, 1965
While draining after degreasing in the solvent, plutonium chips ignited and, when accidentally dropped into the solvent, caused an explosion.
See other METALS: Halocarbons

Water
MCA Case History No. 1212
Plutonium components, normally kept under argon, were accidentally exposed to air and moisture, probably forming plutonium hydride and hydrated oxides. When the plastics containing bag was disturbed, ignition occurred, causing widespread radiation contamination.
See PLUTONIUM HYDRIDE, H_3Pu
See other PYROPHORIC METALS

RUBIDIUM **Rb**

Rubidium is a typical but very reactive member of the series of alkali metals. It is appreciably more reactive than potassium but less so than caesium, and so would be expected to react more violently with those materials which are hazardous with potassium or sodium.

Air,
or Oxygen
Mellor, 1941, Vol. 2, 468; 1963, Vol. 2, Suppl. 2.2, 2172
Rubidium ignites on exposure to air or dry oxygen, largely forming the oxide.

Halogens
Mellor, 1941, Vol. 2, 13; 1963, Vol. 2, Suppl. 2.2, 2174
Ignition occurs in fluorine, chlorine, and bromine or iodine vapours. Contact with liquid bromine would be expected to cause a violent explosion.

Mercury
Mellor, 1941, Vol. 2, 469
Interaction is exothermic and may be violent.

Non-metals
Mellor, 1963, Vol. 2, Suppl. 2.2, 2174–2175
The molten metal ignites in sulphur vapour at 200–300°C, and reacts with various forms of carbon exothermically, one product, rubidium octacarbide, being pyrophoric.

Vanadium trichloride oxide
Mellor, 1963, Vol. 2, Suppl. 2.2, 2176
Interaction is violent at 60°C.

Water
Mellor, 1941, Vol. 2, 469
Contact with cold water is exothermic enough to ignite the hydrogen evolved.
See ALKALI METALS

RHENIUM Re

Oxygen

See OXYGEN (Gas), O_2 : Rhenium

RHENIUM(VII) SULPHIDE Re_2S_7

Sidgwick, 1950, 1299

Readily oxidised in air, sometimes igniting.

See other METAL SULPHIDES

RHODIUM Rh

Sidgwick, 1950, 1513

Metallic rhodium prepared by heating its compounds in hydrogen must be allowed to cool in an inert atmosphere to prevent catalytic ignition of the adsorbed hydrogen on exposure to air.

See also ZINC, Zn: Catalytic metals

Interhalogens

See BROMINE PENTAFLUORIDE, BrF_5 : Acids, etc.
CHLORINE TRIFLUORIDE, ClF_3 : Metals

See other HYDROGENATION CATALYSTS
PYROPHORIC METALS

RUTHENIUM Ru

Aqua regia,
Potassium chlorate

See POTASSIUM CHLORATE, $ClKO_3$: Aqua regia, etc.

SULPHUR S

MCA SD-74, 1959

It may become ignited by frictional heat or incendive sparks, particularly when suspended as dust in air, for which the approximate explosive limits are 35–1400 g/m^3. Handling precautions for the solid and molten element are detailed. In the hazardous reactions given below, the dual capacity of sulphur to oxidise and be oxidised is apparent.

Diethyl ether
Taylor, H. F., *J. R. Inst. Chem.*, 1955, **79**, 43
Evaporation of an ethereal extract of sulphur caused an explosion of great violence. Experiment showed that evaporation of wet, peroxidised ether gave a mildly explosive residue, which became violently explosive on addition of sulphur.

Fluorine
See FLUORINE, F_2: Non-metals

Halogen oxides
Mellor, 1941, Vol. 2, 289, 241, 295; 1956, Vol. 2, Suppl. 1, 540, 542
Sulphur ignites in chlorine dioxide gas, and may cause an explosion, which usually happens on contact of sulphur with the very unstable dichlorine monoxide. Iodine(V) oxide reacts explosively on warming with sulphur. Chlorine trioxide would be expected to oxidise sulphur violently but dichlorine heptaoxide does not react.
See TRIOXYGEN DIFLUORIDE, F_2O_3

Interhalogens
See BROMINE TRIFLUORIDE, BrF_3: Halogens, etc.
BROMINE PENTAFLUORIDE, BrF_5: Acids, etc.
CHLORINE TRIFLUORIDE, ClF_3: Metals, etc.
IODINE PENTAFLUORIDE, F_5I: Metals, etc.

Metal acetylides,
Metal carbides
Mellor, 1946, Vol. 5, 849, 862, 886, 891
Monorubidium acetylide ignites in molten sulphur; barium carbide ignites in sulphur vapour at 150°C and incandesces; while calcium, strontium, thorium and uranium carbides need a temperature around 500°C to ignite.

Metal halogenates
See METAL HALOGENATES: Non-metals
METAL CHLORATES: Phosphorus, etc.

Metal oxides

See SILVER OXIDE, Ag_2O: Non-metals
CHROMIUM TRIOXIDE, CrO_3: Sulphur
MERCURY(II) OXIDE, HgO: Non-metals
'MERCURY(I) OXIDE' Hg_2O: Non-metals
SODIUM PEROXIDE, Na_2O_2: Non-metals
LEAD(IV) OXIDE, O_2Pb: Non-metals
THALLIUM(III) OXIDE, O_3Tl_2: Sulphur

Metals

1. Mellor, 1940, Vol. 4, 268, 480
2. Mellor, 1941, Vol. 3, 639
3. Mellor, 1946, Vol. 5, 210, 393
4. Mellor, 1942, Vol. 15, 149, 527, 627, 696
5. Mellor, 1941, Vol. 7, 208, 328
6. Mellor, 1942, Vol. 12, 31

A mixture of sulphur with powdered zinc explodes on warming, and that with cadmium reacts less vigorously [1]. Mixtures of calcium and sulphur explode on ignition and calcium burns in sulphur vapour at 400°C [2]. Indium [3], palladium and rhodium [4], thorium and tin [5] all ignite and incandesce on heating admixed with sulphur. Uranium [6], osmium and nickel [4] powders ignite and incandesce in boiling sulphur or its vapour at 600°C, although more finely divided (catalytic) forms of nickel should react more readily. Magnesium will react vigorously when red-hot or molten with molten sulphur or its vapour [1] and a mixture of aluminium powder and sulphur will burn violently or explode if ignited with a high-temperature (magnesium) fuse [3].

See LITHIUM, Li: Sulphur
SODIUM, Na: Non-metals
RUBIDIUM, Rb: Non-metals
See also SELENIUM, Se: Cadmium, etc.

Non-metals

1. Mellor, 1946, Vol. 5, 15
2. von Schwartz, 1918, 328
3. Mellor, 1940, Vol. 8, 786
4. Phillips, R., *Org. Synth.*, 1943, Coll. Vol. 2, 579

Boron reacts with sulphur at 600°C, becoming incandescent [1]. Mixtures of sulphur with lamp black or freshly calcined charcoal ignite spontaneously, probably owing to adsorbed oxygen on the catalytic surface [2]. Mixtures of yellow phosphorus and sulphur ignite and/or explode on heating [3]. Ignition of an intimate mixture of red phosphorus and sulphur causes a violent exothermic reaction [4].

Oxidants

See Halogen oxides, above
Interhalogens, above
Metal oxides, above
SILVER CHLORITE, $AgClO_2$: Hydrochloric acid, etc.
SILVER NITRATE, $AgNO_3$: Non-metals
HEPTASILVER NITRATE OCTAOXIDE, Ag_7NO_{11}
POTASSIUM PERCHLORATE, $ClKO_4$: Sulphur
CHROMYL CHLORIDE, Cl_2CrO_2 : Sulphur
LEAD DICHLORITE, Cl_2O_4Pb
AMMONIUM NITRATE, $H_4N_2O_3$: Sulphur
POTASSIUM PERMANGANATE, $KMnO_4$: Non-metals

Phosphorus trioxide

See PHOSPHORUS(III) OXIDE, O_3P_2 : Sulphur

Potassium nitride

Mellor, 1940, Vol. 8, 99

Potassium and other alkali metal nitrides react with sulphur to form a highly flammable mixture, which evolves ammonia and hydrogen sulphide in contact with water.

See TRICAESIUM NITRIDE, Cs_3N

Sodium hydride

Mellor, 1941, Vol. 2, 483

Interaction with sulphur vapour is vigorous.

See other NON-METALS

TIN(II) SULPHIDE SSn

Oxidants

See CHLORIC ACID, $ClHO_3$: Oxidisable materials
DICHLORINE OXIDE, Cl_2O: Oxidisable materials
METAL HALOGENATES: Metals, etc.

STRONTIUM SULPHIDE SSr

Lead dioxide

See LEAD(IV) OXIDE, O_2Pb: Metal sulphides

TIN(IV) SULPHIDE S_2Sn

Oxidants

See CHLORIC ACID, $ClHO_3$: Oxidisable materials
DICHLORINE OXIDE, Cl_2O: Oxidisable materials
METAL HALOGENATES: Metals, etc.

TITANIUM(IV) SULPHIDE S_2Ti

Potassium nitrate

See POTASSIUM NITRATE, KNO_3: Metal sulphides

URANIUM(IV) SULPHIDE S_2U

Nitric acid

See NITRIC ACID, HNO_3: Uranium(IV) sulphide

DIANTIMONY TRISULPHIDE S_3Sb_2

Oxidants

See SILVER OXIDE, Ag_2O: Metal sulphides
HEPTASILVER NITRATE OCTAOXIDE, Ag_7NO_{11}
POTASSIUM CHLORATE, $ClKO_3$: Metal sulphides
DICHLORINE OXIDE, Cl_2O: Oxidisable materials
LEAD DICHLORITE, Cl_2O_4Pb
FLUORINE, F_2: Sulphides
POTASSIUM NITRATE, KNO_3: Metal sulphides
THALLIUM(III) OXIDE, O_3Tl_2: Sulphur, etc.

ANTIMONY Sb

Krebs, H. *et al., Z. Anorg. Chem.*, 1955, **282**, 177
Electrolysis of acidified, stirred antimony halide solutions at low current density produces explosive antimony which contains substantial amounts of halogen.

Alkali nitrates
Mellor, 1947, Vol. 9, 382
Mixtures detonate on heating, forming antimonates.

Halogens
Mellor, 1947, Vol. 9, 379
Antimony ignites in fluorine, chlorine and bromine. With iodine, ignition or explosion may occur if quantities are large enough.

Oxidants
See Halogens, above
BROMINE TRIFLUORIDE, BrF_3: Halogens, etc.
BROMINE PENTAFLUROIDE, BrF_5: Acids, etc.
CHLORINE TRIFLUORIDE, ClF_3: Metals, etc.
PERCHLORIC ACID, $ClHO_4$: Antimony(III) compounds
DICHLORINE OXIDE, Cl_2O: Oxidisable materials
SELENINYL CHLORIDE, Cl_2OSe: Antimony
NITROSYL FLUORIDE, FNO: Metals
IODINE PENTAFLUORIDE, F_5I: Metals
AMMONIUM NITRATE, $H_4N_2O_3$: Metals
POTASSIUM PERMANGANATE, $KMnO_4$: Antimony
POTASSIUM DIOXIDE, KO_2: Metals
SODIUM NITRATE, $NNaO_3$: Antimony
See other METALS

SELENIUM Se

Hexalithium disilicide
See HEXALITHIUM DISILICIDE, Li_6Si_2: Non-metals

Metal acetylides,
Metal carbides
Mellor, 1946, Vol. 5, 862, 886
Metal acetylides incandesce on heating in selenium vapour; those of barium at 150°C, calcium and strontium at 500°C; and thorium carbide at an unstated temperature.
See DIRUBIDIUM ACETYLIDE, C_2Rb_2: Non-metals

Metal chlorates

Mellor, 1956, Vol. 2, Suppl.1, 583

A slightly moist mixture of selenium with any chlorate (except of alkali metals) becomes incandescent.

See METAL HALOGENATES

Metals

1. Mellor, 1942, Vol. 15, 151
2. Mellor, 1947, Vol. 10, 766–777
3. Mellor, 1942, Vol. 12, 31
4. Mellor, 1940, Vol. 4, 480
5. Mellor, 1942, Vol. 16, 158
6. Reisman, A. *et al., J. Phys. Chem.*, 1963, **67**, 22

Nickel and selenium interact with incandescence on gentle heating [1], as also do sodium and potassium, the latter mildly explosively [2]. Uranium [3] and zinc [4] also incandesce when their mixtures with selenium are heated, and platinum sponge incandesces vividly [5]. The particle size of cadmium and selenium must be below a critical size to prevent explosions during synthesis of cadmium selenide by heating the elements together. Similar considerations apply to interaction of cadmium or zinc with sulphur, selenium or tellurium [6].

Nitrogen trichloride

See NITROGEN TRICHLORIDE, Cl_3N: Initiators

Oxidants

See Metal chlorates, above
Oxygen, etc. below
SILVER(I) OXIDE, Ag_2O: Non-metals
BROMINE PENTAFLUORIDE, BrF_5: Acids, etc.
CHLORINE TRIFLUORIDE, ClF_3: Metals, etc.
FLUORINE, F_2: Non metals
SODIUM PEROXIDE, Na_2O_2: Non-metals

Oxygen,
Organic matter

Astin, S. *et al., J. Chem. Soc.*, 1933, 391

During attempted conversion of recovered selenium metal to the dioxide by heating in oxygen, a vigorous explosion occurred. This was attributed to selenium-catalysed oxidation of traces of organic impurities in the selenium. Oxidation of recovered selenium with nitric

acid is rendered vigorous by presence of organic impurities, but is a safe procedure.
See other NON-METALS

SILICON **Si**

Calcium
See CALCIUM, Ca: Silicon

Metal acetylides
See DICAESIUM ACETYLIDE, C_2Cs_2: Non-metals
DIRUBIDIUM ACETYLIDE, C_2Rb_2: Non-metals

Metal carbonates
Mellor, 1940, Vol. 6, 164
Amorphous or crystalline silicon both react exothermically when heated with alkali metal carbonates, attaining incandescence and evolving carbon monoxide.

Oxidants
See SILVER FLUORIDE, AgF: Non-metals
PEROXYFORMIC ACID, CH_2O_3: Non-metals
CHLORINE TRIFLUORIDE, ClF_3: Metals, etc.
CHLORINE, Cl_2: Non-metals
COBALT TRIFLUORIDE, CoF_3: Silicon
NITROSYL FLUORIDE, FNO: Metals, etc.
FLUORINE, F_2: Non-metals
OXYGEN DIFLUORIDE, F_2O: Non-metals
IODINE PENTAFLUORIDE, F_5I: Metals, etc.
LEAD(II) OXIDE, OPb: Non-metals
See other NON-METALS

SAMARIUM **Sm**

1,1,2-Trichlorotrifluoroethane
Anon., *NSC Newsletter, R & D Sect.*, 1970, (8)
An attempt to mill a slurry of the metal in the solvent caused a violent explosion. Experiments showed violent interaction of the freshly ground metal surfaces with the halocarbon solvent.
See other METALS: Halocarbons

TIN Sn

Oxidants

See CHLORINE TRIFLUORIDE, ClF_3: Metals
CHLORINE, Cl_2: Metals
FLUORINE, F_2: Metals
IODINE HEPTAFLUORIDE, F_7I: Metals
AMMONIUM NITRATE, $H_4N_2O_3$: Metals
POTASSIUM DIOXIDE, KO_2: Metals
SODIUM PEROXIDE, Na_2O_2: Metals
SULPHUR, S: Metals

TANTALUM Ta

Although very unreactive in massive forms, the finely divided metal may be pyrophoric.

See PYROPHORIC METAL POWDERS

Bromine trifluoride

See BROMINE TRIFLUORIDE, BrF_3: Halogens, etc.

Fluorine

See FLUORINE, F_2: Metals

TELLURIUM Te

Halogens,
or Interhalogens

See BROMINE PENTAFLUORIDE, BrF_5: Non-metals
CHLORINE FLUORIDE, ClF: Tellurium
CHLORINE TRIFLUORIDE, ClF_3: Metals, etc.
CHLORINE, Cl_2: Non-metals
FLUORINE, F_2: Non-metals

Hexalithium disilicide

See HEXALITHIUM DISILICIDE, Li_6Si_2: Non-metals

Metals

See CADMIUM, Cd: Selenium, etc.
POTASSIUM, K: Selenium, etc.
SODIUM, Na: Non-metals
PLATINUM, Pt: Selenium, etc.
ZINC, Zn: Non-metals

THORIUM Th

Air
The finely divided metal is rather pyrophoric in air.
See Oxygen, below

Halogens
Mellor, 1941, Vol. 7, 207
Thorium incandesces in chlorine, bromine or iodine, and would be expected to ignite readily in fluorine.

Nitryl fluoride
See NITRYL FLUORIDE, FNO_2: Metals

Oxygen
1. Mellor, 1941, Vol. 7, 207
2. Stout, E. L., *Chem. Eng. News,* 1958, **36**(8), 64
Massive metal ignites on heating in oxygen, or in air below red heat. Powdered metal ignites when rubbed or crushed [1], or on pouring as a stream in air, owing to intergranular friction [2]. Storage and handling precautions are discussed.
See other PYROPHORIC METALS

Peroxyformic acid
See PEROXYFORMIC ACID, CH_2O_3: Metals

Phosphorus
See PHOSPHORUS, P: Metals

Sulphur
See SULPHUR, S: Metals
See other PYROPHORIC METALS

THORIUM SALTS Th(4+)

'Cupferron'
See *N*-NITROSOPHENYLHYDROXYLAMINE *O*-AMMONIUM SALT ('CUPFERRON'), $C_6H_9N_3O_2$: Thorium salts

Air
1. Mellor, 1941, Vol. 7, 19
2. Stout, E. L., *Chem. Eng. News,* 1958, **36**(8), 64
3. *NSC Data Sheet 485*
4. Lowry, R. N., *et al., Inorg. Synth.*, 1967, **10**, 3

While massive titanium ignites in air at 1200° [1], the finely divided metal is very pyrophoric and should be stored damp in metal containers [2]. Titanium fires are difficult to extinguish as it burns brilliantly in nitrogen above 800°C, in carbon dioxide above 550°C [3], and may react explosively with metal carbonates [1]. Residual titanium sponge after vapour phase reaction with iodine at 400–425°C is often pyrophoric [4].

See other PYROPHORIC METALS

Aluminium

See ALUMINIUM–TITANIUM ALLOYS, Al–Ti

Bromine trifluoride

See BROMINE TRIFLUORIDE, BrF_3: Halogens, etc.

Carbon dioxide,
or Metal carbonates
1. Stout, E. L., *Chem. Eng. News,* 1958, **36**(8), 64
2. Mellor, 1941, Vol. 7, 20

Titanium powder has ignited as a thin layer under carbon dioxide, and burns in the gas above 550°C [1]. Contact of titanium with fused alkali carbonates causes incandescence [1].

Halocarbons

Pot. Incid. Rep. 39, ASESB, Washington, 1968

Mixtures of powdered titanium and trichlorethylene or 1,1,2-trichlorotrifluoroethane flash or spark under heavy impact.

See other METALS: Halocarbons

Halogens
1. Mellor, 1941, Vol. 7, 19
2. Kirk-Othmer, 1969, Vol. 20, 372

Reaction of heated titanium with the halogens usually causes ignition

and incandescence (chlorine at 350°C, bromine at 360°C, iodine higher, but weak combustion)[1], especially in absence of moisture[2].
See FLUORINE, F_2: Metals

Metal oxides
Mellor, 1961, Vol. 7, 20
Interaction with copper(II) oxide or trilead tetraoxide is violent.

Metal oxosalts
Mellor, 1941, Vol. 7, 20
Contact of powdered titanium with molten potassium chlorate, alkali carbonates or mixed potassium carbonate–nitrate causes vivid incandescence. Heating mixtures of the powdered metal with potassium chlorate, nitrate or permanganate causes explosions.
See POTASSIUM PERCHLORATE, $ClKO_4$: Metal powders

Nitric acid
1. Anon., *Occ. Health and Safety*, 1957, **7**, 214
2. Anon., *Chem. Eng. News*, 1953, **31**, 3320

Investigation showed that commercial titanium alloys, in contact, with acid containing less than 1.34% water and more than 6% dinitrogen tetraoxide may become sensitive to impact, and react explosively with the acid. The possible causes are discussed [1]. The spongy residue formed by prolonged corrosion of titanium–manganese alloys by red fuming nitric acid will explode on exposure to friction or heat [2].

Nitryl fluoride
See NITRYL FLUROIDE, FNO_2: Metals

Oxygen
Kirk-Othmer, 1969, Vol. 20, 372
Ignition may occur in liquid or gaseous oxygen and in gas mixtures containing above 46% oxygen, under impact or abrasion. Mixtures of powdered metal and liquid oxygen are detonable.
See OXYGEN (Gas), O_2: Titanium
OXYGEN(Liquid), O_2: Metals

Silver fluoride
See SILVER FLUORIDE, AgF: Titanium

Water
Mellor, 1941, Vol. 7, 19
NSC Data Sheet 485
Finely divided titanium reacts with steam at 700°C, and in presence of air the evolved hydrogen may ignite or explode.
See other METALS

TITANIUM SALTS **Ti(4+)**

'Cupferron'
See *N*-NITROSOPHENYLHYDROXYLAMINE *O*-AMMONIUM SALT ('CUPFERRON'), $C_6H_9N_3O_2$: Thorium salts

THALLIUM **Tl**

Fluorine
See FLUORINE, F_2 : Metals

URANIUM **U**

MCA Case History No. 1296
Storage of uranium foil in closed containers in presence of air and moisture may produce a pyrophoric surface.
See other PYROPHORIC METALS

Ammonia
Mellor, 1942, Vol. 12, 31
At dull red heat the metal incandesces in ammonia.

Bromine trifluoride
See BROMINE TRIFLUORIDE, BrF_3 : Uranium, etc.

Halogens
Mellor, 1942, Vol. 12, 31
Uranium powder ignites in fluorine at ambient temperature, in chlorine at 150–180°C, in bromine vapour at 210–240°C and in iodine vapour at 260°C.

Nitric acid,
Trichloroethylene
MCA Case History No. 1104
Uranium scrap containing rather finely divided turnings was being pickled, after degreasing with trichlorethylene, by treatment with 4–6M nitric acid solution. During subsequent rinsing with cold water. an explosion ejected the metal and acid liquor from the beaker. It has been reported in the literature that uranium reacts with nitric acid or dinitrogen tetraoxide either explosively or with the formation of an explosive coating or residue. The latter is inhibited by the presence of fluoride ion during pickling. Preventive measures proposed include separate treatment of finely divided uranium scrap, thorough pre-mixing of the acid solution and addition of ammonium hydrogen fluoride. An alternative, or contributory, cause could have been the violent reaction between dinitrogen tetraoxide and residual trichloroethylene.
See DINITROGEN TETRAOXIDE, N_2O_4: Halocarbons

Nitrogen oxide
See NITROGEN OXIDE, NO: Metals

Nitryl fluoride
See NITRYL FLUORIDE, FNO_2: Metals

Selenium,
or Sulphur
Mellor, 1942, Vol. 12, 31
Uranium incandesces in sulphur vapour, and with selenium.
See other METALS

VANADIUM V

Oxidants
See BROMINE TRIFLUORIDE, BrF_3: Halogens, etc.
CHLORINE, Cl_2: Metals
NITRYL FLUORIDE, FNO_2: Metals

TUNGSTEN W

Air,
or Oxidants
Mellor, 1943, Vol. 11, 729

The finely divided metal may ignite on heating in air, or with a range of oxidants.

See BROMINE PENTAFLUORIDE, BrF_5 : Acids, etc.
CHLORINE TRIFLUORIDE, ClF_3 : Metals
NITRYL FLUORIDE, FNO_2 : Metals
FLUORINE, F_2 : Metals
OXYGEN DIFLUORIDE, F_2O: Metals
IODINE PENTAFLUORIDE, F_5I: Metals
HYDROGEN SULPHIDE, H_2S: Metals
LEAD(II) OXIDE, O_2Pb: Metals

See other METALS

ZINC **Zn**

Air

1. Anon., *Chem. Ztg.*, 1940, 289; *Ind. Chem.*, 1944, **20**, 330
2. Robson, S., *School Sci. Rev.*, 1934, **16**(62), 165

A cloud of zinc dust generated by sieving the hot, dried material, exploded violently, apparently after initiation by a spark from the percussive sieve-shaking mechanism. Precautions recommended include use of cold zinc and total enclosure of the process [1]. The possibility of explosions of zinc dust suspended in air is mentioned as a serious hazard [2].

See DUST EXPLOSIONS
METAL DUSTS

Ammonium sulphide

See AMMONIUM SULPHIDE, H_8N_2S: Zinc

Arsenic trioxide

Mellor, 1940, Vol. 4, 486

A mixture with excess zinc filings explodes on heating.

Calcium chloride

See CALCIUM CHLORIDE, $CaCl_2$: Zinc

Carbon disulphide

Mellor, 1940, Vol. 4, 4

Carbon disulphide and zinc powder interact with incandescence.

Chlorinated rubber

See Halocarbons, below

Catalytic metals,
Acids
Sidgwick, 1950, 1530, 1578
Alloys with iridium, platinum, or rhodium, after extraction with acid, leave residues which explode on warming in air owing to the presence of occluded hydrogen (or oxygen) in the catalytic metals.

Halocarbons
1. Berger, E., *Compt. Rend.*, 1920, **170**, 29
2. *ABCM Quart. Safety Summ.*, 1963, **34**, 12
A paste of zinc powder and carbon tetrachloride (with kieselguhr as thickener) will readily burn after ignition by a high-temperature primer [1]. Intimate mixtures of chlorinated rubber with powdered zinc (or its oxide) in presence or absence of hydrocarbon or halocarbon solvents react violently or explosively at about 216°C [2].
See BROMOMETHANE, CH_3Br: Metals
CHLOROMETHANE, CH_3Cl: Metals
See also CHLORINATED RUBBER: Metal oxides
See other METALS: Halocarbons

Halogens,
or Interhalogens
Mellor, 1940, Vol. 4, 476
Warm zinc powder incandesces in fluorine, and zinc foil ignites in cold chlorine if traces of moisture are present.
See BROMINE PENTAFLUORIDE, BrF_5: Acids, etc.
CHLORINE TRIFLUORIDE, ClF_3: Metals

Lead azide
See LEAD(II) AZIDE, N_6Pb: Copper, etc.

Manganese dichloride
Terreil, A., *Bull. Soc. Chim. Fr.* [2], 1874, **21**, 289
Zinc reacts explosively when heated with anhydrous manganese dichloride.

Metal oxides
See POTASSIUM DIOXIDE, KO_2: Metals
TITANIUM DIOXIDE, O_2Ti: Metals
ZINC PEROXIDE, O_2Zn

o-Nitroanisole,
Sodium hydroxide
See *o*-NITROANISOLE, $C_7H_7NO_3$: Sodium hydoxide, etc.

Nitrobenzene
Muir, G. D., private comm., 1968
Zinc residues from reduction of nitrobenzene to *N*-phenylhydroxylamine are often pyrophoric and must be kept wet during disposal.

Nitrogen compounds
See HYDRAZINIUM NITRATE, $H_5N_3O_3$: Metals
HYDROXYLAMINE, H_3NO: Metals

Non-metals
1. Mellor, 1940, Vol. 4, 480, 485
2. Read, C. W. W., *School Sci. Rev.,* 1940, **21**(83), 976

Arsenic, selenium and tellurium all react with incandescence on heating [1]. Interaction of powdered zinc and sulphur on heating is considered to be too violent for use as a school experiment [2].

Oxidants
See Halogens, or Interhalogens, above
Metal oxides, above
PEROXYFORMIC ACID, CH_2O_3 : Metals
POTASSIUM CHLORATE, $ClKO_3$: Metals
NITRYL FLUORIDE, FNO_2 : Metals
NITRIC ACID, HNO_3 : Metals
AMMONIUM NITRATE, $H_4N_2O_3$: Metals

Paint primer base
Muller, K. A., *Spontaneous Combustion of Zinc-rich Paint Mixtures, Sub-committee on Storage & Handling Report,* Washington, API, 1969
Zinc residues from poorly mixed priming paint spontaneously inflamed on prolonged exposure to atmosphere. Aluminium residues may behave similarly.

Pentacarbonyliron,
Transition metal halides
See PENTACARBONYLIRON, C_5FeO_5 : Transition metal halides, etc.

Seleninyl bromide
See SELENINYL BROMIDE, Br_2OSe: Metals

Sodium hydroxide
ABCM Quart. Safety Summ., 1963, **34**, 14
Accidental contamination of a metal scoop with flake sodium hydroxide, prior to its use with zinc dust, caused ignition of the latter.
See Water, below

Water
1. Anon., *Metallwirtschaft,* 1941, **20**, 475
2. Anon., *Chem. Fabrik,* 1940, 362
3. Mellor, 1940, Vol. 4, 474

In contact with atmospheric oxygen and limited amounts of water, zinc dust will generate heat, and may become incandescent. Presence of acetic acid and copper shortens the induction period. Store the metal dry, and keep residues thoroughly wet until disposal [1,2]. Hydrogen is evolved, especially in the presence of acid or alkali [3].
See other METALS

ZIRCONIUM Zr

1. *MCA SD-92,* 1966
2. *Zirconium Fire and Explosion Incidents,* TID-5365, Washington, USAEC, 1956
3. Gleason, R. P., *et al., J. Amer. Soc. Safety Eng.,* 1964, **9**(3), 15
4. Bulmer, G. H. *et al., Rep. AHSB(S) R59,* London, UKAEA, 1964
5. *MCA Case History No. 1234*

The pyrophoric hazards of zirconium powder are well documented [1]; there is a collection of 43 abstracts of unusual zirconium fire and explosion incidents, most of which involved finely divided metal, moisture and friction, occasionally accompanied by the possibility of static sparks from polythene bags [2]. Increasing use of zirconium necessitated a review of methods of controlling the pronounced pyrophoric properties of finely divided zirconium metal. Five considerations for the safe use of zirconium include: exclusion of air or oxygen by use of inert gases; exclusion of water, its vapour, and other contaminants or oxidants; control of particle size; limitation of amount of powder handled or exposed; and limitation of exposure of personnel [3]. Storage and disposal of irradiated metal wastes in various forms has also been discussed [4]. Ignition may not be immediate, however, as fine zirconium dust left uncleared from previous grinding operations was ignited violently by a spark from subsequent grinding operations [5].

See LEAD–ZIRCONIUM ALLOYS, Pb–Zr
METAL DUSTS
PYROPHORIC METALS

See also HAFNIUM, Hf

Carbon tetrachloride

Zirconium Fire and Explosion Incidents, TID-5365, Washington, USAEC, 1956

A mixture of powdered zirconium and carbon tetrachloride exploded violently while being heated.

See other METALS: Halocarbons

Nitryl fluoride

See NITRYL FLUORIDE, FNO_2: Metals

Oxygen-containing compounds

1. Mellor, 1941, Vol. 7, 116
2. Ellern, H., *Military and Civilian Pyrotechnics,* 249, New York, Chem. Publ. Co., 1968

The affinity of zirconium for oxygen is great, particularly when finely divided, and extends to many oxygen-containing compounds. The vigorous reactions with potassium chlorate or nitrate, or with cupric or lead oxides, might be anticipated but there are also explosive reactions with the combined oxygen when zirconium is heated with alkali metal hydroxides or carbonates, or zirconium hydroxide. Hydrated sodium tetraborate behaves similarly [1]. Several alkali metal oxosalts (chromates, dichromates, molybdates, sulphates, tungstates) react violently or explosively when reduced to the parent metals by heating with zirconium [2].

See POTASSIUM CHLORATE, $ClKO_3$: Metals
LITHIUM CHROMATE, $CrLiO_4$: Zirconium
COPPER(II) OXIDE, CuO: Metals
POTASSIUM NITRATE, KNO_3: Metals
LEAD(II) OXIDE, OPb: Metals

Phosphorus

See PHOSPHORUS, P: Metals

Water

1. *MCA SD-92,* 1966
2. Muir, G. D., private comm., 1968
3. *Haz. Chem. Data,* 1972, 253

Powder damp with 5–10% of water may ignite, and although 25% of water is regarded as a safe concentration [1], ignition of a 50% paste on breakage of a glass container has been observed [2]. Although water is used to prevent ignition, the powder, once ignited, will burn under water more violently than in air [3].

See other METALS

ZIRCONIUM SALTS **Zr(4+)**

'Cupferron'

See *N*-NITROSOPHENYLHYDROXYLAMINE *O*-AMMONIUM SALT ('CUPFERRON'), $C_6H_9N_3O_2$: Thorium salts

APPENDIX 1

Source Title Abbreviations used in Handbook References

The abbreviations used in the references for titles of journals and periodicals are those used in BP publications practice and conform closely to the recommendations of the *Chemical Abstracts* system Abbreviations which have been used to indicate textbook and reference book sources of information are set out below with the full titles and publication details.

ABCM Quart. Safety Summ.	*Quarterly Safety Summaries,* London, Association of British Chemical Manufacturers, 1930–1964
ACS 54, 1966	*Advanced Propellant Chemistry,* ACS 54, Washington, American Chemical Society, 1966
ACS 88, 1969	*Propellants Manufacture, Hazards and Testing,* ACS 88, Washington, American Chemical Society, 1969
Alkali Metals, 1957	*Alkali Metals,* ACS 19, Washington, American Chemical Society, 1957
Augustine, 1968	*Reduction, Techniques and Applications in Organic Chemistry,* Augustine, R. L., London, Edward Arnold, 1968
Bahme, 1972	*Fire Officers' Guide to Dangerous Chemicals,* FSP-36, Bahme, C. W., Boston, National Fire Protection Association, 1972
BCISC Quart. Safety Summ.	*Quarterly Safety Summaries,* London, British Chemical Industry Safety Council, 1965 to date
Berufsgenossenschaft	*Zeitshrift fur Unfallversicherung und Betriebssicherheit,* Heidelberg. Berufsgenossenschaft der Chemische Industrie, 1960 to date

Braker, 1971 — *Matheson Gas Data Book,* Braker, W., Mossmann, A. L., East Rutherford, NJ, Matheson Gas Products, 5th Edn., 1971

Brandsma, 1971 — *Preparative Acetylenic Chemistry,* Brandsma, L., Barking, Elsevier, 1971

Brauer, 1963, 1965 — *Handbook of Preparative Inorganic Chemistry,* Brauer, G., (Translation Editor: Riley, R. F.), London, Academic Press, 2nd Edn. Vol 1, 1963; Vol 2, 1965

Castrantas, 1965 — *Fire and Explosion Hazards of Peroxy Compounds,* Special Publication No. 394, Castrantas, H. M., Banerjee, D. K., Noller, D. C., Philadelphia, ASTM, 1965

Castrantas, 1970 — *Laboratory Handling and Storage of Peroxy Compounds,* Special Publication No. 491, Castrantas, H. M., Banerjee, D. K., Philadelphia, ASTM, 1970

Chemiearbeit — *Chemiearbeit, Schutzen und Helfen,* Dusseldorf, Berufsgenossenschaft der Chemische Industrie, 1949–62 (supplement to *Chemische Industrie*)

Cloyd, 1965 — *Handling Hazardous Materials,* Technical Survey No. SP-5032. Cloyd, D. R., Murphy, W. J., Washington, NASA, 1965

Coates, 1960 — *Organometallic Compounds,* Coates, G. E., London, Methuen, 1960

Coates, 1967, 1968 — *Organometallic Compounds,* Coates, G. E., Green, M. L. H., Wade, K., London, Methuen Vol. 1, 1967; Vol. 2, 1968

Dangerous Loads, 1972	*Dangerous Loads,* London, Institute of Fire Engineers, 1972
Dangerous Substances, 1972	*Dangerous Substances: Guidance on Fires and Spillages* (Section 1: 'Inflammable liquids'), London, HMSO, 1972
Davies, 1961	*Organic Peroxides,* Davies, A. G., London, Butterworths, 1961
Davis, 1943	*Chemistry of Powders and Explosives,* Davis, T. L., New York, Wiley, 1943
Dunlop, 1953	*The Furans,* ACS 119, Dunlop, A. P., Peters, F. N., New York, Reinhold, 1953
Emeléus, 1960	*Modern Aspects of Inorganic Chemistry,* Emeléus, H. J., Anderson, J. S., London, Rutledge and Kegan Paul, 1960
Federoff, 1960	*Encyclopaedia of Explosives and Related Compounds,* Federoff, B. T. (Editor), Dover, NJ, Piccatinny Arsenal, 1960
Gaylord, 1956	*Reductions with Complex Metal Hydrides,* Gaylord, N. G., New York, Interscience, 1956
Goehring, 1957	*Ergebnisse und Probleme der Chemie der Schwefelstickstoff-verbindung, Scientia Chemica,* Vol. 9, Goehring, M., Berlin, Akademie Verlag, 1957
Haz. Chem Data, 1971	*Hazardous Chemical Data,* NFPA 49, Boston, National Fire Protection Association, 1971

Houben-Weyl, 1953 to date — *Methoden der Organischen Chemie,* Ed. Müller, E., Stuttgart, Georg Thieme, 4th edn, 16 vols, not yet completed, published unsequentially, 1953 to date

Inorg. Synth., 1939–68 — *Inorganic Syntheses,* Various Editors, London McGraw-Hill. Vol. 1, 1939; Vol. 11, 1968

Karrer, 1950 — *Organic Chemistry,* Karrer, P., London, Elsevier, 1950

Kirk-Othmer, 1963–71 — *Encyclopaedia of Chemical Technology,* Kirk, R. E., Othmer, D. F., London, Interscience. Vol. 1, 1963; Vol. 22, 1970; Supplementary Vol. 1971

Kirshenbaum, 1956 — *Final Report on Fundamental Studies of New Explosive Reactions,* Kirshenbaum, A. D., Philadelphia, Research Institute of Temple University, 1956

Kit and Evered, 1960 — *Rocket Propellant Handbook,* Kit, B., Evered, D. S., London, Macmillan, 1960

MCA Case Histories, 1950 on — *Case Histories of Accidents in the Chemical Industry,* Washington, Manufacturing Chemists' Association Inc. (Published monthly and republished as indexed collected volumes at 2–3 year intervals)

MCA SD-Series — *Safety Data Sheets* (1–99 available), Washington, Manufacturing Chemists' Association Inc. (Published and revised as necessary, 1947–73)

Mackay, 1966 — *Hydrogen Compounds of the Metallic Elements,* Mackay, K. M., London, Spon, 1966

Martin, 1971 — *Dimethylsulphoxid,* Martin, D., Hauthal, H. G., Berlin, Akademie Verlag, 1971

Mellor, 1941-71 — *Comprehensive Treatise on Inorganic and Theoretical Chemistry,* Mellor, J. W., London, Longmans Green, Vol. 1, reprinted 1941; Vol. 16 published 1937; and isolated Supplementary volumes up to Vol. 8 Suppl. 3, 1971

Mellor MIC, 1961 — *Mellor's Modern Inorganic Chemistry,* Parkes, G. D. (Revision Editor), London, Longmans Green, 1961

Miller, 1965–6 — *Acetylene, its Properties, Manufacture and Uses,* Miller, S. A., London, Academic Press. Vol. 1, 1965; Vol. 2, 1966

Org. Synth., 1944 to date — *Organic Syntheses,* Various Editors, New York, John Wiley. Collective Vol. 1, 1944; Collective Vol. 2, 1944; Collective Vol. 3, 1955; Collective Vol. 4, 1962; annual volumes thereafter

Partington, 1967 — *General and Inorganic Chemistry,* 4th Edn, Partington, J. R., London, Macmillan, 1967

Pieters, 1957 — *Safety in the Chemical Laboratory,* Pieters, H. A. J., Creyghton, J. W., London, Academic Press, 2nd Edn, 1957

Pot. Incid. Rep. — *ASESB Potential Incident Report,* Washington, Armed Services Explosives Safety Board

Proc. Nth Combust. Symp. — *Proceedings of Combustion Symposia,* Various Editors,

	Various Publishers, dated year following event. 1st Symposium, 1928; 14th Symposium, 1972
Rodd, 1951–62	*The Chemistry of Carbon Compounds,* Rodd, E. H. (Editor), London, Elsevier. Vol. 1A, 1951; Vol. 5, 1962
Rüst, 1947	*Unfälle beim Chemischen Arbeiten,* Rust, E., Ebert, A., Zurich, Rascher Verlag, 2nd Edn. 1947
Rutledge, 1968	*Acetylenic Compounds,* Rutledge, T. F., New York, Reinhold, 1968
Schmidt, 1967	*Handling and Use of Fluorine and Fluorine–Oxygen Mixtures in Rocket Systems,* SP-3037 Schmidt, H. W., Harper, J. T., Washington, NASA, 1967
Schumacher, 1960	*Perchlorates, their Properties, Manufacture and Uses,* ACS 146, Schumacher, J. C., New York, Reinhold, 1960
Schumb, 1955	*Hydrogen Peroxide,* ACS 128, Schumb, W. C., Satterfield, C. N., Wentworth, R. L., New York, Reinhold, 1955
Shriver, 1969	*Manipulation of Air-sensitive Compounds,* Shriver, D. F., London, McGraw-Hill, 1969
Sichere Chemiearbeit	*Sichere Chemiearbeit,* Heidelberg, Berufsgenossenchaft der Chemische Industrie, 1963 to date (supplement to *Chemiker Zeitung)*

Sidgwick, 1950 — *The Chemical Elements and their Compounds,* Sidgwick, N. V., Oxford, Oxford University Press, 1950

Smith, 1965–6 — *Open-chain Nitrogen Compounds,* Smith, P. A. S., New York, Benjamin. Vol. 1, 1965; Vol. 2, 1966

Steere, 1967 — *Handbook of Laboratory Safety,* Steere, N. V. (Editor), Cleveland, Chemical Rubber Publishing Co., 1967 (2nd Edn, 1971)

Swern, 1970–2 — *Organic Peroxides,* Swern, D. (Editor), London, Wiley-Interscience. Vol. 1, 1970; Vol. 2, 1971; Vol. 3, 1972

Urbanski, 1964–7 — *Chemistry and Technology of Explosives,* Urbanski, T., London, Macmillan. Vol. 1, 1964; Vol. 2, 1965; Vol. 3, 1967

Van Wazer, 1958 — *Phosphorus and its Compounds,* Van Wazer, J. R., New York, Interscience. Vol. 1, 1958; Vol. 2, 1961

Vogel, 1957 — *A Textbook of Practical Organic Chemistry,* Vogel, A. I., London, Longmans Green, 3rd Edn, 1957

von Schwartz, 1918 — *Fire and Explosion Risks,* von Schwartz, E., London, Griffin, 1918

Weast, 1972 — *Handbook of Chemistry and Physics,* Weast, R. C. (Editor), Cleveland, Chemical Rubber Publishing Co., 53rd Edn, 1972

Zabetakis, 1965	*Flammability Characteristics of Combustible Gases and Vapours*, Zabetakis, M. G., Washington, US Bureau of Mines, 1965

APPENDIX 2

Index of Class, Group and Topic Titles used in Section 1

Title	Page